AF443628

CONFIRMATION, EMPIRICAL PROGRESS, AND TRUTH APPROXIMATION

POZNAŃ STUDIES
IN THE PHILOSOPHY OF THE SCIENCES AND THE HUMANITIES

VOLUME 83

EDITORS

Poznań Studies in the Philosophy of the Sciences and the Humanities
is partly sponsored by Adam Mickiewicz University

Address: dr Katarzyna Paprzycka . Instytut Filozofii . SWPS . ul. Chodakowska 19/31
03-815 Warszawa . Poland . fax: +48(0)22-517-9703 . E-mail: drp@swps edu.pl
http://main.amu.edu.pl/~pozn-stu/

MONOGRAPHS-IN-DEBATE

Monographs-in-Debate is a new subseries of the Poznań *Studies in the Philosophy of the Sciences and the Humanities* book series. It publishes monographs that deal with the general area of interest of the *Poznań Studies* series. The special nature of the *Monographs-in-Debate* volumes arises from the recognition of the dialectical nature of philosophical work. Each volume contains a monograph, followed by peer commentaries and the author's replies.

POZNAŃ STUDIES IN THE PHILOSOPHY OF THE SCIENCES AND THE HUMANITIES, VOLUME 83
MONOGRAPHS-IN-DEBATE

Confirmation, Empirical Progress, and Truth Approximation

Essays in Debate with Theo Kuipers
Volume 1

**Edited by
Roberto Festa, Atocha Aliseda
and Jeanne Peijnenburg**

Amsterdam - New York, NY 2005

The paper on which this book is printed meets the requirements of "ISO 9706:1994, Information and documentation - Paper for documents - Requirements for permanence".

ISSN 0303-8157
ISBN: 90-420-1638-8 (Bound)
©Editions Rodopi B.V., Amsterdam - New York, NY 2005
Printed in The Netherlands

CONTENTS

Poznań Studies in the Philosophy of the Sciences and the Humanities, vol. 84

CONTENTS OF THE COMPANION VOLUME

COGNITIVE STRUCTURES IN SCIENTIFIC INQUIRY
Essays in Debate with theo Kuipers, vol. 2

COMPUTATIONAL APPROACHES

THEORIES AND STRUCTURES

SCIENCE AND ETHICS

INTRODUCTION

The present volume, Confirmation, Empirical Progress, and Truth Approximation, and the companion volume, Cognitive Structures in Scientific Inquiry, are Volume 1 and 2, respectively, of Essays in Debate with Theo Kuipers. The subdivision of the latter into two volumes, which can be read independently from each other, is motivated by the fact that they deal with two different, although closely related, clusters of topics and issues.[1]

In the present introduction we will describe the nature of the *Essays* and the structure of this volume (Section 1).[2] Then we will summarize the contents of the relevant book of Kuipers, see below, and the seventeen contributions appearing in this volume (Section 2). Finally, we will make very brief remarks on some 'metaphilosophical problems' raised by the philosophical approach developed by Kuipers in the works debated in these *Essays*, with special reference to the relations between logic and philosophy of science (Section 3).[3]

1. Nature and Structure of the Volume

Essays in Debate with Theo Kuipers deals with the content of two books written by Theo Kuipers and published by Kluwer Academic Publishers in the

---

[1] *Essays in Debate with Theo Kuipers* is the second (two volume) book of the new sub-series *Monographs-in-Debate* (MiD) of *Poznań Studies in the Philosophy of the Sciences and the Humanities*. It has been preceded by the volume devoted to the discussion of Evandro Agazzi, *Right, Wrong, and Science: The Ethical Dimensions of the Techno-Scientific Enterprise* (edited by Craig Dilworth), published in 2004. MiD continues in a systematic form those volumes of *Poznań Studies* that were devoted to important publications in philosophy, which have appeared since 1990. Such volumes include books about the philosophical works of Mario Bunge (vol. 18, 1990), Jonathan Cohen (vol. 21, 1991), Agnes Heller (vol. 37, 1994), Jerzy Kmita (vol. 47, 1996), Ernest Gellner (vol. 48, 1996), and Jaakko Hintikka (vol. 51, 1997).

[2] Section 1 is identical – apart from some minor adaptations – to the corresponding section of the introduction to the companion volume.

[3] Section 3 might be read together with the corresponding section of the Introduction to Volume 2, where further metaphilosophical aspects and implications of Kuipers' approach to philosophy of science are examined.

In: R. Festa, A. Aliseda and J. Peijnenburg (eds.), *Confirmation, Empirical Progress, and Truth Approximation (Poznań Studies in the Philosophy of the Sciences and the Humanities,* vol. 83), pp. 11-20. Amsterdam/New York, NY: Rodopi, 2005.

Synthese Library, viz. *From Instrumentalism to Constructive Realism. On Some Relations Between Confirmation, Empirical Progress, and Truth Approximation* (2000) and *Structures in Science. Heuristic Patterns Based on Cognitive Structures: An Advanced Textbook in Neo-classical Philosophy of Science* (2001). We will refer to these books as ICR and SiS. The two volumes of *Essays* are devoted to a critical discussion of ICR and SiS, respectively, where the division of the work is mirrored by their titles: *Confirmation, Empirical Progress and Truth Approximation* is almost identical to the subtitle of ICR, while *Cognitive Structures in Scientific Inquiry* is essentially a condensed version of the subtitle of SiS. However, the reader should not be surprised to find frequent references also to many other papers by Kuipers: indeed, both ICR and SiS are synthetic monographs, putting much of the earlier work of Theo Kuipers in a revised form together and adding new, formerly missing links.

Essays is intended to provide an occasion for significant dialogue on – and better understanding of – issues and positions involved in ICR and SiS. Each of the thirty-four commentaries included in the *Essays* is directly related to one of the topics dealt with in ICR or SiS: all commentators explicitly discuss Kuipers' approach to the relevant topic and, in many cases, presents her or his own approach. The reader will notice that *Essays* scarcely constitutes a traditional Liber Amicorum. Instead, it provides a genuine and lively debate on theses defended in ICR and SiS. Many articles are critical and polemical, as ought to be the case in any significant debate; see, for instance, the papers of Patrick Maher and Sjoerd Zwart (in Volume 1) and those of Eric R. Scerri and Arno Wouters (in Volume 2). In the spirit of this significant dialogue, each contribution is followed by a substantial reply by Kuipers himself.

The present volume consists of three parts: (1) a synopsis of the target monograph by Kuipers; (2) seventeen commentaries on the monograph (or a related paper), each followed by a reply from Kuipers; (3) a bibliography of Kuipers' work.

Since ICR and SiS are closely related, it is not surprising that the present ICR-related volume includes several references to SiS; moreover, especially in Kuipers' replies, there are references to contributions included in the SiS-related Volume 2 of *Essays*. For a better understanding of these references, a terse Table of Contents of SiS (including only the titles of the parts and the chapters) is enclosed in Appendix 2 of Kuipers' synopsis of ICR, and the complete Table of Contents of Volume 2 of *Essays* is included immediately after the Table of Contents of this Volume.

2. Contents of the Volume

ICR provides an original and convincing synthesis of (adaptations of) the confirmation theory of Carnap and Hempel and the truth approximation theory of Popper. The key element of this synthesis is a sophisticated instrumentalist methodology – called the *evaluation methodology* – which acts as the link between confirmation and truth approximation.

The evaluation methodology is based on the separate and comparative evaluation of theories, *even if already falsified*, in terms of their successes and problems. Besides providing, as one might expect, the straight route for short-term empirical progress in science in the spirit of Laudan, the evaluation methodology has a more surprising feature: it is also effective in achieving truth approximation, and, *pace Popper*, it appears much more efficient than the falsificationist methodology. More precisely, in ICR it is argued that the short-term empirical progress obtained by the application of the evaluation methodology is also functional for *all* kinds of truth approximation: observational, referential, and theoretical.

This sheds new light on the long-term dynamics of science and, hence, on the relation between the main epistemological positions, viz., instrumentalism (Toulmin, Laudan), constructive empiricism (van Fraassen), referential realism (Hacking and Cartwright), and theory realism of a non-essentialist nature or, in ICR-terminology, *constructive realism* (Popper). By investigating the application of the instrumentalist, or evaluation methodology, one can understand that there are good reasons for the epistemological transition from instrumentalism to constructive empiricism, from this to referential realism and, finally, from referential realism to constructive realism. The title of ICR, i.e., "from instrumentalism to constructive realism," precisely points to the possibility of all three epistemological transitions.

While in ICR the above story is presented in great detail, Kuipers' synopsis (this volume) focuses on the main ways of theory evaluation presented in ICR, viz. evaluation in terms of confirmation (or falsification), empirical progress and truth approximation.

The seventeen commentaries included in the present volume discuss the positions and outcomes defended in ICR, as well as their ramifications and implications. For the convenience of the reader, the commentaries have been divided into seven groups. The first three papers, collected under the label CONFIRMATION AND THE HD METHOD, discuss Kuipers' non-standard approach to confirmation (ICR, Part I). The three papers within the second group, EMPIRICAL PROGRESS BY ABDUCTION AND INDUCTION, discuss different aspects of Kuipers' views on empirical progress (ICR, Part II). Since the problems of truthlikeness (what is closer-to-the-truth?) and truth approximation (how can we get closer to the truth?) are the central issues of ICR, it is not surprising that nine of the contributions to this volume – collected in four groups – deal with these issues. In particular, the first three groups – called TRUTH APPROXIMATION BY ABDUCTION, TRUTH APPROXIMATION BY EMPIRICAL AND NONEMPIRICAL MEANS, and TRUTHLIKENESS AND UPDATING – discuss Kuipers' views on "basic truth approximation" (ICR, Part III), while the papers within the fourth group are about Kuipers' notion of REFINED TRUTH APPROXIMATION (ICR, Part IV, especially Ch. 10-12). Finally, under the label METAPHORS AND THEOREMS: METASCIENTIFIC AND METAPHILOSOPHICAL ISSUES, the last group consists of two papers dealing with some metascientific and metaphilosophical issues raised by the account of the scientific enterprise developed by Kuipers in ICR and SiS.[4]

2.1. *Confirmation and the HD Method*

This group includes three papers due to Patrick Maher, John R. Welch, and Gerhard Schurz.

Maher argues that (i) Kuipers' theory of qualitative confirmation, which does not assume the existence of quantitative probabilities (see ICR, Ch. 2.1), is affected by serious flaws, and that (ii) the application of this theory to the solution of two famous paradoxes of confirmation, i.e., the ravens paradox and the paradox of grue emeralds (see ICR, Ch. 2.2) is based on a dubious principle which cannot be derived without making use of Kuipers' quantitative confirmation theory (see ICR, Ch. 3).

The paradoxes in confirmation theory are the main issue also of *Welch*'s paper. Focussing on some aspects of Goodman's "new riddle of induction," Welch extends Kuipers' analysis of Goodman's grue-problem (see ICR, Ch. 2.2) in several directions: (i) he proposes an amplified classification of grue problems, (ii) expands the class of gruesome predicates by incorporating Quine's 'undetached rabbit part', 'rabbit stage', and the like, and (iii) shows how some basic problems can be managed along Bayesian lines, inspired by Kuipers' solution of Goodman's problem.

Schurz argues that, in spite of their intuitive starting points, Kuipers' accounts of hypothetico-deductive confirmation (HD-confirmation) and of truthlikeness lead to counterintuitive consequences. He shows how these unwelcome results can be avoided by adopting Schurz-Weingartner's approaches to confirmation and truthlikeness, both based on the notion of "relevant-element." Moreover, he suggests that his revised definitions of HD-

---

[4] Here 'metaphors' refers to the important role that metaphors about the nature of science play in J.J.A. Mooij's paper on the metascientific issue of scientific realism, and 'theorems' to the important role that the desirability of theorems plays in Roberto Festa's paper on the metaphilosophical issue of the relations between logic and philosophy of science.

confirmation and truthlikeness can be connected in a manner similar to that of Kuipers, by conceiving the "rule of success," leading to empirical progress, as the glue between confirmation and truth approximation.

2.2. *Empirical Progress by Abduction and Induction*

Several papers in this volume (notably those by Aliseda, Batens, van Benthem, Burger and Heidema, Meheus, Mormann, Schurz, and Zwart) and in the companion volume (especially the contribution by Kamps and Ruttkamp) – can be seen as clarifying examples of the abundance of important issues where philosophy of science and logic can cross-pollinate to their mutual benefit. More specifically, the three papers in this group – due to Atocha Aliseda, Joke Meheus, and Diderik Batens – show that interesting logical research can be driven by questions stemming from philosophy of science.

Aliseda takes up the challenge – laid down by Kuipers (1999) – to design a method, a logic or a computer program, for abducing a revised hypothesis that is empirically more successful than a given one. After providing a reformulation of Kuipers' account of empirical progress in the framework of (extended) semantic tableaux, she shows that her general semantic tableaux method for abduction can be used for making empirical progress of a special kind: the identification and subsequent dissolving of lacunae.

Meheus suggests a different way to meet Kuipers' challenge. She approaches the problem of empirical progress by using two ampliative adaptive logics, called LA and LA^k, developed on the basis of Batens' ampliative adaptive logic. While LA is an adaptive general logic for abduction, that enables one to generate explanatory hypotheses from a set of observational statements and a set of background assumptions, LA^k can be seen as an extension of LA, providing a specific logic for empirical progress, which can be used to abduce the maximally successful hypothesis, if any. Among other things, LA^k captures some basic features of Kuipers' notion of empirical progress.

In contrast to the Bayesian approaches to inductive logic – based on the assignment of suitable degrees of inductive probability to hypotheses – *Batens'* "logic of induction" is a purely logical approach to inductive logic. More precisely, such a logic is given by a logical system, called LI, which can be seen as a specialization of his own adaptive version of dynamic logic, applicable to the cases where the premises are restricted to, on the one hand, a set of empirical data and, on the other hand, a set of background generalizations, while the consequences may include generalizations as well as singular statements, some of which may serve as predictions and explanations.

2.3. *Truth Approximation by Abduction*

The first group of papers on truthlikeness includes the contributions by Ilkka Niiniluoto and Igor Douven.

Niiniluoto studies the interplay between the notions of abduction and truthlikeness, and their role in the defence of scientific realism, by focussing on the generalization of abduction to cases where the conclusion states that the best theory is approximately true. After reconstructing Kuipers' proposals within the framework of monadic predicate logic, he applies his own notion of truthlikeness, by showing, among other things, that a theory with higher truthlikeness does not always have greater empirical success than its less truthlike rivals, and that the notion of expected truthlikeness provides a fallible link from the approximate explanatory success of a theory to its truthlikeness.

Douven discusses the rule of inference proposed by Kuipers in ICR under the name Inference to the Best Theory (IBT), and supports Kuipers' turn to IBT as a critically revised version of the standard rule of "inference to the best explanation" (IBE). After arguing that IBT has to be strengthened if it is to serve realist purposes, he describes a method for testing, and perhaps justifying, a suitably strengthened and refined version of it.

2.4. *Truth Approximation by Empirical and Nonempirical Means*

The authors of the three papers in this group are Bert Hamminga, David Miller, and Jesús P. Zamora Bonilla.

Hamminga exploits the structuralist terminology adopted in ICR in defining the relations between confirmation, empirical progress and truth approximation. In his paper, the fundamental problem of Lakatos' classical concept of scientific progress is clarified, and its way of evaluating theories is compared to the real problems of scientists who face the far from perfect theories they wish to improve and defend against competitors. Among other things, Hamminga presents a provocative diagnosis of Lakatos' notion of "novel facts", by arguing that it is not so much related to Popper's notion of "empirical content" of a theory, but rather to its allowed possibilities.

Miller examines the view – advanced by McAllister (1996) and endorsed, with new arguments, by Kuipers (2002) – that aesthetical criteria may reasonably play a role in the selection of scientific theories. After evaluating the adequacy of Kuipers' approach to truth approximation, Miller discusses Kuipers' account of the nature and role of empirical and aesthetic criteria in the evaluation of scientific theories and, in particular, the thesis that "beauty can be a road to truth". Finally, he examines McAllister's doctrine that scientific revolutions are characterized above all by novelty of aesthetic judgments.

Zamora Bonilla argues that the points of view of verisimilitude theorists (who assume that science attempts to reach an increasing truthlikeness), and of radical sociologists of science (who assert that scientists struggle for more mundane goals such as income, power, and fame), can be made compatible by accepting two assumptions: 1) rational individuals only would be interested in engaging in a strong competition (such as that described by radical sociologists) if they knew in advance under what rules their outcomes are to be assessed, and 2) if these rules have to be chosen "under a veil of ignorance", then rules favouring highly verisimilar theories can be preferred to other methodological rules.

2.5. *Truthlikeness and Updating*

The third group of papers on truthlikeness consists of a contribution by Sjoerd Zwart and one by Johan van Benthem.

Zwart argues that Kuipers' choice to represent the applications of a theory by "logical models" leads to the following unwelcome results: (1) in contrast to applications of a theory, logical models are mutually incompatible; (2) an increase of logical strength is paradoxically represented both as increase and as decrease of a set of models; (3) the evidence logically implies the strongest empirical law; (4) a hypothesis and its negation can both be false. He suggests that such problems can be avoided by identifying the applications of a theory not with its logical models, but with *partial* models, that can be extended to the logical models of the language used to formulate the theory.

Addressing one theme raised in Zwart's paper, *van Benthem* discusses Kuipers' view of theories. He starts from the problem of the appropriate formats for formal theories, and then clarifies the issue of what it means to update a theory. Using properties of verisimilitude as a lead, he also provides some connections between formal calculi of theories in the philosophy of science and modal-epistemic logics. Throughout he uses this case study as a platform for discussing more general connections between logic and general methodology.

2.6. *Refined Truth Approximation*

The fourth and last group of papers on truthlikeness includes a contribution by Thomas Mormann and one by Isabella C. Burger and Johannes Heidema.

Following Miller's geometric point of view in the theory of truthlikeness, *Mormann* uses a suitable reformulation of Kuipers' concept of structure-likeness (see ICR, Ch. 10) as the key for the construction of his "geometry of logic." In contrast to the widespread idea that the quantitative and the qualitative approach to truthlikeness are worlds apart, separated by a gap of principle, Mormann argues that, within the general framework of the theory of interval structures, the quantitative approach can be construed in a qualitative way. More precisely, he shows that the qualitative and the quantitative accounts of truthlikeness turn out to be special cases of the interval account.

A key notion in the ICR-analysis of truth approximation is the R/S-ordering of "greater successfulness," where the notion of a more successful theory is defined for couples of theories, w.r.t. the available "empirical data" R and "the strongest established laws" S (see ICR, Ch. 7.3.2). This notion is a main source of inspiration of the paper by *Burger and Heidema*, who argue that philosophy of science is often concerned with comparative orderings on the states of a system, or with theories expressing information about the system. They explain a number of these orderings, study their properties, and unravel some of their interrelationships.

2.7. *Metaphors and Theorems: Metascientific and Metaphilosophical Issues*

The authors of the two papers of the last group are J.J.A. Mooij and Roberto Festa.

Mooij discusses a number of metaphors about the nature of science, introduced in ICR (Ch. 1 and 13) in connection with three types of metaphysical realism: minimal, moderate and essentialistic realism, and argues that the metaphors of the mirror, the net and the map correspond to essentialistic, minimal and moderate realism, respectively. Moreover, he explains why the map-metaphor is by far the most suitable one in the context of the ICR-view of science and – besides sharing Kuipers' preference for this metaphor – gives further and sophisticated arguments for it.

Festa considers the metaphilosophical issue of the relations between logic and philosophy of science, as they appear from the perspective of Kuipers's neo-classical approach to the philosophy of science. Since his paper deals with ideas and outcomes presented *both* in ICR and SiS, it is located at the end of this volume, as a sort of *trait d'union* with the companion SiS-related volume. A brief description of the specific metaphilosophical problems considered in Festa's paper can be found below in Section 3.

3. Philosophy of Science and Logic: An Intimate Relation?

Besides the specific philosophical problems, related to different epistemological and methodological issues – starting from those mentioned in Section 2 and discussed in the contributions to this volume – the views and outcomes presented by Kuipers in ICR and SiS and in his replies raise a number of intriguing metaphilosophical problems about the nature of philosophy of science and its relations to other disciplines. As far as the last problem is concerned, it seems to us that a special relevance, both from an historical and a theoretical point of view, should be attributed to the relations between philosophy of science and logic.

Some aspects of these relations, as they appear from the perspective of Kuipers' neo-classical philosophy of science, are treated in Roberto Festa's contribution to this volume. More precisely, Festa's paper focusses on two couples of issues: (A) the *(dis)similarities* between the goals and methods of logic and those of philosophy of science, w.r.t. (1) the role of theorems within the two disciplines, and (2) the falsifiability of their theoretical claims; and (B) the *interactions* between logic and philosophy of science, w.r.t. (3) the possibility to apply logic in philosophy of science, and (4) the possibility that the two disciplines are sources of challenging problems for each other.

A general conclusion which can be drawn from Festa's paper is that the similarities and interactions between philosophy of science and logic strongly depend on the way in which the two disciplines are conceived and that, in particular, both the similarities and the interactions are deemed to increase if philosophy of science is practised within the neo-classical approach.

At the end of Section 3 of the Introduction to the companion volume, we point out that Kuipers did not write SiS solely for philosophers, but he hoped that his readers will include scientists too. Festa's remarks on the relations between (neo-classical) philosophy of science and logic aim, among other things, to make it sufficiently clear that the neo-classical approach, developed in ICR and SiS, should attract not only the attention of philosophers and scientists, but also the active interest of professional logicians.

ACKNOWLEDGMENTS

The publication of this two volume book of *Essays in Debate with Theo Kuipers* would not have been possible without the support and the cooperation of a number of persons and institutions, whose role and contribution we gratefully acknowledge.

First of all, we wish to express our gratitude to Leszek Nowak and Katarzyna Paprzycka who kindly invited us, to design and edit this book for the new subseries *Monographs-in-Debate*. We are indebted also to Kluwer Academic Publishers, for making available the copies of ICR and SiS for all contributors, and to the Faculty of Philosophy of the University of Groningen, for providing financial support. Finally, we would like to thank four persons who played different but equally crucial roles in the production of the *Essays*, i.e., Hauke de Vries for his unflagging technical assistance, David Atkinson for grammatical corrections and stylistic improvements, Ian Priestnall (Paragraph Services) for his work of linguistic correction, and Lieke Hendriks for her work on the camera ready version.

Roberto Festa, *Trieste* (Italy)
Atocha Aliseda, *Mexico City* (Mexico)
Jeanne Peijnenburg, *Groningen* (The Netherlands)

REFERENCES

Benthem, J. van (1982). The Logical Study of Science. *Synthese* **51**, 431-472.

Benthem, J. van (1999). Wider Still and Wider... Resetting the Bounds of Logic. *European Review of Philosophy* **4**, 21-44.

Kuipers, T.A.F. (2002). Beauty, a Road to The Truth. *Synthese* **131** (3), 291-328.

Kuipers, T.A.F. (1999). Abduction Aiming at Empirical Progress or Even at Truth Approximation, Leading to Challenge for Computational Modelling. In: J. Meheus and T. Nickles (eds.), *Scientific Discovery and Creativity*, special issue of *Foundations of Science* **4** (3), 307-323.

McAllister, J.W. (1996). *Beauty and Revolution in Science*. Ithaca: Cornell University Press.

Theo A.F. Kuipers

THE THREEFOLD EVALUATION OF THEORIES

A SYNOPSIS OF

FROM INSTRUMENTALISM TO CONSTRUCTIVE REALISM. ON SOME RELATIONS BETWEEN CONFIRMATION, EMPIRICAL PROGRESS, AND TRUTH APPROXIMATION (2000)

CONTENTS

Theo A.F. Kuipers

THE THREEFOLD EVALUATION OF THEORIES

A SYNOPSIS OF FROM INSTRUMENTALISM TO CONSTRUCTIVE REALISM. ON SOME RELATIONS BETWEEN CONFIRMATION, EMPIRICAL PROGRESS, AND TRUTH APPROXIMATION (2000)

ABSTRACT. Surprisingly enough, modified versions of the confirmation theory of Carnap and Hempel and the truth approximation theory of Popper turn out to be smoothly synthesizable. The glue between confirmation and truth approximation appears to be the instrumentalist methodology, rather than the falsificationist one.

By evaluating theories separately and comparatively in terms of their successes and problems (hence even if they are already falsified), the instrumentalist methodology provides – both in theory and in practice – the straight route for short-term empirical progress in science in the spirit of Laudan. However, it is argued that such progress is also functional for all kinds of truth approximation: observational, referential, and theoretical. This sheds new light on the long-term dynamic of science and hence on the relation between the main epistemological positions, viz., instrumentalism (Toulmin, Laudan), constructive empiricism (van Fraassen), referential realism (Hacking and Cartwright), and theory realism of a non-essentialist nature (Popper), here called constructive realism.

In *From Instrumentalism to Constructive Realism* (2000) the above story is presented in great detail. The present synopsis highlights the main ways of theory evaluation presented in that book, viz. evaluation in terms of confirmation (or falsification), empirical progress and truth approximation.

Introduction

Over the years I have been working on two *prima facie* rather different, if not opposing, research programs, notably Carnap's confirmation theory and Popper's truth approximation theory. However, I have always felt that they must be compatible, even smoothly synthesizable, for all empirical scientists use confirmation intuitions, and many of them have truth approximation ideas. Gradually it occurred to me that the glue between confirmation and truth approximation was the instrumentalist or evaluation methodology, rather than the falsificationist one. By separate and comparative evaluation of theories in terms of their successes and problems – hence, even if already falsified – the

In: R. Festa, A. Aliseda and J. Peijnenburg (eds.), *Confirmation, Empirical Progress, and Truth Approximation (Poznań Studies in the Philosophy of the Sciences and the Humanities,* vol. 83), pp. 23-85. Amsterdam/New York, NY: Rodopi, 2005.

evaluation methodology provides in theory and practice the straight route for short-term empirical progress in science in the spirit of Laudan. Further analysis showed that this also sheds new light on the long-term dynamic of science and hence on the relation between the main epistemological positions, viz., instrumentalism (Toulmin, Laudan), constructive empiricism (van Fraassen), referential realism (Hacking), and theory realism of a non-essentialist nature, here called constructive realism (Popper). Indeed, thanks to the evaluation methodology, there are good, if not strong reasons for all three epistemological transitions "from instrumentalism to constructive realism."

In this way a clear picture of scientific development arises, with a short-term and a long-term dynamic. In the former there is a severely restricted role for confirmation and falsification, the dominant role being played by (the aim of) empirical progress, and there are serious prospects for observational, referential and theoretical truth approximation. Hence, in regard to this short-term dynamic, the scientist's intuition that the debate among philosophers about instrumentalism and realism has almost no practical consequences can be explained and justified. The long-term dynamic is enabled by (observational, referential and theoretical) inductive jumps, after 'sufficient confirmation', providing the means to enlarge the observational vocabulary in order to investigate new domains of reality. In this respect, a consistent instrumentalist epistemological attitude seems difficult to defend, whereas constructive realism seems most plausible.

In *From Instrumentalism to Constructive Realism* (ICR, 2000) the above story is presented in great detail. The present synopsis highlights the main ways of theory evaluation presented in ICR, viz. evaluation in terms of confirmation (or falsification), empirical progress and truth approximation. It essentially follows the division of ICR in four parts and 13 chapters, here resulting in four parts and 13 sections. However, Part IV (Ch. 10-12), dealing with refined truth approximation, is only briefly sketched. This synopsis is necessarily selective and hence it may be useful to consult from time to time the complete table of contents, including section titles, which is reproduced in Appendix 1.[1] Appendix 2 presents the outline table of contents of the

---

[1] ICR is based on many publications, starting from 1978. The Foreword of ICR (p. x) mentions those 10 papers that have partially been used in writing ICR. This synopsis represents the main lines of ICR from its dominant point of view, viz. theory evaluation. It is supposed to be my last survey of ICR. A number of special topic-oriented surveys have been written before. Their titles indicate their special emphasis: Pragmatic aspects of truth approximation (1998), Abduction aiming at empirical progress or even truth approximation, leading to a challenge for computational modeling (1999), Progress in nomological, explicative and design research (Ch. 9 of SiS, 2001), Beauty, a road to the truth (2002), Empirical and conceptual idealization and concretization: the case of truth approximation (forthcoming), Inference to the best theory: kinds of abduction and induction (2004).

companion volume *Structures in Science* (SiS, 2001), to which occasional reference will be made. Appendix 3 gives a list of acronyms.

I would like to conclude this introduction by referring to Ilkka Niiniluoto's major contribution to the field, viz. his *Truthlikeness* of 1987. Despite our differences regarding the topic of truth approximation, notably his emphasis on a quantitative approach and my emphasis on a qualitative one, the reader will come to understand that I feel much sympathy with his slogan "Popper's voice but Carnap's hands" (Niiniluoto 1987, p. xvi).

1. *General Introduction: Epistemological Positions*

The core of the ongoing instrumentalism-realism debate concerns the nature of theoretical terms and of proper theories using such terms, or rather the attitude one should have towards them. *Prima facie*, the most important epistemological positions in that debate are certainly instrumentalism, constructive empiricism, referential realism and theory realism. They can be characterized and ordered according to the ways in which they answer a number of leading questions, where every subsequent question presupposes the affirmative answer to the previous one. For completeness, I start with two preliminary questions that get a positive answer from the major positions, but a negative one in idealist and extremely relativist postmodern circles:

Question 0: Does a natural world that is independent of human beings exist?
No: ontological idealism; Yes: ontological realism.

Question 1: Can we claim to possess true claims to knowledge about the natural world?
No: epistemological relativism; Yes: epistemological realism.

Question 2: Can we claim to possess true claims to knowledge about the natural world beyond what is observable?
No: empiricism: instrumentalism or constructive empiricism; Yes: scientific realism.

Question 3: Can we claim to possess true claims to knowledge about the natural world beyond (what is observable and) reference claims concerning theoretical terms?
No: entity or, more generally, referential realism; Yes: theory realism.

Question 4: Does there exist a correct or ideal conceptualization of the natural world?
No: constructive realism; Yes: essentialist realism.

Note first that "empiricism" has two variants. They split on the subquestion whether reference of theoretical terms and truth values of theoretical statements even have to be formally denounced, notably as category mistakes by instrumentalists, or not, as constructive empiricism concedes. The splitting of "theory realism" at the end of this question-and-answer game into "constructive realism" and "essentialist realism" suggests that we now have five main positions: instrumentalism, constructive empiricism, and referential, constructive and essentialist realism. The following scheme, starting with Question 2, presents their relation in brief.

```
Q2:  true claims about the natural world      ⇒      empiricism
       beyond the observable?                 no              - instrumentalism
                                                              - constructive empiricism

       yes ⇓ scientific realism


Q3: beyond reference?                         ⇒      referential realism
                                              no              ⇒ entity realism

       yes ⇓ theory realism

Q4: ideal conceptualization?                  ⇒      constructive realism
                                              no
       yes ⇓ essentialist realism
```

The main epistemological positions

Important refinements are obtained when the Questions 2 - 4 are considered from four perspectives on theories. On the one hand, theories supposedly deal only with "the actual world" or primarily with "the nomic world," that is, with what is possible in the natural world. On the other hand, one may only be interested in whether theories are true or false, or primarily in whether they approach "the truth," regarding the world of interest. It should be stressed that "the truth" is always to be understood in a domain-and-vocabulary relative way. Hence, no language-independent metaphysical or essentialist notion of "the truth" is assumed. The four perspectives imply that all (non-relativistic) epistemological positions have an "actual world version" and a "nomic world version" and that they may be restricted to "true-or-false" claims, or emphasize "truth approximation claims." In both cases it is plausible to distinguish between observational, referential, and theoretical claims and corresponding inductions, that is, the acceptance of such claims as true. Instrumentalists, in parallel, speak of theories as "reliable-or-unreliable" derivation instruments or as "approaching the best derivation instrument." All four perspectives occur in particular in their realist versions, but they also make sense in adapted form in most of the other epistemological positions.

ICR is primarily a study of confirmation, empirical progress and truth approximation, and their relations. With the emphasis on their nomic interpretation the five main positions are further characterized and compared in the light of the results of this study, leading to the following conclusions.

There are good reasons for the instrumentalist to become a constructive empiricist; in turn, in order to give deeper explanations of success differences, the constructive empiricist is forced to become a referential realist; in turn, there are good reasons for the referential realist to become a theory realist. The theory realist has good reasons to indulge in constructive realism, since there is no reason to assume that there are essences in the world, the existence of which is a prerequisite for ideal conceptualizations. As a result, the way leads to constructive realism and amounts to a pragmatic argument for this position, where the good reasons mainly deal with the short-term and the long-term dynamics generated by the nature of, and the relations between, confirmation, empirical progress and truth approximation.

The suggested hierarchy of the heuristics corresponding to the epistemological positions is, of course, not to be taken in any dogmatic sense. That is, when one is unable to successfully use the constructive realist heuristic, one should not stick to it, but try weaker heuristics: first the referential realist, then the constructive empiricist, and finally the instrumentalist heuristic. For, as with other kinds of heuristics, although not everything goes all the time, *pace* (the suggestion of) Feyerabend's slogan "anything goes," everything goes sometimes. Moreover, after using a weaker heuristic, a stronger heuristic may become applicable at a later stage: "reculer pour mieux sauter."

Besides epistemological conclusions, there are some general methodological lessons to be drawn. The main one is that there are good reasons for all positions not to use the falsificationist but the instrumentalist or "evaluation(ist)" methodology. That is, empirical (and to some extent perhaps non-empirical) successes and failures should exclusively guide the selection of theories, even if the better theory has already been falsified. This common methodology, directed at the separate and comparative evaluation of theories, is presented in Sections 5 and 6 below.

I. Confirmation

In this part a sketch of the main ideas behind confirmation and falsification of a hypothesis by the so-called HD(hypothetico-deductive) method is followed by a description of the "landscape of qualitative and quantitative confirmation," as I like to call it. Confirmation of a hypothesis, however, has the connotation that the hypothesis has not yet been falsified. Whatever the truth claim associated with a hypothesis, as soon as it has been falsified, the plausibility (or probability) that it is true becomes and remains zero. In the next part I elaborate how theories can nevertheless be evaluated after falsification.

2. *Confirmation by the HD Method*

I start this section with a brief exposition of HD testing, that is, the HD method of testing hypotheses, and indicate the related qualitative explication of confirmation and, in the next section, its quantitative extensions. ICR Ch. 2, moreover, deals in detail with the paradoxes of the ravens and emeralds and some other confirmation problems and solutions. Moreover, it deals briefly with induction, that is, the acceptance of hypotheses.

HD Testing

HD testing attempts to give an answer to one of the questions that one may be interested in, *the truth question*, which may be qualified according to the relevant epistemological position. The HD method prescribes the derivation of test implications and testing them. In each particular case, this may either lead to confirmation or to falsification. Whereas the "language of falsification" is relatively clear, the "language of confirmation" is a matter of great dispute.

According to the leading expositions of the hypothetico-deductive (HD) method by Hempel (1966), Popper (1934/1959) and De Groot (1961/1969), the aim of the HD method is to determine whether a hypothesis is true or false, that is, it is a method of testing. On closer inspection, this formulation of the aim of the HD method is not only laden with the epistemological assumption of theory realism, according to which it generally makes sense to aim at true hypotheses, but it also mentions only one of the realist aims, i.e., answering the "truth question." Applying the HD method to this end will be called HD testing as distinct from HD evaluation, which has other primary aims.

For the moment I will confine my attention to the HD method as a method of testing hypotheses. Though the realist has a clear aim in undertaking HD testing, this does not mean that HD testing is only useful from that epistemological point of view. Let me briefly review the other main epistemological positions as far as the truth question is concerned and recall that claims may pertain to the actual world or to the nomic world (of nomic possibilities). Hypotheses may or may not use the so-called theoretical terms in addition to the so-called observation terms. What is observational is not taken in some absolute, theory-free sense, but depends greatly on the level of theoretical sophistication. Theoretical terms intended to refer to something in the actual or nomic world may or may not in fact successfully refer to something. For the (*constructive*) *empiricist* the aim of HD testing is to investigate whether the hypothesis is observationally true, i.e., has only true observational consequences, or is observationally or empirically adequate, to use van Fraassen's favorite expression. For the *instrumentalist* the aim of HD testing is still more liberal (and essentially part of the aim of HD evaluation):

for which intended applications is the hypothesis observationally true? The *referential realist*, on the other hand, adds to the aim of the empiricist to investigate whether the hypothesis is referentially true, i.e., whether its referential claims are correct. In contrast to the *theory realist*, he is not interested in the question whether the theoretical claims, i.e., the claims using theoretical terms, are true as well.

Methodologies are ways of answering epistemological questions. It turns out that the method of HD testing, the test methodology, is functional for answering the truth question of all four epistemological positions. For this reason, I shall present the test methodology in fairly neutral terms, viz., plausibility, confirmation and falsification.

The expression 'the plausibility of a hypothesis' abbreviates the informal qualification 'the plausibility, in the light of the background beliefs and the evidence, that the hypothesis is true'. Here 'true' may be specified in one of the four main senses: 1) observationally, as far as particular intended applications are concerned, 2) observationally, as far as all intended applications are concerned, 3) and, moreover, referentially, 4) and even theoretically. Admittedly, despite these possible qualifications, the notion of "plausibility" remains necessarily vague, but that is what most scientists would be willing to subscribe to. When talking about "the plausibility of certain evidence," I mean, of course, "the prior plausibility of the (observational!) hypothesis that the test will result in the reported outcome." Hence, here 'observationally true' and 'true' coincide by definition for what can be considered evidential statements.

Regarding the clarity of notions of "confirmation" and "falsification" the situation is rather asymmetric. "Falsification" of a hypothesis simply means that the evidence entails that the hypothesis is observationally false, and hence also false in the stronger senses. However, what "confirmation" of a hypothesis precisely means is not so clear. The explication of the notion of "confirmation" of a hypothesis by certain evidence is here primarily approached from the success perspective on confirmation. This perspective equates confirmation with an increase in the plausibility of the evidence on the basis of the hypothesis, and implies that the evidence increases the plausibility of the hypothesis. However, by a liberalization suggested in debate with Maher (this volume), I now add "or an increase of the plausibility of the hypothesis on the basis of the evidence, if the latter has zero plausibility."

A test of a hypothesis may be experimental or natural. That is, a test may be an experiment, an active intervention in nature or culture, but it may also concern the passive registration of what is or was the case, or what happens or has happened. In the latter case of a so-called natural test, the registration may be a more or less complicated intervention, but is nevertheless supposed to have no serious effect on the course of events of interest.

According to the HD method a hypothesis H is tested by deriving test implications from it, and checking, if possible, whether they are true or false. Each test implication has to be formulated in terms that are considered to be observation terms. A test implication may or may not be general in nature. Usually there is background knowledge B, which is assumed to be true. Moreover, a test implication is usually of a conditional nature, if C then F ($C \rightarrow F$). Here C denotes one or more "initial conditions" and F denotes a potential fact (event or state of affairs) predicted by H and C. If C and F are of an individual nature, F is called an individual test implication, and $C \rightarrow F$ a conditional test implication. When C is artificially realized, it is an experimental test, otherwise it is a natural test.

The basic logic of HD testing can be represented by some (valid) applications of Modus (Ponendo) Ponens (MP), where '$\models$' indicates logical entailment and where 'I' denotes a test implication:

$$B, H \models I \qquad\qquad\qquad B, H \models C \rightarrow F$$
$$\underline{B, H} \qquad\qquad\qquad\qquad \underline{B, H, C}$$
$$I \qquad\qquad\qquad\qquad\qquad F$$

It should be stressed that $B, H \models I$ and $B, H \models C \rightarrow F$ are supposed to be deductive claims, i.e., claims of a logico-mathematical nature.

The remaining logic of hypothesis testing concerns the application of Modus (Tollendo) Tollens (MT). Neglecting complications that may arise, such as that B's or C's truth may be disputed, if the test implication is false, the hypothesis must be false, and therefore has been falsified, for the following arguments are deductively valid ('$\neg$' indicates negation):

$$B, H \models I \qquad\qquad\qquad B, H \models C \rightarrow F$$
$$\underline{B, \neg I} \qquad\qquad\qquad\qquad \underline{B, C, \neg F}$$
$$\neg H \qquad\qquad\qquad\qquad \neg H$$

When the test implication turns out to be true, the hypothesis has of course not been (conclusively) verified, for the following arguments are invalid, indicated by '-/-/-':

$$B, H \models I \qquad\qquad\qquad B, H \models C \rightarrow F$$
$$B, I \qquad\qquad\qquad\qquad\qquad B, C, F$$
$$-/-/- \qquad\qquad\qquad\qquad -/-/-/-/-$$
$$H \qquad\qquad\qquad\qquad\qquad H$$

Since the evidence (I or $C\&F$) is compatible with H, we may at least say that H may still be true. However, we can say more than that. Usually it is said that H has been confirmed. It is important to note that such confirmation by the HD method means more than mere compatibility; it is confirmation in the strong sense that *H* has obtained a *success* of a (conditional) deductive nature. By entailing the evidence, *H* makes the evidence as plausible as possible. This I call *the success perspective* on ((conditional) deductive) *confirmation*.

Falsification and confirmation have many complications, e.g., due to auxiliary hypotheses. I will deal with several complications, related to general and individual test implications, at the end of Section 5. As already indicated, however, there is a great difference between falsification and confirmation. Whereas the "logical grammar" of falsification is not very problematic, the grammar of confirmation, i.e., the explication of the concept of confirmation, has been a subject of much dispute.

Deductive Confirmation

The grammar of confirmation to be presented in this and the following two sections (based on SiS, Subsection 7.1.2, and introducing some new elements relative to ICR) is in many respects a systematic exposition of well-known ideas about deductive, structural, and inductive confirmation. However, these ideas are presented in a non-standard way and refine and revise several standard solutions of problems associated with these ideas.

Here I will only give a sketch of the main lines of the three ICR chapters on confirmation. It is important to note that, although the role of falsification and confirmation will be relativized in many respects in part II, it will also become clear that they remain very important for particular types of hypotheses. Notably, they remain relevant for general observational (conditional) hypotheses, and for several kinds of (testable) comparative hypotheses, e.g., hypotheses claiming that one theory is more successful or (observationally, referentially or theoretically) even more truthlike than another.

This section deals with qualitative (deductive) confirmation that results from applying the HD method, while the next one deals with quantitative, more specifically probabilistic confirmation, including a suitable degree of confirmation. The third section introduces the crucial distinction between structural and inductive confirmation and gives a brief survey of the main systems of inductive confirmation in the Carnap-Hintikka tradition of so-called inductive logic, with a suitable degree of *inductive* confirmation.

The main non-standard aspect is the approach of confirmation from the "success perspective," according to which confirmation is primarily equated with evidential success, more specifically with an increase of the plausibility of the *evidence* on the basis of the hypothesis. Hence, in contrast to standard expositions, confirmation is not directly (at least not in general, see below) equated with an increase of the plausibility of the *hypothesis* by the evidence.

This is merely an additional aspect of confirmation under appropriate conditions and epistemological assumptions.

Contrary to many critics, I believe that the notion of *deductive* (*d-*) *confirmation* makes perfectly good sense, provided the classificatory definition is supplemented with some comparative principles. More specifically, "(contingent) evidence E d-confirms (consistent) hypothesis H, assuming B" is defined by the clause: $B\&H$ (logically) entails E, and further obeys:

> *Comparative principles*:
> P1: if $B\&H$ entails E and E entails E^* (and not vice versa) then E d-confirms H, assuming B, more than E^*.
> P2: if $B\&H$ and $B\&H^*$ both entail E then E d-confirms H and H^*, assuming B, equally.

To be sure, this definition-with-comparative-supplement only makes sense as a *partial* explication of the intuitive notion of confirmation; it leaves room for non-deductive, in particular probabilistic extensions, as we shall see below. However, let us first look more closely at the comparative principles, suppressing the phrase 'assuming B'. They are very reasonable in the light of the fact that the deductive definition can be conceived as a (deductive) *success* definition of confirmation: if H entails E, E clearly is a success of H, if not a predictive success, then at least a kind of explanatory success. From this perspective, P1 says that a stronger (deductive) success confirms a hypothesis more than a weaker one, and P2 says that two hypotheses should equally be praised for the same success. In particular P2 runs counter to standard conceptions. However, in Chapter 2 of ICR I deal extensively with the possible objections and show, moreover, that the present analysis can handle the confirmation paradoxes discussed by Hempel and Goodman.

I would like to conclude this section with the "Confirmation Matrix", i.e., a survey of the four logical relations, with epistemologically plausible names, between a hypothesis and evidence, assuming background knowledge. Recall that 'd-' is short for 'deductive(ly)', '$\neg$' indicates negation and '$\models$' indicates logical entailment. E is assumed to be true.

Conclusion Premises	E (true)	$\neg E$ (false)
B, H	E d-confirms H, assuming B $B \& H \models E$	E falsifies H, assuming B $B \& H \models \neg E$ ($\Leftrightarrow B \& E \models \neg H$)
$B, \neg H$	E d-disconfirms H, assuming B $B \& \neg H \models E$	E verifies H, assuming B $B \& \neg H \models \neg E$ ($\Leftrightarrow B \& E \models H$)

The confirmation matrix

We get conditional notions by also assuming, besides background knowledge *B*, one or more initial conditions *C*, which play(s) logically the same role as *B*. Although *C* may not entail *E*, *E* may or may not be formulated such that it entails *C*.

3. *Quantitative Confirmation, and its Qualitative Consequences*

There is a natural quantitative refinement of deductive confirmation, which will here be characterized. ICR Ch. 3, moreover, presents in detail several qualitative consequences, and also discusses the prospects for quantitative acceptance criteria for hypotheses. In the first appendix the theory of quantitative confirmation is compared with Popper's theory of corroboration, in the second its solution of the ravens paradoxes is compared with the standard Horwich's analysis.

Probabilistic confirmation presupposes, by definition, a probability function, indicated by p, that is, a real-valued function obeying the standard axioms of probability, which may nevertheless be of one kind or another (see Section 4). But first I shall briefly deal with the general question of a probabilistic criterion of confirmation and a degree of confirmation.

The *standard* (or *forward*) *criterion* for probabilistic confirmation is that the *posterior* probability $p(H/E)$ exceeds the (relative to the background knowledge) *prior* probability $p(H)$, that is, $p(H/E) > p(H)$. However, this criterion is rather inadequate for "p-zero" hypotheses. For example, if $p(H) = 0$ and E d-confirms H, this confirmation cannot be seen as an extreme case of probabilistic confirmation, since $p(H/E) = p(H) = 0$. In other words, this criterion has the very undesirable consequence that p-zero hypotheses cannot be confirmed. However, for p-non-zero hypotheses and assuming $0 < p(E) < 1$, the standard criterion is equivalent to the *backward* or *success* criterion, according to which the so-called *likelihood* $p(E/H)$ exceeds the initial probability $p(E)$ of E: $p(E/H) > p(E)$.[2] Now it is easy to check that any probability function respects d-confirmation according to this criterion, since $p(E/H) = 1$ when H entails E, and hence exceeds $p(E)$, even if $p(H) = 0$, in which case it is a matter of a plausible definition. More generally, the success criterion can apply in all p-zero cases in which $p(E/H)$ can nevertheless be meaningfully interpreted.

To be sure, as Maher (this volume) stresses, the success criterion does not work properly for "p-zero evidence," e.g. in the case of verification of a real-

---

[2] This follows directly from the general definition of conditional probability, viz., $p(A/B) =_{\mathrm{def}} p(A\&B)/p(B)$, assuming that $p(B) \neq 0$. Note that this definition creates in general a tension between cases where $p(B) = 0$ while we would also like to say that $p(A/B) = 1$ because of the fact that B entails A.

valued interval hypothesis by a specific value within that interval. However, although this is less problematic than the case of p-zero hypotheses dealt with above (see my reply to Maher), it seems reasonable to accept the standard criterion for p-zero evidence. From now on 'confirmation' will mean forward or backward confirmation when $p(H) \neq 0 \neq p(E)$, backward confirmation when $p(H) = 0$ and $p(E) \neq 0$ and forward confirmation when $p(E) = 0$ and $p(H) \neq 0$; and it is left undefined when $p(H) = 0 = p(E)$.

In sum, we obtain the following survey of deductive confirmation, "general confirmation," or simply "confirmation," and related notions

E d-confirms H, assuming B	$B\&H \vDash E$	and hence $p(E/B\&H) = 1$
E confirms H, assuming B	$p(E/B\&H) > p(E/B)$ or $p(H/B\&E) > p(H/B)$	
E falsifies H, assuming B	$B\&H \vDash \neg E$	and hence $p(E/B\&H) = 0$
or, equivalently,	$B\&E \vDash \neg H$	and hence $p(H/B\&E) = 0$

E is neutral wrt H, assuming B $\qquad p(E/B\&H) = p(E/B)$ and $p(H/B\&E) = p(H/B)$

E d-disconfirms H, assuming B	$B\&\neg H \vDash E$	and hence $p(E/B\&\neg H) = 1$
E disconfirms H, assuming B	$p(E/B\&H) < p(E/B)$ or $p(H/B\&E) < p(H/B)$	
E verifies H, assuming B	$B\&\neg H \vDash \neg E$	and hence $p(\neg E/B\&\neg H) = 1$
or, equivalently,	$B\&E \vDash H$	and hence $p(H/B\&E) = 1$

Finally, we can define 'non-deductive confirmation' as 'confirmation, but no d-confirmation', and 'proper confirmation' as 'confirmation, but no verification'.

I now turn to the definition of a degree of confirmation, suppressing 'assuming B'. I propose, instead of the standard difference measure $p(H/E) - p(H)$, the non-standard ratio measure $p(E/H)/p(E)$ as the degree (or rate) of (backward) confirmation (according to p), indicated by $c_p(H,E)$. This ratio has the following properties. For p-non-zero hypotheses it is equal to the standard ratio measure $p(H/E)/p(H)$, and hence is symmetric ($c_p(H,E) = c_p(E,H)$), for p-non-zero hypotheses, but it leaves room for confirmation (amounting to: $c_p(H,E) > 1$) of p-zero-hypotheses. For p-zero evidence we may turn to the standard ratio measure.

The definition satisfies the comparative principles of deductive (d-) confirmation P1 and P2. Note first that $c_p(H,E)$ is equal to $1/p(E)$ when H entails E, for $p(E/H) = 1$ in that case. This immediately implies P2: if H and H^* both entail E then $c_p(H,E) = c_p(H^*,E)$. Moreover, if H entails E and E^*, and E entails E^* (and not vice versa) then $c_p(H,E) > c_p(H,E^*)$, as soon as we may assume that $p(E) < p(E^*)$. Note that $p(E) \leq p(E^*)$ already follows from the assumption that E entails E^*. The result is a slightly weakened version of P1.

As suggested, there are a number of other degrees of confirmation. Fitelson (1999) evaluates four of them, among them the logarithmic forward version of my backward ratio measure, in the light of seven arguments or conditions of adequacy as they occur in the literature. The ratio measure fails in five cases. Three of them are directly related to the "pure" character of r, that is, its satisfaction of P2.[3] P2 is defended extensively in Chapter 2 of ICR.

However, I also argue there, in Chapter 3, that as soon as one uses the probability calculus, it does not matter very much which "confirmation language" one chooses, for that calculus provides the crucial means for updating the plausibility of a hypothesis in the light of evidence. Hence, the only important point, which then remains, is to always make clear which confirmation language one has chosen.

4. *Inductive Confirmation and Inductive Logic*

I now turn to a discussion of some possible kinds of probability functions and corresponding kinds of probabilistic confirmation. ICR Ch. 4, moreover, deals in detail with Carnap's and Hintikka's theories of inductive confirmation of singular predictions and universal generalizations. It further summarizes some of the main results, following Carnap, that have been obtained with respect to optimum inductive probability functions and inductive analogy by similarity and proximity.

Structural Confirmation

I start with *structural confirmation*, which has an objective and a logical version. Consider first an example dealing with a fair die. Let E indicate the even (elementary) outcomes 2, 4, 6, and H the "high" outcomes 4, 5, 6. Then (the evidence of) an even outcome confirms the hypothesis of a high outcome according to both criteria, since $p(E/H) = p(H/E) = 2/3 > 1/2 = p(H) = p(E)$. I call this the paradigm example of (non-deductive) structural confirmation.

This example illustrates what Salmon (1969) already pointed out in the context of discussing the possibilities of an inductive logic. A probability function may be such that E confirms H in the sense that H *partially* entails E. Here 'partial entailment' essentially amounts to the claim that the relative number of models in which E is true on the condition that H is true is larger than the relative number of models in which E is true without any condition. This general idea can be captured in a quantitative way by defining structural confirmation as (backward) confirmation based on a probability function assigning constant (e.g. equal) probabilities to the elementary outcomes. Such a probability function may represent an *objective* probability process, such as a

---

[3] The P2-related arguments concern the first and the second argument in Fitelson's Table 1, and the second in Table 2. Of the other two, the example of 'unintuitive' confirmation is rebutted in ICR (Chapter 3) with a similar case against the difference measure. The other one is related to the 'grue-paradox', for which Chapters 2 and 3 of ICR claim to present an illuminating analysis in agreement with P2.

fair die. Note that in this paradigm example of structural confirmation, that is, an even outcome of a fair die confirms a high outcome, the corresponding degree of confirmation is $(2/3)/(1/2) = 4/3$. The probability function may also concern the so-called *logical* probability or logical measure function (Kemeny 1953), indicated by m. Kemeny's m-function assigns probabilities on the basis of ((the limit of) the ratio of) the number of structures making a proposition true, that is, its number of models (cf. the random-world or labeled method in Grove, Halpern and Koller 1996). These logical probabilities may or may not correspond to the objective probabilities of an underlying process, as is the case with a fair die. Hence, for structural confirmation, we may restrict the attention to (generalizations of) Kemeny's m-function.

Structural confirmation is a straightforward generalization of d-confirmation. For suppose that H entails E. Then $m(E/H) = (\lim)\ Mod(E\&H)\ /\ Mod(H) = 1 > (\lim)\ Mod(E)\ /\ Mod(Tautology) = m(E)$, where e.g., '$Mod(H)$' indicates the number of models of H. Moreover, as already indicated, it is a probabilistic explication of Salmon's (1969) idea of confirmation by "partial entailment", according to which an even outcome typically is partially implied by a high outcome.

It is important to note that the m-function leads in many cases to "m-zero" hypotheses (cf. Compton 1988). For instance, every universal generalization "for all $x\ Fx$" gets zero m-value for an infinite universe. As we may conclude from the general exposition in Section 3, certain evidence may well structurally confirm such hypotheses by definition, according to the success criterion, but not according to the standard criterion. E.g., a black raven structurally (conditionally deductively) confirms "all ravens are black" according to the success criterion, even if the universe is supposed to be infinite. In this case the m-value of that hypothesis is zero, with the consequence that it is not confirmed according to the standard criterion. However, it is typical for the m-function that it lacks, even from the success perspective, the confirmation property which is characteristic of inductive probability functions.

Inductive Confirmation

Inductive confirmation is (*pace* Popper and Miller 1983) explicated in terms of confirmation based on an inductive probability function, i.e., a probability function p having the general feature of "positive relevance", "inductive confirmation" or, as I like to call it,

instantial confirmation: $p(Fa/E\&Fb) > p(Fa/E)$

where 'a' and 'b' represent distinct individuals, 'F' an arbitrary monadic property and 'E' any kind of contingent evidence compatible with Fa. Note that this definition is easy to generalize to n-tuples and n-ary properties, but I will restrict attention to monadic ones. Since the m-function satisfies the condition $m(Fa/Fb\&E) = m(Fa/E)$, we get for any inductive probability function p:

$$p(Fa\&Fb/E) = p(Fa/E)\cdot p(Fb/E\&Fa) > m(Fa\&Fb/E)$$

Inductive (probability) functions can be obtained in two ways, which may also be combined:

- "inductive priors", i.e., positive prior p-values $p(H)$ for m-zero   hypotheses

  and/or

- "inductive likelihoods", i.e., likelihood functions $p(E/H)$ having the property of instantial confirmation

Note first that forward confirmation of m-zero hypotheses requires inductive priors, whereas backward confirmation of such hypotheses is always possible, assuming that $p(E/H)$ can be interpreted. Below I will give a general definition of inductive confirmation in terms of degrees of confirmation.

In terms of the two origins of inductive probability functions we can characterize the four main theories of confirmation in philosophy of science as follows:

	Inductive priors	Inductive likelihoods
Popper	No	No
Carnap	No	Yes
Bayes[4]	Yes	No
Hintikka	Yes	Yes

Theories of confirmation

Popper rejected both kinds of inductive confirmation, for roughly three reasons: two problematic ones and a defensible one. The first problematic one (Popper 1934/1959) is that he tried to argue, not convincingly (see e.g., Earman 1992, Howson and Urbach 1989, Kuipers 1978), that $p(H)$ could not be positive. The second one is that any probability function has the property "$p(E{\rightarrow}H/E) < p(E{\rightarrow}H)$" (Popper and Miller 1983). Although the claimed property is undisputed, the argument that a proper inductive probability function should have the reverse property, since "$E{\rightarrow}H$" is the "inductive conjunct" in the equivalence "$H \Leftrightarrow (E\vee H)\&(E{\rightarrow}H)$", is not convincing. The indicated reverse property may well be conceived as an unlucky first attempt

---

[4] Here 'Bayes' refers to the Bayesian confirmation theory in the Howson-Urbach style, see below, not to Bayesian statistics.

to explicate the core of (probabilistic) inductive intuitions, which should be replaced by the property of instantial confirmation. The defensible reason is that the latter property merely reflects a subjective attitude and, usually, not an objective feature of the underlying probability process, if there is such a process at all.

Carnap, following Laplace, favored inductive likelihoods, although he did not reject inductive priors. The so-called Bayesian approach in philosophy of science reflects inductive priors (but Bayesian statistics uses inductive likelihoods as well, see Festa 1993). Finally, Hintikka introduced "double inductive" probability functions, by combining the Carnapian and the Bayesian approach.

Degree of (Inductive) Confirmation

I now turn to the problem of defining a *degree of inductive confirmation* such that it entails a general definition of *inductive confirmation*. The present approach is not in the letter but in the spirit of Mura (1990) (see also e.g., Schlesinger, 1995) and Milne (1996) and Festa (1999). The idea is to specify a measure for the degree of inductive influence by comparing the relevant "*p*-expressions" with the corresponding (structural) "*m*-expressions" in an appropriate way. I proceed in two stages.

Stage 1. In the first stage we define, as announced, the degree of inductive influence in this degree of confirmation, or simply the degree of inductive (backward) confirmation (according to p), as the ratio:

$$r_p(H, E) \quad = \quad \frac{c_p(H, E)}{c_m(H, E)} \quad = \quad \frac{p(E/H)/p(E)}{m(E/H)/m(E)}$$

A direct consequence of this definition is that the degree of confirmation equals the product of the degree of structural confirmation and the degree of inductive confirmation.

Stage 2. In the second stage I generally define *inductive confirmation*, that is, E inductively confirms H, of course, by the condition: $r_p(H,E) > 1$. This definition leads to four interesting possibilities for confirmation according to p. Assume that $c_p(H,E) > 1$. The *first* possibility is *purely structural* confirmation, that is, $r_p(H,E)=1$, in which case the confirmation has *no inductive features*. This trivially holds in general for structural confirmation, but it may occasionally apply to cases of confirmation according to some p different from m. The *second* possibility is that of *purely inductive* confirmation, that is, $c_m(H,E) = 1$, and hence $r_p(H,E) = c_p(H,E)$. This condition typically applies in the case of instantial confirmation, since, e.g., $m(Fa/Fb\&E)/m(Fa/E) = 1$. The *third* possibility is that of a combination of structural and inductive confirmation: $c_m(H,E)$ and $c_p(H,E)$ both exceed 1, but the second more than the first. This type of *combined* confirmation typically occurs when a Carnapian inductive probability function is assigned e.g., in the case of a die-like object of which it may not be assumed that it is fair. Starting from equal prior probabilities for the six sides such a function gradually approaches the observed relative frequencies. Now suppose that among the even outcomes a high outcome has been observed more than expected on the basis of equal probability. In this case, (only) knowing in addition that the next throw has resulted in an even outcome confirms the hypothesis that it is a high outcome in two ways: structurally (see above) and inductively.

> *Example*. Let n be the total number of throws so far, let n_i indicate the number of throws that have resulted in outcome i $(1,\ldots,6)$. Then the Carnapian probability that the next throw results in i is $(n_i+\lambda/6)/(n+\lambda)$, for some fixed finite positive value of the parameter λ. Hence, the probability that the next throw results in an even outcome is $(n_2+n_4+n_6+\lambda/2)/(n+\lambda)$, and the probability that it is "even-and-high" is $(n_4+n_6+\lambda/3)/(n+\lambda)$. The ratio of the latter to the former is the posterior probability of a high next outcome given that it is even and given the previous outcomes. It is now easy to check that in order to get a degree of confirmation larger than the structural degree, which is 4/3, as we have noted before, this posterior probability should be larger than the corresponding logical probability, which is 2/3. This is the case as soon as $2n_2 < n_4+n_6$, that is, when the average occurrence of '4' and '6' exceeds that of '2'.

Let me finally turn to the *fourth* and perhaps most surprising possibility: confirmation combined with the "opposite" of inductive confirmation, that is, $r_p(H,E) < 1$, to be called *counter-inductive* confirmation. Typical examples arise in the case of deductive confirmation. In this case $r_p(H,E)$ reduces to $m(E)/p(E)$, which may well be less than 1. A specific example is the following: let E be *Fa&Fb* and let p be inductive then E d-confirms "for all x Fx" in a counter-inductive way. On second thoughts, the possibility of, in particular, deductive counter-inductive confirmation should not be surprising. Inductive probability functions borrow, as it were, the possibility of inductive confirmation by reducing the available "amount" of deductive confirmation.

Further research will have to determine whether deductive and inductive confirmation can ever go together in a meaningful way. For the moment the foregoing completes the treatment of HD testing of a theory in terms of confirmation and falsification. I now turn to HD evaluation, which leaves room for continued interest in theories after their falsification.

II. Empirical Progress

HD testing attempts to give an answer to one of the questions in which one may be interested, *the truth question*, which may be qualified according to the relevant epistemological position. However, *the (theory) realist*, for instance, is not only interested in the truth question, but also in some other questions. To begin with, there is the more refined question of which (individual or general) facts the hypothesis explains (its explanatory successes) and which facts are in conflict with the hypothesis (its failures); *the success question* for short. I show in this part that the HD method can also be used in such a way that it is functional in (partially) answering this question. This method is called HD evaluation, and uses HD testing of test implications. Since the realist ultimately aims to approach the strongest true hypothesis, if any, i.e., *the (theoretical-cum-observational) truth* about the subject matter, the plausible third aim of the HD method is to help answer the question of how far a hypothesis is from the truth, *the truth approximation question*. Here the truth will be taken in a relatively modest sense, viz., relative to a given domain and conceptual frame. In Section 7 I make plausible the contention that HD evaluation is also functional in answering the truth approximation question.

The other epistemological positions are guided by two related, but more modest success and truth approximation questions, and I shall show later that the HD method is also functional in answering these related questions. The *constructive empiricist* may not only be interested in the question of whether the theory is empirically adequate or observationally true; i.e., whether the observational theory implied by the full theory is true. He may also be interested in the refined success question about what its true observational consequences and its observational failures are, and in the question of how far the implied observational theory is from the strongest true observational theory, *the observational truth*. The *referential realist* may, in addition, be interested in the truth of the reference claims of the theory and how far it is from the strongest true reference claim, *the referential truth*. The *instrumentalist* phrases the first question of the empiricist more liberally: for what (sub-)domain is it observationally true? He retains the success question of the empiricist. Finally, he will reformulate the third question as follows: to what extent is it the best (and hence the most widely applicable) derivation instrument?

The method of HD evaluation will turn out, in this part, to be a direct way to answer the success question and, in the next part, an indirect way to answer the truth approximation question, in both cases for all four epistemological positions. This part will again primarily be presented in a relatively neutral terminology, with specific remarks relating to the various positions. The success question will be presented in terms of successes and counterexamples[5]: what are the potential successes and counterexamples of the theory?

In sum, two related ways of applying the HD method to theories can be distinguished. The first one is *HD testing*, which aims to answer the truth question. However, as soon as the theory is falsified, the realist with falsificationist leanings, i.e., advocating exclusively the method of HD testing, sees this as a disqualification of an explanatory success. The reason is that genuine explanation is supposed to presuppose the truth of the theory. Hence, from the realist-falsificationist point of view a falsified theory has to be abandoned and one has to look for a new one.

The second method to be distinguished, *HD evaluation*, keeps taking falsified theories seriously. It tries to answer the success question, the evaluation of a theory in terms of its successes and counterexamples (problems) (Laudan 1977). For the (non-falsificationist) realist, successes remain explanatory successes and, when evaluating a theory, they are counted as such, even if the theory is known to be false.

It is important to note that the term '(HD) evaluation' refers to the evaluation in terms of successes and counterexamples, and not in terms of truth approximation, despite the fact that the method of HD evaluation will nevertheless turn out to be functional for truth approximation. Hence, the method of HD evaluation can be used meaningfully without any explicit interest in truth approximation and without even any substantial commitment to a particular epistemological position stronger than instrumentalism.

Chapters 5 and 6 are pivotal in ICR by providing the glue between confirmation and truth approximation, for which reason they are here relatively extensively summarized. Moreover, they are also included in SiS (as Chh. 7 and 8). In addition to what I will present here, ICR Ch. 5 deals more in detail with "falsifying general hypotheses," which if accepted lead to general problems of theories. Moreover, it briefly deals with statistical test implications. Anticipating Part III, ICR Ch. 6 indicates already why the evaluation methodology can be functional for truth approximation. Moreover it explains and justifies in greater detail than the present synopsis the non-falsificationist practice of scientists, as opposed to the explicit falsificationist view of many of them. This is not only the case in terms of fruitful dogmatism, as discovered by Kuhn and Lakatos, but also in terms of truth approximation, for example, by the paradigmatic non-falsificationist method of idealization and concretization, as propagated by Nowak.

---

[5] If the reader finds that the term 'counterexample' has a realist, or falsificationist, flavor, it may be replaced systematically by 'problem' or 'failure'.

5. *Separate Evaluation of Theories by the HD Method*

In this section it is shown that a decomposition of the HD method applied to theories is possible which naturally leads to an explication of the method of separate HD evaluation, using HD testing, even in terms of three models. Among other things, it will turn out that HD evaluation is effective and efficient in answering the success question. In the next section I use the separate HD evaluation of theories for their comparative HD evaluation.

Evaluation Report

The core of the HD method for the evaluation of theories amounts to deriving from the theory in question, say X, General Test Implication (GTI's) and subsequently (HD) testing them. For every GTI I it holds that testing leads sooner or later either to a counterexample of I, and hence a counterexample of X, or to the (revocable) acceptance of I: a *success* of X. A counterexample, of course, implies the falsification of I and X. A success minimally means a "derivational success"; it depends on the circumstances whether it is a predictive success and it depends on one's epistemological beliefs whether or not one speaks of an explanatory success.

Now, it turns out to be very illuminating to write out in detail what is implicitly well-known from Hempel's and Popper's work, viz., that the HD method applied to theories is essentially a stratified, two-step method, based on a macro- and a micro-argument, with much room for complications. In the macro-step already indicated, one derives GTI's from the theory. In their turn, such GTI's are tested by deriving from them, in the micro-step, with the help of suitable initial conditions, testable individual statements, called Individual Test Implications (ITI's). The suggested decomposition amounts in some detail to the following.

For the macro-argument we get:

Theory: X
Logico-Mathematical Claim (LMC): if X then I
____ Modus Ponens (MP)
General Test Implication (GTI): I

A GTI is assumed formally to be of the form:

I: for all x in D [if $C(x)$ then $F(x)$]

that is, for all x in the domain D, satisfying the initial conditions $C(x)$, the fact $F(x)$ is "predicted." All specific claims about x are supposed to be formulated in observation terms.

Successive testing of a particular GTI I will lead to one of two mutually exclusive results. The one possibility is that sooner or later we get falsification of *I* by coming across a falsifying instance or counterexample of *I*. Although a counterexample of *I* is, strictly speaking, also a counterexample of *X*, I also call it, less dramatically, a *negative instance* of or an *individual problem* for *X*. The alternative possibility is that, despite variations in members of *D* and ways in which *C* can be satisfied, all our attempts to falsify *I* fail, i.e., lead to the predicted results. The conclusion attached to repeated success of *I* is of course that *I* is established as true, i.e., as a general (reproducible) fact. I will call such an *I* a (*general*) *success* of *X*. Finally, it may well be that certain GTI's of *X* have already been tested long before *X* was taken into consideration. The corresponding individual problems and general successes have to be included in the evaluation report of *X* (see below).

Recorded problems and successes are (partial) answers to the success question: what are the potential successes and problems of the theory? Hence, testing GTI's derived in accordance with the macro HD argument is effective in answering this question. Moreover, it is efficient, for it will never lead to irrelevant, neutral results, that is, results that are neither predicted by the theory nor in conflict with it. Neutral results for one theory only come into the picture when we take test results of other theories into consideration, that is, the comparative evaluation of two or more theories (see the next section).

I call the list of partial answers to the success question, which are available at a certain moment *t*, the *evaluation report* of *X* at *t*, consisting of the following two components:

the set of individual problems,
i.e., established counterexamples of GTI's of *X*,
the set of general successes,
i.e., the established GTI's of *X*, that is, general facts derivable from *X*.

Hence, the goal of separate theory evaluation can be explicated as aiming at such an evaluation report.

Models of HD Evaluation

Let us now have a closer look at the testing of a general test implication, the micro-step of the HD method, or, more generally, the testing of a *General Testable Conditional* (GTC). The micro HD argument amounts to:

General Test Conditional (GTC): *G*: for all *x* in *D* [if *C(x)* then *F(x)*]
Relevance Condition: *a* in *D*

____ Universal Instantiation (UI)

Individual Test Conditional: if *C(a)* then *F(a)*
Initial Condition(s) (IC): *C(a)*

____ Modus Ponens (MP)

Individual Test Implication (ITI): *F(a)*

If the specific prediction posed by the individual test implication turns out to be false, then the hypothesis *G* has been falsified. The relevant (description of the) object has been called a *counterexample* or a *negative instance* or an *individual problem* of *G*. If the specific prediction turns out to be true the relevant (description of) the object may be called a *positive instance* or an *individual success* of *G*. Besides positive and negative instances of *G*, we may want to speak of *neutral instances* or *neutral results*. They will not arise from testing *G*, but they may arise from testing other general test implications.

Consequently, the evaluation report of GTC's basically has two sides, like the evaluation reports of theories; one for problems and the other for successes. Again, they form partial answers to the success question now raised by the GTC. However, here the two sides list entities of the same kind: negative or positive instances, that is, individual problems and individual successes, respectively.

It is again clear that the micro HD argument for a GTC *G* is effective and efficient for making its evaluation report: each test of *G* either leads to a positive instance, and hence to an increase of *G*'s individual successes, or it leads to a negative instance, and hence to an increase of *G*'s individual problems. It does not result in neutral instances. Note that what I have described above is the micro HD argument for *evaluating* a GTC. When we confine our attention to establishing its truth-value, and hence stop with the first counterexample, it is the (micro) HD argument for *testing* the GTC.

Concatenation of the macro and micro HD argument gives the full argument for theory evaluation leading to individual problems and individual successes. Instead of the two-step concatenated account, theory evaluation can also be presented completely in terms of contracted HD evaluation, without the intermediate GTI's, leading directly to individual problems and individual successes.

Any application of the HD method (concatenated or contracted) leading to an evaluation report with individual problems and individual successes will be called an application of the *micro-model* of HD evaluation. It is clear that application of the micro-model is possible for all kinds of general hypotheses, from GTC's to theories with proper theoretical terms.

However, as far as theories which are not just GTC's are concerned the macro-step also suggests the model of *asymmetric* HD evaluation of a theory, leading to an evaluation report with individual problems and general successes. In that case, GTI's are derived in the macro-step, and only tested, not evaluated, in the micro-step.In the micro-model of HD evaluation of theories, in particular when contraction is used, the intermediate general successes of theories may disappear from the picture. However, in scientific practice, these intermediate results frequently play an important role. The individual successes of theories are summarized, as far as possible, in general successes. These general successes relativize the dramatic role of falsification via other general test implications. As we shall see in the next section, they form a natural unit of merit for theory comparison, together with counterexamples, as the unit of (individual) problems. In the next section, the model of asymmetric HD evaluation plays a dominant role. The results it reports will then be called counterexamples and (general) successes.

However, individual problems frequently can be summarized in terms of "general problems." They amount to established "falsifying general hypotheses" in the sense of Popper. Hence, there is also room for a *macro-model* of HD evaluation, where, besides general successes, the evaluation report lists general problems as well. In this case, all individual successes and individual problems are left out of the picture as long as they do not fit into an established general success or problem. Note that there is also the possibility of a fourth model of HD evaluation of an asymmetric nature, with individual successes and general problems, but as far as I can see, it does not play a role in scientific practice.

The three interesting models of HD evaluation of theories can be ordered in terms of increasing refinement: the macro-model, the asymmetric model, and the micro-model.

It can be shown that the main lines of the analysis of testing and evaluation also apply when the test implications are of a statistical nature. However, for deterministic test implications there are already all kinds of complications of testing and evaluation, giving occasion to "dogmatic strategies" and suggesting a refined scheme of HD argumentation. Although such problems multiply when statistical test implications are concerned, I shall restrict myself to a brief indication of those in the deterministic case.

Complicating Factors

According to the idealized versions of HD testing and evaluation presented so far there are only cases of evident success or failure. However, as is well known, several factors complicate the application of the HD method. Let us approach them first from the falsificationist perspective. Given the fact that scientists frequently believe that their favorite theory is (approximately) true, they have, on the basis of these factors, developed strategies to avoid the conclusion of falsification. The important point of these dogmatic or conservative strategies is that they may rightly save the theory from falsification, because the relevant factor may really be the cause of the seeming falsification. Although the recognition of a problem for a theory is more dramatic from the falsificationist perspective, when evaluating a theory one may also have good reasons for trying to avoid a problem.

I distinguish five complicating factors, each leading to a standard saving strategy. They show in detail, among other things, how Lakatos' methodology of research programs (Lakatos 1970/1978), saving the hard core, can be defended and effected. Though perhaps less frequently practiced, the same factors may also be used, rightly or wrongly, as point of impact for contesting some success. In this case, there is even one additional factor. All six factors concern suppositions in the concatenated macro and micro HD argument. I do not claim originality with these factors as such; most of them have been mentioned by Lakatos and have been anticipated by Hempel, Popper and others. However, their subsequent systematic survey and localization is made possible by the decomposition of the macro and micro HD argument. It is left to the reader to identify examples of the factors.

In the subjoined, refined schematization of the concatenated HD arguments the six main vulnerable factors or weak spots in the argument have been made explicit and emphasized by the addititon of 'Q', which stands for 'Questionable'. The relevant assumptions have been given suggestive names, such that they may be assumed to be self-explanatory. Some of them have been grouped together by the numbering because of their analogous logical role.

Theory: X
Q1.1: Auxiliary hypotheses: A
Q1.2: Background Knowledge: B
Q2: Logico-Mathematical Claim (LMC): if X, A, B then I

____ Modus Ponens (MP)

General Test Implication (GTI): I: for all x in D [if $C(x)$ then $F(x)$]
Q3: Observation presuppositions: $C = C^*$, $F = F^*$

____

General Test Implication (GTI*): I: for all x in D [if $C^*(x)$ then $F^*(x)$]
Q4.1: Relevance Condition: a in D

____ Universal Instantiation (UI)

Individual Test Conditional: if $C^*(a)$ then $F^*(a)$
Q4.2: Initial Condition(s) (IC): $C^*(a)$

____ Modus Ponens (MP)

Individual Test Implication (ITI): $F^*(a)$
Data from repeated tests
Q5: ____ Decision Criteria

either sooner or later	*or* only positive
a counterexample	instances of GTI*,
of GTI*, leading to the	suggesting inference of GTI*
conclusion not-GTI*	by Inductive Generalization Q6

The consequence of the first five factors (auxiliary hypotheses + background knowledge claims, logico-mathematical claims, observation presuppositions, initial conditions, and decision criteria) is that a negative outcome of a test of a theory only points unambiguously in the direction of falsification under certain conditions. Falsification of the theory only follows when it may be assumed that the auxiliary hypotheses, the background knowledge claims and the observation presuppositions are (approximately) true, that the logico-mathematical claim is valid, that the initial conditions were indeed realized and that the used decision criteria were adequate in the particular case. Hence, it will not be too difficult to protect a beloved theory from threatening falsification by challenging one or more of these suppositions.

If the truth question regarding a certain theory is the guiding question, most points of this section, e.g., the decomposition of the HD method, the evaluation report and the survey of complications, are only interesting as long as the theory has not been falsified. However, if one is also, or primarily, interested in the success question the results remain interesting after falsification. In the next section I will show how this kind of separate HD evaluation can be put to work in comparing the success of theories. Among other things, this application explains and even justifies non-falsificationist behavior, including certain kinds of dogmatic behavior.

6. *Empirical Progress and Pseudoscience*

The analysis of separate HD evaluation has important consequences for theory comparison and theory selection. The momentary evaluation report of a theory immediately suggests a plausible way of comparing the success of different theories. Moreover, it suggests the further testing of the comparative hypothesis that a more successful theory will remain more successful and, finally, the rule of theory selection, prescribing its adoption, for the time being, if it has so far proven to be more successful. The suggested comparison and rule of selection will be based on the asymmetric model of evaluation in terms of general successes and individual problems. However, it will also be shown that the symmetric approach, in terms of either individual or general successes and problems, leads to an illuminating symmetric evaluation matrix, with corresponding rules of selection.

Asymmetric Theory Comparison

A central question for methodology is what makes a new theory better than an old one. The intuitive answer for the new theory being as good as the old is plausible enough. The new theory has at least to save the established strengths of the old one and not to add new weaknesses on the basis of the former tests.

In principle, we can choose any combination of individual or general successes and problems to measure strengths and weaknesses. However, the combination of general successes and individual problems, i.e., the two results of the asymmetric model of (separate) HD evaluation, is the most attractive. First, this combination seems the closest to actual practice and, second, it turns out to be the most suitable one for a direct link with questions of truth approximation. For these reasons I will first deal with this alternative and come back to the two symmetric alternatives.

Given the present choice, the following definition is the obvious formal interpretation of the idea of (*prima facie*) progress, i.e., *increasing success*:

> Theory Y is (at time t) at least as successful as (more successful than or better than) theory X iff (at t)
> - all individual problems of Y are (individual) problems of X
> - all general successes of X are (general) successes of Y
> (– Y has extra general successes or X has extra individual problems)

The definition presupposes, of course, that for every recorded (individual) problem of one theory, it has been ascertained whether or not it is also a problem for the other, and similarly whether or not a (general) success of one is also a success of the other. The first clause may be called the "instantial clause" as appealing and relatively neutral. From the realist perspective it is plausible to call the second clause the "explanatory clause." From other epistemological perspectives one may choose another, perhaps more neutral name, such as, the general success clause. It is also obvious how one should define, in similar terms to those above, the general notion of "the most successful theory thus far among the available alternatives" or, simply, "the best (available) theory."

It should be stressed that the diagnosis that Y is more successful than X does not guarantee that this will remain the case. It is a *prima facie* diagnosis based only on facts established thus far, and new evidence may change the comparative judgment. But, assuming that established facts are not called into question, it is easy to check that the judgement cannot have to be reversed, i.e., that X becomes more successful than Y in the light of old and new evidence. For, whatever happens, X has extra individual problems or Y has extra general successes.

It should be conceded that it will frequently not be possible to establish the comparative claim, let alone that one theory is more successful than all its available alternatives. The reason is that these definitions do not guarantee a constant linear ordering, but only an evidence-dependent partial ordering of the relevant theories. In other words, in many cases there will be "divided success": one theory has successes another does not have, and *vice versa*, and similarly for problems. Of course, one may interpret this as a challenge for refinements, e.g., by introducing different concepts of "relatively maximal" successful theories or by a quantitative approach. However, it will become clear that in case of "divided success" another heuristic-methodological approach, of a qualitative nature, is more plausible.

As a matter of fact, the core of HD evaluation amounts to several heuristic principles. The first principle says that, as long as there is no best theory, one may continue the separate HD evaluation of all available theories. The aim is, of course, to explore the domain further in terms of general facts to be accounted for and individual problems to be overcome by an overall better theory. For the moment, I will concentrate on the second principle, applicable in the relatively rare case that one theory is more successful than another one, and hence in the case that one theory is the best.

Suppose theory Y is at t more successful than theory X. This condition is not yet a sufficient reason to prefer Y in some substantial sense. That would be a case of "instant rationality." However, when Y is at a certain moment more successful than X, this situation suggests the following comparative success hypothesis:

CSH: Y (is and) will remain more successful than X

CSH is an interesting hypothesis, even if Y is already falsified. Apart from the fact that Y is known to have some extra successes or X some extra individual problems at t, CSH amounts at t to two components, one about problems, and the other about successes:

CSH-P: all individual problems of Y are individual problems of X

CSH-S: all general successes of X are general successes of Y

where 'all' is to be read as 'all past and future'.

Although there may occasionally be restrictions of a fundamental or practical nature, these two components concern, in principle, testable generalizations. Hence, testing CSH requires application of the micro HD argument. Following CSH-P, we may derive a GTI from Y that does not follow from X, and test it. When we get a counterexample of this GTI, and hence an individual problem of Y, it may be ascertained if the problem is shared by X. If it is not, we have falsified CSH-P.

Alternatively, following CSH-S, we may derive a GTI from X which cannot be derived from Y, and test it. If it becomes accepted, its acceptance means falsification of CSH-S. Of course, in both cases, the opposite test result confirms the corresponding comparative subhypothesis, and hence CSH, and hence increases the registered success difference. In the following, for obvious reasons, I call (these two ways of) testing CSH *comparative HD evaluation*.

The plausible rule of theory selection is now the following:

Rule of Success (RS)
When *Y* has so far proven to be more successful than *X*, i.e., when CSH has been 'sufficiently confirmed' to be accepted as true, eliminate *X* in favor of *Y*, at least for the time being.

RS does not speak of "remaining more successful," for that would imply the presupposition that the CSH could be completely verified (when true). Hence I speak of "so far proven to be more successful" in the sense that CSH has been "sufficiently confirmed" to be accepted as true; that is, CSH is accepted as a (twofold) inductive generalization. The point at which CSH is "sufficiently confirmed" will be a matter of dispute. Be this as it may, the acceptance of CSH and consequent application of RS is the core idea of *empirical progress*, a new theory that is better than an old one. RS may even be considered as the (fallible) criterion and hallmark of scientific rationality, acceptable for the empiricist as well as for the realist.

As soon as CSH is (supposed to be) true, the relevance of further comparative HD evaluation is diminished. Applying RS, i.e., selecting the more successful theory, then means the following, whether or not that theory already has individual problems. One may concentrate on the further separate HD evaluation of the selected theory, or one may concentrate on the attempt to invent new interesting competitors, that is, competitors that are at least as successful as the selected one.

Given the tension between reducing the set of individual problems of a theory and increasing its (general observational) successes, it is not an easy task to find such interesting competitors. The search for such competitors cannot, of course, be guided by prescriptive rules, like RS, but there certainly are heuristic principles of which it is easy to see that they stimulate new applications of RS. Let me start by explicitly stating the two suggested principles leading to RS. First, there is the *principle of separate HD evaluation* (PSE): "Aim via general test implications to establish new laws which can be derived from your theory (general successes) or, equivalently, aim at new negative instances (individual problems) of your theory". Secondly, the *principle of comparative HD evaluation* (PCE): "Aim at HD testing of the comparative success hypothesis, when that hypothesis has not yet been convincingly falsified". In both cases, a typical Popperian aspect is that one should aim at deriving test implications, which are, in the light of the background knowledge, very unlikely or even impossible. The reason is, of course, that a (differential) success of this kind is more impressive than that of a more likely test implication. In view of the first comparative (confirmation) principle (P1), such a success leads in case of PSE to more confirmation of a theory, assuming that that has not yet been falsified, and in case of PCE to more confirmation of the comparative success hypothesis in general.

As already suggested, RS presupposes previous application of PSE and PCE. But some additional heuristic principles, though not necessary, may also promote the application of RS. To begin with, the *principle of content* (PC) may do so: "Aim at success preserving, strengthening or, *pace* Popper, weakening your theory." A stronger theory is likely to introduce new individual problems but gain new general successes. If the latter arise and the former do not materialize, RS can be applied. Something similar applies to a weaker theory. It may solve problems without sacrificing successes. I would also like to mention the *principle of dialectics* (PD) for two theories that escape RS because of divided success: "Aim at a success preserving synthesis of two RS-escaping theories." In ICR (Section 8.3), I explicate a number of dialectical notions in this direction. Of course, there may come a point at which further attempts to improve a theory and hence to discover new applications of RS are abandoned.

In sum, the asymmetric model of HD evaluation of theories naturally suggests the definition of 'more successful', the comparative success hypothesis, the testing of such a hypothesis, i.e., comparative HD evaluation, and the rule of success (RS) as the cornerstone of empirical progress. Separate and comparative HD evaluation provide the right ingredients for applying first the definition of 'more successful' and, after sufficient tests, that of RS, respectively. In short, separate and comparative HD evaluation are functional for RS, and HD testing evidently is functional for both types of HD evaluation. The method of HD evaluation of theories combined with RS and the principles stimulating the application of RS might well be called the instrumentalist methodology. In particular, it may be seen as a free interpretation or explication of Laudan's problem solving model (Laudan 1977), which is generally conceived as a paradigm specification of the idea of an instrumentalist methodology. However, it will also be called, more neutrally, the *evaluation methodology*. It will be said that RS governs this methodology. The claim is that this methodology governs the short-term dynamic of science, more specifically, the internal and competitive development of research programs.

Note that the evaluation methodology demonstrates continued interest in a falsified theory. The reasons behind it are easy to conceive. First, it is perfectly possible that the theory nevertheless passes other general test implications, leading to the establishment of new general successes. Second, even new tests leading to new individual problems are very useful, because they have to be overcome by a new theory. Hence, at least as long as no better theory has been invented, it remains useful to evaluate the old theory further in order to reach a better understanding of its strengths and weaknesses.

Symmetric Theory Comparison

The symmetric models of separate HD evaluation, i.e., the micro- and the macro-models, suggest a somewhat different approach to theory comparison. Although these approaches do not seem to be in use to the extent of the asymmetric one and can only indirectly be related to truth approximation, they lead to a very illuminating (comparative) evaluation matrix.

A better theory has to be at least as successful as the old one, and this fact suggests general conditions of adequacy for the definitions of 'success', of 'problem' and of 'neutral result'. The asymmetric definition of 'at least as successful' presented above only deals explicitly with individual problems and general successes; neutral results remain hidden, but it is easy to check that they nevertheless play a role. The symmetric models take all three types of results explicitly into account. The macro-model focuses on such results of a general nature, the micro-model on such results of an individual nature.

The notions of general successes and general problems are not problematic. Moreover, general facts are neutral for a theory when they are neither a problem nor a success. A better theory retains general successes as (already tested) general test implications, and does not give rise to new general test implications of which testing leads to the establishment of new general problems. Moreover, general problems may be transformed into neutral facts or even successes, and neutral general facts may be transformed into successes. The notions of individual successes, individual problems and neutral results are not problematic either, as long as we list them in terms of positive, negative and neutral instances, respectively. A better theory keeps the positive instances as such; it does not lead to new negative instances, and neutral instances may remain neutral or become positive. However, if we want to list individual successes and/or individual problems in terms of statements, the situation becomes more complicated, but it is possible (see ICR, pp. 116-7).

Let us now look more specifically at the symmetric micro-model, counting in terms of individual problems, successes and neutral results, that is, negative, positive and neutral instances or (statements of) individual facts. Hence, in total, the two theories produce a matrix of nine combinations of possible instances or individual facts. In order that the matrix can also be made useful for the macro-model, I present it in terms of facts. For the moment, these facts are to be interpreted as individual facts. The entries represent the status of a fact with respect to the indicated theories X and Y.

		X		
		negative	neutral	positive
Y	negative	B4: 0	B2: −	B1: −
	neutral	B8: +	B5: 0	B3: −
	positive	B9: +	B7: +	B6: 0

The (comparative) evaluation matrix

From the perspective of *Y* the boxes B1/B2/B3 represent unfavorable facts (indicated by '−'), B4/B5/B6 (comparatively neutral or) indifferent facts (0), and B7/B8/B9 favorable facts (+). The numbering of the boxes, anticipating a possible quantitative use, was determined by three considerations: increasing number for increasingly favorable results for *Y*, a plausible form of symmetry with respect to the diagonal of indifferent facts, and increasing number for indifferent facts that are increasingly positive for both theories.

It is now highly plausible to define the idea that *Y* is *more successful* than *X* in the light of the available facts as follows: there are no unfavorable facts and there are some favorable facts, that is, B1/2/3 should be empty, and at least one of B7/8/9 non-empty. This state of affairs immediately suggests modified versions of the comparative success hypothesis and the rule of success.

It is also clear that, by replacing individual facts by general facts, we obtain macro-versions of the matrix, the notion of comparative success, the comparative success hypothesis and the rule of success. A general fact may be a general success, a general problem or a neutral general fact for a theory.

In all these variants, the situation of being more successful will again be rare, but it is certainly not excluded. In ICR (Chapter 11) I argue, for instance, that the theories of the atom developed by Rutherford, Bohr and Sommerfeld can be ordered in terms of general facts according to the symmetric definition.

Another set of examples of this kind is provided by the table (adapted from: Panofsky and Phillips 1962[2], p. 282), representing the records in the face of 13 general experimental facts of the special theory of relativity (STR) and six alternative electrodynamic theories, viz., three versions of the ether theory and three emission theories. According to this table, STR is more successful than any of the other; in fact it is maximally successful as far as the 13 experimental facts are concerned. Moreover, Lorentz's contraction version of the (stationary) ether theory is more successful than the contractionless version. Similarly, the ballistic version of the emission theory is more successful than the other two. However, it is also clear that many combinations lead to divided results. For instance, Lorentz's theory is more successful in certain respects (e.g., De Sitter's spectroscopic binaries) than the ballistic theory, but less successful in other respects (e.g., the Kennedy-Thorndike experiments).

                    Theo A. F. Kuipers

Theories	Experimental facts	Light propagation experiments					Experiments from other fields							
		Aberration	Fizeau convection coefficient	Michelson-Morley	Kennedy-Thorndike	Moving sources and mirrors	De Sitter spectroscopic binaries	Michelson-Morley, using sunlight	Variation of mass with velocity	General mass-energy equivalence	Radiation from moving charges	Meson decay at high velocity	Trouton-Noble	Unipolar induction
Ether theories	Stationary ether, no contraction	A	A	D	D	A	A	D	D	N	A	N	D	D
	Stationary ether, Lorentz contraction	A	A	A	D	A	A	A	A	N	A	N	A	D
	Ether attached to ponderable bodies	D	D	A	A	A	A	A	D	N	N	N	A	N
Emission theories	Original source	A	A	A	A	A	D	D	N	N	D	N	N	N
	Ballistic	A	N	A	A	D	D	D	N	N	D	N	N	N
	New source	A	N	A	A	D	D	A	N	N	D	N	N	N
Special theory of relativity		A	A	A	A	A	A	A	A	A	A	A	A	A

Comparison of experimental record of seven electrodynamic theories.
Legend: A: agreement, D: disagreement, N: not applicable

In the present approach it is plausible to define, in general, one type of divided success as a liberal version of more successfulness. *Y is almost more successful* than *X* if, besides some favorable facts and (possibly) some indifferent facts, there are some unfavorable facts, but only of the B3-type, provided there are (favorable) B8- or B9-facts or the number of B3-facts is (much) smaller than that of their antipodes, that is, B7-facts. The provision clause guarantees that it remains an asymmetric relation. Crucial is the special treatment of B3-facts. They correspond to what is called *Kuhn-loss*: the new theory seems no longer to retain a success demonstrated by the old one. The idea behind their suggested relatively undramatic nature is the belief that further investigation may show that and how a B3-fact turns out to be a success after all, perhaps by adding an additional (non-problematic) hypothesis. In this case it becomes an (indifferent) B6-fact. Hence, the presence of B3-facts is first of all an invitation to further research. If this is unsuccessful, such a B3-fact becomes a case of recognized Kuhn-loss. Unfortunately, the table above does not contain an example of an almost more successful theory.

Cases of divided success may also be approached by some (quasi-) quantitative weighing of facts. Something like the following quantitative evaluation matrix is directly suggested by the same considerations that governed the number ordering of the boxes.

		X		
		negative	neutral	positive
Y	negative	B4: −1/−1	B2: −3/+3	B1: −4/+4
	neutral	B8: +3/−3	B5: 0/0	B3: −2/+2
	positive	B9: +4/−4	B7: +2/−2	B6: +1/+1

The quantitative (comparative) evaluation matrix

All qualitative success orderings of electrodynamic theories to which the table gives rise, remain intact on the basis of this quantitative matrix (which is not automatically the case). Moreover, we now of course get a linear ordering, with Lorentz's theory in the second position after STR and far ahead of the other alternatives. Of course, one may further refine such orderings by assigning different basic weights to the different facts, to be multiplied by the relative weights specified in the quantitative matrix.

Like a similar observation in the symmetric case, it is now possible to interpret the qualitative and the quantitative versions of the evaluation matrix as explications of some core aspects of Laudan's (1977) problem-solving model of scientific progress, at least as far as empirical problems and their solutions are concerned.

Scientific and Pseudoscientific Dogmatism

Although the method of HD testing, HD evaluation, and hence the evaluation methodology have a falsificationist flavor, each with its own aim, they are certainly not naïve in the sense in which Popper's methodology has sometimes been construed. Naïve falsificationism in the sense described by Lakatos (1970/1978) roughly amounts to applying HD testing for purposes of theory evaluation and elimination. Its core feature then becomes to further discard (convincingly) falsified theories. Lakatos has also construed a sophisticated version of falsificationism such that, when comparing theories, he takes their "unrefuted content" into account, a practice that allows falsified theories to remain in the game. Moreover, Lakatos has proposed a "methodology of research programs", which operates in a sophisticated falsificationist way. However, it works in such a way that it postpones the recognition of falsifications of the "hard core theory" as long as it is possible to roll off the causes of falsification dogmatically onto auxiliary hypotheses or background theories.

It can be argued that HD evaluation can be seen as an explication of sophisticated falsificationism, leaving room for a dogmatic research program specification. Moreover, it can be argued that the falsificationist and the evaluation methodology may be functional for truth approximation, and that the latter non-falsificationist methodology, ironically enough, is much more efficient for that purpose.

The (naïve) falsificationist methodology amounts to restricting the rule of success (RS) to not-yet-falsified theories in combination with the following rule:

Rule of Elimination (RE)
When a theory has been convincingly falsified, elimination should follow, and one should look for a new theory

The evaluation methodology can be summarized by the

Principle of Improvement (*of theories*) (PI)
Aim at a more successful theory, and successive application of RS

Both methodologies presuppose the

Principle of (*Falsifiability or*) *Testability* (PT)
Aim at theories that can be tested, and hence evaluated, in the sense that test implications can be derived, which can be tested for their truth-value by way of observation

Hence, the relativization of the methodological role of falsification, inherent in the evaluation methodology, should not be construed as a plea to drop falsifiability as a criterion for being an empirical theory. On the contrary, empirical theories are supposed to be able to score successes or, to be precise, general successes. Moreover, PI presupposes the principles of separate and comparative HD evaluation (PSE/PCE) as introduced in Section 5, whereas PE presupposes them only for not yet falsified theories. Finally, it is possible to extend PI to nonempirical features, for example aesthetic features such as simplicity and symmetry. In Sections 8 and 9 I will formally take such features into account in relation to truth approximation. However, it should be clear that their methodological role in theory choice is primarily or even exclusively restricted to cases of equal empirical success (see Kuipers 2002, Section 6, for a detailed treatment of their role).

It is clear that RE may retard empirical progress in the sense of PI. Moreover, it can also be argued that RE affects the prospects for truth approximation. A striking feature of PI in this respect is that the question of whether the more successful theory is false or not does not play a role at all.

That is, the more successful theory may well be false, provided all its counterexamples are also counterexamples of the old theory.

These claims about truth approximation have important methodological consequences. They enable a new explanation, even justification, of the observation of Kuhn, Lakatos and others that there is quite a discrepancy between falsificationist (methodological) theory and non-falsificationist practice. In principle, even with respect to the paradigmatic non-falsificationist method of idealization and concretization, as propagated by Nowak, but this requires "refined" truth approximation (see Section 10). Straightforward (basic or refined) truth approximation may be seen as the primary, conscious or unconscious, motive for non-falsificationist behavior. Dogmatic behavior, in the sense of working within a research program, is only a secondary motive for non-falsificationist behavior. Whatever the main motive, as long as such behavior is directed at theory improvement within the program, it can be distinguished from pseudoscientific behavior.

The following principle expresses the core idea:

Principle of improvement guided by research programs (PIRP)
One should primarily aim at progress within a research program, i.e., aim at a better theory while keeping the hard core of the program in tact. If, and only if, this strategy does not work, try to adapt the hard core, while leaving the vocabulary in tact. If, and only if, this second strategy is also unsuccessful, look for another program with better perspectives on progress

Whereas responsible dogmatic behavior is governed by this refined principle of improvement, leaving room for dogmas, one of the typical marks of pseudoscientific behavior is that one is usually not even aiming at improvement by the first strategy, let alone by the second.

Our notion of comparative evaluation is governed by the notion of being "(almost) more successful". This is a rather strict strategy. In ICR I question the general usefulness of quantitative liberalizations of 'successfulness', and for that matter, of 'truthlikeness', mainly because they need real-valued distances between models, a requirement which is very unrealistic in most scientific contexts. Hence, the applicability of liberal notions may well be laden with arbitrariness. Be this as it may, it is important to stress that the strict strategy does not lead to void or almost void methodological principles. If there is divided success between theories, the Principle of Improvement amounts, more specifically, to the already mentioned recommendation that we should try to apply the Principle of Dialectics: "Aim at a success preserving synthesis of the two RS-escaping theories", of course, with a plausible program-bound version. Hence, the restricted applicability of the strict notion of comparative success does not exclude the possibility of clear challenges being formulated in cases where they do not apply – on the contrary.

III. Basic Truth Approximation

This part introduces and analyzes the theory of naïve or basic truth approximation and its relation to empirical progress and confirmation, first for epistemologically unstratified theories and later for stratified ones.

In Section 7 the qualitative idea of truthlikeness is introduced, more specifically the idea that one theory can be closer to the truth than another, which is called "nomic truthlikeness." Here 'the truth' concerns "the nomic truth," i.e., the strongest true hypothesis, assumed to exist according to the "nomic postulate," about the physical or nomic possibilities, called "the nomic world," restricted to a given domain and, again, as far as can be expressed within a given vocabulary. The Success Theorem is crucial, according to which 'closer to the truth' implies 'being at least as successful', even straightforwardly, if defined in the asymmetric way. It is used to argue that the evaluation methodology is effective and efficient for nomic truth approximation.

ICR Ch. 7, moreover, deals with "actual" truthlikeness and truth approximation, where 'the actual truth' represents the actual possibility or (restricted) world, or their historical succession. The chapter results in a survey of bifurcations of truthlikeness theories and concludes with their plausible methodological and epistemological consequences for the notions of novel facts, crucial experiments, inference to the best explanation and descriptive research programs.

Section 8 argues that "basic" nomic truthlikeness and the corresponding methodology have plausible conceptual foundations, of which the dual foundation will be the most appealing to scientific common sense: 'more truthlike' amounts to 'more true consequences and more correct models', in line with the asymmetric definition of 'more successful'. There is also an indication of how this analysis leaves room for nonempirical considerations in theory evaluation, such as aesthetic ones.

In ICR Ch. 8 a detailed comparison is presented between Popper's original definition of truthlikeness and the basic definition, showing among other things that the latter does not have the generally recognized shortcomings of the former. Moreover, it is also argued that basic truthlikeness suggests a non-standard, viz., intralevel rather than interlevel, explication of the main intuitions governing the so-called correspondence theory of truth. Moreover, it is made clear that the presented cognitive structures suggest logical, methodological and ontological explications of some main dialectical concepts, viz., dialectical negation, double negation, and the triad of thesis-antithesis-synthesis.

Section 9 introduces the first major sophistication, the stratification arising from the (changing) distinction between observational and theoretical terms, leading to the distinction between observational and theoretical truth approximation.

In ICR Ch. 9 this also leads to the idea of "the referential truth," i.e., the truth about which terms of a vocabulary refer and which do not, where 'reference' gets a precise definition on the basis of the nomic postulate. This enables the definition of the referential claim of a theory and hence of the idea of one theory being closer to the referential truth than another. Moreover, the overall analysis is shown to lead to plausible rules of inference to the best theory, viz. as the closest to the observational, the theoretical, and the referential truth.

For readers with a model theoretic background it is important to realize the main divergence between the (dominant) model theoretic view on empirical theories and my favorite so-called "structuralist" perspective on them. According to the former the target of theorizing is one particular "intended application," the actual world, and according to the latter it is a set of "intended applications," the nomic possibilities. Although the suggested model theoretic perspective may be dominant, I would like to leave room for the possibility that an alternative model theory in line with the structuralist perspective will be further developed and become respected. Not in order to replace the dominant one but in order to obtain an alternative that is more suitable for certain purposes. However, I do not see this alternative as a non-Tarskian move in some deep sense. Starting from Tarski's basic definition, which is that of "truth in a structure" (Hodges 1986), and assuming that one has more than one intended application in mind, it is plausible to define that a theory is true if and only if it is true for all intended applications. However, in this case there are at least two coherent ways of defining that a theory is false. In line with the standard approach one may be inclined to call a theory only false when it is false for all intended applications, and indeterminate if it is neither true for all intended applications nor false. However, it is in line with the structuralist approach to call a theory already false if it is false for at least one intended application.

7. Truthlikeness and Truth Approximation

I shall first deal with the logical or conceptual problem of defining '(more) truthlikeness', assuming that we know what "the truth" is. I then turn to the prospects for truth approximation by using the method of HD evaluation. In

                                    Theo A. F. Kuipers

this section we do not yet assume a distinction between theoretical and observational terms, which amounts to assuming that all terms are observational.

Truthlikeness

The starting point of the idea of truthlikeness is a vocabulary and a domain. A conceptual possibility is a situation or state of affairs that can be described in the vocabulary, and is therefore conceivable. Let CP be the set of all *conceptual possibilities* that can be described in terms of the vocabulary, also called the *conceptual frame*. A theory will be associated with a subset of CP. A basic assumption, the *Nomic Postulate*, is that the representation of the chosen domain in terms of the vocabulary results in a unique subset of CP containing the *nomic possibilities*. We can identify this usually unknown subset with the truth T for reasons that will become clear shortly. For the sake of convenience I here assume that we can somehow characterize T in terms of the vocabulary.

The aim of theory formation is the actual characterization of T. Hence, the nomic possibilities constituting T can also be called *desired* possibilities, and the elements in $CP - T$, representing the nomic *impossibilities*, can also be called the *undesired* possibilities. A theory X consists of a subset X of CP, with the strong claim "$X = T$". If X encloses T, X does not exclude desired possibilities. Thus the weaker claim "$T \subseteq X$", meaning that X admits all desired possibilities, is true, in which case we will also say that X is true as a hypothesis. If this weaker claim is false we will also say that X is false as a hypothesis. If $T \subseteq Y \subset X$, Y excludes more undesired possibilities than X and so the claim "$T \subseteq Y$", that goes with it, is stronger than "$T \subseteq X$", but nevertheless true. In this sense theory T itself is the strongest true theory, and I call it *the truth*. It seems useful to call the elements of X (its) *admitted* possibilities and those of $CP - X$ the *excluded* possibilities (of X). Now it is important to note that the elements of $X \cap T$ are the desired possibilities admitted by X, and $X - T$ consists of the undesired possibilities admitted by X. In Figure 1 all four resulting categories are depicted.

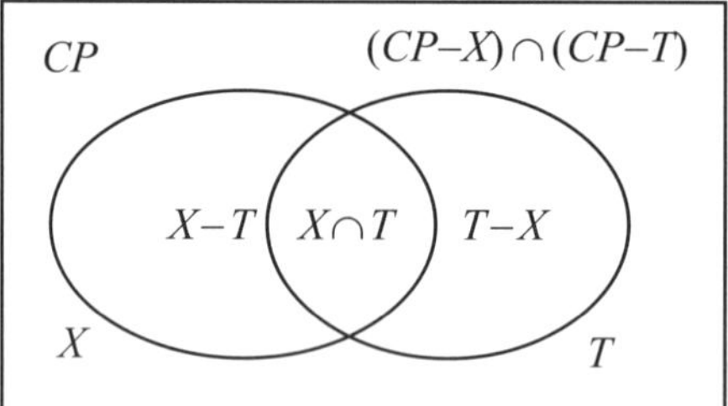

CP: set of conceptual possibilities
T: set of nomic/desired possibilities
X: set of admitted possibilities
$X \cap T$: desired possibilities admitted by X
X–T: undesired possibilities admitted by X
T–X: desired possibilities excluded by X
$(CP - X) \cap (CP - T)$: undesired possibilities excluded by X

*Fig.*1. Four categories of possibilities

This brings us directly to the basic definition of (equal or greater) truthlikeness:

Definitions

Y is *at least as close to T as X* (or: *Y resembles T as much as X*) iff
(DP)   all desired possibilities admitted by *X* are also admitted by *Y*
(UP)   all undesired possibilities admitted by *Y* are also admitted by *X*

Y is (two-sided) *closer to T than X* (or: *Y resembles T more than X*) iff
(DP) & (DP+)   *Y* admits extra desired possibilities
(UP) & (UP+)   *X* admits extra undesired possibilities

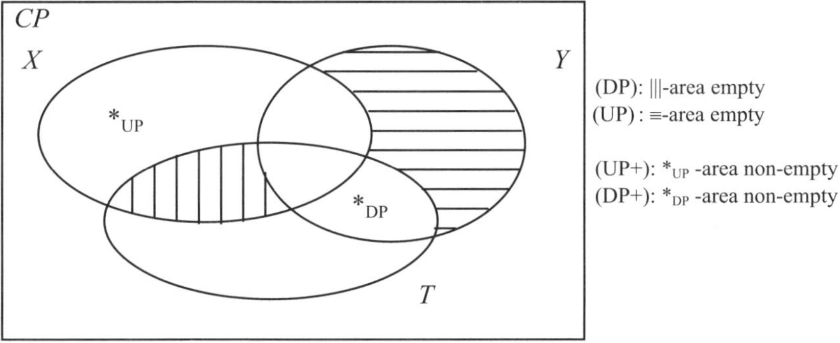

Fig.2. Y is closer to the truth *T* than *X*

Figure 2 indicates which sets must be empty (clause (DP) and (UP): vertical and horizontal shading, respectively) and which sets have to be non-empty (clause (DP+) and (UP+): area *DP and area *UP non-empty, respectively) in the case that *Y* is closer to the truth than *X*.

Truth Approximation

Now we are able to rephrase the notion of empirical progress and sketch its relation to nomic truth approximation, assuming that *T* is unknown. Recall that as far as theories are concerned, we have dealt up to now with the logical problem of defining nomic truthlikeness, assuming that *T*, the set of nomic possibilities, is at our disposal. In actual scientific practice we don't know *T*; it is the target of our theoretical and experimental efforts. I will now explicate the idea that one theory is more successful than another in terms of "realized conceptual possibilities" and show that this can be explained by the hypothesis of nomic truth approximation, that is, the hypothesis that the first theory is closer to the truth than the second.

First it is important to note that you can establish that a certain conceptual possibility is nomically possible by experimentally realizing this possibility, but you cannot establish that a certain conceptual possibility is nomically *im*possible in a direct way, for you cannot realize nomic *im*possibilities. The standard (partially) indirect way to circumvent this problem is by establishing on the one hand nomic possibilities by realizing them, and on their basis establishing (observational) laws on the other. As we have seen in the preceding section, this is precisely what the separate HD evaluation of theories amounts to. That is, theories are evaluated in terms of their capacity to respect the realized possibilities, i.e., to avoid counterexamples, and to entail the observational laws, i.e., to have general successes.

The problems and successes of a theory will have to be expressed in terms of the data to be accounted for. The data at a certain moment t can be represented as follows. Let $R(t)$ indicate the set of realized possibilities up to t, i.e., *the accepted instances (of T)*, which have to be admitted by a theory. Note that there may be more than one realized possibility at the same time, before or at t, with plausible restrictions for overlapping domains.

Up to t there will also be some accepted general hypotheses, *the (explicitly) accepted laws*, which have to be accounted for by a theory. On their basis, *the strongest accepted law* to be accounted for is the general hypothesis $S(t)$ associated with the intersection of the sets constituting the accepted laws. It claims that all nomic possibilities satisfy its condition, i.e., it claims that "$T \subseteq S(t)$". Of course, $S(t)$ is, via the laws constituting it, in some way or other based on $R(t)$; minimally we may assume that $R(t)$ is not in conflict with $S(t)$, that is, $R(t)$ is a subset of $S(t)$. In the following, however, I shall need the much stronger *correct data* (CD-)*hypothesis* $R(t) \subseteq T \subseteq S(t)$, guaranteeing that $R(t)$ only contains nomic possibilities, and that hypothesis $S(t)$ only excludes nomic impossibilities. $S(t)$ is thus (assumed to be) true as a hypothesis and may hence rightly be called a law. Henceforth I assume the CD-hypothesis. $R(t)$ may now be called the set of *established nomic possibilities*, and $S(t)$ *the strongest established law*. In fact, for every superset H of $S(t)$ (but subset of CP), hence $S(t) \subseteq H \subseteq CP$, the claim "$T \subseteq H$" is also true, for which reason H may be called *an (explicitly or implicitly) established law*. Let $Q(S(t))$ indicate the set of all supersets of $S(t)$. Then $Q(S(t))$ represents the set of all *established laws*.

Assuming the data $R(t)$ and $S(t)$, it is now easy to give explications of the notions of individual problems and general successes of a theory X at time t we met in Part II concerning the HD evaluation of theories. The set of *individual problems*, of X at t, is equated with the established members of $R(t) - X$, that is, established nomic possibilities that are not admitted by X. Similarly, the set of *general successes,* of X at t, is equated with the set of established laws that are supersets of X, that is, the members of $Q(S(t)) \cap Q(X)$.

For comparative judgements of the success of theories I shall now explicate the instantial clause of Section 6 in terms of established nomic possibilities, i.e., $R(t)$. Theory Y is *instantially at least as successful as* X if and only if the individual problems of Y form a subset of those of X, that is, Y has no *extra* individual problems. Formally, including some equivalent versions:

$$R(t) - Y \subseteq R(t - X$$
$$\Leftrightarrow X \cap R(t) - Y = \varnothing \Leftrightarrow X \cap R(t) \subseteq Y \cap R(t)$$

On the other hand, for the explanatory (or general success) clause we have two options for explication, one on the (first) level of subsets of CP and one on the (second) levels of sets of such subsets, leading to two equivalent comparative statements. To begin with the second level, the level of consequencés, theory Y is *explanatorily at least as successful as* X if and only if the general successes of X form a subset of those of Y, that is, X has no *extra* general successes. Formally:

$$Q(X) \cap Q(S(t)) \subseteq Q(Y) \cap Q(S(t))$$
$$\Leftrightarrow Q(X) \cap Q(S(t)) - Q(Y) = \varnothing \Leftrightarrow Q(S(t)) - Q(Y) \subseteq Q(S(t)) - Q(X)$$

On the first level, the level of sets, this is equivalent to the condition that the 'established nomic impossibilities' excluded by X form a subset of those of Y. Formally:

$$(CP-S(t)) \cap (CP-X) \subseteq (CP-S(t)) \cap (CP-Y)$$
$$\Leftrightarrow Y-X \cup S(t) = \varnothing \Leftrightarrow Y-S(t) \subseteq X-S(t)$$

The proof of this equivalence is formally the same as that of the (first) 'equivalence thesis' that will be presented in the next section.

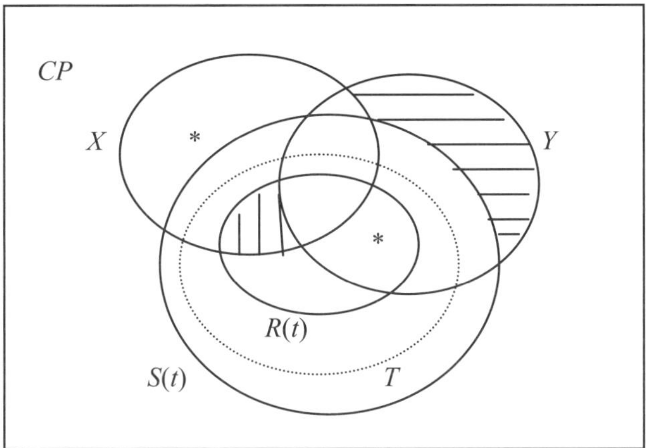

*Fig.*3. Y is two-sidedly more successful than X relative to $R(t)/S(t)$: shaded areas empty, starred areas non-empty. The unknown T is drawn such that the correct data hypothesis is built in.

The conjunction of the instantial and the explanatory clause forms the general definition of the statement that one theory is at a certain time at least as successful as another, relative to the data $R(t)/S(t)$. It will be clear that this definition can be seen as an explication of the "asymmetric" definition given in Section 6. We obtain the strict version, that is, more successful, when in at least one of the two cases proper subsets are concerned. It is called two-sided when proper subsets are involved in both cases. The "two-sided" strict version is depicted on the first level in Figure 3 (in which T is indicated by a dotted ellipse to stress that it is unknown).

Now it is easy to prove the following crucial theorem:

Success Theorem:

If theory Y is at least as close to the nomic truth T as X and if the data are correct then Y (always) remains at least as successful as X.

From this theorem it immediately follows that *success dominance* of Y over X, in the sense that Y is at least as successful as X, can be explained by the following hypotheses: the *truth approximation (TA-)hypothesis*, Y is at least as close to the nomic truth T as X, and the auxiliary *correct data (CD-)hypothesis*.

All notions in the theorem have been explicated, and the proof is, on the first level, only a matter of elementary set-theoretical manipulation, as will be clear from the following presentation of the theorem as an argument:

$$
\begin{array}{llll}
Y\!-\!T \subseteq X\!-\!T & \quad T\!-\!Y \subseteq T\!-\!X & \quad \text{TA-hypothesis} \\
T \subseteq S(t) & \quad R(t) \subseteq T & \quad \text{CD-hypothesis} \\
\hline
Y\!-\!S(t) \subseteq X\!-\!S(t) & \quad R(t)\!-\!Y \subseteq R(t)\!-\!X & \quad \text{success dominance}
\end{array}
$$

As a rule, a new theory will introduce some new individual problems and/or will not include all general successes of the former theory. The idea is that the relative merits can now be explained on the basis of a detailed analysis of the relative "position" to the truth. However, for such cases a general theorem is obviously not possible.

The importance of The Success Theorem is that it can explain that, and how empirical progress is possible within a conceptual frame CP for a given domain. For this purpose, recall first the Comparative Success Hypothesis (CSH) and the Rule of Success (RS), introduced in Section 6:

CSH: Y (is and) remains more successful than X

RS: When Y has so far been proven to be more successful than X, i.e., when CSH has been "sufficiently confirmed" to be accepted as true, then eliminate X, in favor of Y, at least for the time being.

For an instrumentalist, CSH and RS are already sufficiently interesting, but the (theory-)realist will only appreciate it for its possible relation to truth approximation, whereas the empiricist and the referential realist will have intermediate interests.

The Success Theorem shows that RS is functional for approaching the truth in the following sense. Assuming correct data, the theorem suggests that the fact that "Y has so far proven to be more successful than X" may well be the consequence of the fact that Y is closer to the truth than X. For the theorem enables the attachment of three conclusions to the fact that X has so far proven to be more successful than X; conclusions which are independent of what exactly "the nomic truth" is:

- first, it is still possible that Y is closer to the truth than X, a possibility which, when conceived as a hypothesis, the TA-hypothesis, according to the Success Theorem, would explain the greater success in a general way,
- second, it is impossible that Y is further from the truth than X (and hence X closer to the truth than Y), for otherwise, so teaches the Success Theorem, Y could not be more successful,
- third, it is also possible that Y is neither closer nor further from the truth than X, in which case, however, another explanation, now of a specific nature, has to be given for the fact that Y has so far proven to be more successful.

Hence we may conclude that, though "so far proven to be more successful" does not guarantee that the theory is closer to the truth, it provides good reasons to make this plausible. And this is increasingly the case, the more the number and variation of tests of the comparative success hypothesis increase. It is in this sense that I interpret the claim that RS is functional for truth approximation: the longer the success dominance lasts, despite new experiments, the more plausible that this is the effect of being closer to the truth.

In view of the way in which the evaluation methodology is governed by RS, this methodology is, in general, functional for truth approximation. I would like to spell out this claim in more detail. Recall that the separate and comparative HD evaluation of theories was functional for applying RS in the sense that they precisely provide the ingredients for the application of RS. Recall moreover that HD testing of hypotheses is functional for HD evaluation of theories entailing them. Hence, we get a transitive sequence of functional steps for truth approximation:

HD testing of hypotheses
    → separate HD evaluation of theories
        → comparative HD evaluation of theories
            → Rule of Success (RS)
                → Truth Approximation (TA)

Consequently, from the point of view of truth approximation, RS can be justified as a prescriptive rule, and HD testing and HD evaluation as its drive mechanisms. Intuitive versions of the rule and the two methods are usually seen as the hallmark of scientific rationality. The analysis of their truth approximating cooperation can be conceived as an explication of what many scientists are inclined to think, and others are inclined to doubt. To be sure, the understanding is not relevant for the practice. That the practice is functional for truth approximation may be conceived as the cunning of reason in science.

It is important to stress once more that RS does not guarantee that the more successful theory is closer to the truth. As long as one does not dispose of explicit knowledge of T, it is impossible to have a rule of success that can guarantee that the more successful theory is closer to the truth. As we shall see at the end of Section 9, there is only one (near) exception to this claim: purely inductive research.

Another way to summarize the above findings is the following. The TA-hypothesis, claiming that one theory is at least as close to the truth as another, is a perfect example of an empirically testable comparative hypothesis. The Success Theorem says that the TA-hypothesis implies, and hence explains, that the first theory will always be at least as successful as the second.

In terms of an application of HD testing, the Success Theorem amounts to the following claim: the TA-hypothesis has the following two general comparative test implications (assuming the strong, but plausible, auxiliary CD-hypothesis):

all general successes of X are general successes of Y

all individual problems of Y are individual problems of X

Note that these are precisely the two components of the comparative success hypothesis (CSH). Hence, when Y is at least as successful as X, the further HD evaluation, i.e., the further testing of CSH, can indirectly be seen as further HD testing of the TA-hypothesis. When doing so, the latter hypothesis can be falsified, or it can be used to explain newly obtained success dominance.

Recall that I noted in Section 6 that the application of the prescriptive rule RS can be stimulated by several heuristic principles, viz., the *principle of separate HD evaluation* (PSE), the *principle of comparative HD evaluation* (PCE), the *principle of content* (PC), and, finally, the *principle of dialectics* (PD). Of course, we may now conclude that all these principles belonging to the evaluation methodology are indirectly functional for truth approximation.

The Success Theorem is not only attractive from the realist point of view, it is also instructive for weaker epistemological positions, even for the instrumentalist. The Success Theorem implies that a theory that is closer to the truth is also a better derivation instrument for successes, and that the truth (the true theory) is the best derivation instrument. From this not only the self-evident fact follows indirectly that RS and HD evaluation are functional for approaching the best derivation instrument, but also that the heuristic of the realist may also be of help to the instrumentalist. Its core is the Nomic Postulate, according to which, given a domain, each conceptual frame has a unique strongest true hypothesis, i.e., the truth. Intermediate considerations apply both to the constructive empiricist and to the referential realist.

We can go even further. Given the proof of the Success Theorem, the following theorem is now easy to prove:

Forward Theorem

If CSH, which speaks of *remaining* more successful, is true, this implies the TA-hypothesis, that is, if Y is not closer to the nomic truth than X, (further) testing of CSH will sooner or later lead to an extra counterexample of Y or to an extra success of X.

In other words, 'so far proven to be more successful' can only be explained by the TA-hypothesis (Success Theorem) or by assuming that the comparative success hypothesis has not yet been sufficiently tested (Forward Theorem).

It is important to stress once more that the present section was based on the assumption of an observational vocabulary. I withdraw this assumption in the next two sections.

8. *Intuitions of Scientists and Philosophers*

The main point of Ch. 8.1 of ICR is that it is possible to give a "dual foundation" of nomic truthlikeness and the corresponding methodology, which can be seen as an explication of some basic intuitions and practices of scientists. However, it turned out later that this approach can also account for intuitions about the role of nonempirical, e.g. aesthetic, features of theories. Here I present the analysis with the option of explicating this additional intuition of scientists in mind. For this purpose I also leave room for a distinction between observational and theoretical terms.

The definitions of equal and more truthlikeness of Section 7 can be reformulated in terms of desirable and undesirable features of a theory. The starting point consists of properties of possibilities. A feature of a theory will be understood as a "distributed" feature, that is, a property of all the possibilities that the theory admits. This leaves room for empirical features of theories, such as all its possibilities satisfying certain observational laws, but also for nonempirical features. For example, a theory is frequently called symmetric because all its possibilities show a definite symmetry. According to this definition, a feature of a theory can be represented as a set of possibilities, namely as the set of all possibilities that have the relevant property. This set

                      Theo A. F. Kuipers

then contains the set of all possibilities that the theory admits. Note that this means that we could say that a feature of a theory excludes (exactly) all possibilities that do not have that property.

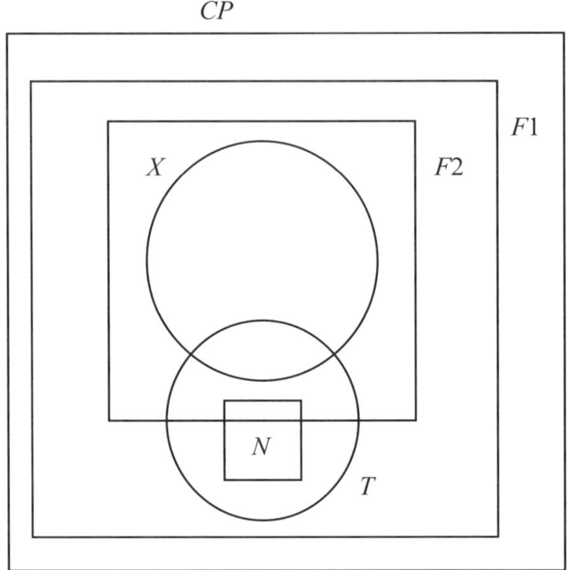

*Fig.*4. Three types of features

It is obvious how we can formulate explicit definitions of desired, undesired, and remaining features in terms of the (logical) exclusion of desired and undesired possibilities: desired features are features that include all desired possibilities or, equivalently, that exclude only undesired possibilities; undesired features are features that include all undesired possibilities or, equivalently, that exclude only desired possibilities. All remaining features, as far as they can be represented as a subset of *CP*, exclude desired and undesired possibilities; that is, they do not include either all desired possibilities or all undesired ones. These are features about which we can be neutral, for which reason I call them neutral features. However, they will play no role in the following analysis.[6] The three types of features are depicted in Figure 4.

---

[6] Popper has given a definition of 'closer to the truth' in terms of more true and fewer false consequences, that was acknowledged later (also by Popper himself) to be unsound. In terms of features Popper's mistake can be rephrased as an exceedingly broad understanding of undesired features: not only the features that are defined as undesired above, but also the neutral features fall under Popper's definition. For further analysis, see ICR, Section 8.1 and also (Zwart 1998/2001, Chapter 2), who has creatively reused part of Popper's intuitions (Chapter 6).

Note that a desired feature F of X is a true feature of X in the sense that not only X but also T is a subset of F, that is, the weak claim that may be associated with F, "$T \subseteq F$", is true. However, not only all undesired features of X are false in this sense but also all neutral features. The undesired features are false in a strong sense: they not only exclude some desired possibilities, but only such possibilities.

The following theorems can now easily be proved [7]:

Equivalence theses

Y is at least as close to T as X iff
(UF)   all undesired features of Y are also features of X (equivalent to (DP))
(DF)   all desired features of X are also features of Y (equivalent to (UP))

Y is two-sidedly closer to T than is X iff
(UF) & (UF+) X has extra undesired features (equivalent to (DP+))
(DF) & (DF+) Y has extra desired features (equivalent to (UP+))

In Figure 5, 'at least as close to the truth' is depicted in terms of features. The rectangle now represents the 'universe' of all possibly relevant, distributed, features, and hence the powerset $P(CP)$ of CP (see Note 6). $Q(X)$ and $Q(Y)$ represent the set of features of X and Y, $Q(T)$ represents the set of desired features (features of T) and $Q(CP - T)$ represents the set of undesired features (the features of $CP - T$). Note that, $Q(T)$ and $Q(CP - T)$ have exactly one element in common, namely the tautology, which can be represented by CP.

Notice the strong analogy between the logical form of (DP) (see the beginning of Section 7) and (DF) and between that of (DP+) and (DF+). The same goes for the logical form of (UP) and (UF), and of (UP+) and (UF+). The equivalences stated in the theorem, though, correspond in the reverse way: (DP) and (UF) are equivalent, as are (DP+) and (UF+), (UP) and (DF), (UP+)

---

[7] Note for readers interested in the technical details. For proving these theorems it is advisable to introduce the set-theoretical interpretation of the universe of features and the set-theoretical characterization of the new clauses in terms of 'powersets' and 'co-powersets'. The powerset $P(X)$ of X is defined as the set of all subsets of X. The rectangle representing the 'universe' of all possibly relevant, distributed, features can now be interpreted as the 'powerset' $P(CP)$ of CP. Like a kind of mirror notion to that of powerset, the co-powerset $Q(X)$ of X is the set of all subsets of CP that include X, also called the supersets of X (within CP). $Q(X)$ then represents the features of X, $Q(T)$ the desired features and $Q(CP - T)$ the undesired features. Note that $Q(T)$ and $Q(CP - T)$ have exactly one set as common element, namely CP, that corresponds with the tautology, and that is of course included in the set of features of every theory. This results in the following formal translations of the four feature clauses:

(UF) $Q(Y) \cap Q(CP - T) \subseteq Q(X) \cap Q(CP - T)$ (UF+) $(Q(X) \cap Q(CP-T)) - Q(Y) \neq \varnothing$

(DF) $Q(X) \cap Q(T) \subseteq Q(Y) \cap Q(T)$ (DF+) $(Q(Y) \cap Q(T)) - Q(X) \neq \varnothing$

Proving the equivalence theses in terms of sets now becomes a nice exercise in 'set calculation'.

                              Theo A. F. Kuipers

and (DF+). None of this is at all surprising, for undesired features could be defined in terms of desired possibilities and vice versa. Therefore it is in principle possible to reproduce the proof of the theses informally, clause by corresponding clause.[8]

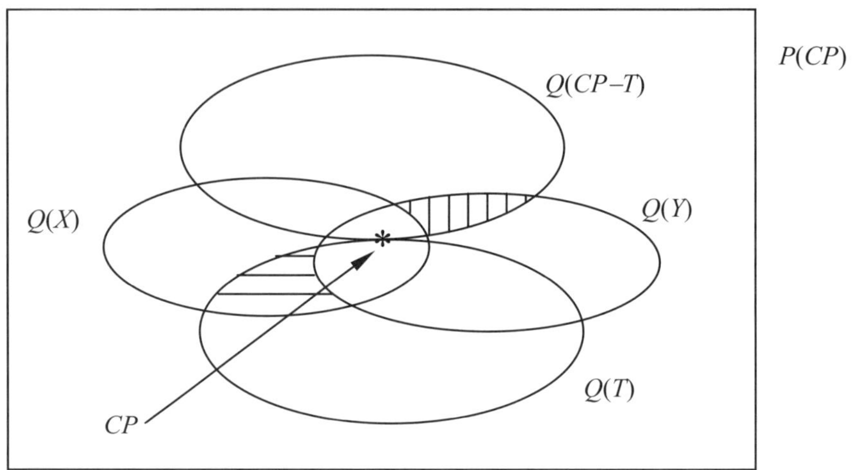

Fig.5. *Y* is at least as close to *T* as *X*, in terms of features.
(UF): |||-area empty, (DF): ≡-area empty

On the basis of the equivalences, it follows that the two principal definitions can also be given in a mixed or "dual" form, in terms of desired possibilities and desired features: (DP) and (DF) for at least as close to the truth, with the addition of (DP+) and (DF+) for (two-sidedly) closer to the truth. Roughly speaking, 'more truthlike' amounts to "more desired possibilities and more desired features." In my opinion, this resonates very much the intuitions of many scientists. We can strengthen this by taking into account that a desired possibility is a "correct model" and a desired feature a "true (general) consequence," where 'general' refers to all nomic possibilities. In this terminology, the dual conceptual foundation for nomic truthlikeness is most appealing to scientific common sense: 'more truthlike' amounts to 'more

---

[8] Another note for readers interested in the technical details. Let me give, by way of example, a proof of the claim that (UF) entails (DP). Assume (UF) and let, contrary to (DP), x be a desired possibility admitted by X, that is, x belongs to $X \cap T$, and let x not be admitted by Y, hence belong to $T - Y$. Now $CP - \{x\}$ is a superset of Y, hence it represents a feature of Y which only excludes desired possibilities, viz. x, (and no undesired ones). Hence it is an undesired feature of Y, which should according to (UF) also be a feature of X, which rules out that x is a member of X. Q.e.d. All proofs are of this elementary nature.

true (general) consequences and more correct models'. Moreover, it is now easy to see that 'more successful' amounts to 'more *established* true (general) consequences, i.e., general successes, and fewer *established* incorrect models, i.e., counterexamples', in line with the asymmetric definition, which is in fact also of a dual nature. Finally, dual nomic truthlikeness and the corresponding dual methodology leave room for nonempirical considerations in theory evaluation, such as aesthetic ones.

9. *Epistemological Stratification of Nomic Truth Approximation*

So far I have discussed objective features of theories in general. Of course there are different kinds of features and corresponding criteria. An obvious classification is the division into empirical and nonempirical features and criteria.

There are two main categories of empirical criteria of a theory, in accordance with the dual design above. I have already mentioned the question whether or not the theory implies a certain established observational law that, if so, can be explained or predicted by the theory. The entailment of an observational law can thus be conceived as an established desired observational feature of the theory. Observational laws are of course established by "object induction" on properties recurring in repeated experiments. Instead of speaking of entailment or explanation and/or prediction of the theory, in what follows I will simply speak of explanation of such laws. Besides the "explanation criterion" there is the "instantial criterion," viz., the admission of an observed possibility, that is, the result of a particular experiment being an example or counterexample of the theory. So an observed possibility can be regarded as an established, desired observational possibility.

Assuming that empirical criteria are primary, relative to their possible aesthetic value, they are the only relevant criteria as long as only observational and no theoretical terms are involved. In other words, nonempirical features are only important if a (relative) distinction between observational and theoretical terms can be made. I suppose that, in the present context, this distinction holds. Of course such a distinction between theoretical and observational terms leads to the distinction between an observational level of conceptual possibilities CPo and a theoretical (cum observational) level of conceptual possibilities $CP = CPt$. This distinction allows a precise definition of empirical versus nonempirical features to be formulated: features of the first kind exclude possibilities on the observational level, features of the second kind do not. Formally, e.g. for the second kind, a subset F of CP represents a nonempirical feature iff for all x in CPo there is at least one y in CPt such that y has x as its "projection" in CPo. This definition may suggest that nonempirical features of theories, in particular aesthetic ones, cannot be indicative of the empirical merits and prospects of a theory. However, by way of meta-induction, that is, inductive extrapolation or even generalization of a feature of certain theories to other ones, such features can come to be conceived as indicative in this respect. In this sense, aesthetic criteria may be seen as indirect empirical criteria, though formally quite different from the two categories of empirical criteria introduced above. From now on I shall speak only of empirical criteria (and features) in the direct sense explained above.

Truth approximation by means of empirical criteria can now be defined and founded on the basis of the following, easy to prove,

> *Combined Projection & Success theorem*
> If Y is closer to T than X then Y is at least as successful as X, almost in the sense of Sections 6 and 7, more precisely:

> (*DF-Success:*) *Explanatory clause*
> All established observational laws explained by X are also explained by Y (or: all established desired observational features of X are also features of Y)

> (*DP-Success:*) *Instantial clause*
> All observed examples of X are also examples of Y "unless X is lucky" (in other words: all observed counterexamples of Y are also counterexamples of X, "unless X is lucky").

The subclause 'unless X is lucky' will be clarified shortly. The underlying assumption for the proof of this theorem is the correctness of the empirical data, that is to say, the observed possibilities and the observational laws that are (through an inductive leap) based on them, are correct.[9]

This theorem permits the functionality argumentation given in Section 7 to be generalized. Assume that theory Y at time t is (two-sidedly) more successful than X in the sense suggested above: not only are the two clauses fulfilled, but also Y explains at least one extra observational law and X has at least one extra observed counterexample (in other words: Y has an extra observed example). This evokes the comparative success hypothesis that Y will be lastingly more successful than X. This hypothesis is a neat empirical hypothesis of a comparative nature that can be tested by deriving and testing new test implications. As soon as this hypothesis has been sufficiently tested, in the eyes of some scientists, the rule of success can be applied, which means that they draw the conclusion that Y will remain more successful than X. It can be proved (recall the Forward Theorem of Section 7) that this is equivalent to concluding that the observational theory that follows from Y is closer to the observational truth To (the strongest true theory that can be formulated with

---

[9] For the set-theoretical formulation of this theorem I refer to ICR Sections 7.3.3 and 9.1.1.

the observational vocabulary, thus as a subset of *CPo*) than X. But this conclusion is in its turn a good argument for the truth approximation (TA-) hypothesis on a theoretical level: Y is closer to the (theoretical) truth $T = Tt$ than X. In other words, the rule of success is functional for truth approximation. For this, three specific reasons have been given in Section 7 (for further details, see ICR, p.162, p.214). They only need some qualification in view of the possibility of lucky hits, to which we now turn.

Whereas the explanatory clause is straightforward, the instantial clause is not, due to the possibility of lucky hits. It is interesting to study the latter in some detail. Let an I(nstantial)-similarity be an observed possibility that is admitted by both or neither theory, and let an I-difference be an observed possibility that is admitted by Y but not by X. Because of the "one-many" character of the relation between the observational and theoretical levels of conceptual possibilities, a theory can have an observational feature on the observational level only if it has one on the theoretical level. The admission of an observational possibility on the theoretical level, though, cannot be based only on the admission of a suitable desired theoretical possibility, but can also be based on a suitable undesired theoretical possibility. If the observed example can be based on some admitted desired theoretical possibility, it may be called a real success of the theory. However, if the observed example can only be based on admitted undesired possibilities, it is some sort of *lucky hit* of that theory. For I-similarities there are all kinds of possibilities for this to occur, but it is not worth the effort of spelling them all out. I-differences, on the other hand, are very interesting. An I-difference can be based on a lucky hit of Y, in which case the DP+-clause, on the theoretical-cum-observational level, will not be verified and so the TA-hypothesis will not be confirmed (and therefore the reversed DP-clause is not falsified). Of course, if it is a real success of Y, the DP+-clause is verified, the TA-hypothesis is confirmed, and the reversed DP-clause is falsified. These possibilities are depicted in Figure 6.

In Figure 6, the observed example a is a real success of Y if there is a theoretical version in area 4 and it is a lucky hit if there is not, in which case there must be versions in 1 and 2. If (DF) [($\Leftrightarrow$ (UP)] holds, area 2 is empty, so a must be a real success. It is clear that whether an extra success of Y is real or only apparent cannot be ascertained on the basis of the observed example. We can say, though, that if TAH (especially (DF)) is true, the example must be real. As said, in that case the DF+-clause is verified and TAH confirmed. Although this is not a completely circular confirmation, it is a "(DF)-laden" and therefore "TAH-laden" confirmation. So an I-difference is not reliable as a (modest) "signpost to the truth," even when it is correctly determined. This nature of I-differences makes it possible for realists who want to defend TAH in a concrete case, to relativize reversed I-differences: after all, an instantial success of *X* that is a counterexample of *Y* could be a lucky hit on the part of *X*. This is the condition mentioned in the instantial clause.

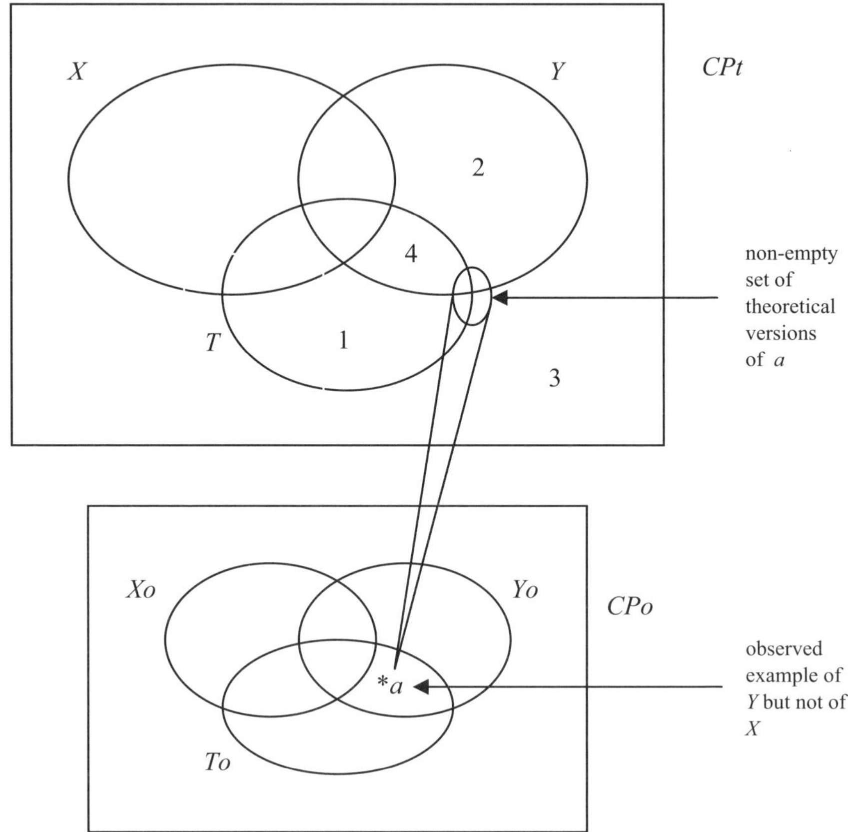

Fig.6. I-difference: observed example of *Y* (but not of *X*) as a real success or as a lucky hit. Further explanation in the text.

As suggested before, not only empirical criteria may play a role in theory choice, but also nonempirical criteria, that is, criteria in terms of logical, conceptual or aesthetic features. In a recent paper (Kuipers 2002) a *formal-cum-naturalistic* analysis is given of the relation between beauty, empirical success, and truth. It supports the findings of James McAllister in his inspiring *Beauty and revolution in science* (1996), by explaining and justifying them. First, scientists are essentially right regarding the usefulness of aesthetic criteria for truth approximation, provided they conceive of them as less hard than empirical criteria. Second, the aesthetic criteria of the time, the "aesthetic canon," may well be based on "aesthetic (meta-) induction" regarding (distributed) nonempirical features of paradigms of successful theories which scientists have come to appreciate as beautiful. Third, they can play a crucial, dividing role in scientific revolutions. Since aesthetic criteria may well be wrong, they may retard empirical progress and hence truth approximation in the hands of aesthetic dogmatists but not in the hands of aesthetically flexible, "revolutionary" scientists.

The truth approximation analysis also affords an opportunity to reconsider the nature of descriptive and explanatory research programs. Such programs presuppose, by definition (see SiS, Ch. 1) a domain, a problem, and a core idea, including a vocabulary, to solve that problem. A *descriptive research program* uses an observational conceptual frame, and may either exclusively aim at one or more true descriptions (as for example in most historiography), or it may also aim at the true (observational) theory in the following specific way. In this nomological type of descriptive program the goal of a true theory is supposed to be achieved exclusively by establishing observational laws. Given that this requires (observational) inductive jumps, it is plausible to call such a descriptive program an *inductive research program*. It is easy to see that such programs "approach the truth by induction". The micro-step of the HD method may be applied for the establishment of observational laws, resulting in true descriptions which either falsify the relevant general observational hypothesis or are partially derivable from it. According to the basic definition of 'more truthlike', assuming that accepted observational laws are true, any newly accepted observational law guarantees a step in the direction of the true theory. For it is easy to verify that if $S(t)$ and $S(t')$ indicate the strongest accepted law at time t and t' later than t, respectively, $S(t')$ is closer to T than $S(t)$. Hence, inductive research programs are relatively safe strategies of truth approximation: as far as the inductive jumps happen to lead to true accepted laws, the approach not only makes truth approximation plausible, it even guarantees it.

Let me now turn to the explication of the nature of explanatory or theoretical programs, which are by definition of a nomological nature. An *explanatory program* may or may not use a theoretical vocabulary. Even (nomic) empiricists can agree that it is directed at establishing the true observational theory. If there are theoretical terms involved, the referential realists will add that it is also directed at establishing the referential truth. The theory realist will add to this that it is even directed at establishing the theoretical truth. Scientists working within such a program will do so by proposing theories respecting the hard core as long as possible, but hopefully not at any price. They will HD evaluate these theories separately and comparatively. RS directs theory choice and is trivially functional for empirical progress. Moreover, although that rule is demonstrably functional for all distinguished kinds of nomic truth approximation, it cannot guarantee a step in the direction of the relevant truth, even assuming correct data. Though the basic notions of successfulness and truthlikeness are sufficient to give the above characterization of the typical features of explanatory research programs, they usually presuppose refined means of comparison, which are presented in Part IV of ICR.

IV Refined Truth Approximation

To keep this synopsis within reasonable limits, I have chosen not to give a detailed impression of the last part of ICR, Ch. 10-12. Brief indications of the chapters will have to suffice. Ch. 10 introduces another sophistication of the basic approach: it accounts for the fact that progress is frequently made by new theories that introduce new mistakes, something which is excluded according to basic truth approximation. In Ch. 11 this refinement allows some real-life illustrations of (potential) truth approximation, one from physics and another from economics. Moreover, in Ch. 12 it is shown that there are also quantitative versions of refined truth approximation, based upon distances between structures.

10. *Refinement of Nomic Truth Approximation*

The study of truthlikeness and truth approximation is completed by introducing a second major sophistication accounting for a fundamental feature of most theory improvement, viz., new theories introduce new mistakes, but mistakes that are in some way less problematic than the mistakes they replace. This refinement is introduced in a qualitative way by taking into account that one incorrect model may be more similar, or "more structurelike" to a target model than another. This leads to refined versions of nomic truthlikeness and truth approximation, with adapted conceptual foundations. It is argued, and illustrated by the Law of Van der Waals, that the frequently and variously applied method of "idealization and successive concretization," propagated by Nowak, is a special kind of (potential) refined nomic truth approximation (in this respect, see also Kuipers, forthcoming). Combining the present sophistication with that of Section 9, one obtains explications of stratified refined nomic truthlikeness and truth approximation.

11. *Examples of Potential Truth Approximation*

Two sophisticated examples illustrate that the final analysis pertains to real-life, theory-oriented, empirical science. The first example shows that the successive theories of the atom, called "the old quantum theory," viz., the theories of Rutherford, Bohr, and Sommerfeld, are such that Bohr's theory is closer to Sommerfeld's than Rutherford's. Here, Bohr's theory is a (quantum) specialization of Rutherford's theory, whereas Sommerfeld's is a (relativistic) concretization of Bohr's theory. This guarantees that the nomic truth, if not caught by the theory of Sommerfeld itself, could have been a concretization of the latter. In both cases, Sommerfeld would have come closer to the truth than Bohr and Rutherford. The second example illustrates a non-empirical use of the idealization and concretization methodology, viz., aiming at (approaching) a provable interesting truth. In particular, it is shown that the theory of the capital structure of firms of Modigliani and Miller is closer to a provable interesting truth than the original theory of Kraus and Litzenberger, of which the former is a "double" concretization.

12. *Quantitative Truthlikeness and Truth Approximation*

Here the prospects for quantitative versions of actual and nomic truthlikeness are investigated. In the nomic refined case there are essentially two different ways of corresponding quantitative truth approximation, a non-probabilistic one, in the line of the qualitative evaluation methodology, and a probabilistic one, in which the truthlikeness of theories is estimated on the basis of a suitable probability function. As stressed in Ch. 3 of ICR, probabilistic methodological reasoning, notably about confirmation, is already rather artificial, although it is used by scientists to some extent. However, quantitatively measuring the distance between theories and between their successes is in most cases even more artificial and, moreover, rare. Hence, the quantitative accounts of (non-) probabilistic truth approximation, notably that of Niiniluoto (1987), are presented with many reservations.

13. *Conclusion: Constructive Realism*

Recall that Section 1 introduces the main epistemological positions, instrumentalism, constructive empiricism, referential realism, constructive (theory) realism and essentialist (theory) realism. In the course of ICR the following conclusion could provisionally be drawn at the end of Part III and could be further strengthened in Part IV:

> The instrumentalist methodology provides good reasons for the transition of
> epistemological positions from instrumentalism to constructive realism. Here, the

intermediate step from constructive empiricism to referential realism turned out to be the hardest one, whereas the step from constructive to essentialist realism had to be rejected.

The rest of ICR Ch. 13 presents the main lines of the resulting favorite epistemological position of *constructive realism*. It is a conceptually relative, hence non-essentialist, nomic truth approximation version of theory realism, accounting for objective truths. They can be approached by an intersubjective method, viz., the evaluation methodology for theories, in which the role of (truth-)testing of hypotheses primarily concerns testing test implications of theories as well as testing comparative success and truth approximation hypotheses of theories.

The term 'constructive realism' has earlier been used by Giere (1985), and my conception of it is rather similar, except that I include in it, of course, truth approximation, whereas Giere still focuses on the true/false dichotomy, but he fully recognizes the nomic aim of theorizing. With respect to truth approximation, my position is rather similar to that of Niiniluoto (1987, see in particular Section 4.3). The main difference between my and his position, besides my primarily qualitative versus his straightforward quantitative approach, is my emphasis on the nomic aim of theorizing. In sum, constructive realism reflects the combination of their deviating strengths by emphasizing *nomic truth approximation* as opposed to the *actual truth-value* of theories.

Ch. 13 deals more in particular with the acceptance (as true) of three types of hypotheses or claims, that is, three types of induction, viz. observational, referential and theoretical induction, and with the formation of observation terms. The resulting metaphysical nature of scientific research is depicted, together with portraits of real and fictitious scientists. It concludes with a discussion of metaphors for empirical science research, and concludes that the map metaphor, rather than the mirror or the net metaphor, is to be preferred, although it is certainly not perfect.

As already mentioned in the introduction, there arises in ICR a clear picture of scientific development, with a short-term and a long-term dynamic. In the former there is a severely restricted role for confirmation and falsification; the dominant role is played by (the aim of) empirical progress, and there are serious prospects for observational, referential and theoretical truth approximation. Hence, regarding this short-term dynamic, the scientist's intuition that the debate among philosophers about instrumentalism and realism has almost no practical consequences can be explained and justified. The long-term dynamic is enabled by (observational, referential and theoretical) inductive jumps, after "sufficient confirmation," providing the means to enlarge the observational vocabulary in order to investigate new domains of reality. In this respect, a consistent instrumentalist epistemological attitude seems difficult to defend, whereas constructive realism seems the most plausible.

I would like to conclude this synopsis by explaining the reason for this last claim, originating from the crucial role played by observational and theoretical induction in the construction and determination of terms and, hence, the long-term dynamic in science. Besides the formation of observation terms by straightforward explicit definition, observational induction may provide the necessary and sufficient empirical conditions, e.g., in the form of existence and uniqueness requirements, for explicitly defining new observation terms. The quantitative notions of pressure and temperature are examples. Such new terms are unambiguously, hence intersubjectively, applicable, and they (may be supposed to) refer if they enable new observational inductions. Besides implying referential induction, theoretical induction may provide the necessary and sufficient conditions, e.g., in the form of existence and uniqueness requirements, for applying theoretical terms, that is, for identifying theoretical entities and for measuring theoretical attributes. For example, the detection of electrons, and the measurement of their mass and charge, are based on such inductions. In this way, theoretical terms (may be supposed to) become referring and unambiguously, hence intersubjectively, applicable. In other words, theoretical terms can be essentially transformed into new observation terms by appropriate theoretical and/or referential induction. Of course, together with earlier or elsewhere accepted observation terms, they can be used for new cases of observational induction. Moreover, with other ones, they will play a crucial role in the separate and comparative evaluation of new theories introducing new theoretical terms, dealing with (partially) new domains, starting another round of the "empirical cycle." For an epistemological instrumentalist it is difficult to account for this long-term dynamic in a consistent way. However, for a constructive realist, focusing on nomic truth approximation, this is easy, ironically enough, in particular when he is prepared to replace the falsificationist methodology by the instrumentalist methodology.

University of Groningen
Department of Theoretical Philosophy
Oude Boteringestraat 52, 9712 GL Groningen
The Netherlands
e-mail: T.A.F.Kuipers@philos.rug.nl
http://www.rug.nl/filosofie/kuipers.html

Appendix 1: Table of Contents of
From Instrumentalism to Constructive Realism: On Some Relations between Confirmation, Empirical Progress, and Truth Approximation

13.3.   Formation of observation terms
13.4.   Direct applicability of terms
13.5.   The metaphysical nature of scientific research
13.6.   Portraits of real and fictitious scientists
13.7.   Reference and ontology
13.8.   Truth definitions and truth criteria
13.9.   Metaphors

Notes, References, Index of Names, Index of Subjects

Appendix 2: Outline Table of Contents of
Structures in Science: Heuristic Patterns based on Cognitive Structures. An advanced textbook in neo-classical philosophy of science

Contents, Foreword

Part I   Units of Scientific Knowledge and Knowledge Acquisition
            1   Research programs and research strategies
            2   Observational laws and proper theories

Part II   Patterns of Explanation and Description
            3   Explanation and reduction of laws
            4   Explanation and description by specification

Part III   Structures in Interlevel and Interfield Research
            5   Reduction and correlation of concepts
            6   Levels, styles, and mind-body research

Part IV   Confirmation and Empirical Progress
            7   Testing and further separate evaluation of theories
            8   Empirical progress and pseudoscience

Part V   Truth, Product, and Concept Approximation
            9   Progress in nomological, design, and explicative research
            10 Design research programs

Part VI   Capita Selecta
            11 Computational philosophy of science
            12 The structuralist approach to theories
            13 'Default-norms' in research ethics

Suggestions for further reading, Exercises, Notes, References, Index of Names, Index of Subjects

Appendix 3: Acronyms

A-difference/-similarity	Aesthetic difference/similarity
CD-hypothesis	Correct Data hypothesis
CP	set of Conceptual Possibilities
CSH	Comparative Success Hypothesis
d-confirmation	deductive confirmation
DF/UF	Desired Features/Undesired Features (clause)
DF-success	Desired Features success (clause)
DP/UP	Desired Properties / Undesired Properties (clause)
DP-success	Desired Properties success (clause)
E-difference/-similarity	Explanatory difference/similarity
GTC	General Testable Conditional
GTI	General Test Implication
HD (evaluation, method, testing)	Hypothetico-Deductive (evaluation, method, testing)
IC	Initial Condition(s)
ICR	*From Instrumentalism to Constructive Realism*
I-difference/-similarity	Instantial difference/similarity
ITI	Individual Test Implication
LMC	Logico-Mathematical Claim
Mod(H)	the set of models of H
MP	Modus (Ponendo) Ponens
MT	Modus (Tollendo) Tollens
PC	Principle of Content
PCE	Principle of Comparative HD evaluation
PD	Principle of Dialectics
PI	Principle of Improvement
PIRP	Principle of Improvement by Research Programs
PSE	Principle of Separate HD evaluation
PT	Principle of (Falsifiability or) Testability
p-(non-)zero	(non-)zero probability
RE	Rule of Elimination
RS	Rule of Success
SiS	*Structures in Science*
STR	Special Theory of Relativity
TA-hypothesis / TAH	Truth Approximation Hypothesis
UI	Universal Instantiation

REFERENCES

Compton, K. (1988). 0-1 Laws in Logic and Combinatorics. In: I. Rival (ed.), *Proceedings* 1987 *NATO Adv. Study Inst. on Algorithms and Order*, pp. 353-383. Dordrecht: Reidel.

Earman, J. (1992). *Bayes or Bust. A Critical Examination of Bayesian Confirmation Theory.* Cambridge: The MIT-Press.

Festa, R. (1993). *Optimum Inductive Methods. A Study in Inductive Probability Theory, Bayesian Statistics and Verisimilitude.* Dordrecht: Kluwer Academic Publishers.

Festa, R. (1999). Bayesian Confirmation. In: M.C. Galavotti and A. Pagnini (eds.), *Experience, Reality, and Scientific Explanation*, pp. 55-87. Dordrecht: Kluwer Academic Publishers.

Fitelson, B. (1999). The Plurality of Bayesian Measures of Confirmation and the Problem of Measure Sensitivity. *Philosophy of Science*, Supplement to Volume **66** (3), S362-S378.

Giere, R. (1985). Constructive Realism. In: P. Churchland and C. Clifford (eds.), *Images of Science*, pp. 75-98. Chicago: The University of Chicago Press.

Groot, A. de. (1961/1969). *Methodologie.* Den Haag: Mouton, 1961. Translated as: *Methodology* (New York: Mouton, 1969).

Grove, A., J. Halpern and D. Koller. (1996). Asymptotic Conditional Probabilities: The Unary Case. *The Journal of Symbolic Logic* **61** (1), 250-275.

Hempel, C. (1966). *Philosophy of Natural Science.* Englewood Cliffs: Prentice-Hall.

Hodges, W. (1986). Truth in a Structure. *Proceedings of the Aristotelian Society, New Series* **86**, 135-151.

Howson, C. and P. Urbach. (1989). *Scientific Reasoning: the Bayesian Approach.* La Salle: Open Court.

Kemeny, J. (1953). A Logical Measure Function. *The Journal of Symbolic Logic* (**18**) 4, 289-308.

Kuipers, T. (1978). *Studies in Inductive Probability and Rational Expectation. Synthese Library*, vol. 123. Dordrecht: Reidel.

Kuipers, T. (1998). Pragmatic Aspects of Truth Approximation. In: P. Weingartner, G. Schurz and G. Dorn (eds.), *The Role of Pragmatics in Contemporary Philosophy.* (Proceedings of the 20th International Wittgenstein-Symposium, August 1997), pp. 288-300. Vienna: Hölder-Pichler-Temsky.

Kuipers, T. (1999). Abduction Aiming at Empirical Progress or Even at Truth Approximation, Leading to a Challenge for Computational Modelling. In: J. Meheus and T. Nickles (eds.), *Scientific Discovery and Creativity*, special issue of *Foundations of Science* **4** (3), 307-323.

Kuipers, T. (2000/ICR). *From Instrumentalism to Constructive Realism. On Some Relations between Confirmation, Empirical Progress, and Truth Approximation. Synthese Library*, vol. 287. Dordrecht: Kluwer Academic Publishers.

Kuipers, T. (2001/SiS). *Structures in Science. Heuristic Patterns Based on Cognitive Structures. An Advanced Textbook in Neo-Classical Philosophy of Science. Synthese Library*, vol. 301. Dordrecht: Kluwer Academic Publishers.

Kuipers, T. (2001). The Logic of Progress in Nomological, Explicative and Design Research. Ch. 9 of Kuipers (2001/SiS), pp. 255-264.

Kuipers, T. (2002). Beauty, a Road to The Truth. *Synthese* **131** (3), 291-328.

Kuipers, T. (2004). Inference to the Best Theory. Kinds of Induction and Abduction, Rather Than Inference to the Best Explanation. In: F. Stadler (ed.), *Induction and Deduction in the Sciences*, pp. 25-51. Dordrecht: Kluwer Academic Publishers.

Kuipers, T. (forthcoming). Empirical and Conceptual Idealization and Concretization. The Case of Truth Approximation. Forthcoming in (English and Polish editions of) Liber Amicorum for Leszek Nowak.

Lakatos, I. (1970/1978). Falsification and the Methodology of Scientific Research Programmes. In: I. Lakatos and A. Musgrave (eds.), *Criticism and the Growth of Knowledge*, pp. 91-196. Cambridge: Cambridge University Press. Reprinted in Lakatos (1978), pp. 8-101.

Lakatos, I. (1978). *The Methodology of Scientific Research Programmes*, eds. J. Worrall and G. Currie. Cambridge: Cambridge University Press.

Laudan, L. (1977). *Progress and Its Problems*. Berkeley: University of California Press.

Maher, P. (2004). Qualitative Confirmation and the Ravens Paradox. This volume.

McAllister, J. (1996). *Beauty and Revolution in Science*. Ithaca: Cornell University Press.

Milne, P. (1996). *Log[P(h|eb)/P(h|b)]* is the one true measure of confirmation. *Philosophy of Science* **63**, 21-26.

Mura, A. (1990). When Probabilistic Support Is Inductive, *Philosophy of Science* **57**, 278-289.

Niiniluoto, I. (1987). *Truthlikeness*. Dordrecht: Reidel.

Panofsky, W. and M. Phillips. (1955/1962^2). *Classical Electricity and Magnetism*. London: Addison-Wesley.

Popper, K. (1934/1959). *Logik der Forschung*. Vienna, 1934. Translated as: *The Logic of Scientific Discovery* (London: Hutchinson, 1959).

Popper, K. and D. Miller. (1983). A Proof of the Impossibility of Inductive Probability. *Nature* **302**, 687-688.

Salmon, W. (1969). Partial Entailment as a Basis for Inductive Logic. In: N. Rescher (ed.), *Essays in honor of Carl G. Hempel*, pp. 47-82. Dordrecht: Reidel.

Schlesinger, G. (1995). Measuring Degrees of Confirmation. *Analysis* **55** (3), 208-212.

Zwart, S. (1998/2001). *Approach to The Truth. Verisimilitude and Truthlikeness*. Dissertation Groningen. Amsterdam: ILLC-Dissertation-Series-1998-02. Revised version: *Refined Verisimilitude, Synthese Library*, vol. 307 (Dordrecht: Kluwer Academic Publishers).

CONFIRMATION AND THE HD METHOD

Patrick Maher

QUALITATIVE CONFIRMATION AND THE RAVENS PARADOX

ABSTRACT. In *From Instrumentalism to Constructive Realism* Theo Kuipers presents a theory of qualitative confirmation that is supposed to not assume the existence of quantitative probabilities. He claims that this theory is able to resolve some paradoxes in confirmation theory, including the ravens paradox. This paper shows that there are flaws in Kuipers' qualitative confirmation theory and in his application of it to the ravens paradox.

Part I of Theo Kuipers' book *From Instrumentalism to Constructive Realism* (Kuipers 2000) is concerned with confirmation. It begins (section 2.1) with what Kuipers calls "a qualitative theory of deductive confirmation." This theory is meant to be qualitative in the sense that it does not assume the existence of quantitative probabilities. It is deductive in the sense that it is concerned with situations in which certain deductive relations hold between the hypothesis and the evidence. Having presented this theory, Kuipers uses it to give solutions to some problems in confirmation theory, the first of which is the ravens paradox (section 2.2.1).

In this paper I will discuss both Kuipers' qualitative theory of deductive confirmation and his application of it to the ravens paradox. The following is an overview of the main claims of this paper.

Section 1: Kuipers' theory of confirmation is founded on a definition of confirmation that differs from standard conceptions in intension and extension. There does not appear to be any cogent motivation for adopting this deviant definition.

Section 2: Kuipers defines what he calls conditional deductive con-firmation, or cd-confirmation. I show that cd-confirmation is a trivial relation that holds between practically all propositions.

Section 3: Kuipers presents two ravens paradoxes but one of these is spurious, being based on fallacious reasoning.

Section 4: Kuipers claims to derive a result that solves the ravens paradox. However, Kuipers' derivation is based on a dubious principle and it appears to require quantitative probabilities.

Section 5: Even if this result is true, it does not solve the paradox.

In: R. Festa, A. Aliseda and J. Peijnenburg (eds.), *Confirmation, Empirical Progress, and Truth Approximation* (*Poznań Studies in the Philosophy of the Sciences and the Humanities*, vol. 83), pp. 89-108. Amsterdam/New York, NY: Rodopi, 2005.

All page references are to Kuipers (2000) unless otherwise indicated. To facilitate comparison I will mostly use the same notation as Kuipers. One difference is that I will consistently use overbars to denote negation; Kuipers only sometimes uses this notation.

1. Definition of Confirmation

Kuipers introduces his definition of confirmation this way:

> The explication of the notion of 'confirmation' of a hypothesis by certain evidence in terms of plausibility will be the main target of [chapters 2 and 3]. It will be approached from the success perspective on confirmation, equating confirmation with an increase of the plausibility of the evidence on the basis of the hypothesis.... (p.18)

Later he formulates what he calls the "success definition of confirmation" as follows:

> E confirms H iff (E is a success of H in the sense that) H makes E more plausible (p.23)

According to etymology, dictionaries, and most works on confirmation theory, "E confirms H" means that E makes H more plausible. Thus Kuipers' success definition of confirmation seems to have things backwards since, according to it, for E to confirm H is for H to make E more plausible.

Kuipers mitigates this discrepancy by enunciating what he calls "the reward principle of plausibility". Kuipers (p. 23) says this principle asserts that E makes H more plausible iff E confirms H. This reward principle appears to imply that Kuipers' definition of confirmation agrees extensionally with the standard definition.

However, in a note attached to his statement of the reward principle Kuipers observes that in his quantitative theory of confirmation the reward principle does not hold for hypotheses with zero probability. Specifically, such hypotheses can be confirmed (according to the success definition of confirmation) but they cannot have their plausibility increased.[1] Hence Kuipers says in this note that "one might refine" the reward principle of plausibility "by imposing the condition that H has some initial plausibility." In this way he acknowledges that his success definition of confirmation differs from the standard definition when the hypothesis has no initial plausibility.

Although Kuipers does not mention it, a similar situation exists when the evidence E has zero prior probability. In this case, E cannot confirm H

---

[1] In the quantitative theory Kuipers identifies plausibility with probability. He allows that $p(E|H)$ may be defined even when $p(H) = 0$, so for such an H it is possible that $p(E|H) > p(E)$ and thus E confirms H according to the success definition. On the other hand, if $p(H) = 0$ and $p(E) > 0$ it follows from the probability calculus that $p(H|E) = 0$, so E does not make H more plausible.

according to the success definition of confirmation but E may nevertheless make H more plausible.[2] So in this case we can have confirmation according to the standard definition but not according to Kuipers' definition.

Thus Kuipers needs to restrict the reward principle of plausibility to cases in which both the evidence E and the hypothesis H have positive prior probability. In other cases the standard and success definitions give different verdicts on whether E confirms H, so these two definitions are not extensionally equivalent. And even if they agreed extensionally, the two definitions would still be very different intensionally. This raises the question of whether there is some merit to defining confirmation Kuipers' way, or whether we should view his success definition of confirmation as merely an idiosyncratic stipulative definition.

Kuipers highlights the fact that his definition differs from the usual one in allowing hypotheses with zero initial probability to be confirmed, so I suppose he believes that this feature is an advantage of his definition. But why is this supposed to be an advantage? Kuipers' answer to that question seems to be given in the following passage:

> Although there may be good reasons (contra Popper ...) to assign sometimes non-zero probabilities to genuine hypotheses, it also occurs that scientists would sometimes assign in advance zero probability to them and would nevertheless concede that certain new evidence is in favor of them. (p. 49)

If this were true then the success definition of confirmation would fit at least some things that scientists say better than the standard definition does. However, Kuipers cites no evidence to support his claim about what scientists would say and I am not aware of any such evidence.

I will concede that if a hypothesis with zero probability entails some evidence then that evidence does seem in some sense favorable to the hypothesis, and this is perhaps Kuipers' point. But an advocate of the standard definition of confirmation can explain this by saying that the hypothesis has made a successful prediction, and perhaps even that the evidence confirms that the hypothesis is close to the truth, while denying that the hypothesis itself has been confirmed. And if a scientist did say that the hypothesis was confirmed, an advocate of the standard definition can easily say that what the scientist said is not strictly true, although it is understandable why someone might say that.

But now suppose that H is a hypothesis with positive probability that is entailed by a proposition E that has zero probability. (For example, E might assert that a continuous random variable has a specific value and H might be that the variable's value is in an interval that contains the specified value.) One

---

[2] E cannot confirm H according to the success definition because if $p(E) = 0$ then $p(E|H) = 0$. Nevertheless E may make H more plausible because it may be that $p(H|E) > p(H)$.

would ordinarily think that verification is an extreme case of confirmation, and hence that observation of E would confirm H to the highest degree possible, but Kuipers must deny that in this case there is any confirmation at all. Thus any advantage that Kuipers' definition might have when the hypothesis has zero probability seems to be more than offset by his definition's unsatisfactory treatment of cases in which the evidence has zero probability.

I conclude that Kuipers' success definition of confirmation differs from the standard concept in both intension and extension and there appears to be no cogent justification for this departure from ordinary usage.

2. Conditional Deductive Confirmation

Kuipers identifies two kinds of deductive confirmation, namely unconditional and conditional deductive confirmation. Kuipers abbreviates these as "d-confirmation" and "cd-confirmation" respectively.

His definition of d-confirmation (p. 22) is that E d-confirms H iff H entails E. Kuipers defines cd-confirmation in two stages (pp. 22f.). He first defines the concept "E deductively confirms H assuming C", which he abbreviates as "E C-confirms H". His definition is that E C-confirms H iff (i) H and C are logically independent, (ii) C does not entail E, and (iii) $H\&C$ entails E. The second definition is that E cd-confirms H iff there exists a C such that E entails C and E C-confirms H.[3]

Kuipers (pp. 22, 36f.) indicates his reason for including condition (i) in the definition of C-confirmation: If condition (iii) was the only requirement then we could take C to be $H \rightarrow E$ and any E would cd-confirm any H. Although Kuipers does not say so, I suppose that his reason for including condition (ii) was similar: Without it we could take C to be E and then any E that is logically independent of H would cd-confirm H. However, these conditions do not prevent the trivialization of cd-confirmation, as the following theorem shows. (Proofs of all theorems are given in Section 7.)

THEOREM 1. If E and H are logically independent, and if there is a proposition that is logically independent of $\bar{E}\,\&\,\bar{H}$, then E cd-confirms H.

As an example of the application of this theorem, let H be that all ravens are black and let E be that chalk effervesces in acids (just to pick something completely irrelevant). There are many propositions that are logically independent of $\bar{E}\,\&\,\bar{H}$, for example, the proposition that Mars has no moon. So by Theorem 1, E cd-confirms H.

---

[3] Kuipers (p. 23) also says that C must not be tautologous; I omit that because condition (i) entails that C is not tautologous.

I will now consider how the concepts of d-confirmation and cd-confirmation are meant to relate to confirmation as defined by either of the definitions discussed in Section 1. (Since those two definitions are extensionally equivalent in ordinary cases, the difference between them will not be important here.)

Kuipers says that d-confirmation "is a paradigm case in which scientists speak of confirmation" (p. 22), so it would seem that Kuipers accepts:

(D) If E d-confirms H then E confirms H.

However, it may be that H entails E but E is maximally plausible without H; in that case, E d-confirms H but E does not confirm H according to either of the definitions in Section 1. Thus (D) cannot be true in general. Kuipers could avoid this problem by adding to his definition of d-confirmation the condition that the evidence is not maximally plausible given background beliefs alone.

I turn now to the relation between cd-confirmation and confirmation. Kuipers' definitions imply the following:

If E cd-confirms H then there exists a C such that E d-confirms H given C.

Here "given C" means "when C is added to the background beliefs." By (D) we can replace "d-confirms" by "confirms", thus obtaining:

(CD$_1$) If E cd-confirms H then there exists a C such that E confirms H given C.

So there seems little doubt that Kuipers is committed to (CD$_1$). This is also an uncontroversial principle that can be given a probabilistic derivation if $p(E|H\&C) < 1$.

A stronger possible relation between cd-confirmation and confirmation is:

(CD$_2$) If E cd-confirms H then E confirms H.

This principle is definitely false. One way to see this is to note that, by Theorem 1, H can be cd-confirmed by a completely unrelated proposition E, and such an E will not confirm H. Furthermore, even in the kinds of cases that Kuipers was trying to capture with his definition of cd-confirmation, (CD$_2$) can still fail. To illustrate this latter possibility, let RH be that all ravens are black, let Ra be that some individual a is a raven, and let Ba be that a is black. Then $Ra\&Ba$ cd-confirms RH, with the condition being Ra. However, given certain kinds of background information, $Ra\&Ba$ may actually disconfirm RH, as Good (1967) showed.

There are passages where Kuipers, speaking of evidence that cd-confirms a hypothesis, says that this evidence "confirms" the hypothesis (p.20), or that it counts as a "success" of the hypothesis (p. 100). These passages suggested to me that Kuipers endorsed the false principle (CD$_2$), but Kuipers has told me that my interpretation was wrong and in these passages what he meant was that the evidence *conditionally* confirmed the hypothesis, or was a *conditional* success of the hypothesis. So I will take it that Kuipers accepts (CD$_1$) but not (CD$_2$). However, in view of this, some aspects of Kuipers' treatment of cd-confirmation are puzzling to me.

As I observed earlier, Kuipers has imposed restrictions on *C*-confirmation that are designed to prevent cd-confirmation being trivial. Theorem 1 shows that these restrictions do not achieve their purpose. But why does Kuipers think it is important to prevent cd-confirmation being trivial? If Kuipers accepted (CD$_2$) then it would be clear that he could not allow cd-confirmation to be trivial; however, since he does not accept (CD$_2$), only (CD$_1$), I do not see any reason why cd-confirmation cannot be trivial. If almost any *E* cd-confirms almost any *H*, what follows from (CD$_1$) is merely that for almost any *E* and *H* there exists some *C* such that *E* confirms *H* given *C,* and this consequence seems perfectly reasonable.

Kuipers (pp. 36f.) appears to take a criticism of hypothetico-deductivism by Gemes as a threat to his account of cd-confirmation. And he seems (p. 37) to identify his account of cd-confirmation with the prediction criterion of Hempel. But the hypothetico-deductive theory that Gemes was criticizing, and the prediction criterion of Hempel, are both criteria for when evidence *E* *confirms H*, not merely criteria for when *E* confirms *H* given some condition. Thus the hypothetico-deductivism that Gemes was criticizing, and Hempel's prediction criterion, correspond to (CD$_2$), not (CD$_1$). Since Kuipers does not accept (CD$_2$), I wonder why he seems to identify his account of cd-confirmation with these other theories that are committed to a principle like (CD$_2$).

3. The Ravens Paradox

The theory of qualitative deductive confirmation that I discussed in the preceding section is presented by Kuipers in section 2.1 of his book. After this, in section 2.2.1, he discusses two "raven paradoxes" due to Hempel.

In their presentations of these paradoxes, Hempel and Kuipers often speak of confirmation as if it was a relation between an object and a proposition. However, an object can be described in different ways that are not equivalent so far as confirmation is concerned, so we should rather view confirmation as a relation between two propositions. With confirmation understood in this way, the paradoxes may be stated as follows.

Two principles of confirmation that seem plausible are:

PRINCIPLE 1. (Nicod's condition). *Ca&Fa* confirms $(x)(Cx \rightarrow Fx)$.

PRINCIPLE 2. (Equivalence condition). If E confirms H, E' is logically equivalent to E, and H' is logically equivalent to H, then E' confirms H'.

As in the previous section, let R be the property of being a raven, B the property of being black, a an arbitrary individual, and RH the raven hypothesis $(x)(Rx \to Bx)$.) Then Principles 1 and 2 imply:

(α) $\overline{R}a \,\&\, \overline{B}a$ confirms RH.

But (α) seems counterintuitive; that is to say, the following principle is plausible:

PRINCIPLE 3. $\overline{R}a \,\&\, \overline{B}a$ does not confirm RH.

What Kuipers calls the "first paradox" can be expressed as being that Principles 1-3, although all plausible, are jointly inconsistent.

What Kuipers calls the "second paradox" is that Principles 1 and 2 imply:

(β) $\overline{R}a \,\&\, Ba$ confirms RH.

Kuipers thinks that (β) is "even more counter-intuitive" than (α) but I disagree. If we find a non-raven to be black then, reasoning by analogy, that is some reason to think ravens are also black, and hence that RH is true. Conversely, finding a non-raven to be non-black is some reason to think there may also be non-black ravens, and hence that RH is false. So I would say that (β) is *less* counter-intuitive than (α).

Furthermore, I am not aware of any valid derivation of (β) from Principles 1 and 2. Hempel (1945, p.15) noted that RH is equivalent to

$$(x)[(Rx \lor \overline{R}x) \to (\overline{R}x \lor Bx)].$$

By Principle 1,

$(Ra \lor \overline{R}a).(\overline{R}a \lor Ba)$ confirms $(x)[(Rx \lor \overline{R}x) \to (\overline{R}x \lor Bx)]$.

So by Principle 2,

(γ) $\overline{R}a \lor Ba$ confirms RH.

However, (γ) does not entail (β).[4]

---

[4] Proof: Let p be a probability function with

$$
\begin{array}{ll}
p(\text{RH} \,\&\, Ra \,\&\, Ba) = .2 & p(\overline{\text{RH}} \,\&\, Ra \,\&\, Ba) = .1 \\
p(\text{RH} \,\&\, Ra \,\&\, \overline{B}a) = 0 & p(\overline{\text{RH}} \,\&\, Ra \,\&\, \overline{B}a) = .2 \\
p(\text{RH} \,\&\, \overline{R}a \,\&\, Ba) = .1 & p(\overline{\text{RH}} \,\&\, \overline{R}a \,\&\, Ba) = .1 \\
p(\text{RH} \,\&\, \overline{R}a \,\&\, \overline{B}a) = .2 & p(\overline{\text{RH}} \,\&\, \overline{R}a \,\&\, \overline{B}a) = .1.
\end{array}
$$

Then $p(\text{RH}) = 1/2$ and $p(\text{RH} \mid \overline{R}a \lor Ba) = 5/8$, so $\overline{R}a \lor Ba$ confirms RH and (γ) is true. However, $p(\text{RH} \mid \overline{R}a \,\&\, Ba) = 1/2$, so $\overline{R}a \,\&\, Ba$ does not confirm RH and (β) is false.

Thus Kuipers' "second paradox" is not established as a paradox at all. We have just one paradox, namely the fact that Principles 1-3, although each intuitive, are jointly inconsistent.

4. Kuipers' Solution

Kuipers proposes solutions to both of his "paradoxes." I will not discuss his solution to the spurious paradox but I will now describe his solution to the genuine paradox.

This solution is based on a principle of comparative confirmation that Kuipers (p.26) states as follows:

> P.1c: If E C-confirms H and E^* C^*-confirms H then E C-confirms H more than E^* C^*-confirms H iff E^* is, given C^*, more plausible than E, given C, in the light of the background beliefs.

Kuipers (p.28) argues that P.1c entails a special principle of conditional confirmation that he calls "$S^{\#}.1c$(-ravens)." The following is my rewording of Kuipers' formulation of this principle. (Following Kuipers, I use $\#R$ and $\#\overline{B}$ to denote the number of individuals that are ravens and that are non-black, respectively.)

> $S^{\#}.1c$: $Ra\&Ba$ Ra-confirms RH more than $\overline{Ra}\,\&\,\overline{Ba}$ $\overline{Ba}$-confirms it iff the background beliefs imply that $\#R < \#\overline{B}$.

Kuipers (p.28) assumes that "the background beliefs" (I take him to mean *our* background beliefs) imply that $\#R$ is much smaller than $\#\overline{B}$. He takes this assumption and $S^{\#}.1c$ to imply a proposition that he labels (4). I will formulate it as:

(4) $Ra\&Ba$ cd-confirms RH more than $\overline{Ra}\,\&\,\overline{Ba}$ does.

According to Kuipers (p.29), (4) solves the paradox "concerning non-black non-ravens", that is, the only genuine paradox.

I will now examine this purported derivation of (4).

4.1. *The Principle P.1c*

I will begin by asking what reason there is to accept P.1c. When Kuipers states P.1c he does not give any justification for it other than that it is the analog for conditional deductive confirmation of another principle, called P.1, that he proposed for unconditional deductive confirmation. Kuipers supported P.1 by claiming (a) that it is intuitive, and (b) that it follows from the quantitative theory of confirmation that he presents in the following chapter (chapter 3). So presumably Kuipers intends P.1c to be justified in the same ways; and in fact Kuipers does in chapter 3 derive P.1c from his quantitative theory of confirmation. I will begin by commenting on this quantitative justification of P.1c.

In the quantitative theory of confirmation, plausibility is identified with probability and E is said to confirm H iff $p(H|E) > p(H)$.[5] More generally, E is said to confirm H given C iff $p(H|E\&C) > p(H|C)$. The question of *how much* E confirms H given C then depends on how one measures the degree to which $p(H|E\&C)$ is larger than $p(H|C)$. Many such measures have been advocated (cf. Festa 1999; Fitelson 1999); the following are among the more popular proposals.

Difference measure:	$d(H, E|C)$	$=$	$p(H|E\&C) - p(H|C)$
Ratio measure[6]:	$r(H, E|C)$	$=$	$p(E|H\&C) / p(E|C)$
Likelihood ratio:	$l(H, E|C)$	$=$	$p(E|H\&C) / p(E|\bar{H}\&C)$.

Some authors favor $\log(r)$ or $\log(l)$.

Kuipers (pp. 50-58) favors r or $\log(r)$ and his quantitative justification of P.1c (p. 58) consists in noting that, if plausibility is measured by a probability function and degree of confirmation is measured by r, then P.1c is true. But:

THEOREM 2. Let plausibility be measured by a probability function and let f and g be any strictly increasing functions defined on $[-1,1]$ and $[0,\infty)$ respectively. Then P.1c is false if degree of confirmation is measured by $f(d)$ or $g(l)$.

Since f and g may be the identity function, this theorem implies that P.1c is false if degree of confirmation is measured by d or l. Also, provided we take $\log(0)$ to be defined with value $-\infty$, the theorem implies that P.1c is false if degree of confirmation is measured by $\log(l)$.

Fitelson (2001) gives good reasons to regard l or $\log(l)$ as a better measure of degree of confirmation than either r or $\log(r)$ or d. Therefore, in view of Theorem 2, I think that quantitative considerations undermine P.1c rather than supporting it.

In any case, reliance on a quantitative justification of P.1c is contrary to Kuipers' claim (p. 43) that "we do not need a quantitative approach" in order to have a qualitative theory of deductive confirmation.

---

[5] This assumes the standard definition of confirmation. On Kuipers' success definition of confirmation the condition would be $p(E|H) > p(E)$. Since these conditions are equivalent when $p(E)$ and $p(H)$ are both positive, I will here for simplicity consider only the condition stated in the text.

[6] I have expressed the ratio measure in the form Kuipers prefers. A more common formulation writes it as $p(H|E\&C) / p(H|C)$. The two formulations are equivalent provided $p(H|C) > 0$ and $p(E|C) > 0$.

I mentioned earlier that Kuipers might also intend for P.1c to be justified by a direct appeal to intuition. But even if one has an initial intuition that P.1c is plausible, I think that intuition should evaporate when one appreciates that the truth of P.1c depends on how one chooses to measure degree of confirmation and there are plausible measures that are inconsistent with P.1c.

Thus P.1c is a dubious principle for which Kuipers has no rationally compelling justification.

4.2. *The Special Principle $S^{\#}.1c$*

I turn now to the second step in Kuipers' solution to the paradox, namely his claim that P.1c entails $S^{\#}.1c$. I begin by observing that:

THEOREM 3.   The following can all hold together:
    (i) P.1c is true.
    (ii) The "only if" part of $S^{\#}.1c$ is false.
    (iii) Plausibility is measured by a probability function.
    (iv) Degree of confirmation is measured by r or $\log(r)$.
    (v) a is known to be selected at random from the population.
    (vi) RH is not certain.

This theorem shows that P.1c does not entail the "only if" part of $S^{\#}.1c$. Furthermore, the theorem shows that even if we assume in addition any or all of conditions (iii)-(vi), still P.1c does not entail the "only if" part of $S^{\#}.1c$.

Things are a bit better with the "if" part of $S^{\#}.1c$, as the following theorem shows.

THEOREM 4.   *If*
    (i) P.1c is true,
    (ii) plausibility is measured by a probability function,
    (iii) a is known to be selected at random from the population, and
    (iv) RH is not certain
then the "if" part of $S^{\#}.1c$ is true.

However, Kuipers is claiming to solve the ravens paradox using only qualitative, not quantitative, confirmation theory. Therefore he cannot assume condition (ii) of Theorem 4.

Kuipers (p. 28) does attempt to show that the "if" part of $S^{\#}.1c$ can be derived from P.1c without assuming quantitative probabilities. I take his argument to be that the "if" part of $S^{\#}.1c$ follows from the following premises:

    (I) P.1c is true.

(II) If F, G, F^*, and G^* are properties, the background beliefs imply that $\#FG/\#G > \#F^*G^*/\#G^*$, and a is known to be selected at random from the population, then Fa is, given Ga, more plausible than F^*a, given G^*a, in light of the background beliefs.

(III) a is known to be selected at random from the population.

(IV) It is known that $\#R < \#\overline{B}$.

(V) It is known that RH is false.

The "if" part of $S^{\#}.1c$ does follow from (I)-(V) but, as I will now argue, (II) and (V) are both objectionable.

Condition (II) is superficially plausible but:

THEOREM 5. If plausibility is measured by a probability function then (II) is false.

Kuipers might respond that his argument does not require (II) but only the following special case of it:

(II-R) If the background beliefs imply that $\#\overline{B}\overline{R}/\#\overline{B} > \#BR/\#R$, and a is known to be selected at random from the population, then $\overline{R}a$ is, given $\overline{B}a$, more plausible than Ba, given Ra, in light of the background beliefs.

However, there seems no *qualitative* reason to believe (II-R) other than that it is a special case of the plausible principle (II), and since (II) is false this is no reason at all.

Turning now to (V): We do not know that RH is false, so this does not represent our background information. Further, if we did know that RH is false then, according to the standard concept of confirmation, $Ra\&Ba$ would not confirm RH, so Nicod's condition would not hold and there would be no paradox. So (V) restricts the case to one that is irrelevant to the paradox.

Thus Kuipers has not shown that the "if" part of $S^{\#}.1c$ can be derived from P.1c without making use of quantitative confirmation theory.

4.3. *The Proposition (4)*

The proposition (4), which Kuipers says solves the ravens paradox, seems to introduce a comparative concept of cd-confirmation. Since Kuipers had not defined such a concept, I considered two ways such a concept could be defined and found that neither was satisfactory for Kuipers' purposes. I therefore came to the conclusion that it would be best to take Kuipers' solution to be not (4) but rather:

(4′) $Ra\&Ba$ *Ra*-confirms RH more than $\overline{R}a\,\&\,\overline{B}a$ *Ba*-confirms it.

Subsequently Kuipers informed me that (4′) is what he meant by (4). So we agree that Kuipers' solution to the ravens paradox should be taken to be (4′).

5. Adequacy of the Solution

In Section 3 I characterized the ravens paradox as the fact that Principles 1-3, although each intuitive, are jointly inconsistent. Maher (1999) proposed that a fully satisfactory solution to this paradox will do three things: (a) identify which of the principles is false, supporting this identification with cogent reasons; (b) for each principle that is deemed false, give an explanation that provides insight into why it is false; and (c) for each principle that is deemed false, identify a true principle that is sufficiently similar to the false one that failure to distinguish the two might explain why the false principle is prima facie plausible. I will now evaluate Kuipers' solution with respect to these criteria.

Kuipers' solution is (4′). However, (4′) is consistent with each of Principles 1-3 and so it does not tell us which of those principles is false. Thus (4′) does not provide part (a) of a solution to the paradox.

After presenting his solution to the paradox, Kuipers does address the question of which principle should be rejected. He writes:

> There remains the question of what to think of Hempel's principles used to derive the paradoxes of confirmation. It is clear that the equivalence condition was not the problem, but Nicod's criterion that a black raven confirms RH *un*conditionally. Whereas Nicod's condition is usually renounced unconditionally, we may conclude that it is (only) right in a sophisticated sense: a black raven is a case of cd-confirmation, viz., on the condition of being a raven, (p.29)

From correspondence with Kuipers I gather that what he has in mind here is this: If we interpret the term "confirms" in the ravens paradox as meaning "d-confirms," then Principle 1 is false and Principles 2 and 3 are true; also, if we interpret "confirms" as meaning "cd-confirms," then Principle 3 is false and Principles 1 and 2 are true. These observations are correct. However, "confirms" does not mean either "d-confirms" or "cd-confirms." Confirmation is the concept whose definition was under dispute in Section 1. On either Kuipers' definition or the standard one, it is plain that d-confirmation is not a necessary condition for confirmation. Also, I showed in Section 2 that cd-confirmation is not a sufficient condition for confirmation. That being the case, it is illegitimate to interpret "confirms" in the ravens paradox as meaning either "d-confirms" or "cd-confirms." Or, if one does interpret "confirms" in either of these ways, then one has changed the subject. Thus, although Kuipers here purports to provide part (a) of a solution to the paradox, he does not in fact do so.

Since Kuipers' qualitative solution does not identify which principle is false, it obviously also does not explain why that principle is false, and thus does not give part (b) of a solution to the paradox.

In Chapter 3 Kuipers gives a quantitative treatment of the ravens paradox. Here he concludes (p. 59) that black ravens do confirm RH, in accordance with Principle 1. Although Kuipers does not mention it, this quantitative treatment of the paradox implies that it is Principle 3 that is false. Here Kuipers is talking about confirmation, as characterized by his definition of confirmation, and not d-confirmation or cd-confirmation.[7]

Turning now to part (c), let us suppose, in accordance with what Kuipers' quantitative treatment implies, that Principle 3 is false. Could Kuipers' qualitative solution (4′) explain why Principle 3 has nevertheless seemed plausible? A basic difficulty is that they are concerned with different situations; Principle 3 concerns a situation in which one learns that an arbitrary object is a non-black non-raven, whereas (4′) is concerned with situations in which an object is known to be non-black and found to be a non-raven, or known to be a raven and found to be black. However, if one thinks that people might not distinguish these different situations, then the truth of (4′), supposing it is true, might at least partially explain why Principle 3 has seemed plausible. On the other hand, I think that Principle 3 can seem plausible even when we are clear that it concerns situations in which the object is not antecedently known to be a raven or non-black.

So at best, Kuipers' qualitative solution to the ravens paradox gives us only part of part (c) of a solution to the paradox. And even this only makes sense on the assumption that part (a) of the solution has been obtained from elsewhere.

6. Conclusion

In the portion of his book that I have discussed, Kuipers attempts to show that there is a substantive qualitative theory of confirmation that does not assume quantitative probabilities. The only part of this theory that I would endorse is his claim that d-confirmation is a sufficient condition for confirmation. (And even there one needs to add the qualification that the evidence must not be

---

[7] Kuipers' quantitative treatment of the ravens paradox is based on assumptions that are consistent with Good's (1967) counterexample to Nicod's condition. But as I have just remarked, Kuipers takes his quantitative analysis to show that black ravens do confirm RH, which is not true in Good's example. This shows that Kuipers' quantitative treatment of the paradox is also fallacious. But that is another topic, outside the scope of this paper.

certainly true and the hypothesis must not be certainly false, relative to the background evidence.) As for the other parts: C-confirmation is just d-confirmation with C added to the background beliefs (plus some restrictive conditions that serve no good purpose) and so adds nothing essentially new. The notion of cd-confirmation is trivial. Kuipers' basic principle for comparing degrees of C-confirmation, namely P.1c, lacks any cogent justification and is inconsistent with some plausible proposed measures of degree of confirmation.

Kuipers tries to demonstrate the power of his qualitative theory by using it to solve the ravens paradox. This solution is based on the dubious P.1c and its derivation from that principle appears to require quantitative probabilities. Furthermore, Kuipers' "solution," even supposing it true, does not solve the paradox.

So Kuipers has not shown that there is a qualitative theory of confirmation, not assuming quantitative probabilities, that can solve the ravens paradox.

7. Proofs

7.1. *Proof of Theorem 1*

Let E and H be logically independent and let D be any proposition that is logically independent of $\bar{E}\,\&\,\bar{H}$. Also let C be $E \vee (\bar{H}\,\&\,D)$. The following observations show that E C-confirms H.

    (i) Suppose H is true. Since E and H are logically independent it is possible that E is true, in which case C is true. Similarly it is possible that E is false, in which case C is false.

Suppose H is false. Since E and H are logically independent it is possible that E is true, in which case C is true. Similarly it is possible that E is false and in addition (since D is logically independent of $\bar{E}\,\&\,\bar{H}$) it is possible that D is false; in that case C is false.

Thus neither the truth nor the falsity of H determines the truth value of C. Hence H and C are logically independent.

    (ii) $\bar{E}\,\&\,\bar{H}\,\&\,D$ is logically possible, in which case C is true and E is false. Thus C does not entail E.

    (iii) $H\&C$ entails E.

Furthermore, E entails C. Thus E cd-confirms H, as claimed.

7.2. *Proof of Theorem 2*

Let H be the hypothesis $(x)(Cx \rightarrow Fx)$. Suppose that the population is known to be distributed in one or other of the following two ways:

$$
\begin{array}{c|cc}
 & F & \bar{F} \\
\hline
C & 3 & 0 \\
\bar{C} & 5 & 2 \\
\end{array}
\qquad
\begin{array}{c|cc}
 & F & \bar{F} \\
\hline
C & 3 & 2 \\
\bar{C} & 2 & 3 \\
\end{array}
$$
$$
\qquad\quad H \qquad\qquad\qquad \bar{H}
$$

Suppose that a is known to be randomly selected from the population, plausibility is measured by probability function p, and $p(H) = 1/2$. Then:

$$p(Fa \,|\, Ca \,\&\, H) = 1 \tag{1}$$
$$p(Fa \,|\, Ca \,\&\, \bar{H}) = 3/5 \tag{2}$$
$$p(\bar{C}a \,|\, \bar{F}a \,\&\, H) = 1 \tag{3}$$
$$p(\bar{C}a \,|\, \bar{F}a \,\&\, \bar{H}) = 3/5 \tag{4}$$

$$p(H \,|\, Ca) = \frac{p(Ca \,|\, H)}{p(Ca \,|\, H) + p(Ca \,|\, \bar{H})} = \frac{3}{8} \tag{5}$$

$$p(H \,|\, \bar{F}a) = \frac{p(\bar{F}a \,|\, H)}{p(\bar{F}a \,|\, H) + p(\bar{F}a \,|\, \bar{H})} = \frac{2}{7} \tag{6}$$

$$p(H \,|\, Fa \,\&\, Ca) = \frac{p(Fa \,\&\, Ca \,|\, H)}{p(Fa \,\&\, Ca \,|\, H) + p(Fa \,\&\, Ca \,|\, \bar{H})} = \frac{1}{2} \tag{7}$$

$$p(H \,|\, \bar{F}a \,\&\, \bar{C}a) = \frac{p(\bar{F}a \,\&\, \bar{C}a \,|\, H)}{p(\bar{F}a \,\&\, \bar{C}a \,|\, H) + p(\bar{F}a \,\&\, \bar{C}a \,|\, \bar{H})} = \frac{2}{5}. \tag{8}$$

In this example, $Fa \,\&\, Ca$ Ca-confirms H and $\bar{F}a \,\&\, \bar{C}a$ $\bar{F}a$-confirms H. Also

$$
\begin{aligned}
p(Fa \,\&\, Ca \,|\, Ca) \;&= p(Fa \,|\, Ca \,\&\, H)p(H \,|\, Ca) + p(Fa \,|\, Ca \,\&\, \bar{H})p(\bar{H} \,|\, Ca) \\
&= (3/8) + (3/5)(5/8), \text{ by } (1), (2), \text{ and } (5) \\
&= 3/4 = 0.75. \\
p(\bar{F}a \,\&\, \bar{C}a \,|\, \bar{F}a) \;&= p(\bar{C}a \,|\, \bar{F}a \,\&\, H)p(H \,|\, \bar{F}a) + p(\bar{C}a \,|\, \bar{F}a \,\&\, \bar{H})p(\bar{H} \,|\, \bar{F}a) \\
&= (2/7) + (3/5)(5/7), \text{ by } (3), (4), \text{ and } (6) \\
&= 5/7 = 0.71 \text{ (to 2 decimal places).}
\end{aligned}
$$

So $p(Fa \,\&\, Ca \,|\, Ca) > p(\bar{F}a \,\&\, \bar{C}a \,|\, \bar{F}a)$ and hence $Fa \,\&\, Ca$ is, given Ca, more plausible than $\bar{F}a \,\&\, \bar{C}a$, given $\bar{F}a$, in light of the background beliefs. But

$$
\begin{aligned}
d(H, Fa \& Ca \mid Ca) \; &= \; p(H \mid Fa \& Ca) - p(H \mid Ca) \\
&= \; (1/2) - (3/8), \text{ by (5) and (7)} \\
&= \; 1/8 = 0.125.
\end{aligned}
$$

$$
\begin{aligned}
d(H, \bar{F}a \& \bar{C}a \mid \bar{F}a) \; &= \; p(H \mid \bar{F}a \& \bar{C}a) - p(H \mid \bar{F}a) \\
&= \; (2/5) - (2/7), \text{ by (6) and (8)} \\
&= \; 4/35 = 0.11 \text{ (to 2 decimal places).}
\end{aligned}
$$

So $d(H, Fa \& Ca \mid Ca) > d(H, \bar{F}a \& \bar{C}a \mid \bar{F}a)$ and hence

$$
f[d(H, Fa \& Ca \mid Ca)] \; > \; f[d(H, \bar{F}a \& \bar{C}a \mid \bar{F}a)].
$$

Thus P.1c is false if degree of confirmation is measured by $f(d)$. Also

$$
l(H, Fa \& Ca \mid Ca) = \frac{p(Fa \mid H \& Ca)}{p(Fa \mid \bar{H} \& Ca)}
$$

$$
= 5/3, \text{ by (1) and (2).}
$$

$$
l(H, \bar{F}a \& \bar{C}a \mid \bar{F}a) = \frac{p(\bar{C}a \mid H \& \bar{F}a)}{p(\bar{C}a \mid \bar{H} \& \bar{F}a)}
$$

$$
= 5/3, \text{ by (3) and (4).}
$$

So $l(H, Fa \& Ca \mid Ca) = l(H, \bar{F}a \& \bar{C}a \mid \bar{F}a)$ and hence

$$
g[l(H, Fa \& Ca \mid Ca)] = g[l(H, \bar{F}a \& \bar{C}a \mid \bar{F}a)].
$$

Thus P.1c is false if degree of confirmation is measured by $g(l)$.

7.3. *Proof of Theorem 3*

Suppose that the population is known to be distributed in one or other of the following two ways:

	B	$\bar{B}$			B	$\bar{B}$
R	2	0		R	3	2
$\bar{R}$	4	4		$\bar{R}$	3	2
RH				$\overline{\text{RH}}$		

Suppose that a is known to be randomly selected from the population, plausibility is measured by probability function p, and $p(\text{RH}) = 1/2$. Then:

$$
p(Ba \mid Ra \& \text{RH}) = 1 \tag{9}
$$

$$
p(Ba \mid Ra \& \overline{\text{RH}}) = 3/5 \tag{10}
$$

$$
p(\bar{R}a \mid \bar{B}a \& \text{RH}) = 1 \tag{11}
$$

$$
p(\bar{R}a \mid \bar{B}a \& \overline{\text{RH}}) = 1/2 \tag{12}
$$

$$
p(\text{RH} \mid Ra) = \frac{p(Ra \mid \text{RH})}{p(Ra \mid \text{RH}) + p(Ra \mid \overline{\text{RH}})} = \frac{2}{7} \tag{13}
$$

$$p(\mathrm{RH} \mid \bar{B}a) = \frac{p(\bar{B}a \mid \mathrm{RH})}{p(\bar{B}a \mid \mathrm{RH}) + p(\bar{B}a \mid \overline{\mathrm{RH}})} = \frac{1}{2}. \tag{14}$$

Let degree of confirmation be measured by r or $\log(r)$. It follows that P.1c is true. Also

$r(\mathrm{RH}, Ra \,\&\, Ba \mid Ra)$

$$= \frac{p(Ba \mid Ra \,\&\, \mathrm{RH})}{p(Ba \mid Ra \,\&\, \mathrm{RH})p(\mathrm{RH} \mid Ra) + p(Ba \mid Ra \,\&\, \overline{\mathrm{RH}})p(\overline{\mathrm{RH}} \mid Ra)}$$

$$= \frac{1}{2/7 + (3/5)(5/7)}, \quad \text{by (9), (10), and (13)}$$

$$= 7/5 = 1.4.$$

$r(\mathrm{RH}, \bar{R}a \,\&\, \bar{B}a \mid \bar{B}a)$

$$= \frac{p(\bar{R}a \mid \bar{B}a \,\&\, \mathrm{RH})}{p(\bar{R}a \mid \bar{B}a \,\&\, \mathrm{RH})p(\mathrm{RH} \mid \bar{B}a) + p(\bar{R}a \mid \bar{B}a \,\&\, \overline{\mathrm{RH}})p(\overline{\mathrm{RH}} \mid \bar{B}a)}$$

$$= \frac{1}{1/2 + (1/2)(1/2)}, \quad \text{by (11), (12), and (14)}$$

$$= 4/3 = 1.33'.$$

So $r(\mathrm{RH}, Ra \,\&\, Ba \mid Ra) > r(\mathrm{RH}, \bar{R}a \,\&\, \bar{B}a \mid \bar{B}a)$ and hence also

$$\log[r(\mathrm{RH}, Ra \,\&\, Ba \mid Ra)] > \log[r(\mathrm{RH}, \bar{R}a \,\&\, \bar{B}a \mid \bar{B}a)].$$

Therefore $Ra \,\&\, Ba$ Ra-confirms RH more than $\bar{R}a \,\&\, \bar{B}a$ $\bar{B}a$-confirms RH. However, the background beliefs do not imply $\#R < \#\bar{B}$; in fact, there is a probability of 1/2 that $\#R > \#\bar{B}$. Thus the "only if" part of $S^{\#}.1c$ is false.

7.4. *Proof of Theorem 4*

Suppose that conditions (i)-(iv) hold and that the background beliefs imply that $\#R < \#\bar{B}$. I will show that it follows that $Ra \,\&\, Ba$ Ra-confirms RH more than $\bar{R}a \,\&\, \bar{B}a$ $\bar{B}a$-confirms it, so that the "if" part of $S^{\#}.1c$ holds.

Let $K_1, \ldots, K_n$ denote the possible states of the world. (For simplicity I here consider only the case in which the number of possible states is finite.) Let the population counts in K_i be symbolized as follows:

$$\begin{array}{c|cc} & B & \bar{B} \\ \hline R & a_i & b_i \\ \bar{R} & c_i & a_i + x_i \end{array}$$

106   *Patrick Maher*

Let p be the probability function that measures plausibility. Then

$$
\begin{aligned}
p(Ra\,\&\,Ba \mid Ra) &= \sum_{i=1}^{n} p(Ba \mid Ra\,\&\,K_i)\,p(K_i \mid Ra) \\[2mm]
&= \sum_{i=1}^{n} p(Ba \mid Ra\,\&\,K_i)\,\frac{p(Ra \mid K_i)\,p(K_i)}{\sum_{j=1}^{n} p(Ra \mid K_j)\,p(K_j)} \\[2mm]
&= \sum_{i=1}^{n} \frac{a_i}{a_i + b_i}\,\frac{(a_i + b_i)\,p(K_i)}{\sum_{j=1}^{n}(a_j + b_j)\,p(K_j)} \\[2mm]
&= \frac{\sum_{i=1}^{n} a_i\,p(K_i)}{\sum_{i=1}^{n}(a_i + b_i)\,p(K_i)}.
\end{aligned}
$$

A similar calculation gives:

$$
p(\overline{Ra}\,\&\,\overline{Ba} \mid \overline{Ba}) = \frac{\sum_{i=1}^{n}(a_i + x_i)\,p(K_i)}{\sum_{i=1}^{n}(a_i + x_i + b_i)\,p(K_i)}.
$$

Thus $p(Ra\,\&\,Ba \mid Ra) < p(\overline{Ra}\,\&\,\overline{Ba} \mid \overline{Ba})$ iff:

$$
\frac{\sum_{i=1}^{n} a_i\,p(K_i)}{\sum_{i=1}^{n}(a_i + b_i)\,p(K_i)} < \frac{\sum_{i=1}^{n}(a_i + x_i)\,p(K_i)}{\sum_{i=1}^{n}(a_i + x_i + b_i)\,p(K_i)}.
$$

Multiplying out the terms in this latter inequality, and then simplifying, we find that it is equivalent to this inequality:

$$
0 < \left[\sum_{i=1}^{n} b_i\,p(K_i) \right]\left[\sum_{i=1}^{n} x_i\,p(K_i) \right].
$$

Since it is not certain that RH is true, $b_i p(K_i)$ is positive for at least one i. Since it is certain that $\#R < \#\overline{B}$, $x_i > 0$ for all i. Hence the latter inequality is true. Thus $p(Ra\,\&\,Ba \mid Ra) < p(\overline{Ra}\,\&\,\overline{Ba} \mid \overline{Ba})$. So by P.1c, $Ra\,\&\,Ba$ Ra-confirms RH more than $\overline{Ra}\,\&\,\overline{Ba}$ $\overline{Ba}$-confirms RH.

7.5. *Proof of Theorem 5*

Let F and G be properties and suppose the population is known to be distributed in one or other of the following two ways:

$$\begin{array}{cc} & G \quad \bar{G} \\ F & \boxed{\begin{array}{cc} 25 & 26 \\ 24 & 25 \end{array}} \\ \bar{F} & \\ & K_1 \end{array} \qquad \begin{array}{cc} & G \quad \bar{G} \\ F & \boxed{\begin{array}{cc} 33 & 1 \\ 64 & 2 \end{array}} \\ \bar{F} & \\ & K_2 \end{array}$$

If K_1 holds then

$$\frac{\#FG}{\#G} = \frac{25}{49} > \frac{26}{51} = \frac{\#F\bar{G}}{\#\bar{G}}$$

and if K_2 holds then

$$\frac{\#FG}{\#G} = \frac{33}{97} > \frac{1}{3} = \frac{\#F\bar{G}}{\#\bar{G}}$$

Thus the background beliefs imply that $\#FG/\#G > \#F\bar{G}/\#\bar{G}$.

Suppose that a is known to be randomly selected from the population, plausibility is measured by probability function p, and $p(K_1) = p(K_2) = 1/2$. Then (II) implies

$$p(Fa \,|\, Ga) > p(Fa \,|\, \bar{G}a). \tag{15}$$

But

$$p(K_1 \,|\, Ga) = \frac{p(Ga \,|\, K_1)}{p(Ga \,|\, K_1) + p(Ga \,|\, K_2)} = \frac{49}{146} \tag{16}$$

$$p(K_1 \,|\, \bar{G}a) = \frac{p(\bar{G}a \,|\, K_1)}{p(\bar{G}a \,|\, K_1) + p(\bar{G}a \,|\, K_2)} = \frac{51}{54} \tag{17}$$

$$p(Fa \,|\, Ga) = p(Fa \,|\, Ga \,\&\, K_1)p(K_1 \,|\, Ga) + p(Fa \,|\, Ga \,\&\, K_2)p(K_2 \,|\, Ga)$$

$$= \frac{25}{49} \cdot \frac{49}{146} + \frac{33}{97} \cdot \frac{97}{146}, \text{ using (16)}$$

$$= \frac{58}{146} = 0.40 \text{ (to 2 decimal places)}$$

$$p(Fa \,|\, \bar{G}a) = p(Fa \,|\, \bar{G}a \,\&\, K_1)p(K_1 \,|\, \bar{G}a) + p(Fa \,|\, \bar{G}a \,\&\, K_2)p(K_2 \,|\, \bar{G}a)$$

$$= \frac{26}{51} \cdot \frac{51}{54} + \frac{1}{3} \cdot \frac{3}{54}, \text{ using (17)}$$

$$= \frac{27}{54} = 0.5.$$

Thus (15) is false and hence (II) is false.

                          Patrick Maher

ACKNOWLEDGMENTS

Theo Kuipers and Roberto Festa provided extensive comments that significantly influenced the final form of this paper.

University of Illinois at Urbana-Champaign
Department of Philosophy
105 Gregory Hall
810 S Wright St
Urbana, IL 61801-3611
USA
e-mail: maher1@uiuc.edu

REFERENCES

Festa, R. (1999). Bayesian Confirmation. In: M. C. Galavotti and A. Pagnini (eds.), *Experience, Reality, and Scientific Explanation*, pp. 55-87. Dordrecht: Kluwer.

Fitelson, B. (1999). The Plurality of Bayesian Measures of Confirmation and the Problem of Measure Sensitivity. *Philosophy of Science* **66**, S362-S378.

Fitelson, B. (2001). A Bayesian Account of Independent Evidence with Applications. *Philosophy of Science* **68**, S123-S140.

Good, I. J. (1967). The White Shoe is a Red Herring. *British Journal for the Philosophy of Science* **17**, 322.

Hempel, C. G. (1945). Studies in the Logic of Confirmation. *Mind* **54**. Page references are to the reprint in Hempel (1965).

Hempel, C. G. (1965). *Aspects of Scientific Explanation.* New York: The Free Press.

Kuipers, T. A. F. (2000/ICR). *From Instrumentalism to Constructive Realism.* Dordrecht: Kluwer.

Maher, P. (1999). Inductive Logic and the Ravens Paradox. *Philosophy of Science* **66**, 50-70.

Theo A. F. Kuipers

THE NON-STANDARD APPROACH TO CONFIRMATION
AND THE RAVENS PARADOXES

REPLY TO PATRICK MAHER

Patrick Maher's (PM, for short) critical paper requires a long reply. His first main point is my non-standard approach to confirmation. The second deals with my notion of conditional deductive confirmation and its application to the ravens paradoxes. In the first part of this reply I defend the non-standard approach extensively in a non-dogmatic way. In the second part I defend the notion of conditional deductive confirmation and its application to both counterintuitive cases dealing with ravens, or rather with black and non-black non-ravens. I am happy to be able to conclude this reply with a survey of the main interesting observations that I learned from Maher's critical exposition.

The Non-Standard Approach, i.e. the Success Definition of Confirmation

On Section 1: Definition of Confirmation

In Section 1, Maher criticizes my success definition of confirmation in a way that demands either retreat or extensive defense. For the moment I opt for the latter. In the introduction to Part I of ICR (p. 15) I announce the three main non-standard aspects of the success definition: its "reversive" (although I did not use that term), its "inclusive" and its "pure" character. That is, it *reverses* the definiens clause from 'E makes H more plausible' into 'H makes E more plausible', it is *pure* in the sense that it is neutral in rewarding hypotheses of different plausibility for the same success, and it *includes* the possibility of confirming "p-zero" hypotheses (i.e. hypotheses with probability zero). I shall deal with these aspects in the reverse order. Whenever the distinction is not relevant, I move freely between qualitative and quantitative, i.e. probabilistic, formulations.

In: R. Festa, A. Aliseda and J. Peijnenburg (eds.), *Confirmation, Empirical Progress, and Truth Approximation* (*Poznań Studies in the Philosophy of the Sciences and the Humanities,* vol. 83), pp. 109-127. Amsterdam/New York, NY: Rodopi, 2005.

Let us start, though, with a relativization by quoting a passage from the introduction to the quantitative chapter in ICR (p. 44):

> Moreover, as in the qualitative case, it will also become clear that there is not one "language of quantitative confirmation", but several, e.g., pure and impure ones, inclusive and non-inclusive ones. As long as one uses the probability calculus, it does not matter which confirmation language one chooses, the only important point is to always make clear which one one has chosen. Although speaking of confirmation languages is hence more appropriate, we will accept the current practice of speaking of confirmation theories.

Unfortunately I did not elaborate the 'as in the qualitative case' in the qualitative chapter itself. But implicitly it is fairly clear in that chapter that I am well aware that there are also different "languages of *qualitative* confirmation*,*" and hence that, if one assumes *the* obvious qualitative plausibility "calculus," viz. the one implied by the (quantitative) probability calculus, "the only important point is to always make clear which one one has chosen." Hence, my defense of the non-standard approach must be seen against this non-dogmatic background. At the end of the following defense I even propose a kind of fusion between my non-standard approach and the pure version of the standard approach.

1. *Zero probabilities*. Maher is right in demanding attention for the fact that a main disadvantage of my approach seems to be that confirmation by "p-zero" evidence (i.e. evidence with probability zero) is indeed impossible. I should have paid explicit attention to this possible objection.

1.1. *Verifying p-zero evidence*. Let us therefore start with Maher's *prima facie* very convincing example of a specific real value as evidence fitting into an interval hypothesis. Maher is right in speaking about verification in this case, but he also wants to see verification, in line with the standard approach, as an extreme, but proper, case of confirmation, which is indeed impossible from my perspective. Inherent in my approach, and hopefully radiating from my (qualitative/deductive) Confirmation Matrix (ICR, p. 22) and the (quantitative/probabilistic) Confirmation Square (ICR, p. 46), is that verification is at most an *im*proper extreme kind of confirmation (see ICR, pp. 46-7). Hence I would indeed like to "deny that in this case there is any [proper] confirmation at all" (PM, p. 4). Instead, it is a straightforward case of verification, not at all made problematic by being due to p-zero evidence. In such a case of verification, E (logically) entails H, and there is nothing more to say about it. For example, whereas in the case of (proper) confirmation it is plausible to distinguish between deductive and non-deductive (i.e. probabilistic) confirmation, a similar distinction is not relevant for verification, nor for falsification for that matter; formally speaking, verification and falsification are unproblematic qualifications. In other words, it is not verification but deductive confirmation that is an extreme proper case of confirmation; verification and deductive confirmation only go together when H and E are logically equivalent.

Historians are well aware of the fundamental distinction between verification and confirmation. In many cases they can just verify their hypotheses of interest. Consider hypotheses about the date and place of birth and death that may have been suggested by some previous evidence. Such hypotheses may subsequently just be verified (or falsified) by consulting the relevant civil records. Of course, such data may occasionally be doubted, but that holds for all types of evidence and will have to be accounted for by the appropriate type of "Jeffrey-conditionalization" or by "globalization," see below. Moreover, if verification is impossible, e.g. a town's civic records might have been lost, historians will of course search for merely confirming evidence. Occasionally this may lead to deductively confirming, but non-verifying, evidence. For example, a more global civic register may survive in the archives of the province or region, containing only the years of birth and death, but not the precise days, let alone the hours. The fact that in the suggested historical cases the evidence may have been assigned some non-zero probability is, of course, not relevant for our arguing for a fundamental distinction between (proper) confirmation and verification

1.2. *Non-verifying p-zero evidence.* As I describe myself (ICR, p. 45) in an example, there are cases of non-verifying p-zero evidence that leave room for defining a meaningful posterior probability $p(H/E)$, despite the fact that the standard definition is not applicable since $p(E) = 0$. The consequence is, as I should have remarked, that this makes confirmation possible in the standard sense but not in my sense, which is technically similar to the way in which my definition leaves room for confirmation of p-zero hypotheses and the standard one does not. However, there is a fundamental difference between the relevance of the counterexamples. When $p(E) = 0$ because E reports a particular real number out of a real interval, it is very likely that one should take measure errors into account or that the relevant parameter, e.g. length, just cannot be said to have such a unique value. For both reasons it is then plausible to rephrase the evidence in terms of a small interval of values, which might be called "globalization" of the evidence, in which case, of course, we get p-non-zero evidence and hence the problem disappears. To be sure, there are cases where this globalization is also possible when dealing with p-zero hypotheses. Take, for example, the hypothesis that a certain die is unbiased. In view of the fact that totally unbiased dice will not exist in the real world, we should assign that hypothesis zero probability. Of course, globalization to a 6-

tuple of very small intervals will make a non-zero assignment plausible. Maher seems to suggest this strategy by the claim "the evidence confirms that the hypothesis is close to the truth" (PM, p. 3).

However, in other at least as typical scientific cases this strategy does not make much sense. Consider Einstein's (general) test implication of (at least a certain degree of) light bending when passing heavy objects. For a Newtonian who assigns probability one to Newton's theory, this test implication might well receive probability zero. Hence, however unproblematic Eddington's data might have been (which they were not, but that is another story), they would not confirm Einstein's specific general test implication according to the standard approach. However, some kind of globalization of the hypothesis, whether or not in the "close to the truth" form, is here out of order. Although Einstein made a much more specific, quantitative prediction, the prediction mentioned is already of a qualitative very global nature, but it nevertheless captures the fundamental surprise and risk of his GTR. Hence, in contrast to the standard approach, according to my theory, a "half-open-minded" Newtonian can see the experimental results as confirming evidence for Einstein's theory. If so, he may well see this as a good reason for a non-Bayesian move, viz. changing his prior distribution such that Einstein's theory receives a positive probability, however small.

1.3. *The counterfactual strategy.* Some version of this move is also suggested by Maher when he states that one may say when a scientist talks about confirmation of a p-zero hypothesis from the standard point of view "that what the scientist said is not strictly true, although it is understandable why someone might say that" (PM, p. 3). Of course, this response is also available for his discussion of confirmation by p-zero evidence. More specifically, in both cases one might interpret his way of talking as some kind of counterfactual personal claim: "if I would have assigned non-zero probability, to the hypothesis respectively the evidence, then the evidence would confirm the hypothesis."

2. *The second comparative principle.* The second main reason for the success definition applies already to the "normal case" of non-zero probabilities for hypotheses and evidence. Maher does not pay attention to my emphasis on comparative principles. In this context, particularly P.2 (ICR, p. 24) and its generalization P.2G (ICR, p. 64) are important. Although P.2G does not entail the non-standard approach, I argue that it provides very good additional reasons for preferring the non-standard approach. Starting from the non-standard definition (*SDC*, ICR, p. 23):

(a) *E* confirms *H* iff (*E* is a success of *H* in the sense that) *H* makes *E* more plausible it is plausible to also have (see ICR, P.2G, p. 64, the core of which is):

(Ca)   *E* confirms *H* more than *H** iff *H* makes *E* more plausible than *H**
         does

         E equally confirms *H* and *H** iff *H* and *H** make *E* equally plausible

The additional definitions for probabilistic versions of both subclauses are obvious: $p(E/H) > p(E/H^*)$ and $p(E/H) = p(E/H^*)$ respectively.

The standard definition:

(b)   *E* confirms *H* iff *E* makes *H* more plausible

suggests in contrast the "double" conditions:

(Cb)   *E* confirms *H* more than *H** iff *E* makes *H* "more more plausible"
         than *H**

         E equally confirms *H* and *H** iff *E* makes *H* "equally more
         plausible" than *H**

which are not so easy to elaborate. In particular, for the probabilistic versions everything depends on whether one chooses the ratio or the difference measure as the degree of confirmation (or a close relative of one of them). Or, more cautiously, for judging "more more plausible" or "equally more plausible" one has to choose between comparing differences of the form $p(H/E) - p(H)$ or ratios of the form $p(H/E)/p(H)$. If one opts for comparing differences one's comparative judgments come very much to depend on the prior probabilities of the hypotheses, my reason for writing in ICR of the impure nature of that approach to confirmation.

At least some philosophers of science seem to subscribe to (Ca), which only leaves room for the ratio measure (or a close relative). For instance, Elliott Sober (2000, p. 5) states the principle (in my symbols):

H is better supported than *H** by *E* iff $p(E/H) > p(E/H^*)$

See also (Sober, 2001, pp. 30-3), where he calls a strong version of it "*E* strongly favors *H* over *H** iff $p(E/H) >> p(E/H^*)$" the Likelihood Principle. To be sure, Sober does not want to talk about 'confirmation' here: "We may ask whether an observation supports one hypothesis better than another. Here we're not interested in whether the one hypothesis has a higher prior probability than the other; we want to isolate what the impact of the observation is" (Sober 2000, p. 5). Although many attempts have been made in the literature to draw such a distinction between confirmation and (evidential) support, I would like to argue that we might well read his principle in terms of confirmation. The reason is that I simply do not believe that scientists would not subscribe to the following general claim:

> *E* better supports *H* than *H** iff *E* confirms *H* more than *H**

And I would claim, in addition, they have good reasons for that, for the only things that really count for the practical purposes of scientists are the unconditional and conditional plausibility or probability of evidence or, for that matter, of hypotheses, and their comparisons. Regarding "diachronic" comparisons, including comparisons of diachronic comparisons, it is rather unclear what other aim we can meaningfully have than "to isolate what the impact of the observation is," that is, the pure perspective. Any other comparison will lead to a mixture of unconditional, conditional and "transitional" aspects, which can be decomposed into purely unconditional, conditional and transitional aspects.

To support this claim I consider cases of deductive and/or non-deductive confirmation of two hypotheses by the same evidence. The upshot will be that many intuitions not only suggest that the impure perspective is problematic, but also that a choice between pure and impure degrees of confirmation does not have to be made, and this only follows from the non-standard definition.

2.1. *Comparing deductive confirmation.* If both *H* and *H** entail *E*, they are equally confirmed according to (Ca), but according to (Cb) we have first to decide whether we want to compare ratios or differences. If we take ratios the same verdict results, but if we take differences we obtain that the resulting verdict totally depends on the relative initial plausibility of the hypothesis: the more plausible the more confirmed. It seems rather strange that for such essentially qualitative judgements one first has to make a choice between quantitative criteria. For example, both Newton and Einstein deductively predict the falling of stones near the surface of the moon. Would somebody who is told about confirming experiments by Neil Armstrong have first to make up his mind about whether he prefers comparing ratios or differences in order to judge whether one of the theories is more confirmed than the other or whether they are equally confirmed? If he were not to do so, he would consider this choice as irrelevant. But that would mean that he can't subscribe to (Cb), for that requires a choice. On the other hand, if he wanted to make up his mind, he would be likely to subscribe to (Cb). If he then came to the conclusion that he would favor comparing differences rather than ratios he would in addition have to make up his mind about which hypothesis he finds the more plausible. On the other hand, if he prefers ratios he comes to the "equal confirmation" conclusion only by a rather technical detour. In sum, (Cb) forces one to consider technicalities of a kind that scientists, usually not very sympathetic to the concerns of philosophers of science, are not inclined to do. On the contrary, scientists are likely to have strong intuitions in the suggested case. In which direction, would essentially have to be tested by psychologists of science, where the third possibility – more confirmation of the less plausible hypothesis – should also be taken into consideration.

2.2. *Comparing deductive and non-deductive confirmation.* Let us quote a long passage from Adam Morton's *Theory of Knowledge* (second edition, 1997, p. 186), with abbreviations between []-brackets added:

> *Evidence supports beliefs that make it more probable.* Suppose a geologist defends a theory [$H1$] which predicts [$P1$] an earthquake somewhere on the Pacific coast of North America sometime in the next two years. Then if an earthquake occurs at a particular place and time [$E1$], the theory is somewhat supported. Suppose, on the other hand, that another geologist defends a theory [$H2$] which predicts [$P2$] an earthquake of force 5 on the Richter scale with its epicentre on the UCLA campus on 14 September (the anniversary of Carnap's death, incidentally) in the year 2,000. If this were to occur [$E2$], it would be very strong evidence for the theory.

In the following formalization I equate $E1$ with $E2 = P2$, neglecting the particular force, and indicate it just by E, because the force is not essential and $E1$ could have been any earthquake verifying $P1$. In this way we get:

$H1$ deductively predicts $P1$ hence, $1 = p(P1/H1)$
$H2$ deductively predicts $P2 = E$ hence, $1 = p(E/H2) = p(P2/H2)$
E logically entails $P1$ and is
even much stronger   hence, $p(P1) > p(E)$, $p(P1/H1) > p(E/H1)$
E obtains

Morton, who, like Sober, also avoids talking about confirmation, concludes that E is very strong evidence for $H2$ and somewhat supports $H1$, but I do not hesitate to claim that scientists would see no problem in also saying:

$H2$ is more confirmed by E than $H1$

From (our formalization of) Morton's description it follows straightforwardly that $p(E/H2) = 1 > p(E/H1)$ and hence the case may well be seen as supporting (Ca).

But assume the (Cb)-perspective for a while. Of course, according to the ratio comparison we get the same verdict, for the denominator does not play a role: $p(E/H2)/p(E) = 1/p(E) > p(E/H1)/p(E)$. According to the difference measure this result obtains iff

$$p(H2)\,(p(E/H2)/p(E) - 1) > p(H1)\,(p(E/H1)/p(E) - 1)$$

and hence iff

$$p(H2)\,(1/p(E) - 1) > p(H1)\,(p(E/H1)/p(E) - 1)$$

which holds of course only under specific conditions. Let us assume that $p(H1) = np(H2) < 1$ and $p(E/H1) = mp(E) < 1$ then   we   get: iff

$P(E) > 1/(1+n(m-1))$. Although we may of course assume that m and n are both fairly large, such that the condition does impose a rather small lower bound for $p(E)$, it is nevertheless perfectly possible that $p(E)$ is smaller, in which case the opposite of Morton's intuition is satisfied: E confirms $H1$ more than $H2$. Suppose, for example, that $m = 11$ and $n = 100$, the lower bound is 1/1001, i.e. 1 promille. Hence, the opposite situation certainly is a realistic possibility. In that case we would have a case where, at least according to Morton, "E is very strong evidence for $H2$ and somewhat supports $H1$," but philosophers of science in favor of the difference measure would nevertheless want to say that "E confirms $H1$ more than $H2$."

2.3 *Comparing non-deductive confirmation.* Let us now turn to the second example suggested by Morton (p. 186):

> Or consider the hypotheses that a coin is fair [$H1$] and that it is biased [$H2$]. Suppose that the coin is tossed and [E] lands heads fourteen times and tails one time. This evidence is consistent with both hypotheses, but it has very low probability on the hypothesis that the coin is fair and much higher probability on the hypothesis that the coin is biased. So it gives much stronger evidence for the hypothesis that the coin is biased.

We may formalize the second example by:

$$0 < p(E/H1) \ll p(E/H2) < 1 \text{ and } E \text{ obtains}$$

According to Morton, E gives much stronger evidence for $H2$ than for $H1$. Again, I would not hesitate to claim that scientists would easily say, in agreement with (Ca) and, only, with the ratio version of (Cb):

E confirms $H2$ much more than $H1$

According to the difference comparison we get this iff

$$p(H2)\,(p(E/H2)/p(E) - 1) > p(H1)\,(p(E/H1)/p(E) - 1)$$

with perfect possibilities for an opposite verdict. Again we would have a case in which, at least according to Morton, E "gives much strong evidence for" $H2$ than $H1$, but philosophers of science favoring differences would nevertheless want to say that "E confirms $H1$ more than $H2$."

In sum, my impression is that scientists easily subscribe to (Ca). If they do that indirectly via (Cb) they need to be aware of a particular choice as degree of confirmation. However, since scientists usually do not express their attitudes in terms of degrees of confirmation (or support), it is more plausible that they directly subscribe to (Ca). Of course, this is my informal prejudice about the reasoning of scientists, which only a systematic research by interviews or questionnaires could decide. What "most works on confirmation theory" say (PM, p. 2), is not decisive because the view of confirmation theorists may well be loaded by their favorite interpretation of what they think about the way scientists reason. Of course, it is likely that I am myself a victim of this, but only meta-empirical research can decide on this. For some further elaboration of this point, see below.

3. *The reverse defeniens clause.* Maher claims that "According to etymology, dictionaries, and most works on confirmation theory, '*E* confirms *H*' means that *E* makes *H* more plausible." (PM, p. 2) Although I certainly agree that this "forward connotation" belongs to "confirmation" (vide my "reward principle"), I am not so sure that the "backward connotation" is not at least as important. My book presentation suggests that "forward confirmation" presupposes a success, that is, a success of a hypothesis is a necessary condition for confirmation of a hypothesis (by that same success). The question is whether it is a sufficient condition, of course, not according to "etymology, dictionaries, and most works on confirmation theory," where the latter may be supposed to be written by philosophers (of science), but according to scientists. If so, the backward definition should be considered more basic. Maybe I should have presented my success definition not by (*SDC*, ICR, p. 23):

(a)   E confirms H iff (*E* is a success of H in the sense that) H makes E more plausible

as quoted by Maher, but in two steps, viz.

(a1)   *E* confirms H iff E is a success of H
(a2)   *E* is a success of H iff H makes E more plausible

which yields:

(a)   *E* confirms H iff H makes E more plausible

We should compare this with the "standard" interpretation:

(b)   *E* confirms H iff E makes H more plausible

which does not seem to be decomposable.

Neglecting extreme cases, I have the strong feeling that scientists will, when asked, agree with all four "absolute" claims as conceptually true statements. Above I argued that they are likely to also agree with the following comparative claims:

(Ca)   *E* confirms H more than H^* iff H makes E more plausible than H^* does

       E equally confirms H and H^* iff H and H^* make E equally plausible

rather than with

(Cb)   *E* confirms H more than H^* iff E makes H "more more plausible" than H^*

E equally confirms *H* and *H** iff *E* makes *H* "equally more plausible" than *H**

The reason is that if they subscribe primarily to (Cb) they not only have to assign some plausibilities or probabilities for any specific application, but they also in general have to decide about which degree of confirmation they prefer. This suggests that it is more likely that they combine (Ca) with the four absolute statements. Apart from the extreme cases, this can be perfectly realized by the ratio measure. To be precise, $p(E/H)/p(E)$ can deal with all five claims (that is, a1, a2 (hence a), b, Ca, and Cb), except when $p(E) = 0$. As I suggested in ICR (pp.50-1), for that case it is plausible to use the form $p(H/E)/p(H)$, which is for normal *p*-values equivalent to $p(E/H)/p(E)$ and which may be defined in this case, assuming that $p(H) > 0$. E.g. in the (improper) extreme case of verification, it becomes $1/p(H)$.

Taking my non-dogmatic attitude seriously, the result is that I could live perfectly happily with the following asymmetric fusion of intuitions:

(ab)   *E* confirms *H* iff
       if *E* has some initial plausibility: *H* makes *E* more plausible
       if *E* has none: *E* makes *H* more plausible

       E is neutral for *H* iff
       if *E* has some initial plausibility: *H* makes *E* neither more nor less plausible
       if *E* has none: *E* makes *H* more nor less plausible

(Cab) *E* confirms *H* more than *H** iff
       if *E* has some initial plausibility: *H* makes *E* more plausible than *H** does
       if *E* has none: *E* makes *H* more more plausible than *H** in the ratio sense

       E equally confirms *H* and *H** iff
       if *E* has some initial plausibility: *H* and *H** make *E* equally plausible
       if *E* has none: *E* makes *H* and *H** equally plausible in the ratio sense

This completes my response to Maher's Section 1.

Conditional Deductive Confirmation and the Ravens Paradoxes

The rest of Maher's paper deals with my notion of cd-confirmation (Section 2) and its application to the ravens paradoxes (Sections 3-5). Although he comes up with a number of interesting formal observations, I reject most of his general critique in the following.

On Section 2: Conditional Deductive Confirmation[8]

In correspondence with Maher I have concluded that it is very important to stress that I see conditional deductive confirmation (cd-confirmation) as a kind of confirmation, but not as a kind of unconditional confirmation. More generally, the following distinctions in Ch. 2 and 3 of ICR are very important:

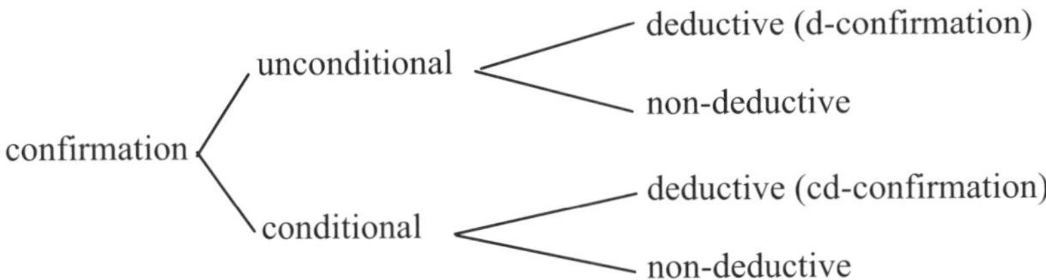

Before I go into a specific criticism of cd-confirmation, it may be helpful to refer to a sentence of Maher in the concluding Section 6 (PM, p. 14). He writes: "C-confirmation [i.e., cd-confirmation with C as condition, TK] is just d-confirmation with C added to the background beliefs … and so adds nothing essentially new." I would not at all object to this claim. My only point is that in the context of cd-confirmation it is very practical to draw a distinction between fixed background beliefs and variable initial conditions as they occur as experimental conditions, such as, look for a raven (and then check its color) or look for a non-black object (and then check its (natural) kind). Note that in my formal presentations in ICR, I omitted, for simplicity, all references to fixed background beliefs, with explicitly saying so.

Let us now return to Section 2. Maher rightly suggests that I give two related but different definitions of cd-confirmation. However, from my presentation it is also clear that their relation is of a type-token or generic-specific kind. I start with (the general structure of) a token of cd-confirmation and then generalize it to a type. The token is of course basic, and there may be problems with the generalization. Maher summarizes both definitions in the second paragraph of Section 2 (with a correct simplification explained in Note 3). I repeat the generic definition of cd-confirmation:

---

[8] Maher is right (PM, p. 5) regarding the desirable exclusion of "maximally plausible evidence" in the definition of deductive confirmation. When summarizing Ch. 2-4 of ICR for SiS (pp. 207-8), I discovered the omission that I had not made explicit that E is supposed to be contingent (and H consistent, in both cases relative to the fixed background beliefs), hence not maximally plausible. Note that my generic definition of cd-confirmation implies that E is non-tautological, by requiring that C is so, and E has to imply C according to that definition.

E cd-confirms *H* iff there exists a *C* such that *E* entails *C* and *E C*-confirms *H*

Maher successfully shows by THEOREM 1 that this generic definition is (still) defective in the sense that almost every *E* cd-confirms *H*. More precisely, he shows that when LI(*E*, *H*) (i.e., *E* and *H* are logically independent) and when there is some *D* LI(*D*, $\neg E$ & $\neg H$), then *Cmaher* = C_m = *Ev*($\neg H$&*D*) is such that *E* C_m-confirms *H*. For all conditions for specific cd-confirmation are now satisfied:

(i) LI(*H*, C_m)
(ii) C_m does not entail *E*
(iii) *H*&C_m entails *E*

Of course, to prevent this trivialization one may either try to define specific cd-confirmation more restrictively or the generic type. In view of the very artificial nature of C_m it is plausible to first think of adapting the generic definition in order to exclude this C_m just because of its artificiality.

The condition C_m = *Ev*($\neg H$ & *D*) is in several respects not very like an initial condition as it occurs in standard examples of explanation of individual events or in my favorite examples of conditional confirmation, e.g. *Cr*: raven (*x*); *Er*: raven (*x*) and black (*x*). First, whatever *E*, *H* and *D* are, C_m can't be of a conjunctive nature, that is, an atomic formula or its negation or a conjunction of such formulas. Second, although C_m is logically independent of *H* in the straightforward sense, it needs *H* for its definition. Third, C_m needs *D* for its definition, although *D* is logically independent of $\neg E$ & $\neg H$. Of course, all three aspects can be used to exclude C_m. At the moment I would favor to requiring a conjunctive nature of *C*, but this may well be too restrictive and/or still leave room for other types of artificial conditions. However, Maher's counterexample does not at all show that it is impossible to prevent trivialization of generic cd-confirmation due to artificially construed conditions. On the contrary, it stimulates the search for an improvement of the generic definition.

On Section 3: The Ravens Paradox[9]

Regarding the object versus propositional form, it is evident that, for example, by giving an example of Nicod's criterion, i.e. Maher's PRINCIPLE 1, in object form, viz. 'a black raven confirms *RH*', where *RH* is short for 'all ravens are black', the propositional form is the intended formal version of the more easy, but somewhat ambiguous, object form. Indeed, Hempel also

---

[9] Unfortunately I speak about "raven paradoxes' and not of "ravens paradoxes". The mistake is due to the fact that in Dutch 'raven' is already the plural form (of 'raaf').

frequently uses the object form, but jumps to the other wherever relevant, and so do I.

More importantly, from my very brief indications in ICR (p. 27) it is clear that I do not claim a new argument for the paradoxes. Hence, as far as the second paradox is concerned, I just intended to refer to Hempel's argument. Maher is certainly right in arguing that there is a gap between deriving (γ) and the claimed derivability of (β) from Nicod's condition and the equivalence condition. To argue, starting from (γ), that "any object which is either no raven or also black" confirms *RH*, in particular, a black non-raven, presupposes what might be called the "converse consequence property with respect to the evidence." This property is indeed problematic and, hence, not defended in ICR. In sum, Maher is right in claiming that Hempel's argument for deriving the second paradox is problematic.

Although I should have paid attention to this problem, my ultimate target would have been the same, namely to argue that, in the context of (conditional) deductive confirmation, a proper explication should not allow the confirmation of *RH* by (the proposition describing) a black non-raven. However, Maher also argues that this confirmation claim is not so counterintuitive as the one dealing with a non-black non-raven, i.e. the first paradox, whereas I suggest the opposite comparative intuition. Maher is certainly right in suggesting that there are contexts in which a black non-raven confirms *RH*. In my quantitative explication I concede this (($1p$), ICR, p. 59) when one is random sampling in the universe of objects, leading to the same degree of confirmation for all three cases. More generally, sampling, randomly or not so randomly, I would subscribe to both questioned confirmation claims as long as the sampling is not among non-ravens or black objects, that is, the context for (conditional) deductive confirmation. Unfortunately, Maher's formulations 'find[ing] a non-raven to be black' and 'finding a non-raven to be non-black' are in this respect rather unclear. In particular, I hesitate to subscribe to the reverse plausibility claim, but I do not exclude types of non-random sampling in which I would agree with this verdict.

On Section 4: Kuipers' Solution

In this section Maher addresses three points with respect to my solution of the first paradox of the ravens hypothesis *RH* (all ravens are black), which amounts to

(4) a black raven cd-confirms *RH* more than a non-black non-raven

or to use Maher's preferred formulation

(4) $Ra \ \& \ Ba$ cd-confirms *RH* more than $\neg Ra \ \& \ \neg Ba$ does

Moreover, the assumption is that the background beliefs (of course, the relevant ones, that is, *our* background beliefs) include or imply that the number of ravens #R is (much) smaller than the number of non-black objects #¬B.

Let us start with *Subsection* 4.3, where he claims that I should have written instead of (4):

(4′) *Ra & Ba Ra*-confirms *RH* more than ¬*Ra* & ¬*Ba* ¬*Ba*-confirms *RH*

However, it is very clear from the context of (4) that this is precisely what I mean more specifically. Let me just quote claim (2), starting just nine lines above (4), and even on the same page, viz. ICR, p. 28.

> (2) a black raven and a non-black non-raven both cd-confirm *RH*, more
> specifically, a black raven on the condition of being a raven and a non-
> black non-raven on the condition of being non-black.

Hence I agree that, strictly speaking, (4′) is my solution of the first paradox.

In *Subsection* 4.1 Maher points out, by THEOREM 2 (PM, pp.10-11), that the suggested *quantitative* rationale of the presupposed underlying *qualitative* conditional principle P.1c, unlike the unconditional version, is sensitive to the degree of confirmation chosen. That is, the ratio measure entails (in Maher's notation):

P.1cq: if E C-confirms H and E^* C^*-confirms H then
$c(H, E/C) > c(H, E^*/C^*)$ iff $p(E^*/C^*) > p(E/C)$

but the difference measure does not.[10] This is an interesting finding. But I am of course inclined to see it as an additional argument in favor of the choice for the ratio measure. Whereas the unconditional and the conditional quantitative version of P.2 are both in favor of the ratio measure, the unconditional quantitative version of P.1 is still satisfied by both measures. However, as soon as we consider the conditional quantitative version, i.e. P.1cq, only the ratio measure satisfies. Instead of seeing it as a case of circular reasoning, as Maher suggests, I see this conclusion more as a case of the so-called "wide reflective equilibrium" between qualitative and quantitative (and simplicity) considerations (cf. Thagard, 1988, adapting the ethical method developed by Rawls and Daniels). Sure, this does not provide a "rationally compelling justification" for P.1c, that is,

P.1c: if E C-confirms H and E^* C^*-confirms H then
E C-confirms H more than E^* C^*-confirms H iff

---

[10] Nor the likelihood ratio measure, which I neglect further, but the same points can be made for that measure. It may be true that Fitelson (2001) gives new arguments in favor of this measure, but in Kuipers (forthcoming, Section 1.2.1) I explain why his arguments in (Fitelson 1999) do not convince me.

> E^* is, given C^*, more plausible than E, given C,
> in the light of the background beliefs

but I am happy with good reason. That is, the main question is, how plausible are P.1c and P.1cq? If they do not hold, there may be cases that a C-experiment is more risky for H than a C^*-experiment but the less surprising evidence E^* would nevertheless confirm H more than the more surprising E. Put in terms of ravens: although investigating ravens may be more risky than investigating non-black objects, RH could be more confirmed by hitting a (non-black) non-raven in the second case than by hitting a black raven in the first case. This sounds rather counterintuitive.

In *Subsection* 4.2 Maher shows, by THEOREM 3, that the 'only if' claim of my specification of P.1c , i.e., $S^\# 1.c$, is not valid, using the ratio measure, for it turns out to leave room for a counterexample. However, I am not really impressed by the counterexample. It amounts to a case of "more cd-confirmation by a black raven (assuming that it is a raven) than by a non-black non-raven (assuming that it is a non-black object)" even if the mentioned condition $\#R < \#\neg B$ is not satisfied in a straightforward sense. As Maher concludes himself, in the example "there is a probability 1/2 that $\#R > \#\neg B$," but one should add that "there is also a probability 1/2 that $\#R < \#\neg B$." Now it is easy to check in the example that the expected value of the ratio (of sizes, not to be confused with the ratio degree of confirmation) $\#R/\#\neg B$ is 7/8. Since this is less than 1 it is a nice case of a sophisticated version of the background belief that $\#R < \#\neg B$. That is, I would already be perfectly happy if all possible counterexamples nevertheless lead to a lower than 1 expectation for the ratio of the sizes. In other words, I would only be impressed, even very impressed, by an example in which this expected ratio is at least 1. In view of my earlier challenge to Maher to provide one, I conclude for the time being that he did not find one.

Maher also discusses the if-side of my claim $S^\# 1.c$. With THEOREM 4 he points out that a sophisticated, probabilistic version of the if-claim obtains. However, I do not see what his objections are to my proof sketch on pp. 28-9 of ICR. I simply point out in terms of percentages that, whatever the numbers of the three types of individuals are that do not falsify RH, for every non-zero number of non-black ravens, hitting a black raven among the ravens is less plausible than hitting among the non-black objects at a non-black non-raven, as soon as the number of ravens is less than the number of non-black objects. This amounts to (II-R) in combination with (I), (III)-(V). Certainly, this is (at most) a case of quasi-quantitative reasoning that can only be made precise in a strictly quantitative sense, but that this is possible is very much suggested by the presented argument. Although I do not want to dispute Maher's THEOREM 5, which is based on the general condition (II), I have only claimed to subscribe to (II-R), in which the starred properties of (II) are limited to the complementary properties of the unstarred ones. For this reason, in contrast to Maher, I find it much easier to call (II-R) a plausible principle than (II). Finally, Maher is right in claiming that my proof sketch is strictly speaking laden with the assumption that *RH* is false. His proof of THEOREM 4 not only makes clear in detail that a quantitative refinement is indeed possible, but also that one only has to assume that *RH* is not certain.

On Section 5: Adequacy of the Solution

In the last substantial Section Maher mentions his criteria for an adequate solution of the (first) ravens paradox in terms of the three, inconsistent, principles mentioned in his Section 3:

> PRINCIPLE 1: *Ra* & *Ba* confirms (x) $(Rx \rightarrow Bx)$ *(RH)*
> i.e. an instance of Nicod's condition

> PRINCIPLE 2 is the equivalence condition and

> PRINCIPLE 3: $\neg RA$ & $\neg BA$ does not confirm *RH*

According to Maher, an adequate solution requires (a) identifying the false principle(s), (b) insight into why they are false, and (c) identifying a true principle that is sufficiently similar to each false one "that failure to distinguish the two might explain why the false principle is *prima facie* plausible."

These criteria sound very reasonable. Let me, therefore, instead of criticizing Maher's evaluation of my solution in detail, summarize my solution in terms of these requirements in combination with my basic distinctions. For it follows from my presentation, whether one likes it or not, that it is important to distinguish between deductive and non-deductive confirmation, and for each, between unconditional and conditional confirmation.

Starting with *unconditional deductive* confirmation, my diagnosis is that (a:) (only) the first principle is false, Nicod's condition, that (b:) it is false because *RH* does not deductively entail the purported confirming instance *Ra* & *Ba*, and that (c:) "*Ra* & *Ba* Ra-confirms *RH*," or equivalently, "*Ra* & *Ba* cd-confirms *RH on the condition Ra*", is sufficiently similar to "*Ra* & *Ba* (d-)confirms *RH*" to explain "that failure to distinguish the two might explain why the false principle is prima facie plausible."

Turning to (specific) *conditional deductive* confirmation in general (a:) the third principle is false, because (b:) *RH* & $\neg Ba$ entails $\neg Ra$, and (c:) which should be distinguished from the claim that *RH* entails $\neg Ba$ & *Ra*.

In terms of non-deductive, probabilistic confirmation, I claim (ICR, pp. 59-60), assuming random sampling in the (finite) universe of objects, regarding *unconditional probabilistic* confirmation (ICR, p. 59, $(1p)$) that (a:) the third principle is false, that (b:) drawing any type of object compatible with *RH* is made more plausible/probable by *RH*, hence also a non-black non-raven, or, if you prefer the standard formulation: the probability of *RH*, if initially positive, increases by such evidence; hence the degree of confirmation for *RH* provided by a non-black non-raven is higher than 1 according to the ratio measure (and positive according to the difference measure), and that (c:) the ratio will be very close to 1: whether we calculate it on the basis of an estimate of the number of non-black ravens (if *RH* is false) or in the sophisticated way indicated in Note 19 of ICR (p. 59, p. 337), as long as the expected number of non-black ravens is a small proportion of the number of objects in the world.

Regarding *conditional probabilistic* confirmation, see (2*p*)-(4*p*) (ICR, pp. 59-60), everything becomes a quantitative version of the corresponding conditional deductive situation.

In sum, according to my analysis, in the unconditional deductive reading the first principle is false and the third true; in all other three readings the opposite is the case. In all four cases the verdict for each principle is explained. Finally, that the verdicts have to be reversed when going from the first reading to one of the other three explains very well why there has been a dispute and why it is so difficult to disentangle the purported paradox. In general: the truth-value of Nicod's condition depends on the precise version of the claim.

Let me finally deal with Note 7, in which Maher criticizes my quantitative treatment of the (first) raven paradox, without going into details. He just claims that the fact that a black raven confirms *RH* (unconditionally) is fallacious because this "is not true in Good's example." Now, in Good's example (Good, 1967), there are very specific and strong background knowledge beliefs. In particular, the number of black ravens is assumed to depend closely on whether or not *RH* is true: if *RH* is true there are 100 black ravens, and a million other birds; if *RH* is false, there are 1000 black ravens, one white, and again a million other birds. Of course, in that case a randomly drawn black raven should disconfirm *RH*, which it does according to all measures. But who wants to take this modeling as merely modeling random sampling in the universe of birds? One plausible way of modeling this, of course, is to assume that there is a fixed (finite, non-zero) number of black ravens and a fixed number of non-ravens, and some equally unknown finite but not necessarily non-zero number of non-black ravens, i.e., 0 or 1 or 2... My detailed unconditional claim (ICR, p. 59 and Note 19) is that when this modeling is adequate a black raven confirms *RH* (as well as a non-raven, black or non-black). For the moment I do not want to rule out that there are more defensible types of modeling random sampling among birds aiming at testing *RH*, but Good's case is not one of them. To put it differently, nobody would see his hypothetical background beliefs as a normal type of case of not knowing the truth-value of *RH*. Of course, and this is Good's point, background beliefs around *RH* may be such that random sampling leads to the conclusion that a black raven disconfirms *RH*. On the other hand, the background beliefs around *RH*, other than those related to (relative) numbers, may be negligible, as I was apparently assuming, by not mentioning other kinds of fixed background beliefs.

Conclusion

This completes my response to Maher's Sections 2-5. In my comments on Section 2, I already referred in a positive sense to his diagnostic statement in his concluding Section 6 regarding the notion of conditional deductive confirmation. For the rest I have already pointed out that I do not agree with his conclusions. However, instead of repeating all disagreements, let me summarize the main interesting observations that I learned from Maher's critical exposition.

Section 1: taking my non-dogmatic attitude to confirmation seriously, I could live perfectly happily with an asymmetric fusion of non-standard and standard intuitions.

Section 2: the generic definition of cd-confirmation needs improvement, in view of THEOREM 1, to prevent it from trivialization.

Section 3: Hempel's derivation of the second ravens paradox is problematic, hence the question is whether it really is a paradox.

Section 4: THEOREM 2 shows that the difference measure for confirmation violates the plausible principle P.1c(q), providing an extra reason for the ratio measure. THEOREM 3 suggests a possible refinement of the formulation of the number condition in my solution of the first ravens paradox: the background beliefs need only to imply that the expected ratio of the number of ravens to the number of non-black objects is (much) smaller than 1. But this should be checked, for both directions. THEOREM 4 shows that a similar weakening of the underlying assumption of the qualitative solution, viz. that the ravens hypothesis is false, is possible: the hypothesis is not certain.

Section 5: It is not yet generally realized that the truth-value of Nicod's condition very much depends on the precise version of the claim.

REFERENCES

Fitelson, B. (1999). The Plurality of Bayesian Measures of Confirmation and the Problem of Measure Sensitivity. *Philosophy of Science*, Supplement to Volume **66**, S362-S378.

Fitelson, B. (2001). A Bayesian Account of Independent Evidence with Applications. *Philosophy of Science* **68**, S123-S140.

Good, I. (1967). The White Shoe is a Red Herring. *The British Journal for the Philosophy of Science* **17**, 322.

Morton, A. (1997). *Theory of Knowledge.* Second Edition. Oxford: Blackwell.

Sober, E. (2000). *Introduction to Bayesian Epistemology*, lecture handout (January 31, 2000). http://philosophy.wisc. edu/sober/courses.htm.

Sober, E. (2001). *Philosophy of Biology.* Second edition. Boulder, CO/Oxford: Westview.

Thagard, P. (1988). *Computational Philosophy of Science.* Cambridge, MA: The MIT Press.

John R. Welch

GRUESOME PREDICATES

ABSTRACT. This paper examines gruesome predicates, the most notorious of which is 'grue'. It proceeds by extending the analysis of Theo A. F. Kuipers' *From Instrumentalism to Constructive Realism* in three directions. It proposes an amplified typology of grue problems, first of all, and argues that one such problem is the root of the rest. Second, it suggests a solution to this root problem influenced by Kuipers' Bayesian solution to a related problem. Finally, it expands the class of gruesome predicates by incorporating Quine's 'undetached rabbit part', 'rabbit stage', and the like, and shows how they can be managed along the same Bayesian lines.

1. Introduction

To classify, in its most primitive sense, is to posit a binary relation between an object and a concept. Objects, paradigmatically, are physical, but we also classify seven as a prime and Shylock as a fiction. Concepts are paradigmatically linguistic, but animals without words plainly cognize some of the same things that we cognize with them.

This note concerns the classification of physical objects with linguistic concepts, and specifically with linguistic concepts of a highly problematic sort: grue and its congeners (Goodman 1979, pp. 74, 79).[1] It focuses, more specifically, on the gruesome predicates that express these concepts. The examination of these predicates in the following pages extends the analysis of Theo A. F. Kuipers' *From Instrumentalism to Constructive Realism* (2000; hereafter referred to as ICR) in three directions. It proposes an amplified typology of grue problems, first of all, and argues that one such problem is the root of the rest (Section 2). Second, it suggests a solution to this root problem influenced by Kuipers' Bayesian solution to a related problem (Section 3). Finally, it expands the class of gruesome predicates by incorporating 'undetached rabbit part', 'rabbit stage', and the like

---

[1] Goodman introduced the concept of grue in his University of London lectures in May of 1953. These lectures were later published as the Project portion of *Fact, Fiction, and Forecast*. The third edition of this work (1979) will be cited throughout. There Goodman stipulates that "'grue' ... applies to all things examined before *t* just in case they are green but to other things just in case they are blue" (p. 74).

In: R. Festa, A. Aliseda and J. Peijnenburg (eds.), *Confirmation, Empirical Progress, and Truth Approximation* (*Poznań Studies in the Philosophy of the Sciences and the Humanities,* vol. 83), pp. 129-137. Amsterdam/New York, NY: Rodopi, 2005.

(Quine 1960, pp. 51-54; 1969, pp. 30-35), and shows how they can be managed along the same Bayesian lines (Section 4).

2. THISGREEN versus THISGRUE

Kuipers treats 'grue' as part of his qualitative theory of confirmation (ICR, pp. 29-36). The predicate can be given more than one reading, as he shows, but for our purposes his temporal reading will serve: 'grue' means "green if examined before 3000, and blue if not examined before 3000" (ICR, p. 29). As Kuipers acknowledges, part of his analysis is inspired by Sober (1994). Sober distinguishes the general hypotheses

    (ALLGREEN)   All emeralds are green.

    (ALLGRUE)   All emeralds are grue.

from the predictive hypotheses

    (NEXTGREEN)   The next emerald I examine will be green.

    (NEXTGRUE)   The next emerald I examine will be grue.

He treats both the generalization problem (ALLGREEN versus ALLGRUE) and the prediction problem (NEXTGREEN versus NEXTGRUE) in two ways: diachronically and synchronically. Helpful though this is, it may encourage the impression that the grue paradox infects hypotheses but not the evidence on which they are based. But if this were so, if the observed emeralds were known to be green and not grue, the prediction problem would reduce to straightforward analogy: from green emeralds as evidence we would infer that the next emerald is green rather than grue. Evidently, though, the problem is not so easily dispatched. For the general and predictive hypotheses above are projected from evidence that has already been interpreted according to the classificatory hypotheses

    (THISGREEN)   This emerald is green.

    (THISGRUE)   This emerald is grue.

In other words, the evidence is just as equivocal as the hypotheses projected from it.[2] But the classification problem presented by THISGREEN and THISGRUE is different from the generalization and prediction problems. It is, indeed, their root.

---

[2] Cf. "Then at time *t* we have, for each evidence statement asserting that a given emerald is green, a parallel evidence statement asserting that that emerald is grue. And the statements that emerald *a* is grue, that emerald *b* is grue, and so on, will each confirm the general hypothesis that all emeralds are grue" (Goodman 1979, p. 74).

Kuipers shows that an asymmetry in the confirmation behavior of 'green' and 'grue' can be created in two different ways (ICR, pp. 33-36). Both involve irrelevance assumptions about time, and both are couched in the language of the generalization problem. The first is based on a strong irrelevance assumption for emeralds that can be stated as follows, where 'E' stands for 'emeralds' and 'M' for 'examined before 3000':

SIA: For all colors C, "all EM are C" implies "all E are C," and hence "all $E\overline{M}$ are C," and vice versa.

The second way is based on a weak irrelevance assumption for emeralds. Where 'E' and 'M' have their previous meanings, 'Q' stands for 'grue (queer)', 'G' for 'green', and 'B' for 'blue', the assumption is:

WIA: For all colors C and C', $C \neq C'$, "all E are C" is (much) more plausible than the conjunction "all EM are C" and "all $E\overline{M}$ are C'" (which is equivalent to "all E are Q" when $C = G$ and $C' = B$).

SIA will be ignored hereafter, for it amounts to denying that the grue hypothesis has any plausibility whatsoever; this is surely too strong, as Kuipers points out (ICR, p. 35). But WIA is far more promising. It does not rule out the grue hypothesis at all, though it does not accord it the same plausibility as the green hypothesis. For our purposes, however, WIA will need to be recast. Unlike WIA, which treats the generalization problem, our modified form of WIA focuses on the classification problem. Where 'E', 'M', 'Q', 'G', and 'B' have the same meanings as before, the weak irrelevance assumption for classifying emeralds is just:

WIA$_c$: For all colors C and C', $C \neq C'$, "This E is C" is (much) more plausible than the conjunction "This EM is C" and "This $E\overline{M}$ is C'" (which is equivalent to "This E is Q" when $C = G$ and $C' = B$).

3. Why Not THISGRUE?

What reasons might one have for adopting WIA$_c$? To lay the groundwork for an answer, let us supply some context. Probability functions such as Kemeny's m (1953) are distinguished from inductive probability functions in that the former lack and the latter have the property of positive instantial relevance (instantial confirmation). That is, for a monadic property F, individuals a and b, and evidence E,

$$p(Fa/E \ \& \ Fb) > p(Fa/E).$$

Kuipers observes that a probability function can be made inductive in one or both of two ways (ICR, p.76). The first is through inductive priors. By contrast with *m*, for instance, which assigns zero probability to universal generalizations in an infinite universe, inductive priors for such *m*-zero hypotheses are nonzero. The second possibility is via inductive likelihoods, which are likelihood functions $p(E/H)$ having positive instantial relevance. Separating these sources of inductive influence enables Kuipers to provide a useful classification of four principal theories of confirmation (ICR, pp. 76-77; 2001, sec. 7.1.2):

	Inductive priors	Inductive likelihoods
Popper	no	no
Carnap	no	yes
Bayes	yes	no
Hintikka	yes	yes

Kuipers' confirmation theory is Bayesian in the standard sense that probabilities are updated by conditionalizing on new evidence. That is, in the face of new evidence E for a hypothesis H, the revised probability of H is its initial conditional probability given E. However, Kuipers' approach is non-standard in two ways. In contrast with the standard approach, "it is inclusive in the sense that it leaves room for a substantial degree of confirmation for '*p*-zero' hypotheses when they are confirmed ..."; in addition, "it is pure in the sense that equally successful hypotheses get the same degree of confirmation, irrespective of their prior probability" (ICR, p. 44).

The approach to the classification problem to be explored in this note is Bayesian as well, though compatible with both standard and nonstandard versions. The basic idea is to tease out the implications of Bayes' theorem for classificatory hypotheses like THISGREEN and THISGRUE. In its simplest form, Bayes' theorem states that, for some hypothesis H and evidence E,

$$p(H/E) = p(H)p(E/H)/p(E).$$

Applying this theorem to hypotheses H_1 and H_2 based on the same evidence E permits the comparative inference

(C) $p(H_1/E) > p(H_2/E)$ if and only if $p(H_1)p(E/H_1) > p(H_2)p(E/H_2)$.

That is, the posterior probability of H_1 may be greater than that of H_2 in three ways: a) the prior probability of H_1 is greater than that of H_2 while its corresponding likelihood is no lower; b) the likelihood of H_1 is greater than that of H_2 while its corresponding prior is no lower; and c) both a) and b) hold at once.

Let us first consider the question of priors. The prior probability of THISGREEN appears to be greater than that of THISGRUE for at least two reasons. The first is the superior fit of THISGREEN with our background knowledge. As Kuipers points out,[3]

> It is surely the case that, as far as we know, there are no types of stones that have changed color at a certain moment in history. However, this does not exclude the possibility that this might happen at a certain time for a certain type of stone, by some cosmic event. To be sure, given what we know, any hypothesis which presupposes the color change is much less plausible than any hypothesis which does not. (ICR, p. 35)

A second reason is the greater simplicity of THISGREEN. It is no accident that the "simplicity solution is among the most popular" of some twenty approaches to the grue problem (Stalker 1994, p. 10). Admittedly, the role of simplicity as a guide to truth is controversial. But however elusive a precise account of simplicity may be, the greater simplicity of THISGREEN with respect to THISGRUE seems clear enough. For any emerald that changes color from green to blue has a more complicated existence, ceteris paribus, than one that stays green from start to finish. For these two reasons, then, the prior probability of THISGREEN is greater than that of THISGRUE.

But what about the relative likelihoods of THISGREEN and THISGRUE? Take the evidence for both hypotheses, first of all, which would normally include how the relevant emerald looks. Though there might be additional evidence based on past observations of emeralds, whether this one or others, that is not essential here. Assume a statement of the evidence E, which for our purposes could be as simple as "This emerald looks both green and grue." Now when we consider the resulting likelihoods, $p(E/\text{THISGRUE})$ would appear to be no greater than $p(E/\text{THISGREEN})$. Hence the higher prior of THISGREEN is not offset by a higher likelihood of THISGRUE.

Let us now bring all this to bear. The prior probability of THISGREEN is, as we have seen, higher than that of THISGRUE. But its likelihood is no lower. If we plug these estimates into (C), the comparative inference above, we find that the posterior probability of THISGREEN is indeed greater than that of THISGRUE. This is in accordance with WIA_c, the weak irrelevance assumption of Section 2.

4. Gruesome Predicates

"Grue" is not an isolated case, of course; it is but the best-known member of a family of similar predicates. Goodman himself coined 'emerose' and 'bleen' (1979, pp. 74 n10, 79), and Quine's 'undetached rabbit part' and 'rabbit stage' turn out to belong to the same family. Though these Quinean predicates have

---

[3] Cf. Sober (1994, pp. 231, 236-237).

already been studied from the standpoint of Goodman's theory of projection (Welch 1984), the following remarks treat them from a different, Bayesian point of view.

Quine's gruesome predicates emerged with his thesis of referential inscrutability (1960, pp. 51-54; 1969, pp. 30-51). The present discussion bears directly on the thesis, of course, but it falls short of a full-fledged treatment for at least two reasons. First of all, though Quine argued that reference was subject to problems of both direct and deferred ostension (1969, pp. 39-40), here we will deal exclusively with problems of direct ostension.[4] In addition, where direct ostension is concerned, some of Quine's gruesome predicates can be safely ignored. 'Rabbit stage', which is true of "brief temporal segments" of rabbits (1960, p. 51), is a case in point. The reason is immediately apparent from Quine's definition of direct ostension:

> The *ostended point*, as I shall call it, is the point where the line of the pointing finger first meets an opaque surface. What characterizes *direct ostension*, then, is that the term which is being ostensively explained is true of something that contains the ostended point. (1969, p. 39)

That is, since rabbit stages are temporal rather than spatial, they cannot have opaque surfaces that contain the ostended point. In what follows, then, we will limit ourselves to gruesome predicates like 'undetached rabbit part', but for expository reasons we will employ the analogous predicate 'undetached emerald part'.

The use of this predicate facilitates continuity with our previous examples. More importantly, however, it points the way to a deeper layer, a problem still more radical than the choice between THISGREEN and THISGRUE. We have already seen that two pairs of rival hypotheses, NEXTGREEN versus NEXTGRUE and ALLGREEN versus ALLGRUE, rely on THISGREEN and THISGRUE as evidence. But THISGREEN and THISGRUE are actually complex hypotheses that presuppose the simple hypothesis "This is an emerald." Yet whenever we can assert "This is an emerald," Quine would claim that we could just as well assert "This is an undetached emerald part." In other words, the deeper layer is the choice between the rival hypotheses

(THISEMERALD)   This is an emerald.
(THISEMERALD-PART)   This is an undetached emerald part.

But this choice is "objectively indeterminate" (Quine 1969, p. 34) – or so goes the thesis of referential inscrutability.

Before discussing this thesis, there are two crucial points to make. The first is that Quine imagined 'undetached rabbit part' and 'rabbit stage' as alternative radical translations of the fictitious native term 'gavagai'. Hence one might object

---

[4] The problems of deferred ostension are treated in Welch (1984, pp. 269-272).

to viewing such terms apart from radical translation. But the Quinean theses on translation and referential inscrutability are not identical; referential inscrutability is one, but only one, reason for translational indeterminacy (Quine 1970). In addition, the problems of referential inscrutability are supposed to arise not only when translating a language unrelated to our own but also within our native language. Reference is inscrutable, according to Quine, when one native speaker talks to another. It is inscrutable even if we talk to ourselves. "[R]adical translation," he insists, "begins at home" (1969, p. 46).

The second point concerns an apparent asymmetry between terms like 'grue' and terms like 'undetached emerald part'. While there is a vivid empirical difference between Goodman's 'grue' and our 'green', the Quinean 'undetached emerald part' and our 'emerald' appear to be empirically indistinguishable, for whenever one points to an emerald, one also points to an undetached emerald part. This difference is merely apparent, however, and the appearance arises from over-reliance on the sense of sight. Suppose that, for whatever reasons, we want to mark the distinction between an emerald and its undetached part. While part-whole distinctions can be visually problematic, they can be made tactilely plain. It is true that whenever one touches a whole emerald, one also touches an undetached emerald part. But the reverse does not hold; to touch just a part is not to touch the whole. Hence the distinction between an emerald and its undetached part makes good empirical sense. One literally grasps it.

To this extent, at least, the choice between THISEMERALD and THISEMERALD-PART is parallel to that between THISGREEN and THISGRUE. In both cases, one is waiting for Godot. It is just that Godot takes the form of visual stimuli for color hypotheses and tactile stimuli for part-whole hypotheses. But could the parallel be extended further? That is, even before the visual data are in, the choice of THISGREEN over THISGRUE is uncertain but hardly arbitrary, as we saw in Section 3. Absent decisive tactile stimuli, then, might the choice between THISEMERALD and THISEMERALD-PART be made in much the same way?

Let us revisit (C), the Bayesian comparative inference of Section 3. It asserts that the posterior probability of one hypothesis may be greater than that of another under three conditions: a) its prior probability is greater while its likelihood is no lower; b) its likelihood is greater while its prior probability is no lower; and c) both previous conditions are satisfied.

Take the question of likelihoods, first of all. The evidence for both THISEMERALD and THISEMERALD-PART includes the observer's sense perceptions (visual, tactile, etc.) of the relevant object. As with THISGREEN and THISGRUE, additional evidence may or may not be at hand. Imagine that the evidence is reported in a statement E such as "This appears to be both an emerald and an undetached emerald part." Then what would we find for the relative likelihoods of THISEMERALD, which is $p(E/\text{THISEMERALD})$, and THISEMERALD-PART, which is $p(E/\text{THISEMERALD-PART})$? I think we would find no reason for one to appear greater than the other.

But what about the priors? Consider, as we did in Section 3, the relative fit of these hypotheses with background knowledge. For THISGREEN and THISGRUE, the relevant background knowledge was of colors. Here, for THISEMERALD and THISEMERALD-PART, the relevant background knowledge concerns a language. Which language? Any one will do, according to Quine; all are subject to referential inscrutability. Then suppose we imagine some native language whose speakers use the term 'esme' as we use 'emerald'. We can assume that, except for 'esme', the language is sufficiently known to translate the native equivalents of THISEMERALD and THISEMERALD-PART. But how would we translate the recalcitrant "esme"?

There are two cases to consider. In both we could secure an empirical foothold for translating native expressions for medium-sized objects in either part- or whole-language. The trick, as we have just seen, is to capitalize on the different feel – quite literally – that parts have relative to wholes. We could, for example, observe native responses to their versions of THISRUBY and THISRUBY-PART under contrasting tactile stimulation.

In the first case, contrasting tactile stimuli have not been used to aid the translation of 'esme', but they have been used for comparable native terms. It would then be reasonable to extrapolate the results of these prior inquiries to the present case. Since these are fictional inquiries into a fictional language, there is no way of really knowing what these results would be. But suppose that they include a preponderance of part-terms. Then we would have reason to think 'undetached emerald part' the more probable translation.

In the second case, not only have contrasting tactile stimuli not been brought to bear on 'esme', but they have also not been employed for comparable native terms. Of course, we might be willing to hazard a guess if we have empirically grounded part- or whole-translations in other languages. But if not? Then there would be no reason to think either 'emerald' or 'undetached emerald part' a better fit with background knowledge, for there is no relevant background knowledge.

But this is where the choice between THISEMERALD and THISEMERALD-PART is considerably more tractable than that between THISGREEN and THISGRUE. The visual data that would finally eliminate one of the color hypotheses will not be available, under Kuipers' interpretation of 'grue', until the year 3000. But the tactile data that would finally eliminate either the part- or the whole-hypothesis need not recede into the distant future. We could, in fact, summon them at will.

In both cases, then, the reference of 'esme' is scrutable. We could directly initiate the relevant touching. Or, before this is done, we could determine that either THISEMERALD or THISEMERALD-PART is a better fit with background knowledge. That would be good reason for assigning one hypothesis a higher initial probability than the other.[5] Consequently, since the likelihood of this hypothesis would be no lower, the higher prior would carry the day. Using (C) as in Section 3, we could draw the inference that this hypothesis's posterior probability is higher as well.

I conclude, therefore, that terms like 'undetached emerald part' are as gruesome as 'grue', and subject to the same Bayesian solution. Quine overstated his case in claiming that the linguist arbitrarily 'leaps to the conclusion' that a native term refers to a whole rather than a part (1960, p. 52). Granted, the evidence might support the opposite conclusion. But the conclusion, whatever it is, can be drawn from the evidence in a reasonable way. Uncertain, yes – arbitrary – no.

Saint Louis University (*Madrid Campus*)
Avenida del Valle, 34
28003 Madrid
Spain

REFERENCES

Goodman, N. (1979). *Fact, Fiction, and Forecast*, Third edition. Indianapolis: Hackett.

Kemeny, J. (1953). A Logical Measure Function. *Journal of Symbolic Logic* **18**, 289-308.

Kuipers, T.A.F. (2000/ICR). *From Instrumentalism to Constructive Realism*. Dordrecht: Kluwer Academic Publishers.

Kuipers, T.A.F. (2001/SiS). *Structures in Science*. Dordrecht: Kluwer Academic Publishers.

Quine, W.V.O. (1960). *Word and Object*. Cambridge, MA: The M.I.T. Press.

Quine, W.V.O. (1969). *Ontological Relativity and Other Essays*. New York: Columbia University Press.

Quine, W.V.O. (1970). On the Reasons for Indeterminacy of Translation. *The Journal of Philosophy* **67**, 178-83.

Sober, E. (1994). No Model, No Inference: A Bayesian Primer on the Grue Problem. In: Stalker (1994), pp. 225-40.

Stalker, D., ed. (1994). *Grue! The New Riddle of Induction*. La Salle, IL: Open Court.

Welch, J.R. (1984). Referential Inscrutability: Coming to Terms without It. *The Southern Journal of Philosophy* **22**, 263-73.

---

[5] It might be argued that the prior probability of THISEMERALD is higher than that of THISEMERALD-PART because it is simpler, but I will not attempt to establish that here.

Theo A. F. Kuipers

'THISGRUE' AND 'THISEMERALD-PART'
REPLY TO JOHN WELCH

I like John Welch's contribution very much, not least because he deals with one of the famous Quinean examples, rabbits versus rabbit parts, in a way that confirms my impression that Quine could have expressed himself much more clearly than he in fact did. To be sure, that would not have stimulated so much exegesis of what he really meant. However, what is at least as important is that Welch demands attention to a problem underlying all "gruesome" problems, the classification problem. Of course, the fact that his solution is inspired by my solution of Goodman's problem with the general grue hypothesis is also something I noted with pleasure. In this reply I first suggest an improved version of the classification problem and solution; I then raise a question about the claimed analogy with the Quinean problem.

The Classification Problem

Let me first state that I very much agree with Welch's claim that the relevant general and predictive hypotheses presuppose a solution of the classification problem as soon as there is some relevant evidence for these hypotheses. Moreover, I also agree with the intuitive a priori claim that "This emerald is green" (THISGREEN) is more plausible than "This emerald is grue" (THISGRUE). But I have some reservations about his explication of this intuition and its defense. To be begin with the former, instead of his WIA_c, at the end of Section 2:

> For all colors C and C', $C \neq C'$, "This E is C" is (much) more plausible than the conjunction "This EM is C" and "This $E\bar{M}$ is C'" (which is equivalent to "This E is Q" when $C = G$ and $C' = B$).

I would prefer:

> For all colors C and C', $C \neq C'$, "This E is C" is (much) more plausible than "This E is MC or $\bar{M}C'$" (which is equivalent to "This E is Q" when $C = G$ and $C' = B$).

In: R. Festa, A. Aliseda and J. Peijnenburg (eds.), *Confirmation, Empirical Progress, and Truth Approximation* (*Poznań Studies in the Philosophy of the Sciences and the Humanities,* vol. 83), pp. 138-139. Amsterdam/New York, NY: Rodopi, 2005.

For, in the standard example, Q is equivalent to the "disjunctive predicate" 'MG or $\overline{MB}$', which makes the whole claim "This E is MG or $\overline{MB}$" easy to interpret, whereas the conjunctive claim suggested by Welch is very difficult to interpret, at least for me.

Happily enough, I do not think that this point really weakens the argumentation in Section 3 in favor of either version of the prior problem. Moreover, regarding the posterior classification problem, the argument with respect to the relative likelihoods really seems to need a similar change. More specifically, the relevant evidential statement "This emerald looks both green and grue" should be replaced. In its present version it is formally equivalent to "This emerald looks both (MG or $\overline{MG}$) and (MG or $\overline{MB}$)," that is, "This emerald looks MG or ($\overline{MG}$ and $\overline{MB}$)," and hence, assuming that B and G are incompatible, "This emerald looks MG," which can hardly be intended. However, the intended version probably is "This emerald looks green or grue," which formally amounts to "This emerald looks (MG or $\overline{MG}$) or (MG or $\overline{MB}$)," that is, "This emerald looks (MG or $\overline{MG}$ or $\overline{MB}$)." Indeed, as Welch attempts to argue for his reading, in my reading there does not seem to be a reason to assign a greater likelihood to THISGRUE relative to this evidence than to THISGREEN.

As an aside, I should warn the reader that Welch's claim early in Section 2 that my theory of confirmation is Bayesian, though non-standard, is correct in the general sense that it fits into a Bayesian probabilistic framework. However his claim is not correct in the specific sense of my classification of principal theories of confirmation, also indicated by Welch, where "Bayes' theory" is a specific kind of enabling inductive confirmation, viz. by inductive priors rather than inductive likelihoods.

'Thisemerald-part' versus 'Thisrabbit-part'

As already suggested, I am pleased with Welch's attempt to construe an analogy between, to be precise, the classificatory grue-problem and Quine's reference problem with 'Gavagai'. Although I think the main line of argument, essentially leading to a strong relativization of Quine's indeterminacy claims, is basically correct, there may be one point of dispute at the very beginning. If a piece of emerald falls apart into a number of pieces (not too many), we would say that we have obtained that number of emeralds. However, rabbits now more viable ways of reproduction than falling apart. Hence the two types of examples may not be as similar as Welch suggests. I leave it to him to find out whether his interesting distinction between visual and tactile impressions strengthens or weakens the relevance of the evident distinction of the two cases.

Gerhard Schurz

BAYESIAN H-D CONFIRMATION AND STRUCTURALISTIC TRUTHLIKENESS:

DISCUSSION AND COMPARISON WITH THE RELEVANT-ELEMENT AND THE CONTENT-PART APPROACH

ABSTRACT. In this paper it is shown that, in spite of their intuitive starting points, Kuipers' accounts lead to counterintuitive consequences. The counterintuitive results of Kuipers' account of H-D confirmation stem from the fact that Kuipers explicates a concept of *partial* (as opposed to *full*) confirmation. It is shown that Schurz-Weingartner's relevant-element approach as well as Gemes' content-part approach provide an account of full confirmation that does not lead to these counterintuitive results. One of the unwelcome results of Kuipers' account of nomic truthlikeness is the consequence that a theory Y, in order to be more truthlike than a theory X (where Y and X are incompatible), must imply the entire nomic truth. It is shown how the relevant-element approach to truthlikeness avoids this result.

1. Introduction

Kuipers' impressive work *From Instrumentalism to Constructive Realism* (Kuipers 2000, hereafter referred to as ICR) is based on some very general philosophical ideas. One of these ideas may be expressed by saying that the instrumentalist and the realist are sitting "in the same boat." For the instrumentalist's strategy of evaluating theories according to their empirical success is at the same time an excellent strategy for the realist's purpose to approximate the truth in a realistic sense. In order to explicate and to prove these ideas, Kuipers' book is based on two central technical notions, the notion of hypothetico-deductive (H-D) confirmation, and the notion of truthlikeness. The precise definition of both notions belongs to the most difficult problems in the Philosophy of Science. In this paper we discuss Kuipers' explications of these two notions. We shall see that, although Kuipers starts from intuitively very appealing ideas, his explications lead to serious counterintuitive consequences.

In: R. Festa, A. Aliseda and J. Peijnenburg (eds.), *Confirmation, Empirical Progress, and Truth Approximation* (*Poznań Studies in the Philosophy of the Sciences and the Humanities,* vol. 83), pp. 141-159. Amsterdam/New York, NY: Rodopi, 2005.

2. Kuipers' Theory of Hypothetico-Deductive (H-D) Confirmation

Kuipers considers H-D confirmation as a special case of probabilistic confirmation. His analysis pursues a straight Bayesian road. He starts from four principles concerning the *unconditional* confirmation of a hypothesis H by an evidence statement E (cf. ICR, subsection 2.1.2):

(1) *Success definition*: E confirms H iff E is a success of H in the sense that H makes E more plausible.

(2) *Reward principle:* E makes H more plausible iff E confirms H.

(3) *Symmetry – implied by* 1 *and* 2: H makes E more plausible iff E makes H more plausible.

(4) *Comparative symmetry*: E^* confirms H^* more than E confirms H iff i) H^* increases the plausibility of E^* more than H increases that of E, iff (following from 3) (ii) E^* increases the plausibility of H^* more than E increases that of H.

Only principles (3) and (4ii) go beyond pure definitions. Assuming that degrees of plausibility can be represented by subjective probability functions p, (3) turns out to be valid, and (4ii) becomes valid if the quantitative degree of confirmation is defined in a "pure" way, i.e., in a way which does *not depend* on the prior probability of H. Kuipers' favorite definition is the so-called *ratio-degree of confirmation* of H by E, $r(H,E)$, defined as (cf. ICR, subsection 3.1.2):

$$(5) \quad r(H,E) =_{df} p(E/H) \,/\, p(E) = p(H/E) \,/\, p(H) = p(H \wedge E) \,/\, p(H).p(E)$$

While the first equality is definitional, the second and third result from probability calculus. Observe that $0 \leq r(H,E) \leq \infty$; $r < 1$ corresponds to disconfirmation (of H by E), $r = 1$ to neutral evidence, and $r > 1$ to confirmation. Note also that p may be any *arbitrary* probability function, merely satisfying the standard (Kolmogorov) axioms or their sentential counterparts. Kuipers mentions it as an additional advantage of the r-degree that if one assumes conditional probability $p(-/-)$ to be axiomatized as a *primitive* concept (as suggested, e.g., by Popper; cf. note 4 to ICR, section 3) such that $p(X/Y)$ is reasonably defined even if Y has zero-probability, then the "success definition" of $r(H,E)$ in terms of $p(E/H) \,/\, p(E)$ is applicable even to p-zero hypotheses.

For the special case of H-D confirmation where $T \Vdash E$ and hence $p(E/H) = 1$ ($\Vdash$ for logical consequence) it results that $r(H,E) = 1/p(E)$, in words, if E is a logical consequence of H then E *always* confirms H, given only that E's *prior* probability $p(E)$ – i.e., E's probability before E was observed – is not already maximal. Moreover, $r(H,E)$ increases with decreasing $p(E)$. If one adds the requirement that only consistent hypotheses are confirmable, then the Bayesian quantitative account gives a straightforward justification of Kuipers' qualitative account of *unconditional* and of *conditional* H-D confirmation (where 'L-true' abbreviates 'logically true', etc.):

(uHD)   *E d-confirms H* iff *H* is consistent, *E* not L-true and $H \Vdash E$.

This definition is extended to *conditional* H-D confirmation as follows (ICR, subsection 2.1.1), where *E* confirms a hypothesis *H* relative to a condition *C*. For example, if *H* is 'all ravens are black', then *E* may be 'this animal is black', or 'this raven is black', and *C* may be 'this animal is a raven'.

(cHD1)   *E C-confirms H* iff (i) $H \wedge C \Vdash E$, (ii) *H* and *C* are L-independent (which means that none of the four entailments "$\pm C \Vdash \pm H$" hold) and (iii) $C \nVdash E$ (while $E \Vdash C$ may hold).

(cHD2)   *E cd-confirms H* iff there exists non-tautological *C* such that $E \Vdash C$ and *E C-confirms H*.

The corresponding r-degree of C-confirmation (cf. ICR, 3.1.3)

$$(6) \quad r(H,E/C) =_{df} p(E/H \wedge C)/p(E/C) = p(H/E \wedge C)/p(H/C)$$
$$= p(H \wedge E/C)/p(E/C) \cdot p(H/C)$$

has similar properties and yields a similar justification as in the unconditional case, because when $H \wedge C \Vdash E$ then $r(H,E;C)$ reduces to $1/p(E/C)$, which is greater than 1 provided that *E*'s prior probability relative to knowledge of *C* is not already maximal.

3. Confirmation of the Improbable?

Kuipers' "pure" version of H-D confirmation implies that confirmation of a hypothesis does not imply anything about its posterior probability, $p(H/E)$. For $p(H/E) = r(H,E) \cdot p(H) = p(H \wedge E) / p(E)$; hence a high degree of d-confirmation of *H* by *E* is compatible with a very low posterior probability of *H*. Therefore, an evidence *E* d-confirms every hypothesis *H* which implies *E*, even if *H* continues to be astronomically improbable after *E* becomes known; and likewise for cd-confirmation. According to this account one would be perfectly right in asserting that, e.g.,

(7) This broken tree cd-confirms that a UFO has broken this tree while it was landing.

(8) The noise in the attic cd-confirms that gremlins are bowling there (this example is due to Sober 1993, p. 32).

(9) That I did not win in the casino cd-confirms that all of its roulette tables are biased.

Even worse, Kuipers' H-D account forces us to say that intuitively reasonable hypotheses – e.g., that a storm has broken this tree in example (7), that the cat in the attic is hunting for mice in example (8), or that simply I was not lucky in example (9) – are *not any more* confirmed by the respective evidences than the weird hypotheses, since $r(H,E) = 1/p(E)$ does *not depend* on the content of H *at all*, if H only L-implies E. Someone who would reason in practical life according to the strange lines exemplified in examples (7-9) would obviously be considered an urgent case for psychological therapy. This shows that the pure account of H-D confirmation is, to say the least, not very close to common-sense intuitions.

Of course, Kuipers may respond that common-sense intuitions are often ambiguous, and confirmation is such a case. We must decide whether we want a "pure" or an "impure" notion of confirmation. If we take the latter choice, then we can no longer say that a single instance of a black raven confirms the raven hypothesis – only a sufficiently large sample would do so in this case. I agree with this hypothetical reply of Kuipers. Yet it is important to be *aware* of the above counterexamples. What makes them really conspicuous is that, intuitively, *even a large collection* of the respective confirmation instances would be regarded as insufficient to confirm them. For example, nobody would say that many instances of broken trees cd-confirm that an UFO had broken them during its landing phase. This observation points to a more fundamental problem, to be treated in the next section.

4. Irrelevance, Redundancy, and Kuipers' Response to these Problems

A qualitative consequence of the fact that $r(H,E)$ does not depend on the content of H is the so-called *converse consequence condition w.r.t.* H which is satisfied by the above account of d-confirmation:

(CC-H):   If E d-confirms H and H^* is consistent and logically entails H, then E d-confirms H^*. In particular, if E d-confirms H then E d-confirms $H^* = H \wedge X$ for any X compatible with H.

For example, this black raven not only confirms the raven hypotheses, but also its conjunction with the mentioned UFO hypothesis. In the case of conditional C-confirmation, one must additionally require that H^* is still logically independent of condition C. This problem of "irrelevant conjunctions," or of "tacking by conjunction" (Glymour), is a well-known one. It has disastrous consequences, provided one accepts the so-called *consequence condition* w.r.t. H:

(C-H): If E d-confirms H, then E confirms every logical consequence of H.

From (CC-H) and (C-H) one obtains the maximal triviality result that every non-tautologous E confirms every hypothesis H which is consistent with E, because E d-confirms $E \wedge H$ and hence confirms H. It is because of problems of this sort that the H-D account set out above is usually called the *naive* one. However, Kuipers' defense of it is anything but naive. Well, he argues, although E which d-confirms H also d-confirms $H^* = H \wedge X$ for every X compatible with H, even as much as H itself, we also know that E does not d-confirm X at all. Hence, d-confirmation remains perfectly localizable – we know perfectly well *which conjunctive part* of H^* has been confirmed by E (ICR, subsection 2.1.2). Kuipers concludes that tacking by conjunction should be accepted because it is harmless. Consequently, Kuipers refutes the consequence condition w.r.t. H: if E confirms H and H' is a logical consequence of H, then according to Kuipers we are *not entitled* to infer that E also confirms H'.

A similar and similarly well-known problem of the naive H-D confirmation account arises from irrelevant disjunctions, or by "tacking by disjunction," in regard to E. For, if E d-confirms H, and $E \vee X$ is not a tautology, then also $E \vee X$ d-confirms H. This consequence is also called the *consequence condition w.r.t. E*. If we join it with the so-called *converse consequence condition w.r.t. E*, which roughly says that if E^* d-confirms H and E entails E^* and no part of E disconfirms H, then also E d-confirms H, then we obtain a second maximal triviality result. Kuipers' reaction to this is again to argue that we should accept the formation of irrelevant E-disjunctions: they are harmless because we can always localize which disjunctive part of E confirms H.

Turning back to irrelevant conjunctions, which is the more important kind of problem in practical terms, let us ask about the situation in which Kuipers' line of argumentation has brought us. He has made clear that his concept of H-D confirmation is one of *partial confirmation*: that E d-confirms H according to (uHD) means merely that E d-confirms *some part* of H's content, but *not* that E confirms all parts of H's content. This is a rather weak claim. Obviously, this weak notion of H-D confirmation *cannot* support the consequence condition w.r.t. H, (C-H). But observe that *for practical purposes the abandonment of the consequence condition w.r.t. H is a severe drawback*. For we want to base our predictions and decisions on those hypotheses which are empirically well confirmed. This means that we draw certain *consequences C* from the well-confirmed hypotheses and conclude that these consequences C are also well confirmed so that we can trust them for prediction and decision purposes. All this presupposes that the condition (C-H) holds. But obviously, condition (C-H) cannot hold for a concept of partial H-D confirmation; it can only hold for a concept of *full* H-D confirmation according to which confirmation of H by E implies not only that some part of H's content is confirmed by E, but rather that *every* part of H's content is confirmed by E, so that we can trust in H's consequences *whatever* they are. How such a conception of full confirmation may be arrived at is discussed in section 5. Before we turn to this problem, we have a look at another small but surprising result.

5. On the Triviality of Bayesian H-D Confirmation: *E* Confirms *H* Because *E* Confirms *H*'s Content Part *E*

Although Kuipers draws a really nice picture of the situation into which his arguments have brought us, a small turnaround suffices for us to recognize that we are on the brink of disaster. Kuipers invites us to *localize* that content part of the hypothesis H which is d-confirmed by E by way of splitting H into some logically equivalent conjunction $H_1 \wedge H_2$ such that one conjunct is not d-confirmed by E – then we may conclude, according to Kuipers, that the confirmed content part of H is localized in the other conjunct. If this were true, then the well-known Popper-Miller objection to the possibility of inductive confirmation would immediately apply to Kuipers' confirmation account. After all, H is logically equivalent to $(E \vee H) \wedge (\neg E \vee H)$, and while the first conjunct is d-confirmed by E, the second conjunct is always probabilistically disconfirmed by E (the proof of this fact is well known). In the case of d-confirmation, $H \Vdash E$ and hence $E \vee H$ is logically equivalent to E. So we would be forced, if we follow Kuipers instructions, to conclude that the only content part of H which really gets confirmed by E is nothing but E itself. If this *were true*, it would mean that *non-trivial* confirmation does not exist *at all*, because it is an essential aspect of (non-trivial) confirmation that it is *ampliative,* or inductive in the broad sense.

Like Kuipers, I do not believe in the validity of the Popper-Miller objection. I think that the mistake of this objection consists in the fact that the conjunctive decomposition of H into $(E \vee H) \wedge (\neg E \vee H)$ is not *logically independent*: the negation of one conjunct implies the other one. So let us try to improve Kuipers' strategy and require that conjunctive decompositions of H must be logically independent. Then the Popper-Miller objection no longer applies. But it is easy to show that even in this refined version the same triviality result is obtained: assume, in the simplest case, that E is the singular statement $E(a)$ (e.g. "if this is a raven then it is black"), and H is the general hypothesis $\forall x E(x)$ (e.g., the raven hypothesis). Then $\forall x E(x)$ is logically equivalent to the conjunction of logically independent conjuncts $E(a) \wedge \forall x (x \neq a \rightarrow E(x))$, and while the left conjunct is trivially d-confirmed by $E(a)$, the right conjunct is not d-confirmed at all. By suitable logical equivalence transformations we can apply this argument to almost all cases of confirmations of general hypotheses by singular observation statements.

So we obtain the same triviality result as before: if E d-confirms H according to (uHD), then E d-confirms merely H's content part E, but nothing which is contained in H and goes beyond E (and likewise for cd-confirmation). This objection reveals a much deeper point, namely the intrinsic weakness of the Bayesian approach to H-D confirmation. Recall that we have mentioned above that the Bayesian H-D confirmation effect is obtained for *every* probability function, even one which does not include any inductive power at all; for example, it is also obtained when one uses Kemeny's logical measure function (cf. ICR, subsection 4.1), which attaches the same probability to every possible model, or world. This explains *why* we have obtained this triviality result: the degree of confirmation which $E(a)$ conveys to $\forall x E(x)$ results from the fact that $E(a)$ raises the probability of $E(a)$ to 1, and $E(a)$'s models are a superset of $\forall x E(x)$'s models, without $E(a)$ needing to raise the probability of $E(b)$ for other instances b (see Figure 1).

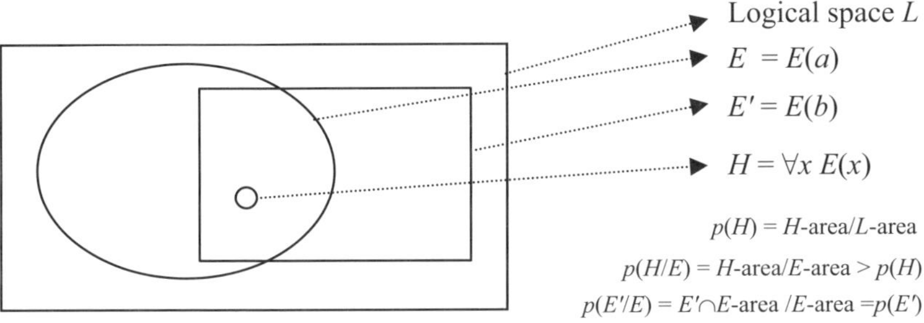

*Fig.*1. Bayesian H-D confirmation

Let us define, with Kuipers (ICR, subsection 4.1), a probability function as *inductive* iff $p(Fa\ /\ Fb \wedge E) > p(Fa\ /\ E)$ for arbitrary a, b and background evidence E about individuals $c_1,\dots,c_n$ distinct from a and b (cf. ICR, subsection 4.1). To avoid misunderstanding, neither the Bayesian account nor Kuipers' account *exclude* inductive probability functions, but their general explication of confirmation does not *rest* on any inductive assumptions[1] – it just rests on the trivial observation that E confirms H because E confirms H's content-part E, and this is the reason why Kuipers' account of *partial* H-D confirmation turns out to be a special case of Bayesian confirmation.

Intuitively, however, if we say that a sample of ravens has confirmed the

---

[1] Many Bayesians are proud of their approach getting around the notorious problem of induction (cf. van Fraassen 1989, ch. 6-7) – but the cost is that their concept of confirmation becomes trivial in the sense explained.

raven hypothesis, we not only mean that it has confirmed what we have observed in the sample, but that it has confirmed that other ravens, *not yet observed*, will also obey the raven hypothesis. This central intuition of confirmation as being *ampliative* is (admitted but) not supported by the Bayesian account of H-D confirmation which Kuipers defends. It follows that if Bayesians are discussing their diverging opinions in the light of a continuous flow of incoming evidence, then all that they are *forced* to agree on is the evidence, while they may continue forever to disagree on that part of their opinion which goes beyond the evidence. This is a disappointing result. The same objection applies to Kuipers' treatment of Goodman's problem (ICR, subsection 2.2.2.): when Kuipers points out that "all observed emeralds are green" does not only *not* d-confirm "all non-observed emeralds are *grue*," but does also *not* d-confirm "all non-observed emeralds are green," he misses the intuitively crucial difference between inductive projectibility of 'green' versus the non-projectibility of 'grue'.

6. Relevant Consequence Elements and H-D Confirmation

The inductive aspect of confirmation can be quantitatively captured by the well-known inductive probability functions in the Carnap-Hintikka tradition which Kuipers discusses in ICR, section 4. However, these systems include many arbitrary features. If one intends to mirror the ampliative (or inductive) aspect of H-D confirmation in a *qualitative* way, one needs an account of *full* H-D confirmation in which E d-confirms H if and only if E d-confirms *every* "essential" content part of H. Such an account has been developed within the theory of relevant consequence elements – *relevant elements*, for short – developed by Schurz and Weingartner (1987) for the purpose of truthlikeness, and it has been applied to H-D confirmation in Schurz (1991, 1994). A brief sketch of the present version of this theory follows (cf. also Schurz 1999). We say that an inference Prem $\Vdash$ Con, for Prem a set of premises and Con a conclusion, is c(onclusion)-relevant iff no predicate of Con is replaceable by a new predicate on *one or several* occurrences of it in C, *salva validitate* of the inference. Prem $\Vdash$ Con is called p(remise)-relevant if no predicate of Prem is replaceable on a *single* occurrence of it in Prem, *salva validitate* of the inference. An inference is called pc-relevant iff it is p- *and* c-relevant. For example $p, p \rightarrow q \Vdash q$ is pc-relevant, $p \wedge q \Vdash p$ is c- but not p-relevant, while $p \Vdash p \vee \underline{q}$ and $p \Vdash (p \rightarrow \underline{q}) \rightarrow \underline{q}$ are p- but not c-relevant (the *salva validitate* replaceable propositional variables are underlined; note that propositional variables are 0-ary predicates). These definitions are not ad hoc, but have very intuitive motivations, for which we refer the reader to Schurz (1991, 1999).

The reason why salva validitate replacements of predicates in the premises must be restricted to single occurrences is that the premises of relevant inferences are allowed to contain (theoretical) concepts which are not contained in the conclusion (while the inverse does not hold); note also that the given definition of premise-relevance is equivalent to another formulation which says that no conjunctive component of the matrix of a certain prenex conjunctive normal form of the premises is redundant in the inference.

The second step is to decompose a statement into its smallest relevant consequences, its so-called *relevant elements*. The resulting decomposition is called an *irreducible representation*. We say that (i) a statement A is *pre-elementary* iff A is not logically equivalent to $A_1 \wedge \ldots \wedge A_n$ ($n \geq 1$) where *each A_i is shorter than A*, (ii) A is *elementary* iff A is a pre-elementary negation normal form and each quantifier-scope in A is a conjunction of pre-elementary formulas; (iii) B is a *relevant element* of A iff B is an elementary relevant consequence of A, and finally (iv) B is an *irreducible representation* of A iff B is a nonredundant conjunction or set of relevant elements of A which is logically equivalent with A, where 'nonredundant' means that no conjunct or element is eliminable salva logical content of A. $\mathbf{I}(A)$ denotes the set of irreducible representations of A. Now we can define H-D confirmation as follows:

(uHDr): E d-confirms H iff there exists $E' \in \mathbf{I}(E)$ and $H' \in \mathbf{I}(H)$ such that $H' \Vdash E'$ is pc-relevant.

(cHDr): E cd-confirms H relative to C iff there exists $E' \in \mathbf{I}(E)$ and $H' \wedge C' \in \mathbf{I}(H \wedge C)$ such that $H' \wedge C' \Vdash E'$ is pc-relevant.

Space limitations mean that we cannot show here how this account solves the problem mentioned above and many others (cf. Schurz 1991, 1996, where irreducible representations are defined in a different way). We only wanted to make clear the essential idea of this account of *full* confirmation of H by E: every "part" of the irreducible representation of H in terms of relevant elements is necessary for deriving E, and no "part" of a similar representation of E is inessential in this derivation.

7. Gemes' Content Parts in Comparison to Relevant Consequence Elements

A closely related approach to a canonical representation of the relevant content parts of statements has been developed by Ken Gemes. According to the simplest formulation of his idea, a logical consequence A of a statement S is a *content part* of S iff S and A are contingent, and for some A^* logically equivalent with A there exists no statement B which logically follows from S and is logically stronger than A but is constructed solely from A's atomic formulas (Gemes 1993, p. 481). To generalize this idea to quantificational logic, Gemes assumes first order quantifiers to be replaced by infinite truthfunctional conjunctions and disjunctions – in other words, he assumes a substitutional interpretation of quantifiers (Gemes 1994). For application to H-D confirmation a second step is necessary in which the given hypothesis is canonically transformed into a so-called *natural axiomatization* of it, defined as follows: $N(H)$ is a natural axiomatization of H iff $N(H)$ is a finite nonredundant set of content parts of H which is L-equivalent with H, where 'nonredundant' means that no content part of some $X \in N(H)$ is logically entailed by some $Y \neq X$, $Y \in N(H)$ (Gemes 1998, p. 9). Where $\mathbf{N}(T)$ denotes the set of natural representations of theory T, Gemes suggests that E confirms axiom H of naturally axiomatized theory $N \in \mathbf{N}(T)$ relative to C iff E and C are content parts of $T \wedge C$ and there exists no $N^* \in \mathbf{N}(T)$ such that for some subset S of N^*, E but not H is a content part of $S \cup \{C\}$ (1993, p. 486); the special case of confirmation of H by E relative to C is obtained by letting $T = H$. Gemes' approach also solves the above-mentioned problem and many others which have to do with irrelevance and redundancy. The following comparison shows some differences between Gemes' content parts and the theory of relevant elements (for further comparisons cf. Gemes 1998; Schurz 1994).

Statement	*Natural Axiomatization*	*Irreducible Representation*
$\forall x(Fx \leftrightarrow Gx)$	$\forall x(Fx \leftrightarrow Gx)$	$\forall x(Fx \rightarrow Gx)$, $\forall x(Gx \rightarrow Fx)$
$p \leftrightarrow (q \leftrightarrow r)$	$p \leftrightarrow (q \leftrightarrow r)$	$p \wedge q \rightarrow r$, $p \wedge r \rightarrow q$
	$[\equiv (p \leftrightarrow q) \leftrightarrow r]$	$q \wedge r \rightarrow p$, $\neg p \wedge \neg q \rightarrow r$
$(s \rightarrow r) \wedge (r \wedge t \rightarrow s)$	$(s \rightarrow r) \wedge (r \wedge t \rightarrow s)$	$(s \rightarrow r)$, $(r \wedge t \rightarrow s)$

The comparison shows that the relevant element approach leads to a more fine-grained decomposition than the content part approach. This more fine-grained quality has advantages in certain cases; as in application to the evaluation of truthlikeness (see below). In the third example, the only natural axiomatization of the theory $(s \rightarrow r) \wedge (r \wedge t \rightarrow s)$ is the set $\{(s \rightarrow r) \wedge (r \wedge t \rightarrow s)\}$ which contains only one axiom. This is so because only $(s \rightarrow r) \wedge (r \wedge t \rightarrow s)$ and $s \rightarrow r$ but not $r \wedge t \rightarrow s$

are content parts. Therefore, Gemes' account has the consequence that r confirms the entire statement $(s{\rightarrow}r){\wedge}(r{\wedge}t{\rightarrow}s)$ relative to the condition s, while according to the relevant element approach, r confirms only the left conjunct $s{\rightarrow}r$ relative to s which is intuitively more appropriate. Another important difference consists in the handling of existential conclusions; for example, both Fa and $\exists x Fx$ are relevant elements of $\forall x Fx$, while only Fa but not $\exists x Fx$ is a content part of $\forall x Fx$ (which results from the substitutional interpretation of quantifiers in Gemes' approach). Again, this is a disadvantage of Gemes' approach because one way of confirming universal hypotheses is by *Popperian basic statements*, which are existential statements. For example, according to Popper's account, $\exists x(Fx{\wedge}Gx)$ confirms $\forall x(Fx{\rightarrow}Gx)$ relative to $\exists x Fx$, which is supported by the relevant element approach, but not by the content part approach. Moreover, Gemes' account does not handle irrelevant parts in $\exists$-scopes; for example, $\exists x Fx$ Gemes-confirms $\exists x(Fx{\wedge}\underline{Gx})$, which is unintuitive since $\underline{Gx}$ is an irrelevant part of the premises and hence, *salva validitate* replaceable. On the other hand, Gemes' content parts have some nice logical properties which relevant elements do not have.

8. Kuipers' Account of Actual and Nomic Truthlikeness

One of the strong points of Kuipers' book is the way in which he brings H-D confirmation and truthlikeness together. From an *instrumentalist* viewpoint, it makes no good sense to eliminate a general theory because it has been falsified by some particular data; the theory may still be an excellent prediction tool for various other purposes. Thus, from an instrumentalist viewpoint, instead of H-D confirmation, which is excessively truth-oriented, *H-D evaluation* is the preferable strategy, where theories are evaluated and compared in terms of their H-D successes and H-D failures – without being eliminated because of falsifications. Kuipers argues convincingly that, at least in the long run, exactly this seemingly "instrumentalist" strategy of H-D evaluation turns out to be functional in terms of getting closer to the truth in the realistic sense. Utilizing a Hegelian metaphor, Kuipers calls this functionality of instrumentalism for realistic purposes the "cunning of reason" (ICR, subsection 6.2).

However, finding a satisfying notion of truthlikeness is a difficult enterprise. According to Popper's original characterization, theory Y is at least as close to the truth T as theory X, abbreviated as $Y \geq_T X$, iff X's true consequences are set-theoretically contained in Y's true consequences, and Y's false consequences are contained in X's false consequences; and Y is closer to the truth than X, $Y >_T X$, if at least one of these two containment relations is proper. Tichý and Miller have shown that Popper's definition of '$Y >_T X$' can impossibly hold if Y is false, i.e., if Y has at least one false consequence (cf. ICR, subsection 8.1.4). Their impossibility proof makes heavy use of irrelevant and redundant consequences. Since then, several alternative definitions of truthlikeness have been developed which are not exposed to Tichý-Miller's knock-down argument. One of these approaches, which is impressively developed but leads a long way from Popper's ideas, is Niiniluoto's quantitative theory of truthlikeness. Another approach is the relevant element approach (Schurz/Weingartner 1987) which replaces classical consequences of theories by relevant elements and is close to Popper's original idea of truthlikeness (Popper 1992 acknowledges this approach in the appendix to his *The Open Society and Its Enemies*). Kuipers has developed an equally impressive but still different approach to truthlikeness. His approach is *structuralist* in that he often prefers to represent propositions not by statements (of a given language), but by classes of models (of an assumed underlying language). Due to well-known results of model theory, his structuralistic conditions may also be expressed in statement terms (cf. ICR, section 8.1).

In his definition of *actual truthlikeness*, Kuipers assumes that the theories being compared, X, Y, ... are *complete* in the sense that they give a complete state description in the underlying language. If this language is propositional, the theories X, Y, ... are sets of propositional variables, each theory containing those variables which are claimed as true by the theory; the variables not contained in theory are automatically claimed to be false (ICR, subsection 7.1.2). For first-order languages, Kuipers takes first-order structures instead of statement theories (ICR, subsection 7.1.3); the statement counterparts of these structures would be the sets of all atomic statements (in an underlying language which contains a name for each individual in the domain) which are claimed to be true by the given theory; the atomic statements not contained in it are claimed to be false. Given this representation of theories X, Y, and letting T be the set of propositional variables, or atomic statements, which are actually true, then Kuipers defines:

(AT) $Y \geq_T X$ iff   (i) $Y - T \subseteq X - T$, and
$\qquad\qquad\qquad\quad$ (ii) $T - Y \subseteq T - X$, or equivalently, $T \cap X \subseteq T \cap Y$.

In words, $Y \geq_T X$ iff (i) Y's false atomic consequences are contained in X's false atomic consequences, and X's true atomic consequences are contained in Y's true atomic consequences.

Kuipers' theory of actual truthlikeness can be regarded as being in the spirit of the Schurz-Weingartner approach insofar as it restricts the set of all consequences of a theory to a special subset of it – the set of atomic consequences. This rather strong restriction only makes sense if the theories being compared are complete. However, most theories in actual practice are not complete, and hence Kuipers' concept of actual truthlikeness cannot be applied to them. For example, there is no way to compare hypotheses such as "all metals are solid and conduct electricity," "all metals conduct electricity and heat," etc., in terms of their actual truthlikeness, because they do not imply any atomic statements. This is the most serious drawback of Kuipers' theory of actual truthlikeness.

However, Kuipers' major purpose is not actual but *nomic* truthlikeness. Roughly speaking, the *nomic truth*, abbreviated by N, is regarded as that proposition which contains all and only those facts (w.r.t. an underlying language) which are true solely because of laws of nature (cf. ICR, subsection 7.2.1). In distinction to the actual truth T, the nomic truth N is not a complete theory – it does not fix the truth value of all statements or in model-theoretic terms, it contains much more than one model – it contains all models which are nomologically possible, and only them. Now, expressing theories as sets of models, Kuipers defines nomic truthlikeness as follows:

(NT) $Y \geq_N X$ iff   (i) $Y - N \subseteq X - N$, and
    (ii) $N - Y \subseteq N - X$, or equivalently, $N \cap X \subseteq N \cap Y$.

In words, Y is nomologically at least as truthlike as X iff (i) Y's mistaken models are contained in X's mistaken models, and (ii) X's correct models are contained in Y's correct models (ICR, subsection 7.2.2). Thereby, a mistaken model is a logically possible but nomologically impossible arrangement of facts which is allowed by the theory, while a correct model is a logically as well as nomologically possible arrangement of facts which is allowed by the theory.

Note that the containment relations in clauses (i) and (ii) can be regarded either as *factual* or as *logical* ones. If they are regarded as factual, then clause (i) would make no sense, because nomologically impossible models are never factually realized, so clause (i) would become trivially true. Clause (ii) would have a factual interpretation, saying that the set of models $(N \cap X) - Y$ is factually empty on *accidental* reasons – but then, nomic truthlikeness would depend on the accidental facts of our worlds, which would also make no good sense. So I assume that Kuipers interprets these containments as logical containments, i.e. containments following from the content of the respective theories – this also accords with Kuipers' interpretations in ICR, section 8. Given that these containments are logical ones, it is useful to reflect their statement counterparts. Assuming Y, X and N are expressed by sets of statements, and utilizing model-theoretic facts, one obtains the following two statement conditions for nomic truthlikeness (as pointed out by Kuipers in ICR, subsection 8.1.2):

(NT-statement): $Y \geq_N X$ iff (i) $Y \Vdash X \vee N$, and (ii) $X \wedge N \Vdash Y$.

Kuipers points out in ICR, subsection 8.1.2, that these two conditions are somehow similar to Popper's original account. If we express them in terms of classical consequences, condition (i) is equivalent to $Cn(Y) \supseteq Cn(X) \cap Cn(N)$, and condition (ii) – which is equivalent to $X \Vdash Y \vee \neg N$ – is equivalent to $Cn(X) \supseteq Cn(Y) \cap Cn(\neg N)$. In words, (i) states that those consequences of X which are nomologically necessary are contained in Y's consequences, and (ii) states that those consequences of Y which are implied by $\neg N$ are contained in X. However, the consequences implied by $\neg N$ are not those which are nomologically impossible (these would be the consequences which imply $\neg N$). So there remains a subtle difference (cf. section 9 of this paper).

9. Counterintuitive Consequences of Nomic Truthlikeness

Although Kuipers' definitions of actual and nomic truthlikeness are formally similar in that they utilize the *symmetric difference* measure, they are not similar in content, because the symmetric difference measure is applied to very different entities (atomic formulas versus models). In particular, actual truthlikeness is not, as one might intuitively expect, the *limiting case* of nomic truthlikeness in which one assumes N to be the actual truth, i.e. to be a complete theory. This is seen as follows: if $N = \{n\}$ = the actual truth, which contains only one model, namely the actual world, and if Y is a false theory, hence $n \notin Y$, then clause (i) reduces to $Y \subseteq X - \{n\}$, and clause (ii) to $X \cap \{n\} \subseteq \emptyset$, which implies that X must also be false, i.e. $n \notin X$, so clause (i) reduces to $Y \subseteq X$ and clause (ii) to $\emptyset \subseteq \emptyset$. This means that if N is the actual truth, then $Y \geq_N X$ iff Y logically entails X, and $Y >_N X$ iff Y is logically stronger than X.

This objection is due to Oddie (1981) and is discussed by Kuipers in ICR (subsection 10.2.2). It shows that Kuipers' concept of nomic truthlikeness is not suited to purposes of actual truthlikeness, for it is clearly counterintuitive that a false theory can be improved merely by logically strengthening it, e.g., by conjoining some other false statements to it. The same aspect of Kuipers' account is reflected in the fact that his definition of nomic truthlikeness blocks only the "first half" (lemma 1) of the Tichý-Miller knockdown argument, but not its "second half" (lemma 2) (cf. ICR, subsection 8.1.4).

However, a similar objection can also be made to a special case of genuine nomic truthlikeness – the case where the theory Y is incompatible with the nomic truth N. This happens in science, for example, when a theory assumes that a certain theoretical entity exists when it in fact does not exist (e.g., the luminiferous ether, or phlogiston); the assumed existence of theoretical entities of this sort will infect all models of the theory, whence all of these models will lie outside the set of physically possible models (because in physically possible models, neither luminiferous ether nor phlogiston exists). In this case, too, it holds that $Y \cap N = \varnothing$, and we get the same counterintuitive result that, given $Y \cap N = \varnothing$, then $Y \geq_N X$ holds iff (i) $Y \subseteq X - N$ and (ii) $X \cap N \subseteq \varnothing$, and therefore, iff Y logically entails X. This shows that Kuipers' theory of nomic truthlikeness cannot handle the truthlikeness of theories which are incompatible with the nomic truth. Kuipers may respond that he can handle this problem with an extension of his account which he calls "refined truthlikeness" (see below) – indeed, Kuipers' argument in ICR, subsection 10.2.2, also applies to the above counterexample. Nevertheless, it is still somewhat disappointing that a purely qualitative problem like the one above cannot be handled within Kuipers' basic approach.

Be that as it may, there exists a more serious problem with Kuipers' approach which concerns the basic intuitions underlying it. This problem becomes clear if we recall what Kuipers' conditions mean in statement terms (cf. NT-statement): $Y \geq_N X$ iff (i) $Y \Vdash N \vee X$, and (ii) $X \wedge N \Vdash Y$. Now, (i) implies that (iii) $Y \wedge \neg X \Vdash N$. This means that if a scientific theory Y, such as Einstein's relativity theory, is closer to the nomic truth than another theory, such as Newtonian mechanics, X, it must be the case that $Y \wedge \neg X$, i.e. Einstein's relativity theory conjoined with the negation of Newtonian mechanics, must logically entail the entire nomic truth. How can this make sense? Even worse: Einstein's relativity theory is incompatible with Newtonian mechanics, and hence $Y \wedge \neg X$ is logically equivalent to Y. So Einstein's relativity theory, in order to be closer to Newtonian mechanics, must entail the entire nomic truth. I do not believe that this makes good sense. It seems that something is wrong with the basic intuitions of this approach.

This impression is supported if Kuipers' account is applied to simple examples. Assume a scenario in which our world is described by only three predicates 'F', 'G' and 'H'; $N = \forall x(Fx \rightarrow Gx \wedge Hx)$, $Y = \forall x(Fx \rightarrow Gx \wedge \neg Hx)$, and $X = \forall x(Fx \rightarrow \neg Gx \wedge \neg Hx)$. Intuitively, Y is closer to the nomic truth than X, because it predicts G's correctly and H's incorrectly, while X predicts both properties incorrectly. But Kuipers' account does not give us this result, because his condition (i) $\forall x \ (Fx \rightarrow Gx \wedge \neg Hx) \Vdash \forall x \ (Fx \rightarrow Gx \wedge Hx) \vee \forall x(Fx \rightarrow \neg Gx \wedge \neg Hx)$ does not hold (while (ii) does in fact hold). It is easy to multiply counterexamples of this sort.

Kuipers extends his basic definition of nomic truthlikeness in many aspects, demonstrating various beautiful properties. When turning to stratified truthlikeness, where the linguistic framework is divided into an observational and a theoretical level, he proves a projection theorem which says that if Y is at least as close to the theoretical truth as X, and X is relatively correct, then Y's observational projection is at least as close to the observational truth as X's observational projection (ICR, subsection 9.1.1). That X is relatively correct means that for every theoretically mistaken but observationally correct model of X there exists an observationally equivalent and theoretically correct X-model. I think that this condition, which Kuipers considers to be weak, is in fact very strong. For example, if a scientific theory contains theoretically mistaken core assumptions, then all of its theoretical models will be infected by these mistakes, but many of its models may still be empirically adequate, whence the condition of relative correctness will not be satisfied. For all such theories, theoretical truthlikeness will not imply observational truthlikeness.

Without being able to go into all details, my main point is that I do not see how Kuipers' refinements can avoid the abovementioned counterexamples. In his account of *refined* truthlikeness (ICR, subsection 10.2.1), Kuipers starts from an assumed similarity relation between structures, or models, $s(x,y,z)$, meaning that structure y is at least as similar to structure z as x, and explicates refined nomic truthlikeness as follows:

(RT): $Y \geq_{Nr} X$ iff (i) for all $y \in Y-(X \cup N)$ there exists $x \in X-N$ and $n \in N-X$ such that $s(x,y,n)$, and (ii) for all $x \in X$ and $n \in N$ (where n is comparable with x) there exists an $y \in Y$ such that $s(x,y,n)$.

$Y >_{Nr} X$ iff $Y \geq_{Nr} X$ but not $X \geq_{Nr} Y$.

In words, clause (i) requires that every mistaken Y-model which is not contained in X is an improvement of some mistaken X-model, i.e. lies between some mistaken X-model and some possible model not allowed by X, and clause (ii) says that every X-model x comparable with some nomologically possible model n is improved by some Y-model such that this Y-model lies between x and n. Apart from the problem that similarity between models is not always a very clear notion, the quantifier clauses in Kuipers' conditions involve some problems of their own. The fact that every X-model is improved by *some Y-model*, and vice versa that every Y-model improves *some X-model*, does not say much about the *majority* of X- and Y-models in terms of their distance to the "true" models. Counterexamples are readily constructed along these lines. For example, assume X and Y are theories about nomologically allowed *positions* of points (or say, point-masses) on a straight line, and the nomic truth contains the truly possible positions of points, and similarity is measured simply by the distance between positions. Then even if *most* of X's positions are much closer to the positions in N than most of Y's positions, Kuipers' conditions for $Y \geq_{Nr} X$ can still be satisfied, because it may be that X contains one worst position (which is improved by every Y-position), and Y contains one best position, which improves every X-position. Moreover, since $X \geq_{Nr} Y$ does not hold, $Y >_{Nr} X$ will result in such a case. The situation is illustrated in *Figure 2*.

$$x_1\, y_1\, y_2 \ldots y_n \qquad\qquad x_2 \ldots x_{n+1}\, y_{n+1} \qquad\qquad n_1 \ldots n_{n+1}$$

Fig. 2. $Y >_{Nr} X$ holds, but most X-points are closer to N-points than most Y-points.

Although we would intuitively consider here X to be closer to the nomic truth than Y, Kuipers' refined approach gives us the inverse result.

10. Truthlikeness and Relevant Elements

The problems of Kuipers' basic account of truth-likeness do not touch the relevant element approach to actual truthlikeness, which is roughly described as follows. Given a theory X, we first form its set of relevant consequences X_r, thereby discounting all singular implications with false antecedent, because they do not count as successes even when they are true (this is an improvement over Schurz/Weingartner 1987). Let X_{tr} be the subset of X_r's true elements, and X_{fr} the subset of X_r's false elements. For the sake of computational feasibility, an irreducible representation of X_{tr} and X_{fr} is chosen, abbreviated as X_{ti} and X_{fi}. Now Y is defined to be at least as close to the actual truth T as X as follows:

(ATr) $Y \geq_T X$ iff (i) $Y_{ti} \Vdash X_{ti}$ and (ii) $X_{fi} \Vdash Y_{fi}$

        (in case of $>_T$, at least one entailment must be proper).

For the above example, where $Y = \forall x(Fx \rightarrow Gx \wedge \neg Hx)$, $X = \forall x(Fx \rightarrow \neg Gx \wedge \neg Hx)$, and the actual truth contains $\exists x Fx \wedge \forall x(Fx \rightarrow Gx \wedge Hx)$, we obtain $Y_{ti} = \{\forall x(Fx \rightarrow Gx)\}$, $Y_{fi} = \{\forall x(Fx \rightarrow \neg Hx)\}$, $X_{ti} = \varnothing$, $X_{fi} = \{\forall x(Fx \rightarrow \neg Gx), \forall x(Fx \rightarrow \neg Hx)\}$. So (ATr) implies the desired result that $Y >_T X$. For further applications of this approach cf. Schurz/Weingartner (1987). The approach can be also extended to truth approximation in the light of established empirical data as indicated in Schurz (1987); a rule of success similar to that given by Kuipers also holds for the relevant element approach. In order to extend this approach to nomic truthlikeness we have to compare the relevant consequent-elements of both theories which are nomologically necessary, and those which are nomologically impossible. (Note that by considering impossible consequences we depart from Kuipers' condition (ii).) Let Y_{Nti} be an irreducible representation of $Y_r \cap Cn(N)$, and Y_{Nfi} be an irreducible representation of $Y_r \cap \{A: N \Vdash \neg A\}$; and likewise for X. By putting these two sets in the place of Y_{ti} and Y_{fi}, respectively (and likewise for X), we obtain the corresponding notion of nomic truthlikeness expressed in terms of relevant elements. Applied to our example, this notion yields the same result as actual truthlikeness (a difference would only arise if Y or X have nomologically indeterminate consequences). The problems of Kuipers' account discussed in section 8 are avoided because we restrict ourselves to relevant elements.

11. Conclusion

Kuipers' general ideas on the relation of confirmation, empirical success and truthlikeness are philosophically appealing and convincing. However, his explications of these notions suffer from several shortcomings. In other places (Schurz/Weingartner 1987; Schurz 1991, 1994, 1999) I have argued that satisfying logical accounts of these and related concepts must refer to a notion of *relevance consequence-element*, which is closely related to Gemes' notion of (relevant) *content-part*, in order to avoid the various traps of logical explication attempts. In this paper I have tried to point out some of the places where these typical traps appear in Kuipers' account, and how the notions of relevant consequence-element or content-part may be helpful in avoiding them.

University of Düsseldorf
Philosophical Institute
Universitätsstrasse 1, Geb. 23.21,
D-40225 Düsseldorf
Germany
e-mail: gerhard.schurz@phil-fak.uni-duesseldorf.de

REFERENCES

Gemes, K. (1993). Hypothetico-Deductivism, Content, and the Natural Axiomatization of Theories. *Philosophy of Science* **54**, 477-487.

Gemes, K. (1994). A New Theory of Content: Basic Content. *Journal of Philosophical Logic* **23**, 595-620.

Gemes, K. (1998). Hypothetico-Deductivism: The Current State of Play; The Criterion of Empirical Significance: Endgame. *Erkenntnis* **49**, 1-20.

Kuipers, T. (2000/ICR). *From Instrumentalism to Constructive Realism.* Dordrecht: Kluwer.

Popper, K. (1992). Die offene Gesellschaft und ihre Feinde. Vol. II. Stuttgart: J.C.B. Mohr (UTB).

Sober, E. (1993). *Philosophy of Biology.* Boulder, CO: Westview Press.

Schurz, G. (1987). A New Definition of Verisimilitude and Its Applications. In: P. Weingartner and G. Schurz (eds.), *Logic, Philosophy of Science and Epistemology*, pp. 177-184. Vienna: Hölder-Pichler-Tempsky.

Schurz, G. (1991). Relevant Deduction. From Solving Paradoxes Towards a General Theory. *Erkenntnis* **35**, 391 - 437.

Schurz, G. (1994). Relevant Deduction and Hypothetico-Deductivism: A Reply to Gemes. *Erkenntnis* **41**, 183 - 188.

Schurz, G. (1999). Relevance in Deductive Reasoning: A Critical Overview. In: G. Schurz and M. Ursic (eds.), *Beyond Classical Logic*, pp. 9-56. St. Augustin: Academia Verlag.

Schurz, G. and P. Weingartner (1987). Verisimilitude Defined by Relevant Consequence-Elements. In: T. Kuipers (ed.), *What Is Closer-To-The-Truth?* (*Poznań Studies in the Philosophy of the Sciences and the Humanities* **10**), pp. 47-78. Amsterdam: Rodopi.

van Fraassen, B. (1989). *Laws and Symmetry*. Oxford: Clarendon Press.

Theo A. F. Kuipers

CONFIRMATION AND TRUTHLIKENESS
REPLY TO GERHARD SCHURZ

The subtitle of ICR reads: "On some relations between confirmation, empirical progress, and truth approximation." Gerhard Schurz's contribution is in fact a critical review of my account of (hypothetico-) deductive (HD) confirmation and of truthlikeness. The former account was intended to be a partial explication of confirmation and the latter is a prerequisite for discussing the prospects of truth approximation by standard procedures, notably the HD method. Assuming revisions of the definitions of HD confirmation and truthlikeness, Schurz suggests he wants to connect them in a similar way as I do by conceiving the "rule of success," leading to empirical progress, as the glue between confirmation and truth approximation. Two of the major claims of ICR are that the rule of success is typically of an instrumentalist nature, for it takes counterexamples into account in a non-falsificationist way, and that this rule is straightforwardly functional for truth approximation. I am very happy with Schurz's support of this general line of argumentation in favor of HD evaluation instead of HD testing. However, I also have to concede that his criticisms of explications of HD confirmation and (basic and refined) truthlikeness are rather severe, so I hasten to respond to them.

As a matter of fact, Schurz's criticism of my account of HD confirmation is of a different nature than that of truthlikeness, although the same notion, viz. "relevant-element," is used for both purposes. My explication of HD confirmation is criticized for being too weak, whereas that of truthlikeness is stated to be on the wrong track. I shall deal with the two notions separately.

Confirmation

The arguments Schurz states to deal with my account of HD confirmation are the following: it leaves room for weird examples, viz. (7)-(9); it doesn't satisfy the "consequence condition" (C-H); and it is not (necessarily) ampliative.

In: R. Festa, A. Aliseda and J. Peijnenburg (eds.), *Confirmation, Empirical Progress, and Truth Approximation* (*Poznań Studies in the Philosophy of the Sciences and the Humanities,* vol. 83), pp. 160-166. Amsterdam/New York, NY: Rodopi, 2005.

Schurz's remedy amounts to a sophisticated strengthening of, let me call it, naïve deductive confirmation, viz., hypothesis H logically entails evidence E, by requiring a number of special logical conditions on the relation between H and E in terms of "relevant elements." He also explains that the strengthening of naïve deductive confirmation in terms of "(the content parts of) a natural axiomatization," proposed by Ken Gemes, is weaker than his own. However, the question of course is whether we need any strengthening at all, so we shall have to consider the three arguments against the "liberal account" put forward by Schurz. I shall briefly comment on these arguments and add an extra argument in favor of the liberal account.

It is indeed very easy to produce weird examples, like (7)-(9), but it should be realized that, historically, similarly weird examples can be given that we have come to appreciate. For example, consider all deductive confirmation of Newton's law of gravitation, assuming the three laws of motion, by (conditionally) deductively well-confirmed empirical laws dealing with freely falling objects, trajectories of projectiles and planetary orbits. This was (naïve) deductive confirmation of a hypothesis that was originally conceived by leading scholars as very occult, notably by Leibniz and Huygens (see Atkinson's contribution to the companian volume, for a concise statement). According to Schurz we should say instead that all this evidence did (almost) not confirm the hypothetical law of gravitation according to Leibniz and Huygens, whereas it did in the eyes of Newton. In sum, the price Schurz wants to pay is that (deductive) confirmation does not apply to hypotheses that are considered as weird and, hence, it is a subjective matter.

Schurz's second argument against naïve deductive confirmation is that it does not satisfy the consequence condition: if E deductively confirms H, it does not automatically deductively confirm any consequence H' of H. Schurz also reports my localization argument relativizing the, probably, counterintuitive fact that naïve deductive confirmation has the opposite "converse consequence condition": if E d-confirms H it d-confirms any strengthening H'' of H. Indeed, if both properties were to hold, any E would confirm any H. However, the *prima facie* desirability of the consequence condition can be explained in another way. Let us consider the previous example of confirmation of the theory of gravitation by the (corrected) law of free fall (LFF). Naïve deductive confirmation leads to the following claims: assuming Newton's theory of motion (TM), LFF not only d-confirms the law of gravitation (LG) but also that law in conjunction with a hypothesis that even now is still considered to be weird (WH). Moreover, assuming LG, LFF d-confirms TM, whereas, even assuming TM, LFF does not d-confirm WH (localization). In other words, when E d-confirms H, irrelevant subhypotheses can be localized and relevant subhypotheses are (at least) conditionally d-confirmed (in the sense indicated). To be sure, in a detailed analysis of a particular case the natural axiomatization of Ken Gemes can play an illuminating role.

Let us finally turn to the quest for an ampliative account. Schurz rightly observes that the probabilistic translation of naïve deductive confirmation leads to Bayesian probabilistic confirmation for any probability function, including the non-inductive or non-ampliative one known as Kemeny's logical measure function. In Chapter 4 of ICR I deal with genuine inductive probability functions, but neglect the non-ampliative type of confirmation. Indeed, naïve deductive confirmation leaves room for the claim that E d-confirms H just because E is a conjunct of H, i.e. $H = E\&H'$, where H' is logically independent of E. Although I would like to agree with Schurz that scientists are inclined to assign an ampliative character to confirmation whenever possible, I strongly disagree with the suggestion that non-inductive confirmation is no confirmation at all. In Section 7.1.2 of SiS, entitled "The landscape of confirmation," I introduce the distinction between structural and inductive confirmation, where structural confirmation is the probabilistic generalization of deductive confirmation, based on the logical measure function. That the latter type of confirmation, and hence the liberal account, should be taken seriously I like to demonstrate with my favorite example, that everybody would agree to say that the evidence of an even outcome (2, 4, 6) of a throw with a fair die confirms the hypothesis of a high outcome (4, 5, 6), whereas this has no ampliative aspect at all. However, typical inductive confirmation functions, like Carnapian or Hintikka systems, essentially combine structural and inductive confirmation, in such a way that the possibility of (purely) inductive confirmation in certain cases, e.g. $E=Fa$ and $H=Fb$, is paid by "counterinductive" confirmation going together with structural confirmation, e.g. deductive confirmation, e.g. $E=Fa\&Fb$ and $H=(x)Fx$. For further details, I refer the reader to SiS or to Kuipers (forthcoming).

Since I find none of the three arguments that Schurz advances against my liberal approach to HD confirmation convincing, I am not inclined to look for far-reaching sophistication, such as the relevant element approach. An extra reason is that this specific sophistication is, however elegant, technically rather complicated, whereas in practice scientists do not seem to use such advanced intuitions, let alone specifications of them.

Truthlikeness

Regarding truthlikeness, Schurz suggests that the number and importance of counterexamples to my basic definition and its refinements are so impressive that we had better look for an alternative definition, viz. again in terms of relevant elements. Although the latter approach may have its advantages, I shall not really go into its separate and comparative evaluation, because most of the reported counterexamples and further counterarguments are either inappropriate or unjustified. But let me begin by saying that Schurz's presentation of my basic (or naïve) and refined account in Sections 7 and 8 are very transparent and largely correct. Just two points. First, Schurz writes in Section 7: "However, most theories of actual practice are not complete [even in the more specific sense of theories not implying atomic statements], and hence, Kuipers' concept of actual truthlikeness cannot be applied to them." (p. 239) However, in my first (1982) paper on truth approximation I extended my treatment of actual truthlikeness to incomplete theories, but since I came to believe that the actual truth is usually not approached by incomplete theories, but by complete or partial descriptions (at the end of ICR, Section 7.3.1 I hint at the easy extension to partial descriptions), I neglected the original extension. Hence, I fail to see it, as Schurz does, as a drawback of my definition to actual truthlikeness. My second point is that Schurz seems to be in doubt whether my definition of nomic truthlikeness might be intended in some factual sense, as opposed to a logical or conceptual sense. I do not understand how my Sections 7.2 and 8.1 could be understood in a factual sense, but I must concede that I state it only once explicitly and specifically for (my general version of) Popper's definition "When the definition applies, this means that certain areas [in Figure 8.1] are empty on logical or conceptual grounds" (ICR, p.180), but it must be clear from the context that the same applies to the empty areas in Figure 8.2 belonging to the basic definition.

Let us start by considering the purported counterexamples in Section 8. After noting that my definition of actual truthlikeness is not a limiting case of that of nomic truthlikeness and that if the latter is nevertheless used for dealing with the problem of actual truthlikeness, Schurz points out that Oddie's well-known child's play objection can be raised. Schurz's claim that, for similar reasons, basic nomic truthlikeness cannot adequately deal with theories that are incompatible with the nomic truth, notably theories using non-referring terms, seems more serious. Let us first turn to the general problem and then to non-referring terms. In general, we can indeed naïvely come closer to the nomic truth when we strengthen a theory that is incompatible with the nomic truth, that is, the nomic version of Oddie's child's play objection. However, in the naïve situation, in which models of theories are simply right or wrong, this is very plausible. Compare it with the situation of intending to grow cabbage plants and no weed, then (even) if you did not yet manage to grow some cabbages it is a step forward when you weed. Similarly, (even) if you do not yet capture correct models, you make progress by eliminating incorrect models, assuming that you cannot improve them. However, the refined version can adequately deal with the nomic version of Oddie's objection (ICR, p. 254), as Schurz rightly remarks, but in the light of the foregoing I do not agree with his final statement "Still, it remains to be somewhat disappointing that a purely qualitative problem like the above one cannot be handled within the Kuipers' basic approach." (p. 241) Schurz is here referring to cases of incompatibility due to non-referring terms, to which I now turn.

Schurz seems to think that the incompatibility of a theory with the nomic truth may straightforwardly be due to the fact that terms may not refer. This, however, is not the case, and it is interesting to explain why. In my treatment of reference in Chapter 9, I make a sharp distinction between reference claims and other claims. To be honest, I consider my definition of reference to be one of the main innovations in ICR relative to my earlier publications. A term is referring iff it has effect on the set of nomic possibilities. In other words, it is a pure fiction when it does not narrow down the set of nomic possibilities. Moreover, a theory is supposed to make a reference claim for a certain term, if its addition to the remaining vocabulary implies a genuine restriction of the set of possibilities allowed by the theory, that is, if not all logical possible extensions of the relevant partial models of the theory become models of the theory. The result is that, for example, the phlogiston theory is compatible with the nomic truth as soon as its observational projection is compatible with it. Of course, we may assume that the phlogiston theory has general observational implications that are incompatible with the observational nomic truth, e.g. net weight decrease of burning material. Hence, the phlogiston theory is incompatible with the nomic truth, not because the term phlogiston does not refer as such, but because of its observational consequences.[2] As we have seen, it remains the case that we can come closer to the (observational and the theoretical) truth by simply strengthening the phlogiston theory, but this is not due to its non-referring nature, but to its false general test implications: in this case we do not eliminate models with non-referring terms, but models with wrong observational features.

I now turn to the second type of counterexample given by Schurz. He is again right in claiming that a new theory (e.g. that of Einstein) that is incompatible with an old one (e.g. that of Newton) can only be naïvely closer to the nomic truth when it entails the (relevant) nomic truth. However, in the "toy world" of naïve truthlikeness (ICR, p. 245) you need to add only correct models if you want to go further than rejecting (all) incorrect models. But the

---

[2] The crucial technical point is that reference claims essentially become meta-claims in my account. Hence, the claim "phlogiston exists" is not directly construed as a part of the phlogiston theory, but that theory, in contrast to the nomic truth, rules out certain states of affairs in such a way that it implies the reference claim with respect to 'phlogiston'. For details of this approach I have to refer to ICR Sections 9.1.2 and 9.2.1.

Newton-Einstein transition certainly is beyond the toy world. The transition is a typical case of idealization and concretization. More specifically, at the end of the refined approach in Chapter 10, 10.4.1, it is argued that the "Double Concretization theorem" holds, according to which a concretization of some idealized point of departure is a refined approximation of (itself and) any further concretization. Hence, if Einstein's theory is a concretization of Newton's theory and if the nomic truth is equivalent to – or a further concretization of – Einstein's theory, then Einstein's theory is closer to the nomic truth than Newton's theory.

For similar reasons, Schurz's third, artificial (*F, G, H*) example plausibly fails to satisfy the naïve condition, but it satisfies the refined condition, for a given domain of objects, all models satisfying *Y* are between models satisfying *X* and *N* and for each combination of (comparable) models in *X* and *N* there is an intermediate in *Y*.

In the criticism of stratified truthlikeness that follows, Schurz mainly contradicts my claim that a certain condition, "relative correctness," is weak. Unfortunately, he does not argue against my three technical general reasons (ICR, pp. 212-3), but his general type of example, viz. "if a scientific theory contains theoretically mistaken core assumptions, then all of its theoretical models will be infected by these mistakes," suggests that he may again have been thinking of non-referring terms. However, on p. 213 I announce and on p. 217 I explain why the condition of relative correctness is even trivial as far as non-referring terms are concerned. This has everything to do with my treating such terms as genuine fictions, that is, terms that do not play any role in shaping the set of nomic possibilities. To be honest, on pp. 212-3, I call a relatively correct theory a theory that is on the right track as far as the theoretical vocabulary (*Vt–Vo*) is concerned. This is a defensible way of speaking as far as referring (theoretical) terms are concerned, but it is of course highly misleading as far non-referring terms are concerned. The better paraphrase in this case would be a negative one, viz., restricting ourselves to a theory with only non-referring theoretical terms; such a theory is relatively correct in the sense that its theoretical terms as such, that is, neglecting the specific claim of the theory, do not exclude nomic observational possibilities.

In Section 8 Schurz finally introduces a quasi-quantitative counterexample to my refined definition of truthlikeness. The counterexample illustrates that the refined definition is, like the basic one, purely qualitative on the level of theories. Hence, further refinements may be desirable. However, the refined one already leaves room for quantitative specifications of the underlying notion of "structurelikeness." For this reason it opens the way to dealing with "truth approximation by concretization," e.g. the Newton-Einstein transition mentioned above.

     Theo A. F. Kuipers

In sum, apart from the last one, all purported counterexamples discussed by Schurz are either mistakenly construed or misinterpreted by calling them counterexamples. Hence, as in the case of HD confirmation, Schurz's arguments do not suggest looking for sophistication of truthlikeness in the direction of the relevant element approach. To be sure, he does not claim that the latter can handle the only counterexample that I find asking for further sophistication. Moreover, although I would now, unlike in the case of confirmation, not claim that my refinements are technically simple, they certainly appeal to procedures used in scientific practice, e.g. idealization and concretization, whereas I do not know of a similar appeal to relevant elements by scientists.

There remains one final remark. This is not the proper place to do full justice to the relevant element approach to confirmation and truthlikeness. Although I expressed my doubts about its need and its relation to scientific practice, Gerhard Schurz makes quite clear in the two sections on that approach that it can be developed in a technically elegant way.

REFERENCE

Kuipers, T.A.F. (forthcoming). Inductive Aspects of Confirmation, Information, and Content. Forthcoming in the "Schilpp volume" dedicated to Jaakko Hintikka.

EMPIRICAL PROGRESS BY ABDUCTION AND INDUCTION

Atocha Aliseda

LACUNAE, EMPIRICAL PROGRESS AND SEMANTIC TABLEAUX

ABSTRACT. In this paper I address the question of the dynamics of empirical progress, both in theory evaluation and in theory improvement. I meet the challenge laid down by Theo Kuipers in Kuipers (1999), namely to operationalize the task of "instrumentalist abduction," that is, theory revision aiming at empirical progress. I offer a reformulation of Kuipers' account of empirical progress in the framework of (extended) semantic tableaux and show that this is indeed an appealing method by which to account for some specific kind of empirical progress, that of lacunae.

1. Introduction

Traditional positivist philosophy of science inherits from logical research not only its language, but also its focus on the *truth question*, that is to say, the purpose of using its methods as means for testing hypotheses or formulae. Hempelian models of explanation and confirmation seek to establish the conditions under which a theory (composed by scientific laws) together with initial conditions, explains a certain phenomenon or whether certain evidence confirms a theory. As for logical research, it has been characterized by two approaches, namely the syntactic and the semantic. The former account characterizes the notion of derivability and aims to answer the following question: given theory H (a set of formulae) and formula E, is E derivable from H? The latter characterizes the notion of logical consequence and responds to the following question: is E a logical consequence of H? (Equivalent to: are the models of H models of E?) Through the truth question we can only get a "yes-no" answer with regard to the truth or falsity of a given theory. Aiming solely at this question implies a static view of scientific practice, one in which there is no place for theory evaluation or change. Notions like derivation, logical consequence, confirmation, and refutation are designed for the corroboration – logical or empirical – of theories.

However, a major concern in philosophy of science is also that of theory evaluation. Since the 1930s, both Carnap and Popper stressed the difference between truth and confirmation, where the last notion concerns theory evaluation as well. Issues like the internal coherence of a theory as well as its

In: R. Festa, A. Aliseda and J. Peijnenburg (eds.), *Confirmation, Empirical Progress, and Truth Approximation* (*Poznań Studies in the Philosophy of the Sciences and the Humanities,* vol. 83), pp. 169-189. Amsterdam/New York, NY: Rodopi, 2005.

confirming evidence, refuting anomalies or even its lacunae, are all key considerations to evaluate a scientific theory with respect to itself and existing others. While there is no agreement about the precise characterization of theory evaluation, and in general about what progress in science amounts to, it is clear that these notions go much further than the truth question. The responses that are searched for are answers to the *success question,* which includes the set of successes, failures and lacunae of a certain theory. This is interesting and useful beyond testing and evaluation purposes, since failures and lacunae indicate the problems of a theory, issues that, if solved, would result in an improvement of a theory to give a better one. Thus, given the success question, the *improvement question* follows from it.

For the purposes of this paper, we take as our basis a concrete proposal which aims at modeling the three questions about empirical progress set so far, namely the questions of truth, success and improvement, the approach originally presented by Theo Kuipers in Kuipers (2000) and in Kuipers (1999). In particular, we undertake the challenge of the latter publication, namely to operationalize the task of theory revision aiming at empirical progress, that is, the task of *instrumentalist abduction.* Our proposal aims at showing that evaluation and improvement of a theory can be modeled by (an extension of) the framework of semantic tableaux. In particular, we are more interested in providing a formal characterization of *lacunae,* namely those phenomena which a scientific theory cannot explain, but that are consistent with it. This kind of evidence completes the picture of successes and failures of a theory Kuipers (1999). The terms 'neutral instance' and 'neutral result' are also used Kuipers (2000), for the cases in which a(n) (individual) fact is compatible with the theory, but not derivable from it. This notion appears in other places in the literature. For example, Laudan (1977) speaks of them as "non-refuting anomalies," to distinguish them from the real "refuting anomalies."[1] These kind of phenomena put in evidence, not the falsity of the theory (as it is the case with anomalies), but its *incompleteness* (Laudan 1977), its incapability of solving problems which it should be able to do.

In contemporary philosophy of science there is no place for *complete theories,* if we take a 'complete scientific theory' to mean that it has the capacity to give explanations – negative or positive – to all phenomena within its area of

---

[1] We also find these kinds of statements in epistemology. Gärdenfors (1988) characterizes three epistemic states of an agent with respect to a statement as follows: (i) acceptance, (ii) rejection, and (iii) undetermination, the latter one corresponding to the epistemic state of an agent in which she neither accepts nor rejects a belief, but simply has no opinion about it. As for myself (Aliseda 1997), I have characterized "novelty abduction" as the epistemic operation that is triggered when faced with a "surprising fact" (which is equivalent to the state of undetermination when the theory is closed under logical consequence). It amounts to the extension of the theory into another one that is able to explain the surprise.

competence.[2] Besides, in view of the recognition of the need for initial conditions to account for phenomena, the incompleteness of theories is just something implicit. Therefore, we should accept the property of being "incomplete" as a virtue of scientific theories, rather than a drawback as in the case of mathematics. A scientific theory will always encounter new phenomena that need to be accounted for, for which the theory in question is insufficient. However, this is not to say that scientists or philosophers of science give up when they are faced with a particular "undecidable phenomenon."

Our suggestion is that the presence of lacunae marks a condition in the direction of progress of a theory, suggesting an extension (or even a revision) of the theory in order to "decide" about a certain phenomenon which has not yet been explained, thus generating a "better" theory, one which solves a problem that the original one does not. Therefore, lacunae not only play a role in the evaluation of a theory, but also in its design and generation.

As for the formal framework for modeling the success and the improvement questions with regard to lacunae, classical methods in logic are no use. While lacunae correspond to "undecidable statements",[3] classical logic does not provide a method for modifyng a logical theory in order to "resolve" undecidable statements. However, extensions to classical methods may provide suitable frameworks for representing theory change. We propose an extension to the method of semantic tableaux, originally designed as a refutation procedure, but used here beyond its purpose in theorem proving and model checking, as a way to extend a theory into a new one which entails the lacunae of the original one. This is a well-motivated standard logical framework, but over these structures, different search strategies can model empirical progress in science.

Section 2 describes Kuipers' empirical progress framework, namely the characterization of evidence types, the account of theory evaluation and comparison, as well as the challenge of the instrumentalist abduction task. Section 3 presents our approach to empirical progress in the (extended) semantic tableaux framework. We start by showing the fundamentals of our method through several examples and then propose a formal characterization

---

[2] It is clear that a theory in physics gives no account to problems in molecular biology. Explanation here is only within the field of each theory. Moreover, in our interpretation we are neglecting all well-known problems that the identification between the notions of explanation and derivability brings about, but this understanding of a "complete scientific theory" highlights the role of lacunae, as we will see later on.

[3] In 1931 Gödel showed that if we want a theory free of contradictions, then there will always be assertions we cannot decide, as we are tied to incompleteness. We may consider undecidable statements as extreme cases of lacunae, since Gödel did not limit himself to show that the theory of arithmetic is incomplete, but in fact that it is "incompletable," by devising a method to generate undecidable statements.

of lacuna evidence type. Section 4 offers our conclusions and section 5 is an appendix that describes formally, though briefly, the classical semantic tableaux method together with the extension we propose.

2. Kuipers' Empirical Progress

Two basic notions are presented by this approach, which characterize the relationship between scientific theories and evidence, namely confirmation and falsification. To this end, Kuipers proposes the *Conditional Deductive Confirmation Matrix*. For the purposes of this paper, we are using a simplified and slightly modified version of Kuipers' matrix. On the one hand, we give the characterization by evidence type, that is, successes and failure, rather than by the notions of confirmation and falsification, as he does. Moreover, we omit two other notions he characterizes, namely (dis)confirmation and verification. On the other hand, we do not take account of the distinction between general successes and individual problems (failures), something that is relevant for his further characterization of theory improvement. Our version is useful since it gives lacunae a special place in the characterization, on a par with successes and failures, and it also permits a quite simple representation in our extended tableaux framework.

2.1. *Success and Failure*

SUCCESS

Evidence E is a success of a theory H relative to an initial condition C whenever:

$$H, C \vDash E$$

In this case, *E confirms H* relative to C.

FAILURE

Evidence E is a failure of a theory H relative to an initial condition C whenever:

$$H, C \vDash \neg E$$

In this case *E falsifies H* relative to C.

This characterization has the classical notion of logical consequence ($\models$) as the underlying logical relationship between a theory and its evidence.[4] Therefore, successes of a theory (together with initial conditions) are its logical consequences and failures are those formulae for which their negations are logical consequences. Additional logical assumptions have to be made. H and C must be logically independent, C and E must be true formulae and neither C nor H give an account of E for the case of confirmation (or of $\neg E$ for falsification) by themselves. That is:

1. Logical Independence:
    $H \not\models C$
    $H \not\models \neg C$
    $\neg H \not\models C$
    $\neg H \not\models \neg C$

2. C and E are true formulae.

3. Assumptions for Conditional Confirmation:
    $H \not\models E$
    $C \not\models E$

4. Assumptions for Conditional Falsification:
    $H \not\models \neg E$
    $C \not\models \neg E$

Notice that the requirement of logical independence assures the consistency between H and C, a requirement that has sometimes been overlooked (especially by Hempel), but it is clearly necessary when the logical relationship is that of classical logical entailment. The assumptions of conditional confirmation and conditional falsification have the additional effect of preventing evidence from being regarded as an initial condition ($H, E \models E$).[5]

2.2. Lacunae

Previous cases of success and failure do not exhaust all possibilities there are in the relationship amongst H, C and E ($\neg E$). It may very well be the case that given theory H, evidence E and all previous assumptions, there is no initial condition C available to account for E ($\neg E$) as a case of success (or failure). In this case we are faced with a *lacuna* of the theory. More precisely:

---

[4] But this need not to be so. Several other notions of semantic consequence (preferential, dynamic) or of derivability (default, etc.) may capture evidence type within other logical systems (cf. Aliseda (1997) for a logical characterization of abduction within several notions of consequence).

[5] These additional assumptions are of course implicit for the case in which E ($\neg E$) is a singular fact (a literal, cf. footnote 8 in appendix), since a set of universal statements H cannot by itself entail singular formulae, but we do not want to restrict our analysis to this case, as evidence in conditional form may also be of interest.

LACUNA

Evidence E is a lacuna of a theory H when the following conditions hold for all available initial conditions C:

$H, C \not\models E$

$H, C \not\models \neg E$

In this case, E neither confirms nor falsifies H.

It is clear that the only case in which a theory has no lacuna is given when it is *complete*. But as we have seen, there is really no complete scientific theory. To find lacunae in a theory suggests a condition (at least from a logical point of view) in the direction of theory improvement. That is, in order to improve a theory (H_1) to give a better one (H_2), we may extend H_1 in such a way that its lacunae become successes of H_2. The existence of lacunae in a theory confront us with its holes, and their resolution into successes indicates progress in the theory. By so doing, we are *completing* the original theory (at least with respect to certain evidence) and thus constructing a better one, which may include additional laws or initial conditions.

The above characterization in terms of successes, failures and lacunae makes it clear that the status of certain evidence is always with respect to some theory and specific initial condition. Therefore, in principle a theory gives no account of its evidence as successes or failures, but only with respect to one (or more) initial condition(s). Lacunae, on the other hand, show that the theory is not sufficient, that there may not be enough laws to give an account of evidence as a case of success or failure. But then a question arises: does the characterization of lacunae need a reference to the non-existence of appropriate initial conditions? In other words, is it possible to characterize lacuna-type evidence in terms of some condition between H and E alone? In section 3 we prove that it is possible to do so in our framework.

2.3. *Theory Evaluation and Comparison*

Kuipers argues for a "Context of Evaluation," as a more appropriate way to refer to the so-called "Context of Justification":

> Unfortunately, the term 'Context of Justification', whether or not specified in a falsificationist way, suggests, like the terms 'confirmation' and 'corroboration', that the truth or falsity of a theory is the sole interest. Our analysis of the HD-method makes it clear that it would be much more adequate to speak of the '*Context of Evaluation*'. The term 'evaluation' would refer, in the first place, to the separate and comparative HD-evaluation of theories in terms of successes and problems. (Kuipers 2000, p. 132)

Kuipers proposes to extend Hempel's HD methodology to account not only for testing a theory, which only gives an answer to the "truth-question," but also for evaluation purposes, in which case it allows one to answer the "success-question," in order to evaluate a theory itself, thus setting the ground for comparing it with others. This approach shows that a further treatment of a theory may be possible after its falsification, it also allows a record to be made of the lacunae of a theory and by so doing it leaves open the possibility of improving it. As Kuipers rightly states, HD-testing leads to successes or problems (failures) of a certain theory "and not to neutral results" (Kuipers 2000, p. 101).

There are, however, several models for HD-evaluation, including the *asymmetric* and the *symmetric* ones (each one involving a micro or a macro argument). For the purposes of this paper we present a simplified version (in which we do not distinguish between individual and general facts) of the symmetric definition, represented by the following *comparative evaluation matrix* (Kuipers 2000, p. 117):

$H_1 \setminus H_2$	Failures	Lacunae	Successes
Failures	0	–	–
Lacunae	+	0	–
Successes	+	+	0

Besides neutral results (0), there are three types of results which are favorable (+) for H_2 relative to H_1, and three types of results which are unfavorable (–) for H_2 relative to H_1.

This characterization results in the following criterion for theory comparison, which allows one to characterize the conditions under which a theory is more successful than another: "Theory H_2 is more successful than theory H_1 if there are, besides neutral results, some favorable results for H_2 and no unfavorable results for H_2." Of course, it is not a question of counting the successes (or failures) of a theory and comparing it with another one, rather it is a matter of inclusion, which naturally leads to the following characterization: (cf. Kuipers 2000, p.112)

Theory Comparison

A theory H_2 is (at time t) as successful as (more successful than) theory H_1 if and only if (at time t):

1. The set of failures in H_2 is a subset of the set of failures in H_1.
2. The set of successes in H_1 is a subset of the set of successes in H_2.
3. At least in one of the above cases the relevant subset is proper.

This model puts forward a methodology in science which is formal in its representation, and it does not limit itself to the truth question. Nevertheless, it is still a static representation of scientific practice. The characterization of a theory into its successes, failures and lacunae is given in terms of the logical conditions that the triple H, C, E should observe. It does not specify, for example, how a lacuna of a theory is identified, that is, how it is possible to determine that there are no initial conditions to give an account of a given evidence as a case of success or failure. Also missing is an explicit way to improve a theory; that is, a way to revise a theory into a better one. We only get the conditions under which two – already given and evaluated – theories are compared with respect to their successes and failures. Kuipers is well aware of the need to introduce the issues of theory revision, as presented in what follows.

2.4. *Instrumentalist Abduction*

In a recent proposal (Kuipers 1999), Kuipers uses the above machinery to define the task of *instrumentalist abduction,* namely theory revision aiming at empirical progress. The point of departure is the *evaluation report* of a theory (its successes, failures and lacunae), and the task consists of the following:

Instrumentalist Abduction Task

Search for a revision H_2 of a theory H_1 such that:

H_2 is more successful than H_1, relative to the available data. [6]

Kuipers sets an invitation for abduction aiming at empirical progress, namely the design of instrumentalist abduction along symmetric or asymmetric lines. More specific, he proposes the following challenge: "*to operationalize in tableau terms the possibility that one theory is 'more compatible' with the data than another*" (Kuipers 1999, p. 320). In this paper we take up this challenge in the following way. We propose to operationalize the above instrumentalist task for the particular case of successes and lacunae (his Task I). We cast the proposal in our extended framework of semantic tableaux (cf. Appendix).

---

[6] In the cited paper this is defined as the first part of Task III (Task III.l). Task III.2 requires a further evaluation task and consists of the following: "H_2 remains more successful than H_1, relative to all future data." Moreover, there are two previous tasks: Task I, Task II, which are special cases of Task III, the former amounts to the case of a "surprising observation" (a potential success or a lacuna) and the task is to expand the theory with some hypothesis (or initial condition) such that the observation becomes a success. This is called novelty guided abduction. The latter amounts to the case of an "anomalous observation" (a failure), and the task is to revise the theory into another one which is able to entail the observation. This one is called anomaly guided abduction. In my own terminology of abduction (Aliseda 1997) I refer to the former as abductive novelty and to the latter as abductive anomaly (cf. footnote 1).

It will be clear that while our proposal remains basically the same as that of Kuipers, our representation allows for a more dynamic view of theory evaluation and theory improvement by providing ways to compute appropriate initial conditions for potential successes, and for determining the conditions under which an observation is a case of lacuna, both leading to procedures for theory improvement. However, this paper does not give all the details of these processes, it only provides the general ideas and necessary proofs.

3. Empirical Progress in (Abductive) Semantic Tableaux

In what follows we show, via a toy example, the fundamentals of our approach, which involves an extension of the tableaux method (cf. Appendix) and provides a proposal for representing Kuipers' account.[7]

3.1. *Examples*

Let $H_1 = \{a \rightarrow b\}$, $E = \{b\}$
A tableau for H_1 is as follows:

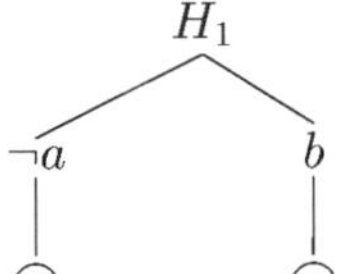

The result is an open tableau. When a formula is added (thereby extending the tableau), some of these possible models may disappear, as branches start closing. For instance, when $\neg b$ is added, the result is the following tableau:

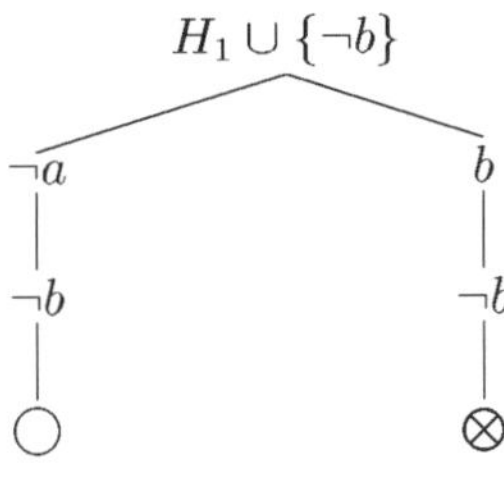

---

[7] Our approach was developed initially for a different purpose, that of implementing abduction in semantic tableaux. We omit here the precise algorithms which produce initial conditions and transform a theory into a better one. These are found in Aliseda (1997). A description of its implementation is found in Vázquez and Aliseda (1999).

Notice that, although the resulting theory remains consistent, one of the two branches has closed. In particular, the previous model is no longer present, as b is no longer true. There is still an open branch though, indicating that there is a model satisfying $H_1 \cup \{\neg b\}$ (a, b false), which indicates that b is not a logical consequence of H_1 ($H_1 \not\models b$).

An attractive feature of the tableaux method is that when E is not a valid consequence *of H_1*, we get all cases in which the consequence fails, graphically represented by the open branches (as shown above, the latter may be viewed as descriptions of models for $H_1 \cup \{\neg b\}$).

This fact suggests that if this *counterexample* were "corrected by amending the theory," through adding more premises, we could perhaps make b a valid consequence of some (minimally but not trivially) extended theory H_1'. For this particular case, we may extend $H_1 \cup \{\neg b\}$ with $\{a\}$ resulting in the following closed tableau:

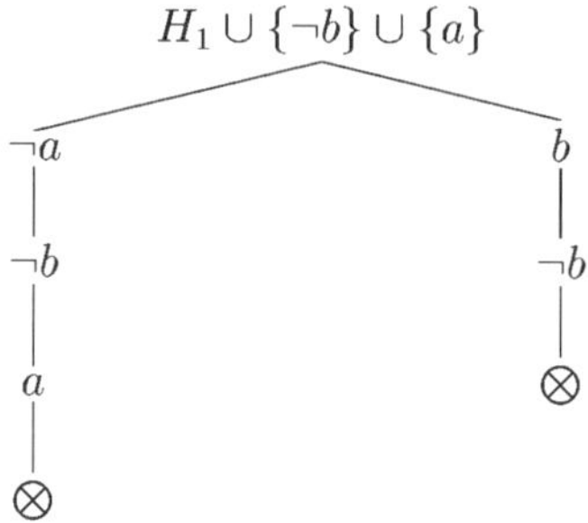

This example shows the basic idea of the process we want to characterize, namely, a mechanism to effect an appropriate *expansion* of a particular tableau (in the original method, once a theory has been represented, no additional expansion or retraction on the tree is performed).

Therefore, in this logical framework, expanding a theory consists of extending the open branches with a formula that closes them (although this may be done carefully by extending with consistent and non-trivial formulae). In our example, H_1 does not account for b as its logical consequence, but $H_1 \cup \{a\}$ does, as it closes the previous open branch.

The mechanism just sketched "implements" Kuipers' characterization of success, showing H_1 as the theory and b as its confirming evidence (a success), with respect to an initial condition a. We note that, viewed within this framework, we may indeed produce the appropriate initial condition a, and not just state the logical conditions that hold, for which evidence b is regarded as a success of the theory H_1. That is, with this framework we suggest a way to compute those initial conditions which, added to a theory, give an account of certain evidence as a case of success.

In a similar way, the tableau above shows that evidence $\neg b$ is a failure of theory H_1 with respect to initial condition a, since its negation b is indeed a success with respect to the same initial condition.

Lacuna

Next, let me show a lacuna-type evidence for this theory.

Let $H_1 = \{a \rightarrow b\}, E = \{c\}$

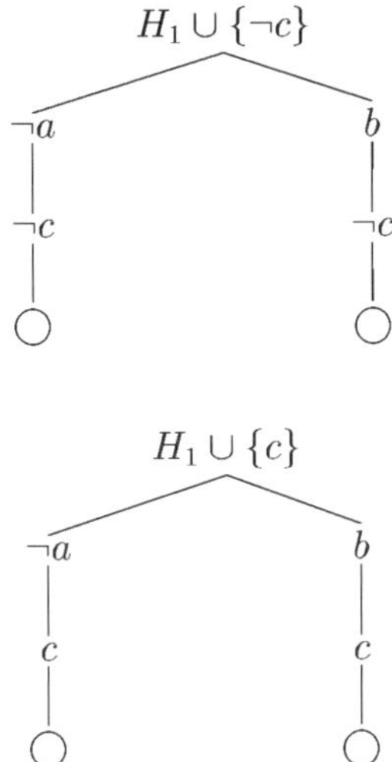

In this case, neither evidence (c) nor its negation ($\neg c$) are valid consequences of the theory H_1, as shown by the two open tableaux above. Moreover, it is not possible to produce any (literal) formula acting as initial condition in order to make the evidence a case of success or failure (unless the trivial case is considered, that is, to add c or $\neg c$). We are then faced with a genuine case of a lacuna, for which the theory holds no opinion about the given evidence, either affirmative or negative: it simply cannot account for it.

Notice that in the previous case of confirming evidence b, its negation $\neg b$ was not a valid consequence of the theory, but as we shall see in the next section, the fact that $\neg b$ closed some branch of the tableau, and in this case $\neg c$ closes no branch, plays an important role in the distinction between successes and lacunae.

In the present case, there is no appropriate initial condition that makes c a success of the theory. For this particular example it is perhaps easy to be aware of this fact (a and $\neg b$ are the only possibilities and neither one closes the whole tableau), but the central question to tackle is, in general, how can we

know when a certain formula qualifies as a case of lacuna with respect to a
certain theory?

As we shall see in a precise way in the next section, we characterize formulae as lacunae of a theory whenever neither one of them nor their negations
close any open branch when the tableau is extended with them. As we have
pointed out, this kind of evidence not only shows that the theory is insufficient
to account for it, but also suggests a condition for theory improvement, thus
indicating the possibility of empirical progress leading to a procedure to
perform the instrumentalist task (cf. Kuipers 1999).

Theory Improvement

In what follows we give an example of the extension of theory H_1 into a theory
H_2, differing in that formula c is a lacuna in the former and a success in the
latter. One way to expand the tableau is by adding the formula $a \rightarrow c$ to H_1
obtaining H_2 as follows:

Let $H_2 = \{a \rightarrow b, a \rightarrow c\}, E = \{c\}$

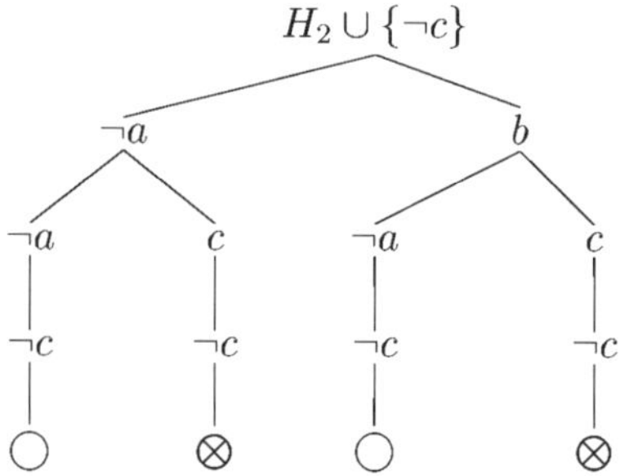

This tableau has still two open branches, but the addition of a new formula
to H_1 (resulting in H_2), converts evidence c into a candidate for success or
failure, since it has closed an open branch (in fact two), and thus its extension
qualifies as *semi-closed* (to be defined precisely in the next section). It is now
possible to calculate an appropriate initial condition to make c a success in H_2,
namely a. Thus, c is a success of H_2 with respect to a.

But H_1 may be extended by means of other formulae, possibly producing
even *better* theories than H_2. For instance, by extending the original tableau
instead with $b \rightarrow c$ the following is obtained:

Let $H_3 = \{a \rightarrow b, b \rightarrow c\}, E = \{c\}$

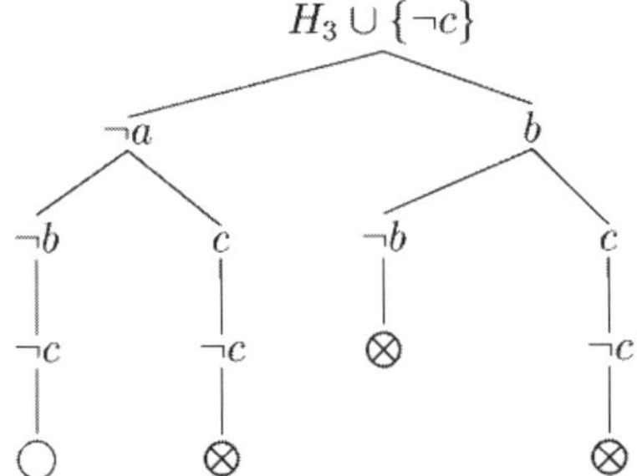

As before, c is a success with respect to initial condition a, but in addition, c is also a success with respect to initial condition b. Formula a as well as b close the open branch in a non-trivial way. Therefore, in H_3 we gained one success over H_2. Shall we then consider H_3 a better theory than both H_1 and H_2?

According to Kuipers' criterion for theory comparison (cf. section 2.3), both H_2 and H_3 are more successful theories than H_1, since the subset of successes in H_1 is a proper subset of both the set of successes in H_2 and of H_3. Moreover, H_3 is more successful than H_2, since in the former case c is a success with respect to two initial conditions and only with respect to one in the latter case.

We leave here our informal presentation of the extended method of semantic tableaux applied to empirical progress leading to the instrumentalist task, as conceived by Kuipers.

3.2 Lacuna Evidence Type Characterization

In what follows we aim to characterize lacuna evidence in terms of the extension type it effects over a certain tableau for a theory when it is expanded with it. The main motivation for this characterization is to answer the question raised in the previous section: can we know in general when a certain formula E qualifies as a case of lacuna with respect to a theory H?

Let me begin by giving a translation of lacuna evidence (cf. section 2.2) into tableaux, denoted here as lacuna* in order to distinguish it from Kuipers' original usage:

LACUNA*

Given a tableau $T(H)$ for a theory, evidence E is a lacuna* of a theory H, when the following conditions hold for all available initial conditions C:

$T(H) \cup \{\neg E\} \cup \{C\}$ is an open tableau
$T(H) \cup \{E\} \cup \{C\}$ is an open tableau

In this case, neither E confirms nor falsifies H.

Although this is a straightforward translation, it does not capture the precise type of extension that evidence E (and its negation) effects on the tableau. To this end, we propose a distinction between "open," "closed" and "semi-closed" extensions. While the first two are characterized by those formulae which when added to the tableau do not close any open branch, or close all of them respectively, the last one are exemplified by those formulae which close some but not all of the open branches. (For the formal details of the extension characterization in terms of semantic tableaux see the Appendix).

As we are about to prove that, when a formula E (and its negation) effects an open extension over a tableau for a certain theory H, it qualifies as a lacuna*. More precisely:

LACUNA* CLAIM

Given a theory H and evidence E:

IF $T(H) + \{\neg E\}$ is an open extension and $T(H) + \{E\}$ is also an open extension,

THEN E is a lacuna* of H.

Proof

(i) To be proved: $T(H) \cup \{\neg E\} \cup \{C\}$ is an open tableau for all C.

(ii) To be proved: $T(H) \cup \{E\} \cup \{C\}$ is an open tableau for all C.

- Let $T(H) + \{\neg E\}$ be an open extension ($H \nvDash E$).
  Suppose there is a C ($C \neq E$ is warranted by the second assumption of conditional confirmation), such that $T(H) + \{\neg E\} + \{C\}$ is a closed extension.
  Then $T(H) + \{C\}$ must be a closed extension, that is, $H \nvDash \neg C$, but this contradicts the second assumption for logical independence.
- Therefore,
  $T(H) + \{\neg E\} + \{C\}$ is not a closed extension, that is, $H, C \nvDash E$, thus concluding that $T(H) \cup \{\neg E\} \cup \{C\}$ is an open tableau.
- **(ii)** Let $T(H) + \{E\}$ be an open extension ($H \nvDash \neg E$).
- Suppose there is a C ($C \nvDash \neg E$ is warranted by the second assumption of conditional falsification), such that $T(H) + \{E\} + \{C\}$ is a closed extension.
  Then $T(H) + \{C\}$ must be a closed extension, that is, $H \nvDash \neg C$, but this contradicts the second assumption of logical independence.
- Therefore,
  $T(H) + \{E\} + \{C\}$ is not a closed extension, that is, $H, C \nvDash \neg E$, thus concluding that $T(H) \cup \{E\} \cup \{C\}$ is an open tableau.

Notice that in proving the above claim we used two assumptions that have not been stated explicitly for lacunae; one of conditional confirmation in (i) $(C \nvDash E)$ and the other one of conditional falsification in (ii) $(C \nvDash \neg E)$. Therefore, we claim these assumptions should also be included in Kuipers' additional logical assumptions as restrictions on lacunae. This fact gives us one more reason to highlight the importance of lacunae as an independent and important case for theory evaluation.

Finally, our characterization of lacuna** evidence type in terms of our proposed tableaux extensions is as follows:

LACUNA**

Given a tableau for a theory $T(H)$, evidence E is a lacuna** of a theory H whenever:

$T(H) + \{\neg E\}$ is an open extension

$T(H) + \{E\}$ is an open extension

In this case, E neither confirms nor falsifies H.

This account allows one to characterize lacuna type evidence in terms of theory and evidence alone, without having to consider any potential initial conditions. This result is therefore attractive from a computational perspective, since it prevents the search of appropriate initial conditions when there are none.

The proof of the above lacuna claim assures that our characterization implies that of Kuipers, that is, open extensions imply open tableaux (lacuna** implies lacuna*), showing that our characterization of lacuna** is stronger than the original one of lacuna*. However, the reverse implication (lacuna* implies lacuna**) is not valid. An open tableau for $T(H) \cup \{\neg E\} \cup \{C\}$ need not imply that $T(H) + \{\neg E\}$ is an open extension. Here is a counterexample:

Let $H = \{a \rightarrow b, a \rightarrow d, b \rightarrow c\}$

$E = \{b\}$ and $C = \{\neg d\}$

Notice that $T(H) \cup \{\neg b\} \cup \{\neg d\}$ is an open tableau, but both $T(H) + \{\neg b\}$ and $T(H) + \{\neg d\}$ are semi-closed extensions.

3.3. *Success and Failure Characterization?*

The reader may wonder whether it is also possible to characterize successes and failures in terms of their extension types. Our focus in this paper has been on lacunae, so we limit ourselves to presenting the results we have found so far, omitting their proofs:

Success IF E is a Success of H THEN

> $T(H) + \{\neg E\}$ is a semi-closed extension and there is an initial condition available C for which $T(H) + \{C\}$ is a semi-closed extension.

Failure IF E is a failure of H THEN

> $T(H) + \{E\}$ is a semi-closed extension and there is an initial condition C available for which $T(H) + \{C\}$ is a semi-closed extension.

For these cases it turns out that the corresponding valid implication is the reverse of the one with respect to that of lacunae. This means that closed tableaux imply semi-closed extensions, but not the other way around. In fact, the above counterexample is also useful to illustrate this failure: both $T(H) + \{\neg b\}$ and $T(H) + \{\neg d\}$ are semi-closed extensions, but E is not a success of H, since $T(H) \cup \{\neg b\} \cup \{\neg d\}$ is an open tableau.

4. Discussion and Conclusions

The main goal of this paper has been to address Kuipers' challenge in Kuipers (1999), namely to operationalize the task of *instrumentalist abduction*. In particular, the central question raised here concerns the role of lacunae in the dynamics of empirical progress, both in theory evaluation and in theory improvement. That is, the relevance of lacunae for the success and the improvement questions.

Kuipers' approach to empirical progress does take account of lacunae for theory improvement, by including them in the evaluation report of a theory (together with its successes and failures), but it fails to provide a full characterization of them, in the same way it does for successes and failures (the former via the notion of confirmation, and the latter via the notion of falsification). Moreover, this approach does not specify precisely how a lacuna of a theory is recognized, that is, how it is possible to identify the absence of initial conditions which otherwise give an account of certain evidence as a case of success or failure.

Our reformulation of Kuipers' account of empirical progress in the framework of (extended) semantic tableaux is not just a matter of a translation into a mathematical framework. There are at least two reasons for this claim. On the one hand, with tableaux it is possible to determine the conditions under which a phenomenon is a case of lacuna for a certain theory, and on the other hand, this characterization leads to a procedure for theory improvement. Regarding the first issue, we characterize formulae as lacunae of a theory whenever neither one of them nor their negations close any open branch (when the tableau is extended with them), they are *open extensions* in our terminology, and we prove that our formal characterization of lacuna implies that of Kuipers. As for the latter assertion, lacunae type evidences not only show that the theory is insufficient to account for them, but also suggests a condition for theory improvement, thus indicating possibilities of empirical progress leading to a procedure for performing the instrumentalist task.

We have suggested that a fresh look into classical methods in logic may be used not only to represent the corroboration of theories with respect to evidence, that is, to answer the truth question, but also to evaluate theories with respect to their evidence as well as to generate better theories than the original ones. To this end, we have presented (though informally) the initial steps to answer the success and improvement questions, showing that given a theory and some evidence, appropriate initial conditions may be computed in order to make that evidence a success (failure) with respect to a theory; and when this is not possible, we have proposed a way to extend a theory and thus improve it in order to account for its lacunae.

Universidad Nacional Autónoma de México (UNAM)
Instituto de Investigaciones Filosóficas
Circuito Mario de la Cueva, s/n
México, 04510, D.F.

APPENDIX

1. *Semantic Tableaux: The Classical Method*

The logical framework of Semantic Tableaux is a refutation method introduced independently by Beth and Hintikka in the 1950s. A more modern version is found in Smullyan (1968) and that is the one used here. The main idea of this framework is the construction of a tableau (a tree-like structure) for a theory H (a finite set of formulae) and a formula $\neg E$, which we denote as $\mathcal{T}(H \cup \{\neg E\})$, in order to prove whether $H \nvDash E$. Here is an example, in which $H = \{a \rightarrow b, a\}$ and $E = \{b\}$:

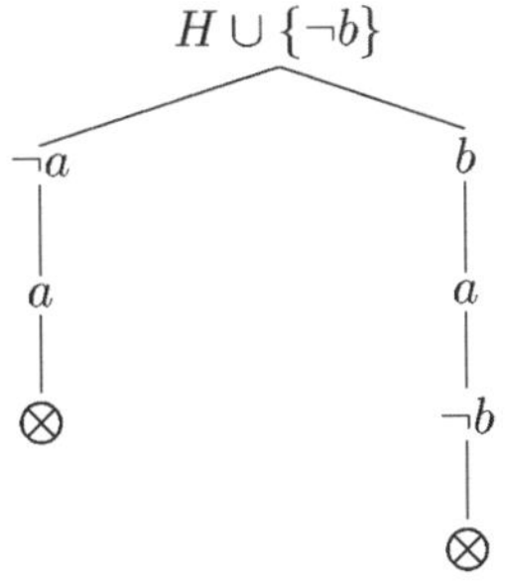

This tableau is closed, since both branches contain a formula and its negation (for clarity, each closed branch ends with the symbol $\otimes$). Therefore, H and $\neg E$ are unsatisfiable, showing that $H \not\models E$ ($\{a \rightarrow b, a\} \not\models \{b\}$).

Otherwise, one or more counterexamples have been constructed and $H \not\models E$ is concluded, as the following tableau example illustrates ($H = \{a \rightarrow b, a\}$ and $E = \{c\}$):

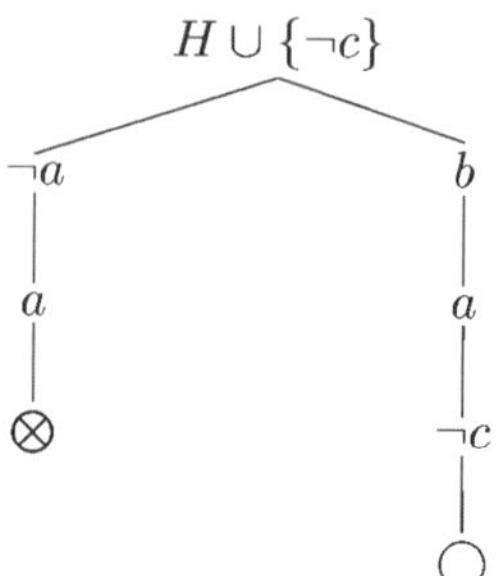

An open tableau indicates that the theory H is consistent and each open branch (which ends with the symbol d) corresponds to a verifying model. In this case, the second branch (from left to right) indicates that a model for H is given by making a and b true. Generally speaking, when constructing the tableau, the possible valuations for the formulae are depicted by the branches.

Semantic Tableaux are a *sound and complete* system:

$$H \not\models E \qquad \text{iff} \qquad T(H \cup \{\neg E\}) \text{ is a closed tableau.}$$

Furthermore, they constitute a decision method for propositional logic. This is different from predicate logic, where quantifier rules may lead to unbounded repetitions. In the latter case, the tableaux method is only semi-decidable. (If the initial set of formulae is unsatisfiable the tableau will close in finitely many steps, but if it is satisfiable the tableau may become infinite, without terminating, recording an infinite model.) In this paper, we only consider the propositional case.

Here is a quick reference list of some major notions concerning tableaux.

Closed Branch : A branch of a tableau is closed if it contains some formula and its negation.

Atomically Closed Branch : A branch is atomically closed if it is closed by a literal (an atomic formula or a negation thereof).

Open Branch : A branch of a tableau is open iff it is not closed.

Complete Branch : A branch B of a tableau is complete if for every formula in conjunctive form which occurs in B, both its conjuncts occur in B, and for every formula in disjunctive form, at least one of its disjuncts occurs in B.

Completed Tableau: A tableau T is completed if every branch of T is either closed or complete.

Proof of $H \not\models E$: A proof of $H \not\models E$ is a closed tableau for $H \cup \{\neg E\}$.

2. *Semantic Tableaux: Our Extended Method*

Below we present our proposal to extend the framework of semantic tableaux in order to account for issues of empirical progress in science. We describe a way to represent a tableau for a theory as the set of its branches, and operations to extend and characterize tableau extension types.

Tableau Representation

Given a theory H, we represent its corresponding *completed tableau* $T(H)$ by the set formed by its branches:

$$T(H) = \{\Gamma_1, \ldots, \Gamma_k\} \text{ where each } \Gamma_i \text{ may be open or atomically closed.}$$

Tableaux are widely used in logic, and they have many further interesting properties. These can be established by simple analysis of the rules and their motivation, as providing an exhaustive search for a counter-example. This presentation incurs no loss of generality. Given a completed tableau for a theory H, that is, $T(H)$:

- If $T(H)$ has open branches, H is consistent. Each open branch corresponds to a verifying model.

- If $T(H)$ has all branches closed, H is inconsistent.

Another, more computational feature is that, given some initial verification problem, the order of rule application in a tableau tree does not affect the result. The structure of the tree may be different, but the outcome is the same as to consistency.

Tableau Extension

A tableau is extended with a formula via the usual expansion rules. An extension may modify a tableau in several ways. These depend both on the form of the formula to be added and on the other formulae in the theory represented in the original tableau. If an atomic formula is added, the extended tableau is just like the original with this formula appended at the bottom of its open branches. If the formula has a more complex form, the extended tableau may look quite different (e.g., disjunctions cause every open branch to split into two). In total, however, when expanding a tableau with a formula, the effect on the open branches can be only one of three types: (i) the added formula closes no open branch; or (ii) it closes all open branches; or (iii) it may close some open branches while leaving others open. In order to compute appropriate initial conditions and characterize the lacuna case, we need to clearly distinguish these three ways of extending a tableau. We label them as *open, closed,* and *semi-closed* extensions, respectively. In what follows we define these notions more precisely.

Branch Extension

Given a branch Γ_i the addition of a formula $\{\gamma\}$ to it is defined by the following $+$ operation[8]:

- $\Gamma_i + \{\gamma\} = \Gamma_i \cup \{\gamma\}$ iff Γ_i is open.

- $\Gamma_i + \{\gamma\} = \Gamma_i$ iff Γ_i is closed.

Branch Extension Types

Given a branch Γ_i the addition of a literal formula $\{\gamma\}$ may result in an open or a closed extension:

- $\Gamma_i + \{\gamma\}$ is an open extension iff $\Gamma_i \cup \{\gamma\}$ results in an open branch.

- $\Gamma_i + \{\gamma\}$ is a closed extension iff $\Gamma_i \cup \{\gamma\}$ results in a closed branch.

The operation $+$ is defined over branches, but it easily generalizes to tableaux as follows:

Tableau Extension (with literals)

$$\mathcal{T}(H) + \{\gamma\} = \{\Gamma_i + \{\gamma\} \mid \Gamma_i \in \mathcal{T}(H)\}$$

Next we characterize the tableau extension types as follows:

Open Extension $\mathcal{T}(H) + \{\gamma\}$ is an open tableau extension iff $\Gamma_i + \{\gamma\}$ is an open branch extension for all $\Gamma_i \in \mathcal{T}(H)$.

Closed Extension $\mathcal{T}(H) + \{\gamma\}$ is a closed tableau extension iff $\Gamma_i + \{\gamma\}$ is a closed branch extension for all $\Gamma_i \in \mathcal{T}(H)$.

Semi-Closed Extension $\mathcal{T}(H) + \{\gamma\}$ is a semi-closed tableau extension iff $\Gamma_i + \{\gamma\}$ is an open branch extension and $\Gamma_j + \{\gamma\}$ is a closed branch extension for some i, j with $i \neq j$.

---

[8] For the purposes of this paper we only give the definition of the addition of a *literal*. Cf. Aliseda (1997) for the general definition covering all kinds of formulae.

REFERENCES

Aliseda, A. (1997). *Seeking Explanations: Abduction in Logic, Philosophy of Science and Artificial Intelligence.* Dissertation Stanford. Amsterdam: ILLC Dissertations Series (1997-04).

Gärdenfors, P. (1988). *Knowledge in Flux: Modeling the Dynamics of Epistemic States.* Cambridge: The MIT Press.

Kuipers, T.A.F. (2000/ICR). *From Instrumentalism to Constructive Realism. On Some Relations between Confirmation, Empirical Progress and Truth Approximation. Synthese Library*, vol. 287, Dordrecht: Kluwer.

Kuipers, T.A.F. (1999). Abduction Aiming at Empirical Progress or Even at Truth Approximation, Leading to Challenge for Computational Modelling. In: J. Meheus and T. Nickles (eds.): *Scientific Discovery and Creativity*, special issue of *Foundations of Science* **4** (3), 307-323.

Laudan, L. (1977). *Progress and Its Problems.* Berkeley: University of California Press.

Smullyan, R.M. (1968). *First Order Logic.* New York: Springer-Verlag.

Vázquez, A. and A. Aliseda. (1999). *Abduction in Semantic Tableau: Towards an Object-Oriented Programming Implementation.* In: J.M. Ahuactzin (ed.), Encuentro Nacional de Com putación, Taller de Lógica. Sociedad Mexicana de Ciencia de la Computación. ISBN: 968-6254-46-3. Hidalgo, México.

Theo A. F. Kuipers

THE INSTRUMENTALIST ABDUCTION TASK AND THE NATURE OF EMPIRICAL COUNTEREXAMPLES

REPLY TO ATOCHA ALISEDA

After the 1998 Ghent conference on scientific discovery and creativity, organized by Joke Meheus and Thomas Nickles, Atocha Aliseda was the first to take up my challenge (published in 1999) for computational modeling of instrumentalist abduction, with Joke Meheus following, witness the next paper. I greatly appreciate the clear and convincing way in which Aliseda shows that her general semantic tableau method for abduction can be used to make empirical progress of a special kind: the identification and subsequent resolution of lacunae. An especially nice feature of her method is that it does not favor revisions by adding initial conditions, but also generates proper theory revisions, as her leading example in Section 3 (from H1 to H3) essentially illustrates by adding a conditional statement.

To be honest, I do not rule out that Patrick Maher will be able to find some holes in this paper, in particular in Section 3.2 with the crucial transition from Aliseda's first explication of my rather intuitive notion of lacuna, lacuna*, to her second explication, lacuna**, a transition in which reference to the available initial conditions is removed. However, I am also fairly sure that such problems can either be solved by some further refinement or that they are typically artificial in nature, such that it is difficult, if not impossible, to imagine real-life instantiations of them.

In the rest of this reply I first deal with the remaining task of instrumentalist abduction, in particular the general task of theory revision in the face of remaining counterexamples. For this purpose it seems important to explicate the nature of empirical counterexamples, in particular by comparing them with logical counterexamples, which is my second aim.

Toward the General Instrumentalist Abduction Task

As Aliseda explains in Note 6, her paper is restricted to the first special task of abduction that I discerned in my 1999-paper, viz. "novelty guided abduction."

In: R. Festa, A. Aliseda and J. Peijnenburg (eds.), *Confirmation, Empirical Progress, and Truth Approximation* (*Poznań Studies in the Philosophy of the Sciences and the Humanities,* vol. 83), pp. 190-192. Amsterdam/New York, NY: Rodopi, 2005.

A novelty, that is, a lacuna in the face of the available theory(-cum-initial-conditions) and background knowledge, is transformed into a success of a revised theory. As her paper and Aliseda (1997) make clear, she would be able to deal in a similar way with the other special task, "anomaly guided abduction," that is, transforming a counterexample of a theory (together with the background knowledge), into a success, or at least a lacuna, of a revised theory. However, in the first case it is assumed that the total available evidence reports no counterexample and, in the second case, there is at most one, the target counterexample. However, general instrumentalist abduction will also have to operate in the face of the concession that other counterexamples have to remain in the game, at least for the time being. It is not yet entirely clear to me that this can be done, nor how, with Aliseda's adapted tableau method. This is not to say that I think that it is impossible; I just have no clear view on whether it can be done in a similar way. One reason is that the method will now have to deal with the problem of which evidence is taken into account and which evidence is set aside. Another reason is that, from a tableau perspective, whether of some standard kind or of Aliseda's abductive kind, the notion of counterexamples and logical entailment are intimately related, but it is not at all clear what such (logical) counterexamples have to do with empirical counterexamples, as they occur in the empirical sciences. This is an interesting question independent of whether one assigns dramatic consequences to empirical counterexamples, as Popper is usually supposed to do, or the more modest role they play in my "comparative evaluation matrix" and subsequent theory of truth approximation. Incidentally, the latter theory is abductively related to the matrix (Kuipers 2004).

Logical and Empirical Counterexamples

For the moment I would like to confine myself to the major similarities and differences between logical and empirical counterexamples, and their relation. As is well-known, the standard form of a logical counterexample pertains to an argument with a (set of) premise(s) P and a purported conclusion C. It is a model of P, that is, a relevant structure on which P is true, which is a countermodel of C, that is, a structure on which C is false. Such a type of counterexamples is typically sought by the standard tableau methods. Of course, one may also say that such a model is a countermodel to "If P then C" as a purported logical truth.

An empirical counterexample typically is a counterexample *of a certain empirical theory*, say X. The explication of this is less standard. Certainly, an empirical counterexample of X is a countermodel of X and hence a logical counterexample of X when X is taken as a purported logical truth. However, nobody will normally claim this for an empirical theory. Hence we will have to look more closely at what an empirical theory is or claims. In my favorite "nomic structuralistic approach" (see the Synopsis of ICR), theorizing is trying to grasp the unknown "set of nomic possibilities" that is, uniquely determined, according to the Nomic Postulate, given a domain of reality and a vocabulary. This set may well be called "the (nomic) truth" (see below), indicated by T. Assuming that the vocabulary is a first order language in which (X and) T can be characterized, an empirical counterexample of X must represent a "realized (nomic) possibility," that is, a piece of existing or experimentally realized reality, for realizing an empirical *im*possibility is of course by definition impossible, assuming that no representation mistakes have been made. Hence, an empirical counterexample of a theory X is not only a countermodel of X but also a model of T, and therefore it is a logical counterexample to the claim that T logically entails X, and hence to the claim that "if T then X" is a logical truth. The latter I call the (weak[1]) claim of a theory X: all models of T are models of X or, equivalently, X is true for all nomic possibilities. I call a X true when its claim is true. In this sense, (the statement) T is the (logically) strongest true statement. For this reason, T is called "the truth." In sum, from the nomic structuralist perspective there is a clear relation between empirical and logical counterexamples: an empirical counterexample of a theory X is also a special type of logical counterexample, viz. to the claim that T logically entails X.

REFERENCES

Aliseda, A. (1997). *Seeking Explanations: Abduction in Logic, Philosophy of Science and Artificial Intelligence.* Dissertation Stanford. Amsterdam: ILLC Dissertations Series (1997-04).

Kuipers, T.A.F. (1999). Abduction Aiming at Empirical Progress or Even at Truth Approximation, Leading to Challenge for Computational Modelling. In: J. Meheus and T. Nickles (eds.): *Scientific Discovery and Creativity*, special issue of *Foundations of Science* **4** (3), 307-323.

Kuipers, T.A.F. (2004). Inference to the Best Theory, Rather Than Inference to the Best Explanation. Kinds of Induction and Abduction. In: F. Stadler (ed.), *Induction and Deduction in the Sciences*, pp. 25-51. Dordrecht: Kluwer Academic Publishers.

---

[1] The strong claim is that X and T are equivalent.

Joke Meheus[†]

EMPIRICAL PROGRESS AND AMPLIATIVE ADAPTIVE LOGICS

ABSTRACT. In this paper, I present two ampliative adaptive logics: **LA** and **LA**^{k}. **LA** is an adaptive logic for abduction that enables one to generate explanatory hypotheses from a set of observational statements and a set of background assumptions. **LA**^{k} is based on **LA** and has the peculiar property that it selects those explanatory hypotheses that are empirically most successful. The aim of **LA**^{k} is to capture the notion of empirical progress as studied by Theo Kuipers.

1. Introduction

In his *From Instrumentalism to Constructive Realism*, Kuipers presents a detailed analysis of the notion of empirical progress (the replacement of an old theory by a new and better one), and its relation to confirmation on the one hand and to truth approximation on the other. On Kuipers' account, a theory X should be eliminated in favor of a theory Y if and only if Y has so far proven to be more successful than X. Intuitively, a theory Y is more successful than a theory X if and only if Y saves the empirical successes of X, faces at most the same empirical problems as X, and is better in at least one of these respects.

In order to evaluate the empirical successes and the empirical problems of a theory (or hypothesis), Kuipers relies on the hypothetico-deductive (HD-) method. Central to this method is the derivation of observational statements from the hypotheses under consideration together with a set of background assumptions $\mathcal{B}$. The logic that is used for these derivations is Classical Logic (henceforth **CL**). Thus, according to the HD-method, an observational statement O is an empirical success for a hypothesis H iff $\mathcal{B} \cup \{H\} \vdash_{\mathbf{CL}} O$, and O is true; O is an empirical problem for H iff $\mathcal{B} \cup \{H\} \vdash_{\mathbf{CL}} O$, and $\neg O$ is true.[1] Where $\mathcal{S}$ and $\mathcal{S}'$ refer to the sets of empirical successes of H and H', and $\mathcal{P}$ and $\mathcal{P}'$ to their sets of empirical problems, the notion of "empirically more successful" is defined by

---

[†] Postdoctoral Fellow of the Fund for Scientific Research – Flanders (Belgium).

[1] In Kuipers (2000), Kuipers presents a semantic account of the notion of empirical progress. In view of the purposes of the present paper, I shall rely on a syntactic account.

In: R. Festa, A. Aliseda and J. Peijnenburg (eds.), *Confirmation, Empirical Progress, and Truth Approximation (Poznań Studies in the Philosophy of the Sciences and the Humanities,* vol. 83), pp. 193-217. Amsterdam/Atlanta, New York, NY: Rodopi, 2005.

H' is empirically more successful than H iff
$\mathcal{S}' \supset \mathcal{S}$ and $\mathcal{P}' \subseteq \mathcal{P}$, or
$\mathcal{S}' \supseteq \mathcal{S}$ and $\mathcal{P}' \subset \mathcal{P}$.

Three remarks are important. First, the conclusion that H' is more successful than H is relative to the set of background assumptions $\mathcal{B}$, and the set of observational statements $\mathcal{O}$. For instance, discovering new empirical problems for H' may lead to the withdrawal of the conclusion that H' is more successful than H. Next, according to Kuipers' definition, hypotheses may be incomparable with respect to their empirical success. For instance, if $\mathcal{S} \not\subseteq \mathcal{S}'$, and $\mathcal{S}' \not\subseteq \mathcal{S}$, then H and H' are "equally successful." To avoid confusion, I shall say that a hypothesis H is *maximally successful* with respect to some set $\mathcal{B} \cup \mathcal{O}$ iff no hypothesis H' is more successful with respect to $\mathcal{B} \cup \mathcal{O}$ than H. Finally, to decide that O is an empirical success (respectively empirical problem) for H, one does not need to reason *from H*. Indeed, if $\mathcal{B} \cup \{H\} \vdash_{\mathbf{CL}} O$, then $\mathcal{B} \vdash_{\mathbf{CL}} H \supset O$. Hence, O can be considered as an empirical success for H if $\mathcal{B} \vdash_{\mathbf{CL}} H \supset O$ and O is true.

This last remark suggests a method for identifying the hypotheses that are maximally successful (given a set of background assumptions and the available evidence) that is different from the method followed by Kuipers. Kuipers derives consequences from the hypotheses under consideration (together with some appropriate set of background assumptions), and next confronts these consequences with the relevant observational statements. This confrontation enables him to compare the successes and problems of the individual hypotheses, and to identify those hypotheses that are maximally successful. An alternative would be to *infer* the maximally successful hypotheses from the set of background assumptions and the set of observational statements. Evidently, the logic behind this method would not be **CL**, but some ampliative logic that enables one to infer formulas of the form H from formulas of the form $H \supset O$ and O.

In the present paper, I shall explore this alternative method, and discuss the logic $\mathbf{LA}^k$ on which it is based. One of the requirements for $\mathbf{LA}^k$ is to capture the notion of empirical progress as defined by Kuipers. Thus, $\mathbf{LA}^k$ should enable one to infer from a set of background assumptions $\mathcal{B}$ and a set of observational statements $\mathcal{O}$ those hypotheses H that are maximally successful according to Kuipers' definition. As one may expect, the inference relation of $\mathbf{LA}^k$ is *non-monotonic*. This is in line with the fact that, on Kuipers' analysis, adding new empirical evidence may lead to a revision of what the maximally successful hypotheses are.

A central observation for the design of the logic $\mathbf{LA}^k$ is that 'maximally successful with respect to' is (at the predicative level) not only undecidable, but that there even is no positive test for it.[2] This has several important consequences. One consequence is that the reasoning process by which one arrives at the conclusion that H is maximally successful with respect to $\mathcal{B} \cup \mathcal{O}$ is not only non-monotonic, but also *dynamic*.[3] A simple example may clarify this. Suppose that, at some moment in time, one derived from $\mathcal{B}$ the following consequences:

(1) $H_1 \supset O_1$
(2) $H_2 \supset O_2$
(3) $H_1 \supset O_3$
(4) $H_2 \supset O_3$

Suppose further that $\mathcal{O}$ consists of:

(1) O_1
(2) O_2
(3) $\neg O_3$

If all consequences of $\mathcal{B}$ that link H_1 and H_2 to observational statements follow from (1)-(4), one may conclude that both H_1 and H_2 are maximally successful with respect to $\mathcal{B} \cup \mathcal{O}$. However, for undecidable fragments, it may be impossible to establish that some statement, say $H_1 \supset O_2$, is *not* derivable from $\mathcal{B}$. In view of this, no reasoning can *warrant* that both H_1 and H_2 are maximally successful with respect to $\mathcal{B} \cup \mathcal{O}$.

In the absence of such an absolute warrant, the only rational alternative is to derive conclusions on the basis of one's *best insights*. It immediately follows from this that a deepening of one's insights may lead to the withdrawal of previously derived conclusions. Thus, in the example above, it seems justified to conclude that H_1 as well as H_2 are maximally successful with respect to $\mathcal{B} \cup \mathcal{O}$. If, however, at a later moment in time, $H_1 \supset O_2$ is derived from $\mathcal{B}$, one has to reject the conclusion that H_2 is maximally successful.

The lack of a positive test has another important consequence. Kuipers' definition of 'more successful' is static: it is relative to a given body of knowledge, but does not depend on the evolving *understanding* of that body. Even for undecidable contexts, such a static definition is indispensable: it determines what the *final conclusions* are that should be reached or "approached." However, for practical and computational purposes we also need a definition of

---

[2] In **CL**, 'to follow from' is undecidable, but it has a positive test. The latter means that if one constructed a proof of A from Γ, one may be sure that A follows from Γ, and if A follows from Γ, there is bound to "exist" a proof of A from Γ, even if we may never find it. This does not obtain for 'is maximally successful with respect to'. Even if H is maximally successful with respect to $\mathcal{B} \cup O$, there need not exist any finite construction that establishes this.

[3] I say that a reasoning process is dynamic if the mere analysis of the premises may lead to the withdrawal of previously derived conclusions.

196

'more successful at some moment in time'. The latter definition should refer to the *understanding* of the relevant knowledge.[4]

The dynamic definition of "maximally successful at some moment in time" is not only important in view of the absence of a positive test. As Kuipers himself observes (see Kuipers 2000, p. 112), his (static) definition presupposes that one checks, for every registered problem that one hypothesis faces, whether it is also a problem for any of the other hypotheses, and similarly for the successes. This presupposition is unrealistic, even for decidable contexts. Due to the fact that resources are limited, scientists often have to decide which hypotheses are maximally successful without being able to check all known problems and successes for every hypothesis under consideration. So, also here one needs the notion of 'maximally successful at some moment in time'.

The logic that I present in this paper has a dynamic proof theory. A line that is added at some stage of the proof may at a later stage be marked (as no longer derivable) in view of the understanding of the premises offered by the proof at that stage. At a still later stage, the line may be unmarked again. Formulas that occur on non-marked lines will be considered as derived at that stage. These formulas will enable us to define the notion of 'maximally successful at some moment in time'. In addition to this, the logic will enable us to define Kuipers' (static) notion of 'maximally successful'. The latter definition is needed to guarantee that different dynamic proofs lead "in the end" to the same set of conclusions.

Thanks to its dynamic character, **LA**^{k} nicely captures Kuipers' notion of empirical progress. If at some moment in time a hypothesis H' turns out to be more successful than a hypothesis H – for instance, because new empirical successes are added (or discovered) for H' – then H will be withdrawn in favour of H'. Importantly, this withdrawal is governed by the logic itself, and hence, does not depend on a decision of the user.

The techniques that led to the logic **LA**^{k} derive from the adaptive logic programme. The first adaptive logic was designed by Diderik Batens around 1980 (see Batens 1989), and was meant to handle in a sensible and realistic way inconsistent sets of premises. This logic was followed by other inconsistency-adaptive systems (see, for instance, Priest 1991 and Meheus 2000), and the idea of an adaptive logic was generalized to other forms of

---

[4] Also Kuipers refers to 'moment of time' in his definition of 'more successful'. However, as his definition is relative to the body of knowledge that is given at some moment t, and not to the (evolving) understanding of that body of knowledge, I shall use 'more successful with respect to $B \cup O$' (and where no confusion is possible also simply 'more successful') instead of Kuipers' 'more successful at time t'. The phrases 'more successful at some moment in time' and 'maximally successful at some moment in time' will in this paper always refer to the *understanding* of a given body of knowledge at some moment in time.

logical abnormalities, such as negation-completeness, and ambiguity (see Batens 1999 for an adaptive logic that can handle all **CL**-abnormalities). An important new development concerns the design of ampliative adaptive logics (see Meheus 1999 for an informal introduction). These logics are designed to handle various forms of ampliative reasoning. At the moment, (formal) results are available on adaptive logics for compatibility (Batens and Meheus 2000), for pragmatic truth (Meheus 2002), for the closed world assumption and negation as failure (Vermeir forthcoming), for diagnostic reasoning (Weber and Provijn forthcoming and Batens, Meheus, Provijn, and Verhoeven 2003), for induction (Batens 2004), for abduction (Batens and Meheus forthcoming; 2000; Meheus *et al.* 2002), for question generation (Meheus 2001), and for the analysis of metaphors (D'Hanis 2000; 2002). An informal discussion of adaptive logics for analogies can be found in Meheus (1999). Also **LA**^{k} is an example of an ampliative adaptive logic.

As some readers may have noticed, there are resemblances between the method proposed in this paper (to *infer* the maximally successful hypotheses from a set of background assumptions and observational statements) and abduction (in the sense of Modus Ponens in the reversed direction). As mentioned in the previous paragraph, adaptive logics are available for this particular form of reasoning. Like **LA**^{k}, these logics enable one to generate explanatory hypotheses for a set of observational statements. However, they do not take into account the success of these hypotheses. **LA**^{k} differs from them in that it only delivers those explanatory hypotheses that are maximally successful.

LA^k is based on a new adaptive logic for abduction **LA**. **LA** exhibits several differences with the adaptive logics of abduction presented in Batens and Meheus (forthcoming) and Meheus and Batens (forthcoming). One is that **LA** validates inferences of the form $A \supset B, B \mathbin{/} A$, whereas the logics from Batens and Meheus (forthcoming) and Meheus and Batens (forthcoming) only validate inferences of the form $(\forall \alpha)(A(\alpha) \supset B(\alpha)), B(\beta) \mathbin{/} A(\beta)$. Another difference is that, in **LA**, abduced hypotheses are not rejected when they are falsified. As we shall see, both properties are needed for a reconstruction of Kuipers' notion of empirical progress.[5]

I shall proceed as follows. First, I briefly discuss the basic ideas of adaptive logics (Section 2). Next, I present the logic **LA** that constitutes the basis for **LA**^{k} (Section 3), and show how it can be transformed into the logic **LA**^{k} (Section 4). Finally, I present some concluding remarks and some open problems (Section 5).

---

[5] The logics for abduction that are presented in Meheus *et al.* (2001) also validate inferences of the form $A \supset B, B \mathbin{/} A$. However, they do not enable one to infer hypotheses that are empirically falsified.

198

2. Some Basics of Adaptive Logics

The enormous strength of adaptive logics is that they provide a unified frame-work for the formal study of reasoning processes that are non-monotonic and/or dynamic.[6] One of the main characteristics of such processes is that a specified set of inference rules is applied in a conditional way: they are applied on the condition that one or more formulas are *not* derived. If this condition is no longer satisfied, then previously drawn conclusions may be rejected. This also holds true for the application context that interests us here: an explanatory hypothesis H is derived from a set $\mathcal{B} \cup \mathcal{O}$ *on the condition* that H is maximally successful with respect to $\mathcal{B} \cup \mathcal{O}$. If this condition is no longer satisfied – an explanatory hypothesis H' is inferred that is more successful with respect to $\mathcal{B} \cup \mathcal{O}$ than H – H as well as all inferences that rely on H are rejected.

Adaptive logics capture this dynamics. The mechanism by which this is realized is actually very simple, both at the semantic and the syntactic level. In this paper, however, I shall restrict myself to the proof theory.[7]

The general idea behind the proof theory of an adaptive logic is that there are two kinds of inference rules: *unconditional rules* and *conditional rules*. If a formula is added by the application of a conditional rule, a "condition" that is specified by the rule is written to the right of the line. If a formula is added by the application of an unconditional rule, no condition is introduced, but the conditions (if any) that affect the premises of the application are conjoined for its conclusion. At each stage of the proof – with each formula added – one or more "marking criteria" are invoked: for each line that has a condition attached to it, it is checked whether the condition is fulfilled or not. If it is not, the line is *marked*. The formulas derived at a stage are those that, at that stage, occur on non-marked lines.

All adaptive logics available today are based on **CL**. They can, however, be divided into two categories: corrective and ampliative. Corrective adaptive logics are obtained by turning some of the inference rules of **CL** into conditional rules. Inconsistency-adaptive logics are typical examples in this category. Am-pliative adaptive logics are obtained by *adding* some conditional rules to **CL**. They thus lead (in general) to a richer consequence set than **CL**. The logics presented in this paper are ampliative.

The proof theory for an adaptive logic may be direct or indirect. The proof theory of an adaptive logic that is based on some logic **L** is *direct* if it proceeds

---

[6] Dynamic reasoning processes are not necessarily non-monotonic. In Batens (2001) it is shown, for instance, that the pure logic of relevant implication can be characterized by a dynamic proof theory.

[7] For most adaptive logics available today the proof theory as well as the semantics have been designed, and the soundness and completeness proofs have been presented. In view of these results, the design of the semantics for **LA** and **LA^k** is rather straightforward.

in terms of **L**, and indirect if it proceeds in terms of some other system, for instance a modal one. Especially in the case of ampliative adaptive logics, an indirect proof theory is usually much more attractive. The reason for this is not difficult to understand. Ampliative inferences typically lead to conclusions that are compatible[8] with (a subset of) the premises.[9] These conclusions, however, are not necessarily *jointly* compatible with the premises. For instance, both A and $\neg A$ may be inferred as explanatory hypotheses for the same set of explananda.

In view of this, restrictions are needed to avoid that mutually inconsistent conclusions lead to triviality. Formulating these restrictions in terms of **CL** leads to a proof theory that is quite complex and not very transparent. They can, however, easily be formulated in terms of a modal logic.

The transition to a modal approach is actually very simple. If $\Gamma \vdash_{\mathbf{CL}} A$, then A is true in *all* models of Γ. What this comes to, in modal terms, is that $\neg A$ is *impossible* ($\neg\Diamond\neg A$), or, in other words, that A is *necessary* ($\Box A$). If A is compatible with Γ, then A is true in *some* model of Γ, but not necessarily in all of them. In line with all this, it seems sensible to consider the members of Γ as necessarily true, and the sentences that are compatible with Γ as possibly true.

This idea is used in Batens and Meheus (2000) to design an adaptive logic for compatibility that is called **COM**. **COM** is based on **S5**, and is defined with respect to sets of premises of the form $\Gamma^\Box = \{\Box A \mid A \in \Gamma\}$. As is shown in Batens and Meheus (2000), **COM** has the interesting property that $\Gamma^\Box \vdash_{\mathbf{COM}} \Diamond A$ iff $\Gamma \nvdash_{\mathbf{CL}} \neg A$, and hence, iff A is compatible with Γ. Note that this transition to a modal approach immediately solves the problem concerning mutually inconsistent conclusions (in view of $\Diamond A, \Diamond\neg A \nvdash B$).

The logics presented here share their general format with **COM**: premises are treated as necessarily true, and conclusions arrived at by ampliative steps as possibly true. So, the proof theory will not rely on

(8) $A \supset B, B \,/\, A$

but on its modal translation, namely

(9) $\Box(A \supset B), \Box B \,/\, \Diamond A$

There is, however, a small complication. This is related to the fact that $\Box B \vdash \Box(A \supset B)$, for arbitrary A, and that $\Box\neg A \vdash \Box(A \supset B)$, for arbitrary B. In view of this, it has to be avoided that the application of (9) leads to arbitrary explanations.

---

[8] A formula A is said to be compatible with Γ iff $\Gamma \nvdash_{\mathbf{CL}} \neg A$.

[9] In all currently available ampliative adaptive logics, the conclusions of ampliative inferences are compatible with the set of premises. The only exceptions are the logics presented in this paper: here the conclusions are compatible with the background assumptions, but not necessarily with the union of the observational statements and the background assumptions.

200

There are several options to solve this difficulty. One option is to formulate a suitably restricted form of (9). Another option is to presuppose that there is a clear distinction between the set of observational statements Γ_1 and the set of background assumptions Γ_2 and to require that (9) can only be applied if $\Box(A \supset B)$ is **S5**-derivable from Γ_2^W, whereas $\Box B$ and $\Box \neg A$ are not. In order to obtain a logic for abduction that is as general as possible, I followed the second option in the design of **LA**.[10]

The easiest way to realize this option is to rely on a bimodal version of **S5** – let us call it $\mathbf{S5^2}$. The language of $\mathbf{S5^2}$ includes two necessity operators ('$\Box_1$' and '$\Box_2$') and two possibility operators ('$\Diamond_1$' and '$\Diamond_2$'). To simplify things, I shall only consider modal formulas of first degree – a modal formal formula is said to be of first degree if it contains one or more modal operators, but none of these is inside the scope of any other modal operator. A formula that does not contain any modal operator will be said to be of degree zero.

The operator '$\Box_1$' will be used in the formalization of the members of Γ_1 (the premises that are observational statements), and '$\Box_2$' in the formalization of the members of Γ_2 (the premises that are background assumptions). $\mathbf{S5^2}$ will be defined in such a way that $\Box_1 A$ is derivable from $\Box_2 A$, but not *vice versa*. This ensures, on the one hand, that the information from both sets of premises can be conjoined (to derive predictions, for instance), and, on the other hand, that it remains possible to recognize which sentences are derivable from the background assumptions alone.

So, the general idea is this. **LA** and $\mathbf{LA^k}$ are ampliative adaptive logics based on **CL**. To avoid the derivation of arbitrary explanations, their consequence relation is defined with respect to couples $\Sigma = \langle \Gamma_1, \Gamma_2 \rangle$, in which Γ_1 and Γ_2 are sets of closed formulas of the standard predicative language. It is assumed that Γ_1 is the set of observational statements, and Γ_2 the set of background assumptions. The proof theories of **LA** and $\mathbf{LA^k}$ are defined with respect to modal adaptive logics based on $\mathbf{S5^2}$. These will be called **MA** and $\mathbf{MA^k}$. It will be stipulated that A is an **LA**-consequence (respectively $\mathbf{LA^k}$-consequence) of some theory Γ iff $\Diamond A$ is an **MA**-consequence (respectively $\mathbf{MA^k}$-consequence) of the modal translation of Γ.

---

[10] The first option is followed in Batens and Meheus (forthcoming) and Meheus and Batens (forthcoming). Unlike **LA**, the logics presented there only validate inferences of the form $(\forall \alpha)(A(\alpha) \supset B(\alpha))$, $B(\beta) \ / \ A(\beta)$. The advantage, however, is that no distinction has to be made between the set of background assumptions and the set of observational statements.

3. A Simple Adaptive Logic for Abduction

Let $\mathcal{L}$ be the standard predicative language of **CL**, and let $\mathcal{L}^M$ be obtained from $\mathcal{L}$ by extending it with '$\square_1$', '$\square_2$', '$\Diamond_1$', and '$\Diamond_2$'. Let the set of wffs of $\mathcal{L}^M$, $\mathcal{W}^M$, be restricted to wffs of degree zero and first degree.

Syntactically, $\mathbf{S5}^2$ is obtained by extending an axiomatization of the full predicative fragment of **CL** with every instance (for $i \in \{1, 2\}$) of the following axioms, rule, and definition:

Al	$\square_i A \supset A$
A2	$\square_i(A \supset B) \supset (\square_i A \supset \square_i B)$
A3	$\square_2 A \supset \square_1 A$
NEC	if $\vdash A$ then $\vdash \square_i A$
D$^{\Diamond}$	$\Diamond_i A =_{\mathrm{df}} \neg\square_i\neg A$

Axioms A1-A2 are the usual axioms for **S5** adapted to the bimodal case. As $\mathcal{W}^M$ contains only wffs of degree zero and first degree, no axiom is needed for the reduction of iterated modal operators. Axiom A3 warrants that '$\square_2$' is stronger than '$\square_1$': $\square_2 A$ implies $\square_1 A$, but not *vice versa*. Note that, in view of A3 and D$^{\Diamond}$, $\Diamond_1 A$ implies $\Diamond_2 A$, but that the converse does not hold.

The relation between **MA** and **LA** is given by Definition 1. Where $\Sigma = (\Gamma_1, \Gamma_2)$, $\Sigma^{\square}$ refers to $\{\square_1 A \mid A \in \Gamma_1\} \cup \{\square_2 A \mid A \in \Gamma_2\}$.

Definition 1 $\Sigma \vdash_{\mathbf{LA}} A$ *iff* $\Sigma^{\square} \vdash_{\mathbf{MA}} \Diamond_2 A$.

Let us now turn to the proof theory for **MA**. As is usual for adaptive logics, **MA**-proofs consist of lines that have five elements: (i) a line number, (ii) the formula A that is derived, (iii) the line numbers of the formulas from which A is derived, (iv) the rule by which A is derived, and (v) a condition. The condition has to be satisfied in order for A to be so derivable.

The condition will either be $\varnothing$ or a couple of the form $\langle \Phi, \Theta \rangle$ in which Φ contains one closed formula and Θ is a set of closed formulas. Thus, if the fifth element of a line in a proof is not empty, the line will have the form

$$i \quad A \quad j_1, ..., j_n \quad \text{RULE} \quad \langle \{B\}, \{C_1, ..., C_n\} \rangle$$

Intuitively, a line of this form will be read as 'A provided that B is a good explanation for $C_1,...,C_n$'. Evidently, the phrase 'is a good explanation for' may be interpreted in different ways. Here, I shall choose for a minimal interpretation that is most adequate in view of the design of $\mathbf{LA}^k$. As we shall see below, this interpretation does not require that a good explanation is not falsified. It does require, however, that it is compatible with the background assumptions – the reason for this will become clear below.[11]

The proof format of **MA** follows the generic proof format for adaptive logics that is presented in Batens, De Clerq and Vanackere (forthcoming). In addition to a premise rule RP, I shall introduce an unconditional rule RU, a conditional rule RC, and three marking criteria. The latter determine when a line should be marked (as no longer derivable). A wff will be said to be derived unconditionally iff it is derived on a line the fifth element of which is empty.

Intuitively, a line is marked if its condition is not (no longer) satisfied. Thus, in line with the interpretation of the condition (see above), the marking criteria determine which requirements a hypothesis A and a set of explananda $B_1,\ldots, B_m$ should meet so that A is a *good* explanation for $B_1,\ldots, B_m$. I shall come back to the marking criteria later.

Let us first look at the three generic rules that govern **MA**-proofs from $\Sigma^\square$. After listing them, I shall briefly comment on each of them.

RP If $A \in \Sigma^\square$, then one may add to the proof a line consisting of

    (i)   the appropriate line number,
    (ii) A,
    (iii)   '–',
    (iv)   'RP', and
    (v) $\varnothing$.

RU If

    (a) $A_1, \ldots, A_n \vdash_{\mathbf{S5^2}} B$ $(n \geq 0)$,
    (b) $A_1, \ldots, A_n$ occur in the proof, *and*
    (c)   at most one of the A_i occurs on a non-empty condition Δ,

then one may add to the proof a line consisting of:

    (i)   the appropriate line number,
    (ii) B,
    (iii)   the line numbers of the A_i if any, and '–' otherwise,
    (iv)   'RU', and
    (v)   if all A_i occur unconditionally, and Δ otherwise.

RC If

    ( ) $\square_1 B_1, \ldots, \square_1 B_m$ $(m \geq 1)$ occur in the proof, and
    ( )   also $\square_2(A \supset (B_1 \wedge \ldots \wedge B_m))$ occurs in it,

---

[11] The minimal interpretation that is followed here may be strengthened in several ways. The systems presented in Batens and Meheus (forthcoming), for instance, require that a good explanation, among other things, is not falsified.

then one may add to the proof a line consisting of:

   (i)   the appropriate line number,
   (ii) $\Diamond_2 A$,
   (iii)   the line numbers of the formulas mentioned in (a) – (b),
   (iv)   'RC', and
   (v) $\langle \{A\}, \{B_1,\ldots, B_m\} \rangle$.

As it should be, the rule RP warrants that all premises are introduced unconditionally (on a line the fifth element of which is empty).

As for the rule RU, note that it enables one to conjoin formulas that are conditionally derived to formulas that are unconditionally derived. This is important, because it warrants that explanatory hypotheses can be used to derive predictions from the background assumptions. The rule RU does not enable one, however, to conjoin explanatory hypotheses to one another (it cannot be applied if more than one of the A_i occurs on a non-empty condition). By using a somewhat more complex format for the condition, the rule RU can easily be generalized to such inferences. However, as we are not interested in **MA** itself, and as the generalization would not lead to a richer consequence set for **LA** (in view of $\Diamond_i A, \Diamond_i B \nvdash_{\mathbf{S5^2}} \Diamond_i(A \wedge B)$) and Definition 1), this would only complicate matters.

The rule RC warrants that the conclusion of an abductive inference is always preceded by '$\Diamond_2$'. Also this is important: if explanatory hypotheses would be preceded by '$\Diamond_1$', their falsification would lead to triviality (see also the example below).

From RU and RC, several rules can be derived that make the proofs more interesting from a heuristic point of view. The following one, the predicative version of RC, will be useful in the examples below:

RD If

   (a) $\Box_1 B_1(\beta),\ldots, \Box_1 B_m(\beta)$ $(m \geq 1)$ occur in the proof, and also
   (b) $\Box_2 (\forall \alpha)\, (A(\alpha) \supset (B_1(\alpha)\ldots B_m(\alpha))$ occurs in it,

then one may add to the proof a line consisting of:

   (i)   the appropriate line number,
   (ii) $\Diamond_2 A(\beta)$,
   (iii) the line numbers of the formulas mentioned in (a) – (b),
   (iv) 'RD', and
   (v) $\langle \{A(\beta)\}, \{B_1(\beta),\ldots, B_m(\beta)\} \rangle$.

Let me illustrate the proof theory with a very simple example. Suppose one observes a new heavenly body (call it 'a') that seems too large to be a star, but that at the same time does not seem to move around the sun. Suppose further one

204

believes, on the one hand, that a star appears as a small heavenly body that does not move around the sun and does not have a tail, and on the other hand, that a comet appears as a large heavenly body that moves around the sun and that has a tail. Let us agree on the following letters to formalize our explananda and our background assumptions:

S 'is a star'
C 'is a comet'
H 'is a heavenly body'
L 'appears large'
M 'moves around the sun'
T 'has a tail'

This is how the result could look like – I omit brackets in the case of continuous conjunctions:

1   $\Box_1 Ha$ –   RP $\varnothing$
2   $\Box_1 La$ –   RP $\varnothing$
3   $\Box_1 \neg Ma$ –   RP $\varnothing$
4   $\Box_2 (\forall x)\,(Sx \supset (Hx \wedge \neg Lx \wedge \neg Mx \wedge \neg Tx))$ –   RP $\varnothing$
5   $\Box_2 (\forall x)\,(Cx \supset (Hx \wedge Lx \wedge Mx \wedge Tx))$ –   RP $\varnothing$

From 4 we can derive by RU that stars are heavenly bodies that do not move around the sun:

6   $\Box_2 (\forall x)\,(Sx \supset (Hx \wedge \neg Mx))$ 4   RU $\varnothing$

The rule RD can now be applied to infer *Sa on the condition* that *Sa* is a good explanation for *Ha* and $\neg Ma$:

7   $\Diamond_2 Sa$ 1, 3, 6   RD$\langle \{Sa\}, \{Ha, \neg Ma\} \rangle$

In an analogous way, *Ca* can be inferred on the condition that *Ca* is a good explanation for *Ha* and *La:*

8   $\Box_2 (\forall x)\,(Cx \supset (Hx \wedge Lx))$ 5   RU $\varnothing$
9   $\Diamond_2 Ca$ 1, 2, 8   RD $\langle \{Ca\}, \{Ha, La\} \rangle$

From 2 and 4, it can be inferred that $\neg Sa$ holds true:

10   $\Box_1 \neg Sa$ 2, 4   RU $\varnothing$

What this comes to is that the explanatory hypothesis that occurs on line 7 is empirically falsified. Thanks to the bimodal approach, however, this does not lead to problems – I leave it to the reader to check that neither $\Box_2 \neg Sa$ nor $\Diamond_1 Sa$ is derivable. In view of 3 and 5, it can be inferred that also the hypothesis on line 9 is falsified:

11   $\Box_1 \neg Ca$ 3, 5   RU $\varnothing$

Although they are empirically falsified, both hypotheses can be used to derive predictions:

12	$\Diamond_2 Ta$	5, 9 RU $\langle \{Ca\}, \{Ha, La\} \rangle$
13	$\Diamond_2 \neg Ta$	4, 7 RU $\langle \{Sa\}, \{Ha, \neg Ma\} \rangle$

Note that the derivation of predictions does not lead to the introduction of a new condition, but that the conditions of the lines to which RU is applied are preserved. This will guarantee that the predictions are no longer derivable when the explanatory hypotheses on which they rely are withdrawn. Note also that the predictions are mutually inconsistent, but that this cannot lead to arbitrary conclusions (in view of $\Diamond_i A, \Diamond_i B \nvdash_{\mathbf{S5^2}} \Diamond_i (A \wedge B)$).

Let us now turn to the marking criteria – the criteria that determine when the condition of a line is satisfied, and thus, fix the meaning of 'is a good explanation for'. As mentioned above, I shall not demand that a good explanation is not empirically falsified. I shall demand, however, that it satisfies three basic requirements without which the very notion of explaining would become superfluous.

The first requirement is that the explanation should not be trivial – in view of $\Diamond_i B \vdash_{\mathbf{S5^2}} \Box_i (A \supset B)$, for arbitrary A, and $\Box_i \neg A \vdash_{\mathbf{S5^2}} \Box_i (A \supset B)$, for arbitrary B (see above). The second is that the explanatory hypothesis should not be derivable from one of the explananda. This is needed to rule out (partial) self-explanations. For instance, we want neither B nor $A \vee B$ as an explanation for B. Cases like this are ruled out by requiring that the truth of the explanatory hypothesis is not warranted by the truth of one of the *explananda* (or, that the explanans is not derivable from one of the explananda). The third and final requirement is that the explanatory hypothesis should be as parsimonious as possible. This is especially important in view of the fact that $\Box_i (A \supset B) \vdash_{\mathbf{S5^2}} \Box_i ((A \wedge D) \supset B)$, and hence, that one needs to prevent that $A \wedge D$ can be abduced, whenever A can.

To each of these requirements corresponds a marking criterion – respectively called T-marking, S-marking and P-marking. The first of these, that for ruling out trivial explanations, is straightforward:

CMT   A line that has $\langle \{A\}, \Theta \} \rangle$ as its fifth element is T-marked iff
    (i)   for some $B \in \Theta$, $\Box_2 B$ occurs in the proof, or
    (ii) $\Box_2 \neg A$ occurs in the proof.

Note that, if a line is T-marked, it remains T-marked at any further stage of the proof. Note also that, in view of (ii) of CMT, it is required that an explanatory hypothesis is compatible with the background assumptions.

The criterion for ruling out (partial) self-explanations involves only a small complication, namely that one should be able to recognize, *in the proof*, that

206

some explanatory hypothesis A is entailed by some explanandum B. I shall assume that this is done by deriving $\Box_i(B \supset A)$ on a line j such that, at line j, the path of $\Box_i(B \supset A)$ does not include any premise.[12] Expressions of the form $\pi_j(A)$ will refer to the path of a formula A as it occurs on line j. To keep things simple, I first define a set $\pi_j^{\circ}(A)$:

Definition 2 *Where A is the second element of line j, $\pi_j^{\circ}(A)$ is the smallest set Λ that satisfies:*

(i) $j \in \Lambda$, *and*

(ii)   *if $k \in \Lambda$, and $l_1,\dots, l_n$ is the third element of line k, then $l_1,\dots, l_n \in \Lambda$.*

Definition 3 $\pi_j(A) = \{B \mid B$ *occurs as the second element of line i and $i \in \pi_j^{\circ}(A)\}$.*

Here is the second marking criterion:

CMS   A line that has $\langle\{A\}, \Theta\rangle$ as its fifth element is S-marked iff, for some $B \in \Theta$, a line j occurs in the proof such that

    (i) $\Box_i(B \supset A)$ is its second element, and

    (ii) $\pi_j(\Box_i(B \supset A)) \cap (\Gamma_1 \cup \Gamma_2) = \varnothing$.

The final criterion requires a bit more explanation. As mentioned above, it has to be prevented that $A \wedge D$ can be abduced, whenever A can – in view of $\Box_i(A \supset B) \vdash_{\mathbf{S5^2}} \Box_i((A \wedge D) \supset B)$. This can be solved by requiring that the explanatory hypothesis is as parsimonious as possible. However, selecting the most parsimonious explanatory hypotheses can be realized in different ways. The most obvious one is to warrant that, whenever A and C explain the same set of explananda, and C is (according to one's best insights) logically weaker than A, A is rejected in favour of C. This, however, raises a further problem. If one would simply select the logically weakest explanatory hypothesis, it would be impossible to generate alternative explanations for the same set of explananda – in view of $\Box_i(A \supset C), \Box_i(B \supset C) \vdash_{\mathbf{S5^2}} \Box_i((A \vee B) \supset C)$.

In some application contexts, this is exactly what one wants. For instance, in the context of medical diagnosis, one is interested in the weakest explanation: whenever two or more explanations can be abduced for the same set of symptoms, one will only accept their disjunction.[13] In other contexts, however, one wants to be able to generate alternative explanations for the same set of

---

[12] This is in line with the general philosophy behind adaptive logics, namely that the application of inference rules and criteria is relative to the distinctions that, at a certain stage, are made by the reasoner. The easiest way to realize this is to refer to formulas that are actually written down in the proof.

[13] Evidently, one will try to strengthen this explanation by asking further questions, or doing further tests. The fact remains, however, that at any moment in time one only accepts the weakest explanation.

explananda. This holds true, for instance, for the case that interests us here: comparing the empirical success of different hypotheses only makes sense if one is able to generate alternative explanations for the same set of explananda.

In this paper, I shall present a very simple solution to the problem that relies on the complexity of formulas – by the complexity of a formula I shall refer to the number of *binary* connectives that occur in it (if a formula contains no binary connectives, its complexity will be said to be zero).[14] The basic idea is this: if two explanations A and C are available for a set of explananda $B_1, ..., B_m$, and A entails C, then A should be withdrawn in favour of C, *provided* that the complexity of the former is greater than that of the latter. Thus, if $p \wedge q$ and p are alternative explanations for the same set of *explananda*, the former should be withdrawn in favor of the latter (because $p \wedge q$ entails p, and moreover the complexity of the former is greater than that of the latter). If, however, the alternative explanations are $p \wedge q$ and $(p \wedge q) \vee r$, then $p \wedge q$ should not be withdrawn (although $p \wedge q$ entails $(p \wedge q) \vee r$, its complexity is smaller than that of $(p \wedge q) \vee r$).

Let $c(A)$ denote the complexity of A. The criterion for P-marking can now be formulated as follows:

CMP   A line that has $\langle \{A\}, \Theta \rangle$ as its fifth element is P-marked iff
- (i)   some line in the proof that is neither T-marked nor S-marked has$\langle \{C\}, \Theta \rangle$ as its fifth element (for some C),
- (ii) $\Box_i(A \supset C)$ occurs on a line i such that $\pi_i(\Box_i(A \supset C)) \cap (\Gamma_1 \cup \Gamma_2) = \varnothing$, and
- (iii) $c(A) > c(C)$.

I immediately add four remarks to this. First, to keep the proof theory as realistic as possible, also this criterion refers to distinctions that have been made in the proof (this is why (ii) does not refer to the fact that A entails C, but to the fact that *it has been recognized* that A entails C).

Secondly, formulas that are P-marked at some stage may at a later stage be unmarked (for instance, because, at some stage, A is P-marked in view of C which is itself, at a later stage, T-marked or S-marked).

Thirdly, as the criterion for P-marking refers to the other two criteria (and as P-marked lines may at a later stage be unmarked), the easiest way to perform the marking is as follows: first, remove all P-marks; next, check which lines have to be marked according to CMT and CMS; finally, check which lines have to be marked according to CMP.[15]

---

[14] For a different solution, see Meheus *et al.* (2001). The solution presented there is more complex, but has the advantage that it offers a better insight in the amount of information contained in a formula.

[15] Evidently, this procedure has to be repeated whenever a line is added to the proof.

208

Finally, if a formula A contains some "irrelevant" letters,[16] A may be marked in view of another formula C that is logically equivalent to it. For instance, as soon is it established that $(p \wedge q) \vee (p \wedge \neg q)$ entails p, the former will be marked in view of the latter.[17] This seems justified in view of the fact that one should try to formulate an explanatory hypothesis in the simplest way possible.

Let me illustrate the marking criteria with a simple prepositional example. Suppose that $\Gamma_1 = \{p, q\}$, and $\Gamma_2 = \{r \supset p, s \wedge p, p \supset \neg t, t \supset q, u \supset q\}$. In view of RP, both the observational statements and the background assumptions may be introduced on an empty condition, the former preceded by '$\square_1$', the latter by '$\square_2$':

1	$\square_1 p$	–	RP $\varnothing$
2	$\square_1 q$	–	RP $\varnothing$
3	$\square_2 (r \supset p)$	–	RP $\varnothing$
4	$\square_2 (s \wedge p)$	–	RP $\varnothing$
5	$\square_2 (p \supset \neg t)$	–	RP $\varnothing$
6	$\square_2 (t \supset q)$	–	RP $\varnothing$
7	$\square_2 (u \supset q)$	–	RP $\varnothing$

In view of RC, $\lozenge_2 r$ may be derived from 1 and 3, on the condition that r is a good explanation for p[18]:

...

8	$\lozenge_2 r$	1, 3 RC$\langle \{r\}, \{p\}\rangle$

However, from 4, one may derive:

...

9	$\square_2 p$	4 RU $\varnothing$

which indicates that p is derivable from Γ_2. Hence, to avoid trivial explanations, line 8 is marked in view of (i) of CMT[19]:

...

1	$\lozenge_2 r$	1, 3 RC $\langle \{r\}, \{p\}\rangle$	$\sqrt{}_{T9}$
2	$\square_2 p$	3, 4 RU $\varnothing$	

At this stage of the proof, the formula on line 8 is no longer considered as derived.

---

[16] I say that a letter occurs irrelevantly in A iff A is logically equivalent to a formula C in which that letter does not occur.

[17] This is a further difference with the solution proposed in (Meheus, Verhoeven, Van Dyck, and Provijn forthcoming): there, logically equivalent formulas are never marked in view of one another.

[18] To illustrate the dynamical character of the proof, I shall each time give the complete proof, but, for reasons of space, omit lines that are not needed to see the dynamics.

[19] The numbers in the marking sign refer to the lines on the basis of which the marking is performed.

It is easily observed that, because of line 9, it no longer makes sense to derive explanatory hypotheses for p. Still, one may try to derive an explanation for q. For instance, from lines 2 and 6, one may derive $\Diamond_2 t$ on the condition that t is a good explanation for q:

$$\ldots$$

| 3 | $\Diamond_2 t$ | | 2, 6 RC $\langle \{t\}, \{q\} \rangle$ |

But, as is clear from the following line

| 4 | $\Box_2 \neg t$ | | 5, 9 RU $\varnothing$ |

t is not compatible with Γ_2. Hence, in view of (ii) of RMT, also line 10 has to be marked:

$$\ldots$$

| 1 | $\Diamond_2 t$ | | 2, 6 RC $\langle \{t\}, \{q\} \rangle \checkmark_{T^{11}}$ |
| 2 | $\Box_2 \neg t$ | | 5, 9 RU $\varnothing$ |

It is possible, however, to derive an alternative explanation for q, for which it is easily observed that it will not be marked by any of the marking criteria:

$$\ldots$$

| 3 | $\Diamond_2 u$ | | 2, 7 RC $\langle \{u\}, \{q\} \rangle$ |

Now, suppose one continues the proof as follows:

$$\ldots$$

| 4 | $\Box_2((u \vee q) \supset q)$ | | 7 RU $\varnothing$ |

In that case, it becomes possible to generate $u \vee q$ as an "explanation" for q:

$$\ldots$$

| 5 | $\Diamond_2(u \vee q)$ | | 2, 13 RC $\langle \{u \vee q\}, \{q\} \rangle$ |

However, as soon as one recognizes that $u \vee q$ is entailed by q, the former is withdrawn in view of CMS:

$$\ldots$$

| 1 | $\Diamond_2(u \vee q)$ | | 2, 13 RC $\langle \{u \vee q\}, \{q\} \rangle \ \checkmark_{S^{15}}$ |
| 2 | $\Box_2(q \supset (u \vee q))$ | | $-$ RU $\varnothing$ |

In a rather similar vein, one may also try to derive, say $u \wedge w$, as an explanation for q:

$$\ldots$$

| 3 | $\Box_2((u \wedge w) \supset q)$ | | 7 RU $\varnothing$ |
| 4 | $\Diamond_2(u \wedge w)$ | | 2, 16 RC $\langle \{u \wedge w\}, \{q\} \rangle$ |

However, as soon as it is established that $u \wedge w$ entails u (which is an alternative, but less complex, explanation for q), $\Diamond_2(u \wedge w)$ is withdrawn in view of CMP:

$$\ldots$$

| 1 | $\Diamond_2(u \wedge w)$ | | 2, 16 RC$\langle \{u \wedge w\}, \{q\} \rangle \checkmark_{P^{12,18}}$ |
| 2 | $\Box_2((u \wedge w) \supset u)$ | | $-$ RU $\varnothing$ |

210

In view of the marking criteria, two forms of derivability can be defined: derivability at a stage and final derivability. I say that a line is *marked* iff it is *T*-marked, *S*-marked or *P*-marked. If the mark of a line is removed, I say that it is unmarked.

Definition 4 *A is* derived at a stage *in an* **MA***-proof from* $\Sigma^{\square}$ *iff A is derived in the proof on a line that is not marked.*

Definition 5 *A is* finally derived *in an* **MA***-proof from* $\Sigma^{\square}$ *iff A is derived on a line i that is not marked, and any extension of the proof in which line j is marked may be further extended in such a way that line j is unmarked.*

It is easily observed that, in the above example, the formula on line 12 is finally derived. The formula on line 10, however, is derived at stage 10, but is no longer derived at stage 11.

As is usual for adaptive logics, the consequence relation of **MA** is defined with respect to final derivability:

Definition 6 $\Sigma^{\square} \vdash_{\mathbf{MA}} A$ *(A is finally derivable from* $\Sigma^{\square}$*) iff A is finally derived in an* **MA***-proof from* $\Sigma^{\square}$.

4. An Adaptive Logic for Empirical Progress

As was explained in the previous section, the logic **LA** enables one to generate explanatory hypotheses on the basis of a set of background assumptions and a set of observational statements. It does, however, not take into account the empirical success of the hypotheses. I shall now show what changes are needed to obtain the logic $\mathbf{LA}^k$ that only generates the maximally successful hypotheses. Like for **LA**, the proof theory of $\mathbf{LA}^k$ is defined with respect to some modal adaptive logic that is based on $\mathbf{S5}^2$. This logic will be called $\mathbf{MA}^k$. The relation between $\mathbf{MA}^k$ and $\mathbf{LA}^k$ is as that between **MA** and **LA**:

Definition 7 $\Sigma \vdash_{\mathbf{LA}^k} A$ *iff* $\Sigma^{\square} \vdash_{\mathbf{MA}^k} \Diamond_2 A$.

Three changes are needed to obtain the proof theory for $\mathbf{MA}^k$ from that for **MA**. The first concerns the format and the interpretation of the fifth element of the lines in a proof. If the fifth element of a line is not empty, it will be a triple of the form $\langle \{A\}, \Theta, \Pi \rangle$ in which Θ and Π are sets of closed formulas. Intuitively, Θ stands for the set of empirical successes of A, and Π for its set of empirical problems; the union $\Theta \cup \Pi$ will be said to be *the empirical record* of A. A line of the form

$$i \quad A \quad j_1, \ldots, j_n \quad \text{RULE} \quad \langle \{B\}, \Theta, \Pi \rangle$$

will be read as 'A provided that B is a good explanation for Θ, $\Theta \cup \Pi$ is the empirical record of B, and no hypothesis is empirically more successful than B'. This new format will enable us to compare the empirical success of alternative explanations, and to mark lines in accordance with this comparison.

The second change is related to the conditional rule RC – the premise rule and the unconditional rule are as for **MA**. The difference with the conditional rule for **MA** is that one not only "keeps track" of the empirical successes (confirming instances) but also of the empirical problems (falsifying instances). The basic idea is actually very simple, and best illustrated by means of a prepositional example. Consider the following set of premises:

$$\{\Box_1 p, \Box_1 q, \Box_2(r \supset (p \wedge \neg q)), \Box_1(s \supset p)\}$$

In this case, the explanatory hypothesis s is empirically more successful than the hypothesis $r - p$ is a confirming instance for both, but q is a falsifying one for r.

In **MA**, this difference will not show: both hypotheses will be derived on a similar condition, namely that they adequately explain p. As a consequence, **MA** does not offer an easy way to compare the empirical success (in Kuipers' sense) of alternative hypotheses. Things change, however, if it is allowed that $\Diamond_2 r$ is introduced on the condition $\langle \{r\}, \{p\}, \{q\} \rangle$, and $\Diamond_2 s$ on the condition $\langle \{s\}, \{p\}, \varnothing \rangle$.[20] Thanks to this difference in condition, it becomes possible to observe that r and s share the same empirical success (namely p), but that r moreover faces an empirical problem not faced by s (namely q).

The new conditional rule is a generalization of this basic idea to sets of confirming instances and of falsifying instances:

RC^k If

    (a) $\Box_1 B_1, ..., \Box_1 B_m$ $(m \geq 1)$,
    (b) $\Box_1 C_1, ..., \Box_1 C_n$ $(n \geq 0)$, and
    (c) $\Box_2(A \supset ((B_1 \wedge ... \wedge B_m) \wedge (\neg C_1 \wedge ... \wedge \neg C_n)))$

    occur in the proof, then one may add to the proof a line consisting of:

    (i)   the appropriate line number,
    (ii) $\Diamond_2 A$,
    (iii)   the line numbers of the formulas mentioned in (a) – (c),
    (iv)   'RC^k', and
    (v) $\langle \{A\}, \{B_1, ..., B_m\}, \{C_1, ..., C_n\} \rangle$.

---

[20] As indicated above, the first two elements of these conditions are interpreted in the same way as for **MA** – namely, that r, respectively s, are good explanations for p (for instance, that they are not self-explanations).

212

As in the case of **MA**, it is possible to formulate derived rules that make the proof theory more interesting from a heuristic point of view. Here is the predicative version of RC^k that will prove useful for the examples below:

RD^k If

    (a) $\Box_1 B_1(\beta)$, ..., $\Box_1 B_m(\beta)$ $(m \geq 1)$,
    (b) $\Box_1 C_1(\beta)$, ..., $\Box_1 C_n(\beta)$ $(n \geq 0)$, and
    (c) $\Box_2(\forall \alpha)\, (A(\alpha) \supset ((B_1(\alpha) \wedge ... \wedge B_m(\alpha)) \wedge (\neg C_1(\alpha) \wedge ... \wedge \neg C_n(\alpha))))$

occur in the proof, then one may add to the proof a line consisting of:

    (i)   the appropriate line number,
    (ii) $\Diamond_2 A(\beta)$,
    (iii)   the line numbers of the formulas mentioned in (a) – (c),
    (iv)   'RD^k',and
    (v) $\langle \{A(\beta)\}, \{B_1,(\beta), ..., B_m(\beta)\}, \{C_1(\beta), ..., C_n(\beta)\} \rangle$.

The last change is that two additional marking criteria are needed – I shall call the first R-marking and the second M-marking.[21] The criterion for R-marking warrants that a line i, that has $\langle \{A\}, \Theta, \Pi \rangle$ as its fifth element, is marked if, according to one's best insight in premises, the empirical record of A at line i (that is, $\Theta \cup \Pi$) is not complete – put more precisely, if there is a line j in the proof such that the empirical record of A at line j is a (real) superset of that at line i[22]:

CMR   A line that has $\langle \{A\}, \Theta, \Pi \rangle$ as its fifth element is R-marked iff
        (i)   some line in the proof, that is not T-, S-, or P-marked has $\langle \{A\},$ $\Theta', \Pi' \rangle$ as its fifth element, and
        (ii) $\Theta' \cup \Pi' \supset \Theta \cup \Pi$.

The criterion for M-marking warrants that a line i, that has $\langle \{A\}, \Theta, \Pi \rangle$ as its fifth element, is marked if, according to one's best insight in premises, A is not maximally successful – that is, if an alternative explanation C has been derived for which it is established that it saves at least the same successes as A, faces at most the same problems as A, and fares better in at least one of these respects[23]:

CMM   A line that has $\langle \{A\}, \Theta, \Pi \rangle$ as its fifth element is M-marked iff
        (i)   some line in the proof, that is not T-, S-, P- or R-marked has $\langle \{C\}, \Theta', \Pi' \rangle$ as its fifth element (for some C), and
        (ii) $(\Theta' \supset \Theta$ and $\Pi' \subseteq \Pi)$ or $(\Theta' \supseteq \Theta$ and $\Pi' \subset \Pi)$.

---

[21] The criteria for T-, S-, and P-marking are as for **MA**, except for the evident change that '$\langle \{A\}, \Theta \rangle$' and '$\langle \{C\}, \Theta \rangle$' have to be replaced systematically by '$\langle \{A\}, \Theta, \Pi \rangle$' and '$\langle \{C\}, \Theta, \Pi \rangle$'.

[22] The marking criterion will be illustrated by an example below.

[23] Compare (ii) of the marking criterion CMM with Kuipers' definition of 'empirically more successful' presented in the first section.

As for **MA**, I shall say that a line is marked iff it is marked according to one of the above criteria.

At first sight, CMR may seem redundant in view of CMM. Note, however, that the former is needed to ensure that, for a given hypothesis, not only the empirical successes are taken into account, but also the empirical problems. This is illustrated by the following example:

1	$\Box_1 Qa$	–	RP $\varnothing$
2	$\Box_1 Ra$	–	RP $\varnothing$
3	$\Box_2 (\forall x)\,(Px \supset Qx)$	–	RP $\varnothing$
4	$\Box_2 (\forall x)\,(Px \supset \neg Rx)$	–	RP $\varnothing$
5	$\Diamond_2 Pa$	1, 3	$\mathrm{RD}^k \langle \{Pa\}, \{Qa\}, \varnothing \rangle\ \surd_{R7}$
6	$\Box_2 (\forall x)\,(Px \supset (Qx \wedge \neg Rx))$	3, 4	RU $\varnothing$
7	$\Diamond_2 Pa$	1, 2, 6	$\mathrm{RD}^k \langle \{Pa\}, \{Qa\}, \{Ra\} \rangle$

As the empirical record for the hypothesis Pa is not complete at line 5 (in view of line 7), the former is R-marked. Without this marking, line 7 would be M-marked in view of line 5. The criterion for M-marking will be illustrated by the examples below.

As in the case of **MA**, lines that are marked at some stage may at a later stage be unmarked, and some criteria refer to others. Hence, also here the marking has to be performed in some specified way – for instance, by first removing all marks, and, next, checking which lines have to be marked according to, *in this order*, CMT, CMS, CMP, CMR and CMM.

The definitions of derivability at a stage and final derivability are as for **MA**, and so is the definition of the consequence relation:

Definition 8 *A is* derived at a stage *in an* **MA**[k]*-proof from* $\Sigma^\Box$ *iff A is derived in the proof on a line that is not marked.*

Definition 9 *A is* finally derived *in an* **MA**[k]*-proof from* $\Sigma^\Box$ *iff A is derived on a line i that is not marked, and any extension of the proof in which line j is marked may be further extended in such a way that line j is unmarked.*

Definition 10 $\Sigma^\Box \vdash_{\mathbf{MA}^k} A$ *(A is finally derivable from* $\Sigma^\Box$*) iff A is finally derived in an* **MA**[k]*-proof from* $\Sigma^\Box$.

I promised in the first section of this paper to enrich Kuipers' static notion of 'maximally successful' with a notion of 'maximally successful *at some moment in time*'. The latter notion is captured by the definition of *derivability at a stage*: explanatory hypotheses that are derived at some stage in an **MA**[k]-proof are maximally successful *relative to the insight in the premises at that stage*. Gaining a better insight in the premises (by further analysing them) may lead to a revision of what the maximally successful hypotheses are. Note that this at once gives an additional dimension to Kuipers' notion of empirical progress:

explanatory hypotheses are replaced by new and better ones, not on the basis of some absolute criterion that is unrealistic for all interesting cases, but on the basis of one's best available insights.

The dynamic proofs of $\mathbf{MA}^k$ nicely capture the dynamics of this process of empirical progress. Still, one needs to guarantee that the dynamics is sensible: that different dynamic proofs lead, "in the end," to the same set of explanatory hypotheses. This is warranted by the definition of *final derivability* (which corresponds to Kuipers' static notion of 'maximally successful').

In order to illustrate the proof theory for $\mathbf{MA}^k$, and to compare it with that for $\mathbf{MA}$, let us return to the astronomy example from the previous section.

As we have seen, the logic $\mathbf{LA}$ enabled us to derive, from the set of observational statements and the set of background assumptions, two explanatory hypotheses (Sa and Ca) and two predictions (Ta and $\neg Ta$). The logic $\mathbf{LA}^k$ leads to the same result (lines 1-5 are as in the previous section):

...

1	$\Box_2(\forall x)\,(Sx \supset (Hx \wedge \neg Lx \wedge \neg Mx))$		4	RU $\varnothing$
2	$\Diamond_2 Sa$		1-3,6	$RD^k\langle\{Sa\},\{Ha,\neg Ma\},\{La\}\rangle$
3	$\Diamond_2\neg Ta$		4, 7	RU $\langle\{Sa\},\{Ha,\neg Ma\},\{La\}\rangle$
4	$\Box_2(\forall x)\,(Cx \supset (Hx \wedge Lx \wedge Mx))$		5	RU $\varnothing$
5	$\Diamond_2 Ca$		1-3, 9	$RD^k\langle\{Ca\},\{Ha,La\},\{\neg Ma\}\rangle$
6	$\Diamond_2 Ta$		5, 10	RU $\langle\{Ca\},\{Ha,La\},\{\neg Ma\}\rangle$

The hypotheses are derived on lines 7 and 10, the predictions on lines 8 and 11. That neither of the hypotheses is withdrawn in view of the other seems justified: their empirical records are incomparable. Suppose, however, that we replace the third premise by Ma. In that case, we obtain the following:

...

3′	$\Box_1 Ma$	−	RP $\varnothing$

...

1	$\Box_2(\forall x)\,(Sx \supset (Hx \wedge \neg Lx \wedge \neg Mx))$		4	RU $\varnothing$
2	$\Diamond_2 Sa$		1-3′, 6	$RD^k\langle\{Sa\}, \{Ha\}, \{La, Ma\}\rangle$
3	$\Diamond_2\neg Ta$		4, 7	RU $\langle\{Sa\}, \{Ha\}, \{La, Ma\}\rangle$
4	$\Box_2(\forall x)\,(Cx \supset (Hx \wedge Lx \wedge Mx))$		5	RU $\varnothing$
5	$\Diamond_2 Ca$		1-3′, 9	$RD^k\langle\{Ca\}, \{Ha, La, Ma\}, \varnothing\rangle$

At stage 10, it becomes clear that the empirical successes for hypothesis Ca are a superset of those for Sa, and that the empirical problems for the former are a subset of those for the latter (compare the fifth element of lines 7 and 10). Hence, all lines that are derived on the condition that Sa is maximally successful are M-marked:

...

1	$\Diamond_2 Sa$		1-3′, 6	$RD^k\langle\{Sa\}, \{Ha\},\{La, Ma\}\rangle \checkmark_{M^{10}}$
2	$\Diamond_2\neg Ta$		4, 7	RU $\langle\{Sa\}, \{Ha\}, \{La, Ma\}\rangle \checkmark_{M^{10}}$

...

1 $\Diamond_2 Ca$ 1-3′, 9 RD^k $\langle \{Ca\},\ \{Ha, La, Ma\},\ \varnothing \rangle$

What this comes to is that the hypothesis *Sa* is withdrawn in view of the hypothesis *Ca*, and hence, that also all predictions based on *Sa* are withdrawn (remember that only formulas on non-marked lines are considered as derived). Note that all this is in line with Kuipers' analysis. According to his definitions, *Ca* is empirically more successful than *Sa*, and hence, the latter should be eliminated in favor of the former.

5. In Conclusion

In this paper, I presented the ampliative adaptive logic $\mathbf{LA}^k$ that is based on the modal adaptive logic $\mathbf{MA}^k$. I showed that $\mathbf{LA}^k$ captures the notion of empirical progress as studied by Theo Kuipers. One of the central characteristics of $\mathbf{LA}^k$ is that it enables one to generate, from a set of observational statement and a set of background assumptions, the most successful explanations.

One of the main results of this paper concerns the proof theory for $\mathbf{LA}^k$. Kuipers' definitions adequately capture the concept of empirical progress. However, the latter did not yet have a proof theory that does justice to these definitions. This proof theory is especially important as, at the predicative level, the notion of 'most successful' is not only undecidable, there even is no positive test for it. This raises the problem how to come to justified conclusions concerning the empirical success of explanatory hypotheses in undecidable contexts. The proof theory presented here solves that problem: it warrants that the conclusions one derives at a certain stage are justified in view of one's insights in the premises at that stage.

In this paper, I restricted the presentation of $\mathbf{LA}^k$ to the syntactic level. Evidently, the semantics should be designed, and the soundness and completeness proofs should be formulated. Another important problem concerns the design of alternative systems. $\mathbf{LA}^k$ only enables one to generate (the most successful) explanations for novel facts (facts not entailed by, but consistent with the theory). However, as Atocha Aliseda has convincingly argued in her [1], we also need systems to generate explanations for anomalous facts (facts not consistent with the theory). In Meheus *et al.* (2001), an adaptive logic for abduction is presented that is adequate for this. By adjusting this logic along the lines followed in the present paper, a logic for empirical progress may be obtained that is adequate for both novel facts and anomalies.[24]

---

[24] Unpublished papers in the reference section (and many others) are available from the internet address `http://logica.rug.ac.be/centrum/writings/`.

ACKNOWLEDGMENTS

I am indebted to Atocha Aliseda and to Theo Kuipers for many helpful comments and suggestions.

Ghent University
Centre for Logic and Philosophy of Science
Belgium
e-mail: joke.meheus@ugent.be

REFERENCES

Aliseda, A. (1997). *Seeking Explanations: Abduction in Logic, Philosophy of Science and Artificial Intelligence*. Dissertation Stanford. Amsterdam: ILLC Dissertations Series (1997-04).

Batens. D. (1989). Dynamic Dialectical Logics. In: G. Priest, R. Routley, and J. Norman (eds.), *Paraconsistent Logic. Essays on the Inconsistent*, pp. 187-217. München: Philosophia Verlag.

Batens, D. (1999). Zero Logic Adding up to Classical Logic. *Logical Studies* **2**, 15. (Electronic Journal: http://www.logic.ru/LogStud/02/LS2.html).

Batens, D. (2001). A Dynamic Characterization of the Pure Logic of Relevant Implication. *Journal of Philosophical Logic* **30**, 267-280.

Batens, D. (2004). On a Logic of Induction. In: R. Festa, A. Aliseda, J. Peijnenburg (eds.), *Confirmation, Empirical Progress, and Truth Approximation* (*Poznań Studies in the Philosophy of the Sciences and the Humanities*, vol. 83), pp. 193-219. Amsterdam/New York, NY: Rodopi.

Batens, D., K. De Clercq and G. Vanackere (forthcoming). Simplified Dynamic Proof Formats for Adaptive Logics.

Batens, D. and J. Meheus (2000). The Adaptive Logic of Compatibility. *Studia Logica* **66**, 327-348.

Batens, D. and J. Meheus (forthcoming). Adaptive Logics of Abduction.

Batens, D., J. Meheus, D. Provijn, and L. Verhoeven (2003). Some Adaptive Logics for Diagnosis. *Logic and Logical Philosophy* **11/12**, 39-65.

D'Hanis, I. (2000). Metaforen vanuit een taalfilosofisch, wetenschapsfilosofisch en logisch perspectief. Master's thesis. Gent: Universiteit Gent.

D'Hanis, I. (2002). A Logical Approach to the Analysis of Metaphors. In: Magnani *et al.* (2002), pp. 21-37.

Kuipers, T.A.F. (2000/ICR). *From Instrumentalism to Constructive Realism*. Dordrecht: Kluwer Academic Publishers.

Magnani, L., N.J. Nersessian and C. Pizzi, eds. (2002). *Logical and Computational Aspects of Model-Based Reasoning*. Dordrecht: Kluwer.

Meheus, J. (1999). Deductive and Ampliative Adaptive Logics as Tools in the Study of Creativity. *Foundations of Science* **4** (3), 325-336.

Meheus, J. (2000). An Extremely Rich Paraconsistent Logic and the Adaptive Logic Based on It. In: D. Batens, C. Mortensen, G. Priest, and J. P. Van Bendegem, (eds.), *Frontiers of Paraconsistent Logic*, pp. 189-201. Baldock, UK: Research Studies Press.

Meheus, J. (2001). Adaptive Logics for Question Evocation. *Logique et Analyse* **173-174-175**, 135-164. Appeared 2003.

Meheus, J. (2002). An Adaptive Logic for Pragmatic Truth. In: W. A. Carnielli, M. E. Coniglio, and I.M. Loffredo D'Ottaviano (eds.), *Paraconsistency. The Logical Way to the Inconsistent*, pp. 167-185. New York: Marcel Dekker.

Meheus, J. and D. Batens (forthcoming). Dynamic Proof Theories for Abductive Reasoning.

Meheus, J., L. Verhoeven, M. Van Dyck, and D. Provijn (forthcoming). Ampliative Adaptive Logics and the Foundation of Logic-Based Approaches to Abduction. In: L. Magnani *et al.* (2002), *Logical and Computational Aspects of Model-Based Reasoning*, pp. 39-71. Dordrecht: Kluwer.

Priest, G. (1991). Minimally Inconsistent **LP**. *Studia Logica* **50**, 321-331.

Vermeir, T. (forthcoming). Two Ampliative Adaptive Logics for the Closed World Assumption.

Weber, E., and D. Provijn (1999). A Formal Analysis of Diagnosis and Diagnostic Reasoning. *Logique et Analyse* **165-166**, 161-180. Appeared 2002.

Theo A. F. Kuipers

ANOTHER START FOR ABDUCTION AIMING AT EMPIRICAL PROGRESS

REPLY TO JOKE MEHEUS

As mentioned already in my reply to Aliseda, Joke Meheus was the second one to take up the challenge that I presented in 1998 and published in 1999, viz. to design a method, a logic or a computer program, for abducing a revised hypothesis that is empirically more successful than a given one. Whereas Aliseda starts from Beth's semantic tableaux method, Meheus starts from Batens' adaptive logic program. In this reply I would like to evaluate the question to what extent the specific logic developed by Joke Meheus meets the challenge. But let me start by stressing that, although her logic is in many respects incomplete, I appreciate it very much, for it seems a very promising start. She shows at least that the Ghentian style of ampliative adaptive logic enables one separately and comparatively to evaluate abductive individual hypotheses. More precisely, given a set of (general) background beliefs and (individual) observations, explanatory hypotheses can be derived by using a set of rules, consisting of the classical rules, amplified with some general and some specific ones, in a stepwise, adaptive way, that is, in the course of a proof, a previously derived conclusion may have to be withdrawn. In fact, it is a two-level construction; the adaptive, first-order, logic itself and a modal proof theory for it. The result is that hypotheses and predictions appear as possibilities in view of the background knowledge and the given observations. Besides the general rules and marking criteria for a general logic for abduction (**LA**), some specific rules and criteria are needed to get a specific logic for empirical progress (**LA**^{k}), abducing the maximally successful hypothesis, if any.

If I see it correctly, **LA**^{k} still has some severe restrictions, which might be withdrawn later. To begin with, as Meheus remarks herself, it is essentially restricted to hypotheses explaining surprising or novel events, that is, events that are not only not entailed by the background knowledge but also compatible with it. Moreover, it seems to be restricted to singular explanatory hypotheses. Last but not least, it essentially deals with the evaluation of explanatory hypotheses, not with their generation. In the rest of this reply, I first deal with the restriction

In: R. Festa, A. Aliseda and J. Peijnenburg (eds.), *Confirmation, Empirical Progress, and Truth Approximation* (*Poznań Studies in the Philosophy of the Sciences and the Humanities,* vol. 83), pp. 218-220. Amsterdam/New York, NY: Rodopi, 2005.

to singular hypotheses explaining surprising events, before turning to the generation issue.

Singular Hypotheses Explaining Novel Events

Let me start by noting that given the restriction to novel events, an explanatory hypothesis may be seen as a candidate for empirical progress relative to the background beliefs alone. The "old" hypothesis, to be replaced by a "new" one, may just be the tautology. However, as becomes clear from the final example (1)-(17), with (3) replaced by (3'), the method may also first lead to a hypothesis (11) that is later replaced by a better one (16). Hence, both separate and comparative evaluation is covered by the method. This seems to suggest how to proceed with anomalous observations, that is, observations in conflict with the background beliefs, at least as soon as the conflicting background beliefs can be shown to be a proper subset. In that case, the natural question is whether the conjunctive hypothesis of these beliefs can first be derived in $\mathbf{LA}^k$, possibly with using older observations, and then be replaced by a better one.

Let us now turn to the apparent restriction to singular hypotheses. If I see it correctly, hypotheses can only come in the game by RC in $\mathbf{LA}$ and, in addition, by RC^k in $\mathbf{LA}^k$. The question is whether, in both cases, the modally hypothesized A can be of the same conditional logical form as (the non-modal versions of) the background beliefs are apparently assumed to have. This form is essential for general explanatory hypotheses. In its simplest form, the question is whether $\mathbf{LA}^{(k)}$ can deal with (conditional) inductive generalizations. Be this as it may, my impression is that, if not, it will not be too difficult to adapt the method for this purpose. In both cases, a toy example might be very helpful.

There remains the question of the generalization of the method to the general instrumentalist abduction task, that is, the generation and evaluation of theory revision in the face of remaining counterexamples. For the evaluative side I should like to refer to my reply to Aliseda, whose method is in a similar position in this respect. However, regarding the generation side, the situation seems to be different.

Generation

As Aliseda (1997) has pointed out, abduction in the sense of Peirce essentially covers the generation and evaluation of explanatory hypotheses. However, $\mathbf{LA}^{(k)}$ does not generate a hypothesis, but evaluates it, in the sense that there may be routes of reasoning such that the hypothesis may be (conditionally) derived and not yet have to be withdrawn. The crucial rule $RC^{(k)}$ presupposes that one introduces the formula '*A*' oneself. Hence, the question is whether there is such a construction method for one or more of such hypotheses. In this respect, the tableau method of Aliseda and the one suggested by Urbanski (2001) seem to have an advantage. However, I do not want to rule out that Meheus might give her method a constructive turn. To be sure, a decent method to prove that a certain hypothesis may be abduced as the most successful one, relative to the background beliefs and the available evidence, of those that have been considered so far, is of independent, substantial value.

REFERENCES

Aliseda, A. (1997). *Seeking Explanations: Abduction in Logic, Philosophy of Science and Artificial Intelligence*. Dissertation Stanford. Amsterdam: ILLC Dissertations Series (1997-04).

Kuipers, T. (1999). Abduction Aiming at Empirical Progress or Even at Truth Approximation, Leading to Challenge for Computational Modelling. In: J. Meheus and T. Nickles (eds.): *Scientific Discovery and Creativity*, special issue of *Foundations of Science* **4** (3), 307-323.

Urbanski, M. (2001). Remarks on Synthetic Tableaux for Classical Propositional Calculus. *Bulletin of the Section of Logic* **30** (4), 195-204.

Diderik Batens

ON A LOGIC OF INDUCTION

ABSTRACT. In this paper I present a simple and straightforward logic of induction: a consequence relation characterized by a proof theory and a semantics. This system will be called LI. The premises will be restricted to, on the one hand, a set of empirical data and, on the other hand, a set of background generalizations. Among the consequences will be generalizations as well as singular statements, some of which may serve as predictions and explanations.

1. Prelude

I published my first paper in English a long time ago. In the paper (Batens 1968) I compared Carnap's and Popper's approach to induction, and basically assigned each approach a context of application, except that a modification was proposed for Popper's corroboration function. I had sent the paper to Carnap, Popper, Hempel, Kemeny, and several other famous people. With one exception, all had returned a few encouraging lines. Not long thereafter, I received a letter, in Dutch, by someone I immediately recognized as Dutch because he used an impressive number of middle initials – the Flemish use them in official documents only. The letter contained some questions and suggestions; a brief correspondence ensued.

I left the field later. However, for the sake of an old friendship, I dedicate this first logic of induction to Theo.

2. Aim of This Paper

It is often said that there is no logic of induction. This view is mistaken: this paper contains one. It is not a contribution to the great tradition of Carnapian inductive logic – see (Kuipers 2000, Ch. 4); it is a logic of induction in the most straightforward sense of the term, a logic that, from a set of empirical data and possibly a set of background generalizations, leads to a set of consequences that comprises generalizations and their consequences. Incidentally, the underlying

In: R. Festa, A. Aliseda and J. Peijnenburg (eds.), *Confirmation, Empirical Progress, and Truth Approximation* (*Poznań Studies in the Philosophy of the Sciences and the Humanities,* vol. 83), pp. 221-247. Amsterdam/New York, NY: Rodopi, 2005.

ideas oppose the claims that were widespread in Carnap's tradition – see, for example, (Bar-Hillel 1968).

LI is characterized by a proof theory and a semantics. Some people will take these properties to be insufficient for calling **LI** a logic. I shall not quarrel about this matter, which I take to be largely conventional. As far as I am concerned, any further occurrence of 'logic' may be read as 'giclo'. The essential point is that **LI** is characterized in a formally decent way, that its metatheory may be phrased in precise terms, and, most importantly, that **LI** may serve to explicate people's actual inductive reasoning.

LI takes as premises descriptions of empirical data as well as background generalizations that are formulated in the language of standard predicative logic. Its consequences follow either deductively or inductively from the premises. By deductive consequences I mean statements that follow from the premises by Classical Logic (**CL**). The main purpose of **LI** obviously concerns the inductive consequences. In this respect the proof of the pudding will be in the eating: the reader will have to read this paper to find out whether he or she considers **LI** as sensible with respect to the intended domain of application. For now, let me just mention that the inductive consequences of a set of empirical data and a set of background knowledge will, first and foremost, be empirical generalizations, and next, the deductive consequences of the empirical generalizations and the premises, including singular statements that may serve the purposes of prediction and explanation.

LI is only one member of a family of logics. It is severely restricted by the standard predicative language. This rules out statistical generalizations as well as quantitative predicates (lengths, weights, etc.). **LI** will not take account of degrees of confirmation or the number of confirming (and disconfirming) instances. **LI** will not deal with serious problems, usually connected to discovery and creativity, such as the genesis of new concepts and other forms of conceptual dynamics. Nor will **LI** deal with the historically frequent case of inconsistent background knowledge – see (Brown 1990), (Norton 1987), (Norton 1993), (Smith 1988), (Nersessian 2002), (Meheus 1993), (Meheus 2002), … **LI** is a bare backbone, a starting point.

More sophisticated inductive logics may be designed by modifying LI. Some of the required modifications are straightforward. But given the absence of any logic of induction of the kind, it seems advisable to present a simple system that applies in specific (although common) contexts. Incidentally, I shall also keep my remarks in defense and justification of **LI** as simple as possible. As most people reading the present book will be familiar with the literature on induction, they will easily see further arguments. It also seems wise, in defending a logic of induction, to refrain from siding with one of the many parties or schools in the research on induction. The logic **LI** is intended to please most of these parties. It should serve as a point of unification: this bit at least we all agree about, even if each explains it in his or her own way.

When working on this paper I wondered why a system as simple and clarifying as **LI** had not been presented a long time ago.[1] However, although **LI** is simple and straightforward to understand, its formulation presupposes familiarity with the adaptive logic programme. I shall not summarize this programme here because several easy introductions to its purpose and range are available, such as (Batens 2000) and (Batens forthcoming). Rather, I shall introduce the required adaptive elements as the paper proceeds. However, it is only fair to the reader to mention that the ideas underlying adaptive logics and dynamic proof theories have some pedigree and are by no means the outcome of the present research.

3. Tinkering with the Dynamic Proof Theory

Children have a profound tendency to generalization. This tendency has a clear survival value. In a sense, our present scientific (and other) knowledge is the result of a sophistication of this tendency. Of course, we know today that all simple empirical generalizations are false – compare (Popper 1973, p.10). This insight is a result of experience, of systematization, of free inquiry, and of systematic research. Our present knowledge, however, is neither the result of an urge that is qualitatively different from children's tendency to systematization, nor the outcome of a form of reasoning that is qualitatively different from theirs.

Let us for a moment consider the case in which only a set of empirical data is available – I shall remove this utterly unrealistic supposition in the present section. Where these empirical data are our only premises, what shall we want to derive from them? Apart from the **CL**-consequences of the premises, we shall also want to introduce some general hypotheses. Only by doing so may we hope to get a grasp of the world – to understand the world and to act in it. And from our premises and hypotheses together we shall want to derive **CL**-consequences (to test the hypotheses, to predict facts, and to explain facts).

LI should define a consequence relation that connects the premises with their **CL**-consequences, with the generalizations, and with their common **CL**-consequences. Is there such a consequence relation? Of course there is. The

---

[1] In the form of a formal logic, that is. Mill's canons come close. There are also clear connections with Reichenbach's straight rule, if restricted to general hypotheses, and with Popper's conjectures and refutations. Articulating the formal logic is worthwhile, as we shall see.

consequence relation is obviously non-monotonic[2] – inductive reasoning is the oldest and most familiar form of non-monotonic reasoning.

Generalizations that are inductively derived from the set of premises, Γ, should be compatible with Γ. A further requirement on inductively derived statements is that they should be *jointly* compatible with Γ. The latter requirement is the harder one. The logic of compatibility – see (Batens and Meheus 2000) – provides us with the set of all statements that are compatible with Γ. The problem of induction is, in its simplest guise, to narrow down this set in such a way that the second requirement is fulfilled. And yet, as I shall now explain, this problem is easy to solve.

Consider an (extremely simple) example of a **CL**-proof of the usual kind – for the time being, just disregard the $\varnothing$s at the end of the lines. As stated before, all premises will be singular statements.

1	$(Pa \wedge Pb) \wedge Pc$		PREM	$\varnothing$
2	$Rb \vee \sim Qb$		PREM	$\varnothing$
3	$Kb \supset \sim Pb$		PREM	$\varnothing$
4	$(Sa \wedge Sb) \wedge Qa$		PREM	$\varnothing$
5	Pa	1	RU	$\varnothing$
6	Pb	1	RU	$\varnothing$
7	Qa	4	RU	$\varnothing$
8	Sa	4	RU	$\varnothing$
9	Sb	4	RU	$\varnothing$

The rule applied in lines 5-9 is called **RU**. This name refers to the generic "unconditional rule." For the moment, just read it as: formula 5 is **CL**-derivable from formula 1, etc.

Suppose that our data comprise 1-4, and that we want to introduce an empirical generalization, for example $(\forall x)(Px \supset Sx)$. Obviously, this formula is not **CL**-derivable from 1-4. However, we may want to accept it *until and unless* is has been shown to be problematic – for example, because some P are not S. In other words, we may want to consider $(\forall x)(Px \supset Sx)$ as conditionally true in view of the premises. By a similar reasoning, we may want to consider $(\forall x)(Px \supset Qx)$ as conditionally true. This suggests that we add these universally quantified formulas to our proof, but attach a condition to them, indicating that the formulas will not be considered as derived if the condition shows false. So we extend the previous proof as follows[3]:

---

[2] A consequence relation '⊢' is non-monotonic iff a consequence of a set of premises need not be a consequence of an extension of this set. Formally: there is a Γ, a Δ, and an A such that $\Gamma \vdash A$ and $\Gamma \cup \Delta \nvdash A$.

[3] The superscript L_{14} on line 11 is explained below.

10	$(\forall x)(Px \supset Sx)$	RC	$\{(\forall x)(Px \supset Sx)\}$
11	$^{L}_{14}\ (\forall x)(Px \supset Qx)$	RC	$\{(\forall x)(Px \supset Qx)\}$

The set $\{(\forall x)(Px \supset Sx)\}$ will be called the *condition* of line 10. If some member of this set is contradicted by the data, the formula derived at line 10, which happens to be $(\forall x)(Px \supset Sx)$, should be withdrawn (considered as not derived). Conditionally derived formulas may obviously be combined by **RU**. As expected, the condition of the derived formula is the union of the conditions of the formulas from which it is derived. Here is an example:

12	$^{L}_{14}\ (\forall x)(Px \supset (Qx \wedge Sx))$	10, 11 RC	$\{(\forall x)(Px \supset Qx)\}$

The interpretation of the condition of line 12 is obviously that $(\forall x)(Px \supset (Qx \wedge Sx))$ should be considered as not derived if either $(\forall x)(Px \supset Sx)$ or $(\forall x)(Px \supset Qx)$ turns out to be problematic.

Logicians not familiar with dynamic proofs will complain that the negation of 11 *is* derivable from 1-4. Let me first show them to be right:

13	$\sim Q$	2, 3, 6 RU	$\varnothing$
14	$\sim(\forall x)(Px \supset Qx)$	6, 13 RU	$\varnothing$

As $(\forall x)(Px \supset Qx)$ is shown to be contradicted by the data, lines 11 and 12, which rely on the presupposition that $(\forall x)(Px \supset Qx)$ is not problematic, have to be *marked*. Formulas that occur in marked lines are considered as not being inductively derivable from the premises.[4]

Some logicians may still complain: 14 is **CL**-derivable from 1-4, and hence, they might reason, it was simply a mistake to add lines 11 and 12 to the proof. Here I strongly disagree. Moreover, the point touches an essential property of dynamic proofs; so let me explain the matter carefully.

Suppose that Γ is a finite set. In view of the restrictions on generalizations and on Γ, it is decidable whether a generalization (in the sense specified below) is or is not derivable, and hence it is decidable whether some singular statement is or is not derivable. So, indeed, one may avoid applications of RC that are later marked (if Γ is finite). However, nearly any variant of **LI** that overcomes some of the restrictions on **LI** – see earlier as well as subsequent sections – will be undecidable and, moreover, will lack a positive test for derivability.[5]

---

[4] When a line is marked I shall sometimes say that the formula that is its second element is marked. We shall see later that there are two kinds of marks, L and B. At stage 14, lines 11 and 12 have to be L-marked. Normally, one would just add an L to those lines. In order not to avoid repeating the proof at each stage, I add L_{14} to indicate that the lines are L-marked at stage 14 of the proof.

[5] A logic is decidable iff there is an algorithm to find out, for any set of premises and for any formula, whether the formula is derivable from the premises. There is a positive test for derivability iff there is an algorithm that leads, for any set of premises and for any formula, to the answer 'Yes' in case the formula is derivable from the premises. **CL**-derivability is decidable for the propositional

In view of this, and in preparation for those more fascinating variants, it seems rather pointless to try circumventing a dynamic proof theory for **LI**. There is a second argument and it should not be taken lightly. It is the purpose of the present paper to explicate actual inductive reasoning. Quite obviously, humans are unable to see all the relevant consequences of the available information. Given our finite brains it would be a bad policy to make inductive hypotheses contingent on complete deductive certainty. To do so would slow down our thinking and often paralyse it. This does not mean that we neglect deductive logic. It only means that we often base decisions on incomplete knowledge, including incomplete deductive knowledge – see (Batens 1995) for a formal approach to the analysis of deductive information. The third (and last) argument is of a different nature. I shall show in this paper that the dynamic proof theory of **LI** is formally sound and leads, within the bounds described in Section 2, to the desired conclusions. All this seems to offer a good reason to continue our journey.

To find out whether the sketched proof procedure holds water, we should also have a look at its weirder applications. Let us consider a predicate that does not occur in our premises, and see what happens to generalizations in which it occurs.

15   $^{L}_{17}$ $(\forall x)(Px \supset Tx)$ RC $\{(\forall x)(Px \supset Tx)\}$

Obviously, 1-4 do not enable one to contradict $(\forall x)(Px \supset Tx)$. However, we may moreover add:

16   $^{L}_{17}$ $(\forall x)(Px \supset {\sim}Tx)$ RC $\{(\forall x)(Px \supset {\sim}Tx)\}$

And now we see that we are in trouble, as the proof may obviously be continued as follows:

17    ${\sim}(\forall x)(Px \supset Tx) \vee {\sim}(\forall x)(Px \supset {\sim}Tx)$ 5 RU $\varnothing$

Although neither 15 nor 16 is contradicted by the empirical data, their conjunction is. The thing to do here is obvious (and well known from the Reliability strategy of adaptive logics). As 15 and 16 are on a par, both of them should be considered as *unreliable*, and hence hence lines 15 and 16 should both be marked in view of their conditions.[6]

Let me straighten this out and introduce some useful terminology. We suppose that generalizations are not problematic until and unless they are shown to

---

fragment of **CL** and undecidable for full **CL**. Nevertheless, there is a positive test for **CL**-derivability. A standard reference for such matters is (Boolos and Jeffrey 1989).

[6] In comparison to the Reliability strategy, the Minimal Abnormality strategy leads to a slightly richer consequence set in some cases. I return to this point later.

be contradicted by the empirical data. So the normal case will be that a generalization is compatible with the data. In view of this, the (derivable) negation of a generalization will be called an *abnormality*. Sometimes abnormalities are connected. Line 17 is a good example: the disjunction of two abnormalities is derivable, but neither of the abnormalities is. Derivable abnormalities and derivable disjunctions of abnormalities will be called *Dab*-formulas – an abnormality is itself a disjunction with one disjunct only. Where Δ is a finite set of generalizations, $Dab(\Delta)$ is a handy abbreviation for $\vee\{\sim\!A \mid A \in \Delta\}$ (the disjunction of the negations of the members of Δ).

In view of the derivability of 17, both $(\forall x)(Px \supset Tx)$ and $(\forall x)(Px \supset \sim\!Tx)$ are unreliable. But the fact that $Dab(\Delta)$ is derivable does not indicate that all members of Δ are unreliable. Indeed,

$$\sim(\forall x)(Px \supset Qx) \vee (\forall x)(Px \supset Sx)$$

is derivable from 14, but adding this formula to the proof does not render $(\forall x)(Px \supset Sx)$ unreliable. The reason is that, even if the displayed formula were added to the proof, it would not be a minimal *Dab*-formula in view of 14 (in the sense that a formula obtained by removing one of its disjuncts has been derived). *A* is unreliable at some stage of a proof, iff there is a Δ such that $A \in \Delta$ and $Dab(\Delta)$ is a *minimal Dab*-formula that is unconditionally derived in the proof at that stage.[7] Here is a further illustration:

18	$\sim(\forall x)(Px \supset Sx) \vee \sim(\forall x)(Px \supset \sim\!Sx)$	5 RU	$\varnothing$
19	$\sim(\forall x)(Px \supset Sx)$	5, 8 RU	$\varnothing$

At stage 18 of the proof, $(\forall x)(Px \supset Sx)$ is unreliable, and hence line 10 is marked. However, at stage 19, $(\forall x)(Px \supset Sx)$ is again reliable – 19 is a minimal *Dab*-formula at this stage, whereas 18 is not – and hence line 10 is unmarked.[8] This nicely illustrates both sides of the dynamics: formulas considered as derived at one stage may have to be considered as not derived at a later stage, and *vice versa*. All this may sound unfamiliar, or even weird. And yet, as we shall see in subsequent sections, everything is under control: ultimately the dynamics is bound to lead to stability, the stable situation is determined only by the premises (as the semantics illustrates), and there are heuristic means to speed up our journey towards the stable situation.

---

[7] The reason for considering only unconditionally derived formulas is straightforward. Indeed, from 17 one may derive $\sim(\forall x)(Px \supset \sim\!Tx)$ on the condition $\{(\forall x)(Px \supset Tx)\}$, but this obviously does not render (Vx)(Px D Tx) reliable. The disjunction 17 follows from the premises by **CL**, and neither of its disjuncts does.

[8] So, at stage 18 of the proof, an L is attached to line 10, but at stage 19 the L is removed.

228   *Diderik Batens*

Having made this precise – formal definitions follow in Section 4 – I leave it
to the reader to check that the introduction of 'wild' hypotheses leads nowhere.
As the predicates U and V do not occur in the premises 1-4, applying RC to add
formulas such as $(\forall x)(Ux \supset Vx)$ to the proof, will lead to lines that are bound to
be marked sooner or later – and very soon if some simple heuristic instructions
are followed.

Before moving on to background knowledge, let me add some important
comments. We have seen that $(\forall x)(Px \supset Qx)$ was not inductively derivable
from 1-4. However, $(\forall x)((Px \wedge Rx) \supset Qx)$ is. Indeed, line 20 below is not
marked in the present proof. In some artificial and clumsy extensions of the
proof, line 20 may be marked. But it is easy enough to further extend the proof in
such a way that line 20 is unmarked. This is an extremely important remark to
which I return later.

$\qquad$ 20 $\quad (\forall x)((Px \wedge Rx) \supset Qx)$ $\qquad\qquad$ RC $\ \{(\forall x)((Px \wedge Rx) \supset Qx)\}$

The next comment concerns the *form* of formulas derived by RC. All that
was specified before is that these formulas should be universally quantified.
However, a further restriction is required. Suppose that it is allowed to add

$\qquad$ [21] $\quad (\forall x)((Qx \vee \sim Qx) \supset \sim Sc)$ $\qquad$ RC $\ \{(\forall x)((Qx \vee \sim Qx) \supset \sim Sc)\}$

to the proof. As

$$\sim(\forall x)(Px \supset Sx) \ \vee \ \sim(\forall x)((Qx \vee \sim Qx) \supset \sim Sc)$$

is derivable from 1, not only line [21] but also line 10 would be marked in view
of this formula. Similar troubles arise if it is allowed to introduce such
hypotheses as $(\forall x)((Qx \vee \sim Qx) \supset (\exists x)(Px \wedge \sim Sx))$.

The way out of such troubles is simple enough. RC should not allow one to
introduce singular statements or existentially quantified statements in disguise.
Hence, we shall require that the generalizations introduced by RC consist of a
sequence of universal quantifiers followed by a formula of the form $A \supset B$ in
which no constants, propositional letters or quantifiers occur. From now on,
'*generalization*' will refer to such formulas only.[9] Some people will raise a
historical objection to this restriction. Kepler's laws explicitly refer to the sun,
and Galileo's law of the free fall to the earth. This, however, is related to the fact
that the earth, the sun, and the moon had a specific status in the Ptolemaic
worldview, and were slowly losing that status in the days of Kepler and Galileo.

---

[9] It is possible to devise a formal language in which the effect of this restriction is reduced to nil.
This is immaterial if such languages are not used in the empirical sciences to which we want to apply
LI. But indeed, the formal restriction hides one on content: all predicates should be well entrenched,
and not abbreviate identity to an individual constant.

In the Ptolemaic worldview, each of those three objects was taken, just like God, to be the only object of a specific kind. So those generalizations refer to kinds of objects, rather than to specific objects – by Newton's time, any possible doubt about this had been removed.[10]

Any *generalization* may be introduced by RC. This includes such formulas as 21 and 22, that are **CL**-equivalent to 23 and 24 respectively. So the implicative form of generalizations may be circumvented.

21	$(\forall x)((Qx \vee \sim Qx) \supset Px)$		RC	$\{(\forall x)((Qx \vee \sim Qx) \supset Px)\}$
22	$(\forall x)(Rx \supset (Qx \wedge \sim Qx))$		RC	$\{(\forall x)(Rx \supset (Qx \wedge \sim Qx))\}$
23	$(\forall x)Px$	21	RU	$\{(\forall x)((Qx \vee \sim Qx) \supset Px)\}$
24	$(\forall x)\sim Rx$	22	RU	$\{(\forall x)(Rx \supset (Qx \wedge \sim Qx))\}$

Is the dynamics of the proofs bound to stop at some finite point? The answer to this question is not simple, but nevertheless satisfactory. However, I postpone the related discussion until we have gained a better grasp of **LI**.

Let us now move on to situations in which *background knowledge* is available. Clearly, background knowledge cannot be considered as unquestionable. For one thing, the empirical data might contradict it. If they do, we face an inconsistent set of premises, which leaves us nowhere on the present approach.[11] So we shall consider background knowledge as defeasible. It is taken for granted *unless and until* it is shown to be problematic.

This being settled, it is simple enough to integrate background knowledge in the dynamic proof format. Background knowledge is the result of inductive inferences made in the past, by ourselves our by our predecessors.[12] For this reason, I shall restrict background knowledge to background generalizations – another simplification – and introduce them as *conditional premises*. Here is an example:

25	$(\forall x)(Qx \supset Rx)$		BK	$\{(\forall x)(Qx \supset Rx)\}$
26	Ra	7, 25	RU	$\{(\forall x)(Qx \supset Rx)\}$

---

[10] Even in the Ptolemaic era, those objects were identified in terms of well entrenched properties – properties relating to their kind, not to accidental qualities. The non-physical example is even more clear: God has *no* accidental properties.

[11] Scientists may justifiably stick to hypothetical knowledge that is falsified by the empirical data, for example because no non-falsified theory is available. Including such cases in the picture requires that we move to a paraconsistent logic. Although I have worked in this domain for some thirty years now, I fear that I would lose most of the readers if I were to open that Pandora's box. So let me stress that there is absolutely no problem in including the paraconsistent case in the logic of induction, but that I leave it out for reasons of space as well as for pedagogical reasons.

[12] Or rather, background knowledge is so interpreted for present purposes. This is a simplification. Humanity did not start from scratch at some point in time, not even with respect to scientific theories – see also Section 7.

The central difference between background generalizations and other generalizations – the latter will be called *local generalizations* from now on – is that the former are retained whenever possible. If $Dab(\Delta)$ is unconditionally derived, and each member of Δ is a background generalization, then, in the absence of further information, we have to consider all members of Δ as unreliable. So we shall mark all lines the condition of which overlaps with Δ. This includes the lines on which the background generalizations are introduced as conditional premises.[13]

If, however, we unconditionally derive $\sim A_1 \vee \ldots \vee \sim A_n \vee \sim B_1 \vee \ldots \vee \sim B_m$, and each A_i is a reliable background generalization (in the sense of the previous paragraph), then we should consider the local generalizations $B_1,\ldots, B_m$ as unreliable, and retain the background knowledge $A_1,\ldots, A_n$. Here is a simple example:

27	$^{L}{}_{29}$ $(\forall x)(Qx \supset \sim Rx)$		RC	$\{(\forall x)(Qx \supset \sim Rx)\}$
28	$^{L}{}_{29}$ $\sim Ra$	7, 27	RC	$\{(\forall x)(Qx \supset \sim Rx)\}$
29	$\sim(\forall x)(Qx \supset \sim Rx) \vee \sim(\forall x)(Qx \supset \sim Rx)$	7	RU	$\varnothing$

$(\forall x)(Qx \supset Rx)$ is a background generalization and has not been shown to be an unreliable background generalization.[14] But the local generalization $(\forall x)(Qx \supset \sim Rx)$ is unreliable in view of 29. Hence, lines 27 and 28 are marked, but lines 25 and 26 are not, as desired.

In view of the asymmetry between background hypotheses and local hypotheses, **LI** is a *prioritized* adaptive logic. This means that the members of one set of defeasible formulas, the background hypotheses, are retained at the expense of the members of another set, the local generalizations.

Before moving on to the precise formulation of the dynamic proof theory, let me intuitively explain some peculiarities of the proof format. Traditionally, a *proof* is seen as a list of formulas. This is not different for **LI**-proofs: the line numbers, the justification of the line (a set of line numbers and a rule), the conditions, and the marks are all introduced to make the proof more readable, but are not part of the proof itself. However, there is a central difference in this connection. In the dynamic case, one writes down a list of formulas, but the proof consists only of the *unmarked* formulas in the list. This does not make the marks part of the proof itself: which formulas are marked is determined by the

---

[13] Here is a simple example. If $(\forall x)(Px \supset Qx)$ is a background generalization, one may introduce it by the rule BK. However, this background generalization (and anything derived from it) would be B-marked in view of line 14. See the next section for the precise definition.

[14] This agrees with the above criterion: there is no set of background generalizations Δ such that $(\forall x)(Qx \supset Rx) \in \Delta$ and $Dab(\Delta)$ is a minimal Dab-formula at stage 29 of the proof.

empirical data, the background generalizations, and the list of formulas written down. Let us now continue to speak in terms of the annotated proof format.

What we are interested in are formulas that are *finally* derivable. On our way toward them, we have to go through the stages of a proof. Some formulas derived *at a stage* may not be finally derivable. As formulas that come with an empty condition (fifth element of the line) cannot possibly be marked at a later stage, they are sometimes called *unconditionally* derived. These formulas are deductively derived (by **CL**) from the empirical data. Formulas that have a non-empty condition are called *conditionally* derived. These formulas are inductively derived *only*. Of course, the interesting formulas are those that are inductively derived only, but nevertheless finally derived. In the present paper I offer a correct definition of final derivability, but cannot study the criteria that are useful from a computational point of view.

A last comment concerns the rules of inference. The *unconditional rules* of **LI** are those of Classical Logic, and they carry the conditions from their premises to their conclusion. The *conditional rules* BK and RC add a *new* element to the condition, and hence start off the dynamics of the proofs. As far as their structure is concerned, however, they are of the same type as the standard premise and axiom rules.

4. The Dynamic Proof Theory

Our language will be that of predicative logic. Let $\forall A$ denote A preceded by a universal quantifier over any variable free in A. A *generalization* is a formula of the form $\forall(A \supset B)$ in which no individual constant, sentential letter or quantifier occurs in either A or B.

A dynamic proof theory consists of (i) a set of unconditional rules, (ii) a set of conditional rules, and (iii) a definition of *marked* lines. The rules allow one to add lines to a proof. Formulas derived on a line that is marked at a stage of the proof are considered as not inductively derived at that stage (from the premises and background generalizations).[15]

Lines in an annotated dynamic proof have five elements: (i) a line number, (ii) the formula derived, (iii) a set of line numbers (of the lines from which the formula is derived), (iv) a rule (by which the formula is derived), and (v) a set of conditions.

---

[15] That background generalizations may be B-marked themselves illustrates that they are defeasible premises.

The logic **LI** operates on ordered sets of premises, $\Sigma = \langle \Gamma, \Gamma^* \rangle$, in which Γ is a set of *singular formulas* (the empirical data) and Γ^* is a set of *generalizations* (the background generalizations).

The rules of **LI** will be presented here in generic format. There are two unconditional rules, PREM and RU, and two conditional rules, BK and RC:

PREM   If $A \in \Gamma$, one may add a line comprising the following elements: (i) an appropriate line number, (ii) A, (iii) –, (iv) PREM, and (v) $\varnothing$.

BK   If $A \in \Gamma^*$, one may add a line comprising the following elements: (i) an appropriate line number, (ii) A, (iii) –, (iv) BK, and (v) $\{A\}$.

RU   If $A_1,\ldots, A_n \vdash_{\mathbf{CL}} B$ and each of $A_1,\ldots, A_n$ occur in the proof on lines $i_1,\ldots, i_n$ that have conditions $\Delta_1,\ldots, \Delta_n$ respectively, one may add a line comprising the following elements: (i) an appropriate line number, (ii) B, (iii) $i_1,\ldots, i_n$, (iv) RU, and (v) $\Delta_1 \cup \ldots \cup \Delta_n$.

RC   Where A is a *generalization*, one may add a line comprising the following elements: (i) an appropriate line number, (ii) A, (iii) –, (iv) RC, and (v) $\{A\}$.

A proof constructed by these rules will be called an **LI**-proof from Σ. In such a proof, a formula is *unconditionally derived* iff it is derived at a line of which the fifth element is empty. It is conditionally derived otherwise.

An *abnormality* is the negation of a generalization. *Dab*-formulas are formulas of the form $Dab(\Delta) = \bigvee\{\sim A \mid A \in \Delta\}$, in which Δ is a finite set of generalizations.[16] $Dab(\Delta)$ is a *minimal Dab*-formula at stage s of a proof iff $Dab(\Delta)$ is unconditionally derived in the proof at stage s and there is no $\Delta' \subset \Delta$ such that $Dab(\Delta')$ is unconditionally derived in the proof at that stage.

Definition 1   *Where $Dab(\Delta_1)$, ..., $Dab(\Delta_n)$ are the minimal Dab-formulas at stage s of a proof from $\Sigma = \langle \Gamma, \Gamma^* \rangle$, $U_s^*(\Gamma) = \bigcup\{\Delta_i \subseteq \Gamma^* \mid 1 \leq i \leq n\}$.*

Definition 2   *Where Δ is the fifth element of line i, line i is B-marked iff $\Delta \cap U_s^*(\Gamma) \neq \varnothing$.*

$U_s^*(\Gamma)$ comprises the background generalizations that are *unreliable* at stage s of the proof. As lines that depend on unreliable background generalizations are B-marked, these generalizations are themselves removed from the proof. This is interpreted by not considering them as part of the background knowledge at that stage of the proof. What remains of the background knowledge at stage s will be denoted by $\Gamma_s^* = \Gamma^* - U_s^*(\Gamma)$.

---

[16] Note that $Dab(\Delta)$ refers to any formula that belongs to an equivalence class that is closed under permutation and association.

Now we come to an important point. In order to determine which local generalizations are unreliable, we have to take the reliable background knowledge for granted. A *Dab*-formula *Dab* (Δ) will be called a *minimal local Dab-formula* iff no formula $Dab(\Delta')$ occurs in the proof such that $(\Delta' - \Gamma_s^*) \subset (\Delta - \Gamma_s^*)$.

Definition 3 *Where* $Dab(\Delta_1)$, ..., $Dab(\Delta_n)$ *are the minimal localDab-formulas at stage s of a proof from* $\Sigma = \langle \Gamma, \Gamma^* \rangle$, $U_s^o(\Gamma) = \bigcup \{\Delta_i - \Gamma_s^* \mid 1 \leq i \leq n\}$.

Definition 4 *Where* Δ *is the fifth element of a line i that is not B-marked, line i is L-marked iff* $\Delta \cap U_s^o(\Gamma) \neq \varnothing$.

$U_s^o(\Gamma)$ comprises the unreliable local generalizations at stage s. These generalizations may have been introduced by RC, they may be unreliable background generalizations, or they may be generalizations that do not occur in the proof (or occur as derived formulas only). Let me briefly clarify Definition 3.

Given the *B*-marks, we have to assess the hypotheses introduced by RC. Which of these are unreliable at stage s of the proof? The key to the answer to this question lies in the following theorem, the proof of which is obvious:

Theorem 1 $Dab(\Delta \cup \Delta')$ *is a minimal Dab-formula at stage s of a proof, iff a line may be added that has* $Dab(\Delta)$ *as its second, RC as its fourth, and* Δ' *as its fifth element.*

Suppose that, in a proof at stage s, Δ' contains only reliable background generalizations, whereas no such background generalization is a member of Δ — that is, $\Delta' \subseteq \Gamma_s^*$ and $\Delta \cap \Gamma_s^* = \varnothing$. That $Dab(\Delta)$ is derivable on the condition Δ' indicates that some member of Δ is unreliable *if* the background generalizations in Δ' are reliable. Moreover, we consider the background generalizations to be more trustworthy than the local generalizations. So from the occurrence of the minimal local *Dab*-consequence $Dab(\Delta \cup \Delta')$ we should conclude that the members of Δ are unreliable.

Incidentally, an equivalent (and also very intuitive) proof theory is obtained by defining $U_s^o(\Gamma)$ in a different way. Let $Dab(\Delta_1),..., Dab(\Delta_n)$ be the minimal (in the usual, simple sense) *Dab*-formulas that have been derived at stage s on the conditions $\Theta_1, ..., \Theta_n$ respectively, and for which $(\Delta_1 \cup ... \cup \Delta_n) \cap \Gamma_s^* = \varnothing$ and $\Theta_1 \cup ... \cup \Theta_n \subseteq \Gamma_s^*$. $U_s^o(\Gamma)$ may then be defined as $\Delta_1 \cup ... \cup \Delta_n$.[17] But let us stick to Definition 3 in the sequel.

Definition 5 *A formula A is derived at stage s of a proof from* Σ, *iff A is the second element of a non-marked line at stage s.*

---

[17] This alternative definition need not lead to the same results with respect to a specific proof at a stage, but it determines the same set of finally derivable consequences (see below in the text) in view of Theorem 1.

Definition 6 $\Sigma \vdash_{\mathbf{LI}} A$ (*A is finally* **LI**-*derivable from* Σ) *iff A is derived at a stage s of a proof from* Σ, *say at line i, and, whenever line i is marked in an extension of the proof, it is unmarked in a further extension of the proof.*

This definition is the same as for other dynamic proof theories. The following theorem is helpful to get a grasp of **LI**-proofs. The formulation is somewhat clumsy because the line may be marked, in which case A cannot be said to be derivable.

Theorem 2 *To an* **LI**-*proof from* $\Sigma = \langle \Gamma, \Gamma^* \rangle$ *a (marked or unmarked) line may be added that has A as its second element and* Δ *as its fifth element, iff* $\Gamma \vdash_{\mathbf{CL}} A \vee Dab(\Delta)$.

The proof of the theorem is extremely simple. Let the **CL**-transform of an **LI**-proof from $\Sigma = \langle \Gamma, \Gamma^* \rangle$ be obtained by replacing any line that has B as its second and Θ as its fifth element, by an unconditional line that has $B \vee Dab(\Theta)$ as its second element. To see that this **CL**-transform is a **CL**-proof from Γ, it is sufficient to note the following: (i) the **CL**-transform of applications of PREM are justified by PREM, (ii) the **CL**-transform of applications of BK and RC are justified in that they contain a **CL**-theorem of the form $A \vee {\sim}A$, and (iii) the **CL**-transform of applications of RU are turned into applications of the **CL**-derivable (generic) rule "If $A_1, ..., A_n \vdash_{\mathbf{CL}} B$, then from $A_1 \vee C_1, ..., A_n \vee C_n$ to derive $B \vee C_1 \vee ... \vee C_n$." This establishes one direction of the theorem. The proof of the other direction is immediate in view of the **LI**-derivable rule: "Where all members of Δ are generalizations, to derive A on the condition Δ from $A \vee Dab(\Delta)$."

So, in a sense, **LI**-proofs are **CL**-proofs in disguise. We interpret them in a specific way in order to decide which generalizations should be selected.

In order to obtain a better grasp of final derivability, I first define the sets of unreliable formulas with respect to Γ, independently of the stage of a proof. First we need: $Dab(\Delta)$ is a *minimal Dab-consequence* of Γ iff $\Gamma \vdash_{\mathbf{CL}} Dab(\Delta)$ and, for no $\Delta' \subset \Delta$, $\Gamma \vdash_{\mathbf{CL}} Dab(\Delta')$.

Definition 7 *Where* Ω^* *is the set of all minimal Dab-consequences of* Γ *in which occur only members of* Γ^*, $U^*(\Gamma) = \bigcup(\Omega^*)$.

This defines the set of background generalizations that are unreliable with respect to the empirical data Γ. The set of *retained* background generalizations is $\Gamma^*_\Gamma = \Gamma^* - U^*(\Gamma)$.

$Dab(\Delta)$ is a *minimal local Dab-consequence* of Γ iff $\Gamma \vdash_{\mathbf{CL}} Dab(\Delta)$ and, for no Δ', $\Gamma \vdash_{\mathbf{CL}} Dab(\Delta')$ and $(\Delta' - \Gamma^*_\Gamma) \subset (\Delta - \Gamma^*_\Gamma)$.[18]

---

[18] As $\Gamma \vdash_{\mathbf{CL}} Dab(\Delta)$, $\Delta \neq \varnothing$ and, in view of Definition 7, $\Delta' - \Gamma^*_\Gamma \neq \varnothing$; similarly for Δ'.

Definition 8 *Where Ω is the set of all minimal local Dab-consequences of Γ,*
$$U^{\circ}(\Gamma) = \bigcup(\Omega) - \Gamma_{\Gamma}^{*}.$$

This defines the set of local generalizations that are unreliable with respect to the empirical data Γ.

Given that **LI**-proofs are **CL**-proofs in disguise, the proofs of the following theorems can safely be left to the reader:

Theorem 3 *Where $\Sigma = \langle \Gamma, \Gamma^{*} \rangle$, $\Gamma_{\Gamma}^{*} = \{A \in \Gamma^{*} \mid \Sigma \vdash_{\mathbf{LI}} A\}$.*

Theorem 4 *Where $\Sigma = \langle \Gamma, \Gamma^{*} \rangle$, $\Sigma \vdash_{\mathbf{LI}} A$, A is finally **LI**-derivable from Σ, iff there is a (possibly empty) Δ such that (i) $\Gamma \cup \Gamma_{\Gamma}^{*} \vdash_{\mathbf{CL}} A \vee Dab(\Delta)$, and (ii) $(\Delta - \Gamma_{\Gamma}^{*}) \cap U^{\circ}(\Gamma) = \varnothing$.*

This sounds much simpler in words. A is an **LI**-consequence of Σ iff A is **CL**-derivable from Γ together with the reliable background generalizations, or, for some set A of reliable local generalizations,[19] $A \vee Dab(\Delta)$ is **CL**-derivable from Γ together with the reliable background generalizations.

The **LI**-consequence relation may be characterized in terms of compatibility – where Δ is compatible with Δ' iff $\Delta \cup \Delta'$ is consistent (iff no inconsistency is **CL**-derivable from this set).[20] The characterization is remarkably simple, as appears from the following three theorems. The proof of the theorems is obvious in view of Definition 7 and Theorem 4.

Theorem 5 $A \in \Gamma_{\Gamma}^{*}$ *iff* $A \in \Gamma^{*}$ *and* $\Delta \cup \{A\}$ *is compatible with* Γ *whenever* $\Delta \subseteq \Gamma^{*}$ *is compatible with* Γ.

A background generalization A is retained iff, whenever a set Δ of background generalizations is compatible with the data, then A and Δ are jointly compatible with the data.

Theorem 6 *Where $\Sigma = \langle \Gamma, \Gamma^{*} \rangle$ and A is a generalization, $\Sigma \vdash_{\mathbf{LI}} A$ iff $\Delta \cup \{A\}$ is compatible with $\Gamma \cup \Gamma_{\Gamma}^{*}$, whenever a set of generalizations Δ is compatible with $\Gamma \cup \Gamma_{\Gamma}^{*}$.*

A generalization A is inductively derivable iff, whenever a set Δ of generalizations is compatible with the data and retained background generalizations, then A and Δ are jointly compatible with the data and retained background generalizations. Let Σ^{G} be the set of generalizations that are inductively derivable from Σ.

Theorem 7 *Where $\Sigma = \langle \Gamma, \Gamma^{*} \rangle$, $\Sigma \vdash_{\mathbf{LI}} A$ iff $\Gamma \cup \Gamma_{\Gamma}^{*} \cup \Sigma^{G} \vdash_{\mathbf{CL}} A$.*

---

[19] Remark that $\Gamma \cup \Gamma_{\Gamma}^{*} \vdash_{\mathbf{CL}} A \vee Dab(\Delta)$ iff $\Gamma \cup \Gamma_{\Gamma}^{*} \vdash_{\mathbf{CL}} A \vee Dab(\Delta - \Gamma_{\Gamma}^{*})$.

[20] This definition presupposes that nothing is compatible with an inconsistent set – see also (Batens and Meheus 2000), for an alternative.

A is inductively derivable from a set of data and background generalizations iff it is **CL**-derivable from the data, the reliable background generalizations, and the inductively derivable generalizations.

Let me finally mention, without proofs, some properties of the **LI**-consequence relation: Non-Monotonicity, Proof Invariance (any two proofs from Γ define the same set of final consequences), **CL**-Closure ($Cn_{\mathbf{CL}}(Cn_{\mathbf{I}}(\Sigma)) = Cn_{\mathbf{I}}(\Sigma)$),[21] Decidability of $\langle \Gamma, \Gamma^* \rangle \vdash_{\mathbf{LI}} A$ whenever Γ and Γ^* are finite and A is either a generalization or a singular formula. Cautious cut with respect to facts: where A is a singular statement, if $\langle \Gamma, \Gamma^* \rangle \vdash_{\mathbf{LI}} A$ and $\langle \Gamma \cup \{A\}, \Gamma^* \rangle \vdash_{\mathbf{LI}} B$, then $\langle \Gamma, \Gamma^* \rangle \vdash_{\mathbf{LI}} B$. Cautious monotonicity with respect to facts: where A is a singular statement, if $\langle \Gamma, \Gamma^* \rangle \vdash_{\mathbf{LI}} A$ and $\langle \Gamma, \Gamma^* \rangle \vdash_{\mathbf{LI}} B$, then $\langle \Gamma \cup \{A\}, \Gamma^* \rangle \vdash_{\mathbf{LI}} B$. By the last two: that inductively derivable predictions are verified, does not lead to new inductive consequences. Cautious cut with respect to generalizations: where A is a generalization, if $\langle \Gamma, \Gamma^* \rangle \vdash_{\mathbf{LI}} A$ and $\langle \Gamma, \Gamma^* \cup \{A\} \rangle \vdash_{\mathbf{LI}} B$, then $\langle \Gamma, \Gamma^* \rangle \vdash_{\mathbf{LI}} B$. Cautious monotonicity with respect to generalizations: where A is a generalization, if $\langle \Gamma, \Gamma^* \rangle \vdash_{\mathbf{LI}} A$ and $\langle \Gamma, \Gamma^* \rangle \vdash_{\mathbf{LI}} B$, then $\langle \Gamma, \Gamma^* \cup \{A\} \rangle \vdash_{\mathbf{LI}} B$. By the last two: if inductively derivable generalizations are accepted as background knowledge, no new inductive consequences follow.

5. The Semantics

The previous sections merely considered the dynamic proof theory of **LI**. This proof theory is extremely important, as it enables us to explicate actual inductive reasoning – humans reach conclusions by finite sequences of steps. A logical semantics serves different purposes. Among other things, it provides insights into the conceptual machinery. Such insights increase our understanding of a logic, even if they are not directly relevant for the computational aspects.

Let $\mathcal{M}_\Gamma$ denote the set of **CL**-models of Γ. The **LI**-models $\Sigma = \langle \Gamma, \Gamma^* \rangle$, will be a subset of $\mathcal{M}_\Gamma$. This subset is defined in terms of the abnormal parts of models – see (Batens 1986) for the first application of this idea (to a completely different kind of logic). The abnormal part of a model (the set of abnormalities verified by a model) is defined as follows. Let $\mathcal{G}$ denote the set of generalizations.

Definition 9 $Ab(M) = \{\forall(A \supset B) \mid M \nvDash \forall (A \supset B); \forall(A \supset B) \in \mathcal{G}\}.$

In words: the abnormal part of a model is the set of generalizations it falsifies. Obviously, $Ab(M)$ is not empty for any model M. For example, either $(\forall x)((Px \vee \sim Px) \supset Qx) \in Ab(M)$ or $(\forall x)((Px \vee \sim Px) \supset \sim Qx) \in Ab(M)$ And if

---

[21] $Cn_{\mathbf{L}}(\Gamma) = \{A \mid \Gamma \vdash_{\mathbf{L}} A\}$ as usually.

$M \models Pa$, then either $(\forall x)(Px \supset Qx) \in Ab(M)$ or $(\forall x)(Px \supset {\sim}Qx) \in Ab(M)$. However, in some models of Pa both $(\forall x)(Px \supset Qx)$ and $(\forall x)(Px \supset {\sim}Qx)$ belong to $Ab(M)$, whereas in others only one of them does.

Given that **CL** is sound and complete with respect to its semantics, $Dab(\Delta)$ is a *minimal Dab-consequence* of Γ iff all $M \in \mathcal{M}_\Gamma$ verify $Dab(\Delta)$ and no $\Delta' \subset \Delta$ is such that all $M \in \mathcal{M}_\Gamma$ verify $Dab(\Delta')$.

This semantic characterization of the minimal *Dab*-consequences of Γ immediately provides a semantic characterization of $U^*(\Gamma)$, of $U^{\circ}(\Gamma)$, and of Γ_Γ^*. This is sufficient to make the first required selection. The proof of Theorem 8 is obvious.

Definition 10 $M \in \mathcal{M}_\Gamma$ *is* background-reliable *iff* $(Ab(M) \cap \Gamma^*) \subseteq U^*(\Gamma)$.

Theorem 8 $M \in \mathcal{M}_\Gamma$ is background-reliable iff $M \models \Gamma_\Gamma^*$.

In words, the retained background knowledge consists of the members of Γ^* that are verified by all background-reliable models of Γ. So a model M of Γ is background-reliable iff it verifies all reliable background generalizations. For any consistent Γ and for any set of background generalizations Γ^*, there are background-reliable models of Γ.[22] This is warranted by the compactness of **CL**: Γ is compatible with Γ_Γ^* iff it is compatible with any finite subset of Γ_Γ^*.

I now proceed to the second selection of models of Γ.

Definition 11 $M \in \mathcal{M}_\Gamma$ *is* reliable (*is an* **LI**-*model of* Σ)[23] *iff* $Ab(M) \subseteq U^{\circ}(\Gamma)$.

Since, in view of Definitions 7 and 8, $U^*(\Gamma) = U^{\circ}(\Gamma) \cap \Gamma^*$, it follows that:

Theorem 9 All reliable models of Σ are background reliable.

One should not be misled by this. $Ab(M) \subseteq U^{\circ}(\Gamma)$ only warrants $(Ab(M) \cap \Gamma^*) \subseteq U^*(\Gamma)$ because the definition of $U^{\circ}(\Gamma)$ refers to the definition of $U^*(\Gamma)$.

Definition 12 *Where* $\Sigma = \langle \Gamma, \Gamma^* \rangle$, $\Sigma \models_{\mathbf{LI}} A$ *iff all reliable models of* Γ *verify* A.

Theorem 10 $\Sigma \vdash_{\mathbf{LI}} A$ *iff* $\Sigma \models_{\mathbf{LI}} A$. (*Soundness and Completeness*)

The proof is longwinded, especially its right-left direction, but follows exactly the reasoning of the proofs of Theorems 5.1 and 5.2 from (Batens 1999). The present proof is simpler, however, as it concerns **CL**.

Some further provable properties: Strong Reassurance (if a **CL**-model M of Γ is not an **LI**-model of Σ, then some **LI**-model M' of Σ is such that $Ab(M') \subset$

---

[22] In the worst case, all background generalizations are unreliable, and hence all models of Γ are background-reliable.

[23] As **LI** is an adaptive logic, it does not make sense to say that M is or is not an **LI**-model, but only that M is or is not an **LI**-model of some Σ.

Ab(*M*)), and Determinism of final derivability (the co-extensive semantic consequence relation defines a unique consequence set for any Σ).

Although it is important to semantically characterize final **LI**-derivability in terms of a set of models of Σ, some might complain that the dynamics of the proofs does not appear in the semantics. However, there is a simple method to obtain a dynamic semantics for adaptive logics. This method, exemplified in (Batens 1995), offers a dynamic semantics that is characteristic for derivability at a stage.

A slightly different (and richer) result would be obtained by applying the Minimal Abnormality strategy. I skip technicalities and merely mention the central difference from the Reliability strategy. In the presence of an instance[24] of *Px* and in the absence of instances of both $Px \wedge Qx$ and $Px \wedge {\sim}Qx$, the Reliability strategy leads to the rejection of both $(\forall x)\,(Px \supset Qx)$ and $(\forall x)\,(Px \supset {\sim}Qx)$ – if any of these formulas occurs in the fifth element of a line, the line is marked. It follows that even the disjunction of both generalizations will be marked. On the Minimal Abnormality strategy, both generalizations are marked but their disjunction is not. This supplementary consequence seems weak and pointless. Moreover, the Minimal Abnormality strategy, while leading to a very simple semantics, terribly complicates the proof theory. For this reason I shall not discuss it further here.

6. Heuristic Matters and Further Comments

Some people think that all adaptive reasoning (including all non-monotonic reasoning) should be explicated in terms of heuristic moves rather than in terms of logic proper. For their instruction and confusion, I shall first spell out some basics of the heuristics of the adaptive logic **LI**. I leave it to the reader to compare both conceptions.

Suppose that one applies RC to introduce, on line *i*, the generalization $\forall(A \supset B)$ on the condition $\{\forall(A \supset B)\}$. As (1) is a **CL**-theorem, it may be derived in the proof and causes $\forall(A \supset B)$ to be *L*-marked.

$${\sim}\forall(A \supset B) \vee {\sim}\forall(A \supset {\sim}B) \vee {\sim}\forall({\sim}A \supset B) \vee {\sim}\forall({\sim}A \supset {\sim}B) \tag{1}$$

So, in order to prevent $\forall(A \supset B)$ from being *L*-marked, one needs to uncondi-tionally derive

$${\sim}\forall(A \supset {\sim}B) \vee {\sim}\forall({\sim}A \supset B) \vee {\sim}\forall({\sim}A \supset {\sim}B)$$

---

[24] An instance of the open formula *A*, is any closed formula obtained by replacing each variable free in *A* by some constant.

or a "sub-disjunction" of it. How does one do so? An instance of A enables one to derive

$$\sim\forall(A \supset B) \lor \sim\forall(A \supset \sim B) \tag{2}$$

whereas an instance of $\sim A$ enables one to derive

$$\sim\forall(A \supset B) \lor \sim\forall(A \supset \sim B) \tag{3}$$

An instance of $A \land B$ enables one to derive

$$\sim\forall(A \supset \sim B) \tag{4}$$

and so on.

In view of this, it is obvious how one should proceed. Suppose that one is interested in the relation between A and B. It does not make sense to introduce by RC, for example, the generalization $\forall(A \supset B)$, if falsifying instances (instances of $A \land \sim B$) are derivable – if there are, the generalization is marked and will remain marked forever. Moreover, in order to prevent $\forall(A \supset B)$ becoming marked in view of (1) or (2), one needs a confirming[25] instance (an instance of $A \land B$) and one needs to derive (4) from it. So two aims have to be pursued: (i) search for instances of $A \land \sim B$ in order to make sure that one did not introduce a falsified generalization, and (ii) search for instances of $A \land B$ in order to make sure that the generalization is not marked.

To see that the matter is not circular, note that it does not make sense, with respect to (ii) from the previous paragraph, to derive, say $B(a)$ from $A(a)$ together with the generalization $\forall(A \supset B)$ itself. Indeed, $B(a)$ will then be derived on the condition $\{\forall(A \supset B)\}$. (4) is derivable from $B(a)$, but again only on the condition $\{\forall(A \supset B)\}$. The only *Dab*-formula that can be unconditionally derived from (4) on the condition $\{\forall(A \supset B)\}$ is (2) – compare Theorem 2. In view of this, the line at which $\forall(A \supset B)$ was introduced by RC will still be marked.

But suppose that $A(a)$ and $C(a)$ occur unconditionally in the proof and that the generalization $\forall(C \supset B)$ was introduced by RC. If we derive $B(a)$ from these, it will be derived on the condition $\{\forall(C \supset B)\}$. So we are not able to unconditionally derive (4) from $A(a)$ and $B(a)$. All we can unconditionally derive along this road is

$$\sim\forall(A \supset \sim B) \lor \sim\forall(C \supset B) \tag{5}$$

and, in view of this, both $\forall(C \supset B)$ and $B(a)$ will be marked.

The reader might find this weird. There may be unconditional instances of $C \land B$ in the proof, and hence $\sim\forall(C \supset \sim B)$ may be unconditionally derived. This seems to warrant that $\forall(C \supset B)$ is finally derived, but obviously it does not. If

---

[25] Obviously, 'confirming' is meant here in the qualitative sense – see (Kuipers 2000, Ch. 2).

such unexpected dependencies between abnormalities obtain, are we not losing control? Nothing very complicated is actually going on here. Control is provided by the following simple and intuitive fact:

> (†) If the introduction a local generalization G entails a falsifying instance of another generalization $\forall(A \supset B)$, and no falsifying instance of the latter is derivable from the empirical data together with the reliable background knowledge, then $\sim\!G \lor \sim\!\forall(A \supset \sim\!B)$ is unconditionally derivable.

What does all this teach us? If we introduce a generalization, we want to find out whether it is finally derived in view of the present data. In order to do so, we should look for falsifying as well as for confirming instances, and we should look for falsifying instances of other generalizations, as specified in (†).[26] These instances may be derived from the union of the empirical data, the reliable background generalizations, and the reliable local generalizations. There is a clear bootstrapping effect here. At the level of the local generalizations the effect is weak, in that wild generalizations will not be finally derivable. At the level of the background generalizations, the effect is very strong – it is only annihilated by falsifying instances. However, at the level of the local generalizations, the bootstrapping effect does *not* reduce to a form of circularity.

So in order to speed up our journey towards the stable situation we need to look for the instances mentioned in the previous paragraph. As this statement may easily be misunderstood let me clarify it. Let the generalization introduced by RC be $\forall(A \supset B)$. (i) We need to find a confirming instance – if there are none, the generalization is bound to be marked.[27] (ii) We need to search for falsifying instances of the generalization and for falsifying instances of other generalizations that are novel with respect to the empirical data and reliable background generalizations – if there are falsifying instances of either kind, the generalization is bound to be marked. As a result of the search for falsifying instances (of either kind), we may find more confirming instances as well as a number of undetermined cases – individual constants for which there is an instance of A but not of either B or $\sim\!B$. When new empirical data become available, objects about which we had no information, or only partial information, may turn out to be falsifying, and so may objects about which we can only derive *conditionally* that they are confirming. So, (iii) we need to collect further data, by observation and experiment. At this point, confirmation theory enters

---

[26] It is unlikely that effects like the one described by (†) will be discovered if one does not handle induction in terms of a logic. I have never seen such effects mentioned in the literature on induction, and they certainly are not mentioned in (Kuipers 2000).

[27] This should be qualified. If there are instances of $\sim\!A$ and none of A, then $\forall(A \supset B)$ may be derivable and unmarked because $\forall A$ is so.

the picture. Although **LI** does not take into account the number of confirming instances, only well-established hypotheses will convincingly eliminate potential falsifiers. Incidentally, I tend to side with Popper in this respect: what is important is not the number of confirming instances, but rather the strength of the tests to which a generalization has been subjected. Whether this concept may be explicated within the present qualitative framework is dubious.

Although the heuristics of **LI** depends on confirmation theory in the sense described above, **LI** in itself enables us to spell out quite interesting heuristic maxims. Given a set of empirical data and a set of background generalizations, it is clear how we should proceed. Most of what was said above relates to that. If the given data and background knowledge do not allow one to finally derive any generalization concerning the relation between A and B because there is insufficient information, **LI** clearly instructs one about the kind of data that should be gathered to change the situation. In this sense, **LI** *does* guide empirical research. This guidance may be considered somewhat unsophisticated, but it is the basic guidance, the one that points out the most urgent empirical research.

I now turn to a different kind of heuristic maxims. In order to speed up our journey towards stability with respect to *given* empirical data and background generalizations, it is essential to derive as soon as possible the minimal *Dab*-consequences of Γ and to derive as soon as possible the minimal local *Dab-consequences* of Γ. Some **LI**-derivable rules are extremely helpful in this respect, and are related to deriving inconsistencies – the techniques to do so are well-known from the **CL**-heuristics. I mention only two examples. Suppose that, in an Li-proof from Σ, A is unconditionally derived, and that $\sim\!A$ is derived on the condition Δ. Then $Dab(\Delta)$ is unconditionally derivable in the proof. Similarly, if an inconsistency is derived on the condition Δ, $Dab(\Delta)$ is unconditionally derivable in the proof.

An equally helpful derivable rule was exemplified before (and is warranted by Theorem 2). If a *Dab*-formula $Dab(\Delta)$ is derived on the condition Δ', then $Dab(\Delta \cup \Delta')$ is unconditionally derivable. Similarly, if an instance of A is derived on the condition Δ and an instance of B is derived on the condition Δ', then $\sim\!\forall(A \supset \sim\!B) \lor Dab(\Delta \cup \Delta')$ is unconditionally derivable – either or both of Δ and Δ' may be empty.

A very rough summary reads as follows: derive all singular statements that lead to instances of formulas no instances of which have been derived, and derive *Dab*-formulas that change the either minimal *Dab*-formulas or the minimal local *Dab*-formulas. The first instruction requires the derivation of a few formulas only. The second may be guided by several considerations, (i) Whenever $Dab(\Delta)$ has been derived, one should try to unconditionally derive $Dab(\Delta')$ for all $\Delta' \subset \Delta$. This is a simple and decidable task, (ii) One should only try to derive

Dab(Δ) when Δ consist of background generalizations, generalizations introduced by the rule RC, or "variants" of such generalizations – the variants of ∀(*A* ⊃ *B*) being the four generalizations that occur in (1). This instruction may be further restricted. Given a background generalization or local generalization ∀(*A* ⊃ *B*), one should first and foremost try to derive the *Dab*(Δ) for which Δ contains variants of ∀(*A* ⊃ *B*). The only cases in which it pays to consider other *Dab*-formulas is the one described in (†).

Up to now I have considered the general heuristic maxims that apply to **LI**. However, **LI** has distinct application contexts, in which different aims are pursued and specific heuristic maxims apply. I shall consider only two very general application contexts.

If one tries to derive *Dab*-formulas that result in some lines being marked or unmarked, one basically checks whether the introduced generalizations are compatible with and confirmed by the available empirical data. However, one might also, after introducing a generalization, concentrate on its consequences by deriving singular statements from it. These singular statements will be derived conditionally. As said before, this may be taken to be a good reason to invoke observation and experiment in order to test them. This leaves room for a "Popperian" application of **LI**. Even if a generalization may be marked in view of derivable *Dab*-formulas, and even if it *is* marked in view of derived *Dab*-formulas, we may try to gather novel data that cause the generalization to be unmarked. Incidentally, the "stronger" generalizations in the sense of (Popper 1935) and (Popper 1963) are those from which a larger number of weaker generalizations are derivable, and hence have more potential falsifiers. Popper was quite right, too, to stress that it is advisable to infer the most general (the bolder) generalizations first. If they become marked, we may still retract to less general generalizations. As long as these are not marked, the less general generalizations are available for free because they are **CL**-consequences of the more general ones.

A distinction is useful in the present context. If an instance of *Px* is derivable from the empirical data together with the reliable background knowledge, but no instances of either *Px* ∧ *Qx* or *Px* ∧ ~*Qx* are so derivable, then both (∀*x*)(*Px* ⊃ *Qx*) and (∀*x*)(*Px* ⊃ ~*Qx*) may be marked *because we have no means to choose between them*. If instances of both *Px* ∧ *Qx* and *Px* ∧ ~*Qx* are **CL**-derivable from the empirical data together with the reliable background knowledge, then both (∀*x*)(*Px* ⊃ *Qx*) and (∀*x*)(*Px* ⊃ ~*Qx*) may be marked *because both are falsified*. The transition from the first situation to the second clearly indicates an increase in knowledge. Moreover, in the second situation it does not make sense to look for further confirming instances of either generalization. What does make sense in the second situation, and not in the first, is that one looks for less general hypotheses, for example $(\forall x)((Px \wedge Rx) \supset Qx)$ that may still be derivable.

This at once answers the objection that **LI** too severely restricts a scientist's freedom to launch hypotheses. **LI** does not in any way restrict the freedom to introduce generalizations. Rather, **LI** points out, if sensibly applied, which generalizations cannot be upheld, and which empirical research is desirable. A scientist's "freedom" to launch hypotheses is not a permission for dogmatism – to make a claim and stick to it. If it refers to anything, then it is to the freedom to break out of established conceptual schemes. Clearly, the introduction of new conceptual schemes goes far beyond the present simple logic of induction – I return to this in Section 7. Given the limits of **LI**, the set of **LI**-consequences of a given Σ should be determined by Σ and should be independent of any specific line of reasoning. In this respect the rule RC differs drastically from such rules as Hintikka's bracketing rule – see, for example, (Hintikka 1999) and (Hintikka forthcoming).

A very different application context concerns predictions derived in view of actions. It makes sense, in the Popperian context, to derive predictions from a generalization A, even before checking whether the proof can be extended in such a way that A is marked. In the present context, it does not. It would be foolish to act on the generalization $(\forall x)(Px \supset Qx)$ in the absence of confirming instances – such actions would be arbitrary. In action contexts, one should play the game in a safer way by introducing only well-confirmed generalizations, not bold ones. Thus $(\forall x)(Px \supset Qx)$ should be *derived* from safe generalizations, for example, $(\forall x)((Px \wedge Rx) \supset Qx)$ and $(\forall x)((Px \wedge \sim Rx) \supset Qx)$ if both of these happen to be safe.

In both contexts,[28] **LI** suggests a specific heuristic procedure. This procedure differs from one context to the other, and may be justified in view of the specific aims.

Some people may find it suspect that applications of the rule RC do not require the presence of any formulas in the proof. RC is a positing rule rather than a deduction rule. This is no reason to worry. **LI** has a dynamic proof theory. A proof at a stage should not be confused with a proof of a logic that has a (static) proof theory of the usual kind. The central question in an **LI**-proof is not whether a generalization can be introduced, but whether it can be retained – the aim is final derivability, not derivability at a stage. The preceding paragraphs make it sufficiently clear that final derivability is often difficult to reach, and that one needs to follow a set of heuristic rules in order even to obtain a sensible estimate of final derivability – see also below. In this connection, it is instructive

---

[28] This distinction between action contexts and contexts concerning theoretical inquiry was one of the points made in my (Batens 1968).

to see that $Cn_I(\langle\varnothing, \varnothing\rangle) = Cn_{CL}(\varnothing)$, that $Cn_I(\langle\{Pa, \sim Pa\}, \varnothing\rangle) = Cn_{CL}(\{Pa, \sim Pa\})$, and hence that neither of these comprises a non-tautological generalization.

A final comment concerns the nature of an adaptive logic. It would be foolish to build a logic that allows for some mistakes. Obviously, adaptive logics do not allow for mistakes: $Cn_I(\Sigma)$ is a well-defined set that leaves no room for any choice or arbitrariness. The dynamic proof theory constitutes a way to search for $Cn_I(\Sigma)$. A proof at a stage merely offers an estimate of $Cn_I(\Sigma)$ – an estimate that is determined by the insights in the premises that are provided by the proof. We have seen that there are heuristic means to make these insights as rich and useful as possible. There also are criteria to decide, in some cases, whether a formula is finally derived in a proof – see (Batens 2002). In the absence of a positive test, that is the best one can do in a computational respect.

For large fragments of the language, **LI**-derivability is decidable. This includes all generalizations, and hence all predictions and explanations. But even for undecidable fragments of the language, dynamic proofs at a stage offer a sensible estimate of $Cn_I(\Sigma)$, the best estimate that is available from the proof – see (Batens 1995). This means that an **LI**-proof at a stage is sufficient to take justified decisions: decisions that may be mistaken, but are justified in terms of our present best insights.

7. Further Research

As announced, **LI** is very simple – only a starting point. In the present section I briefly point to some open problems. Some of these relate to alternatives for **LI**, others to desirable sophistication.

With respect to background generalizations, an interesting alternative approach is obtained by not introducing members of Γ^* but rather generalizations that belong to $Cn_{CL}(\Gamma^*)$. Suppose that $(\forall x)(Px \supset Qx) \in \Gamma^*$, and that Pa, Ra and $\sim Qa$ are **CL**-consequences of Γ. According to **LI**, $(\forall x)(Px \supset Qx)$ is falsified, and hence not retained. According to the alternative, $(\forall x)((Px \wedge \sim Rx) \supset Qx)$ would, for all that has been said, be a retained background generalization. This certainly deserve further study, both from a technical point of view and with respect to application contexts.

LI is too empiricist, even too positivistic. Let me just mention some obvious sophistication that is clearly desirable. Sometimes our background knowledge is inconsistent and sometimes falsified generalizations are retained. As there is room for neither in **LI**, this logic is unfit to explicate certain episodes from the history of the sciences. It is not difficult to modify **LI** in such a way that both inconsistent background knowledge and the application of falsified generalizations are handled. Available (and published) results on inconsistency adaptive logics make this change a rather easy exercise.

Another weakness of **LI**, or rather of the way in which **LI** is presented in the present paper, is that there seems to be only room for theories in the simple sense of the term: sets of generalizations. This weakness concerns especially background theories – the design of new theories is not a simple inductive matter anyway. Several of the problems listed above are solved in (Batens and Haesaert forthcoming); this paper contains also a variant of **LI** that follows the standard format for adaptive logics.

LI does not enable one to get a grasp of conceptual change or of similar phenomena that are often related to scientific creativity and discovery. This will be the hardest nut to crack. That it is not impossible to crack it will be obvious to readers of such papers as (Meheus 1999a), (Meheus 1999b), and (Meheus 2000).

Let me say no more about projected research. The basic result of the present paper is that there is now a logic of induction. It is simple, and even a bit old-fashioned, but it exists and may be applied in simple circumstances.[29]

ACKNOWLEDGMENTS

Research for this paper was supported by subventions from Ghent University and from the Fund for Scientific Research – Flanders, and indirectly by the the Flemish Minister responsible for Science and Technology (contract BIL98/37). I am indebted to Atocha Aliseda, Theo Kuipers, Dagmar Provijn, Ewout Vansteenkiste, and Liza Verhoeven for comments on previous drafts.

University of Ghent
Centre for Logic and Philosophy of Science
Blandijnberg 2
B-9000 Ghent
Belgium
e-mail: diderik.batens@rug.ac.be

---

[29] Unpublished papers by members of our research group are available from the internet address `http://logica.rug.ac.be/centrum/writings/`.

Diderik Batens

REFERENCES

Bar-Hillel, Y. (1968). The Acceptance Syndrome. In: I. Lakatos (ed.), *The Problem of Inductive Logic*, pp. 150-161. North-Holland, Amsterdam.

Batens, D. (1968). Some Proposals for the Solution of the Carnap-Popper Discussion on Inductive Logic. *Studia Philosophica Gandensia* **6**, 5-25.

Batens, D. (1986). Dialectical Dynamics within Formal Logics. *Logique et Analyse* **114**, 161-173.

Batens, D. (1995). Blocks. The Clue to Dynamic Aspects of Logic. *Logique et Analyse* **150-152**, 285-328. Appeared 1997.

Batens, D. (1999). Inconsistency-Adaptive Logics. In: E. Orłowska (ed.), *Logic at Work. Essays Dedicated to the Memory of Helena Rasiowa*, pp. 445-472. Heidelberg/New York: Physica Verlag Springer.

Batens, D. (2000). A Survey of Inconsistency-Adaptive Logics. In: D. Batens, C. Mortensen, G. Priest, and J.P. Van Bendegem (eds.), *Frontiers of Paraconsistent Logic*, pp. 49-73. Baldock, UK: Research Studies Press.

Batens, D. (2002). On a Partial Decision Method for Dynamic Proofs. In: H. Decker, J. Villadsen, and T. Waragai (eds.), *PCL 2002. Paraconsistent Computational Logic*, pp. 91-108. *Datalogiske Skrifter* vol. 95. Also available as cs.LO/0207090 at `http://arxiv.org/archive/cs/intro.html`.

Batens, D. (2004). Extending the Realm of Logic. The Adaptive-Logic Programme. In: P. Weingartner (ed.), *Alternative Logics. Do Sciences Need Them?* Berlin, Heidelberg: Springer Verlag.

Batens, D. and L. Haesaert (2001). On Classical Adaptive Logics of Induction. *Logique et Analyse* **173-175**, 255-290. Appeared 2003.

Batens, D. and J. Meheus (2000). The Adaptive Logic of Compatibility. *Studia Logica* **66**, 327-348.

Boolos, G.S. and R.J. Jeffrey (1989). *Computability and Logic.* Third edition. Cambridge: Cambridge University Press.

Brown, B. (1990). How to Be Realistic about Inconsistency in Science. *Studies in History and Philosophy of Science* **21**, 281-294.

Hintikka, J. (forthcoming). Argumentation in a Multicultural Context.

Hintikka, J. (1999). *Inquiry as Inquiry: A Logic of Scientific Discovery.* Dordrecht: Kluwer.

Kuipers, T. (2000). *From Instrumentalism to Constructive Realism. Synthese Library*, vol. 287. Dordrecht: Kluwer.

Meheus, J. (1993). Adaptive Logic in Scientific Discovery: The Case of Clausius. *Logique et Analyse* **143-144**, 359-389. Appeared 1996.

Meheus, J. (1999a). Deductive and Ampliative Adaptive Logics as Tools in the Study of Creativity. *Foundations of Science* **4**, 325-336.

Meheus, J. (1999b). Model-Based Reasoning in Creative Processes. In: L. Magnani, N. Nersessian and P. Thagard (eds.), *Model-Based Reasoning in Scientific Discovery*, pp.199-217. Dordrecht: Kluwer/Plenum.

Meheus, J. (2000). Analogical Reasoning in Creative Problem Solving Processes: Logico-Philosophical Perspectives. In: F. Hallyn (ed.), *Metaphor and Analogy in the Sciences*, pp. 17-34. Dordrecht: Kluwer.

Meheus, J. (2002). Inconsistencies in Scientific Discovery. Clausius's Remarkable Derivation of Carnot's Theorem. In: H. Krach, G. Vanpaemel and P. Marage (eds.), *History of Modern Physics*, pp. 143-154. Brepols: Brepols.

Nersessian, N. (2002). Inconsistency, Generic Modeling, and Conceptual Change in Science. In: J. Meheus (ed.), *Inconsistency in Science*, pp.197-211. Dordrecht: Kluwer.

Norton, J. (1987). The Logical Inconsistency of the Old Quantum Theory of Black Body Radiation. *Philosophy of Science* **54**, 327-350.

Norton, J. (1993). A Paradox in Newtonian Gravitation Theory. *PSA 1992* **2**, 421-420.

Popper, K.R. (1935). *Logik der Forschung.* Wien: Verlag von Julius Springer.

Popper, K.R. (1959). *Logic of Scientific Discovery.* London: Hutchinson. English translation with new appendices of Popper (1935).

Popper, K.R. (1963). *Conjectures and Refutations.* London: Routledge & Keagan.

Popper, K.R. (1973). *Objective Knowledge.* Oxford: Clarendon.

Smith, J. (1988). Inconsistency and Scientific Reasoning. *Studies in History and Philosophy of Science* **19**, 429-445.

Theo A. F. Kuipers

A BRAND NEW TYPE OF INDUCTIVE LOGIC
REPLY TO DIDERIK BATENS

The correspondence to which Diderik Batens refers dates from the autumn of 1971, and resulted in my very first publication in English, albeit a very short one (Kuipers 1972). Ever since, he has been for me one of the few role models as a philosopher trying to bridge the gap between logic and philosophy of science. Although he certainly is much more of a logician than I am, in many cases, as in the present one, he remains driven by questions stemming from philosophy of science. I am not the only Dutch speaking philosopher influenced by this role model. In Belgium, notably Ghent, he shaped the interests of Jean Paul Van Bendegem, Erik Weber, Helena de Preester and Joke Meheus, to mention only those who have contributed to one of the present two volumes. Certainly the great example in the Netherlands is Evert Willem Beth. Unfortunately I was too young to ever meet him. Although Beth exerted a powerful influence on a whole generation of Dutch philosophers, their emphasis was even more on (mathematical or philosophical) logic and, later, its computational and linguistic applications. Happily enough, Hans Mooij is one of the few exceptions. He was the first supervisor of my dissertation and has now contributed to the present volume. At one time, Johan van Benthem, Beth's indirect successor, seemed to become the great example from and for my own generation. However, after his review-like programmatic paper "The logical study of science" (Van Benthem, 1982) on general philosophy of science, he, unfortunately for my field, directed his logical skills to other areas. But times seem to change, witness his contribution to the present volume.

Batens' contribution is a typical example of doing logic in the service of philosophy of science. Since his contribution is already an impressive logical system, it may be seen as the idealized point of departure for a really rich logic of induction and so I would like to focus my reply on some points that may be relevant for further concretization. However, before doing that, I would like to situate Batens' project in the realm of different approaches to inductive logic.

In: R. Festa, A. Aliseda and J. Peijnenburg (eds.), *Confirmation, Empirical Progress, and Truth Approximation* (*Poznań Studies in the Philosophy of the Sciences and the Humanities,* vol. 83), pp. 248-252. Amsterdam/New York, NY: Rodopi, 2005.

Kinds of Inductive Logic

It is interesting to see how Batens deviates from the old approaches to a logic of induction or an inductive logic. Basically, I mean the two approaches initiated by Carnap, the first being based on the idea of first assigning degrees of inductive probability to hypotheses, prior and posterior relative to the evidence, and then basing rules of inference on them that avoid paradoxes, notably the lottery paradox. Hintikka and Hilpinen made serious progress along these lines, although at the price of assigning non-zero prior probabilities to genuine generalizations. Carnap was not willing to pay this price, which makes him a dogmatic skeptic, to use Niiniluoto's (1999) apt phrase for this attitude. Be that as it may, Carnap made the decision-theoretic move by restricting the task of inductive logic to the probability assignments to be used in decisions, taking relevant utilities into account. As can be derived from Ch. 4 of ICR, even this restricted program of inductive logic, despite its dogmatic skeptic nature, was certainly successful, internally and externally, falsifying Lakatos' premature claim that it was a degenerating program.

It is true that the general idea of an "inductive logic" has several other elaborations. Bayesian philosophy of science is sometimes described this way. As a matter of fact, its standard version can be seen as one of the three basic approaches in the second sense indicated above (see Section 4 of the Synopsis of ICR, and more extensively, SiS, Section 7.1.2), viz. the one rejecting dogmatic skepticism, that is, by taking "inductive priors" into account, but also rejecting "inductive (or adaptive) likelihoods." Carnap, in contrast, rejected inductive priors in favor of inductive likelihoods. Finally, Hintikka has chosen the "double inductive" approach, that is, inductive priors and inductive likelihoods. The common feature of these three approaches is that they aim at realizing the property of instantial confirmation or positive instantial relevance: another occurrence of a certain outcome increases its probability for the next trial.

Besides these (restricted or unrestricted) probabilistic approaches to inductive logic, there are a number of totally different approaches. Besides that of Batens, three of them should be mentioned, all of a computational nature. The first one is that of Thagard c.s. (Holland *et al* 1986, Thagard 1988), leading to the computer program **PI** (Processes of Induction). The second operates under the heading of "inductive logic programming" (see Flach and Kakas 2000) and the third under "abductive logic programming" (see Kakas *et al* 1998). Whereas the first is not so much logically inspired, but connectionistic, the other two typically are. Batens' approach is, at least so far, a purely logical one and hence is rightly called a "logic of induction." It is a specialization of his own adaptive version of dynamic logic aiming at deriving (inductive) generalizations of the type: for all x, if Ax then Bx.

Points for Concretization

I shall not concentrate on technical matters regarding Batens' logic of induction. Although it is presented in a very transparent way by first giving a more informal description of the main means and ends, I do not want to suggest that I have grasped all the details. Incidentally, readers will find in Meheus' paper another nice entry into adaptive logic. Although Batens writes of modifications rather than concretizations, his contribution, like several others, nicely illustrates that not only the sciences but also philosophy can profit greatly from the idealization & concretization (I&C) strategy.[1] I shall concentrate on some points of concretization that are desirable from the point of view of philosophy of science.

A first point is the restriction to generalizations not referring to individual constants. In my opinion Batens defends this idealization in Section 3 too strongly by referring – as such correctly – to the history of the laws of Galileo and Kepler according to which the reference to the earth and the sun, respectively, disappeared in a way in light of Newton's theory (see also his Notes 9 and 10). Typically of inductive methods, rather than hypothetico-deductive ones, I would suggest that in particular in the heuristic phase of inductive research programs (see ICR, 7.5.4) reference to individual objects seems very normal. Indeed, the work of Galileo and Kepler may well be seen from this perspective, whereas Newton indeed saw earth and sun merely as objects of a kind. Moreover, in many areas, e.g. in the humanities, many (quasi-) generalizations seem only to make sense when linked to individuals. More precisely, dispositions of human beings are frequently bound to one individual. People may have more or less unique habits. Hence, a realistic logic of induction should be able to deal with generalizations that merely hold for individual objects. Happily enough, Batens claims, also in his Note 9, that it is at least possible to reduce the effect of the relevant restriction to zero.

A second possible concretization is leaving room for falsified background knowledge. In Note 11 Batens explains that it would be possible to do so by moving to paraconsistent logic. To be sure, Batens is the leading European scholar is this enterprise. Although his formulation might suggest otherwise, I am fairly sure that he does not want to suggest that this paraconsistent move

---

[1] In Kuipers (forthcoming) I illustrate this conceptual version of I&C, as a variant of the empirical version, in terms of the theory of (confirmation, empirical progress, and) truth approximation presented in ICR. In this illustration the two versions of I&C meet each other: revised truth approximation is a conceptual concretization of basic truth approximation, accounting for empirical concretization, e.g. the transition from the ideal gas law to the law of Van der Waals.

requires a complete departure from the present adaptive dynamic approach. What is at stake here seems to be a matter of the order of concretization. The concretization to paraconsistent adaptive logic is a general concretization of that logic, not specifically related to inductive ends. Hence, the question that intrigues me is how important the concretization to paraconsistency is from my philosophy of science point of view. In this respect it is important to note first that I fully subscribe to Batens' first sentence of Note 11: "Scientists may justifiedly stick to hypothetical knowledge that is falsified by the empirical data, for example because no non-falsified theory is available." (p. 203) In a way, this sentence could be seen as the main point of departure of ICR. However, ICR develops an explication of this observation that, at least at first sight, completely differs from the paraconsistent move. In this respect it may be interesting to note that paraconsistent logic is still very much "truth/falsity" oriented, whereas ICR is basically "empirical progress and truth approximation" oriented. (See ICR, Ch. 1, for this distinction.) The strange thing, however, is that although "being falsified" of a theory becomes from my perspective a meaningful but non-dramatic event for a theory, the falsification of a hypothetical inductive generalization (or a first order observational induction, ICR, p. 65) is a crucial event. Since the data at a certain moment (t) are composed of (partial) descriptions of realized possibilities $R(t)$ and inductive generalizations based on them, summarized by $S(t)$, a falsification of one of the latter means that the "correct data" assumption is no longer valid. In other words, we have to weaken $S(t)$ in a sufficient way, preferably such that it is just sufficient. Note that this not a concretization move. Note moreover, that it not only holds for the basic approach but also for the refined approach (ICR, Ch. 10). To be sure, one may argue in particular that taking falsifications of $S(t)$ into account in some sophisticated way might further concretize the refined approach. However, I submit that scientists will be more inclined to adapt $S(t)$ as suggested. Hence, from my point of view, the concretization to paraconsistency is not particularly urgent or even relevant for the role of inductive generalizations in aiming at empirical progress and truth approximation. This attitude seems to be supported by Batens and Haestert (forthcoming) where they extend and improve upon Batens' present contribution. Of course, when genuinely inconsistent theories are at stake the paraconsistent move may become unavoidable.

Another possibility for concretization intrigues me very much. Batens argues at the beginning of Section 6 that it becomes relevant to search for confirming and falsifying instances of "for all x if $A(x)$ then $B(x)$" of the type $A(x)$ & $B(x)$ and, of course, $A(x)$ & *non-B(x)*, respectively. Although he refers in Note 25 to qualitative confirmation in the sense of Ch. 2 of ICR, it remains unclear whether my analysis of kinds of non-falsifying instances in terms of two types of confirming instances ($A(x)$ & $B(x)$ and *non-A(x)* & *non-B(x)*)) and one type of neutral instances (*non-A(x)* & *B(x)*) plays any role. More specifically, from that perspective one would expect, in line with general dynamic logic intuitions, that one starts either with *A*-cases, and finds out whether they are *B* or *non-B*, or with *non-B*-cases, and find out whether they are *A* or *non-A*. All this in order to avoid searching for neutral cases. If I am right that this selective search does not yet play a role, a concretization in this direction would certainly lead to a more realistic and more efficient logic.[2]

Let me conclude with a point that has nothing to do with concretization, but that puzzles me a lot. Although I think I can follow why (†) holds in the logic, I do not understand why it is a "simple and intuitive fact" (p. 214) of which it is "unlikely that [it] will be discovered if one does not handle induction in terms of logic." The combination seems implausible, but knowing Batens, he must have something serious in mind.

REFERENCES

Batens, D. and L. Haesert (forthcoming). On Classical Adaptive Logics of Induction. Forthcoming in *Logique et Analyse*.

Benthem, J. van (1982). The Logical Study of Science. *Synthese* **51**, 431-472.

Flach, P.A. and A.C. Kakas, eds. (2000). *Abduction and Induction*. Dordrecht: Kluwer Academic Publishers.

Holland, J., K. Holyoak, R. Nisbett and P. Thagard (1986). *Induction. Processes of Inference, Learning and Discovery*. Cambridge MA: The MIT Press.

Kakas, A., R. Kowalski and F. Toni (1998). The Role of Abduction in Logic Programming. In: D. Gabbay, C. Hogger, and J. Robinson (eds.), *Handbook of Logic in Artificial Intelligence and Logic Programming*, vol. 5, pp. 235-324. Oxford: Oxford University Press.

Kuipers, T. (1972). A Note on Confirmation. *Philosophica Gandensia* **10**, 76-77.

Kuipers, T. (forthcoming). Empirical and Conceptual Idealization and Concretization. The Case of Truth Approximation. Forthcoming in (English and Polish editions of) Liber Amicorum for Leszek Nowak.

Niiniluoto, I. (1999). *Critical Scientific Realism*. Oxford: Oxford University Press.

Thagard, P. (1988). *Computational Philosophy of Science*. Cambridge MA: The MIT Press.

---

[2] Unfortunately, I had difficulties in understanding precisely the core of the paragraph starting with "So, in order to speed up our journey towards the stable situation …" (p. 214). Maybe this paragraph entails selective search.

TRUTH APPROXIMATION BY ABDUCTION

Ilkka Niiniluoto

ABDUCTION AND TRUTHLIKENESS

ABSTRACT. This paper studies the interplay between two notions which are important for the project of defending scientific realism: abduction and truthlikeness. The main focus is the generalization of abduction to cases where the conclusion states that the best theory is truthlike or approximately true. After reconstructing the recent proposals of Theo Kuipers within the framework of monadic predicate logic, I apply my own notion of truthlikeness. It turns out that a theory with higher truthlikeness does not always have greater empirical success than its less truthlike rivals. It is further shown that the notion of expected truthlikeness provides a fallible link from the approximate explanatory success of a theory to its truthlikeness. This treatment can be applied also in cases where even the best potential theory is an idealization that is known to be false.

Abduction and truthlikeness are two concepts which are often mentioned in debates concerning scientific realism. Many realists think that the best reasons for scientific theories are abductive, or must appeal to what is also called inference to the best explanation (IBE), while some anti-realists have argued that the use of abduction in defending realism is question-begging, circular, or incoherent. Many realists claim that even our strongest theories in science are at best truthlike, while their critics have urged that the explication of this notion has so far been a failure. In order to clarify these debates, it is desirable to study the interplay of these key notions. This paper takes up this task by considering the recent proposals by Theo Kuipers (1999, 2000). Even though the connections between abduction and truthlikeness seem to be generally more complex than Kuipers suggests, I shall illustrate his basic thesis in the special case of monadic first-order languages.

1. Peirce on Abduction

The term 'abduction' was introduced into philosophical discussions by Charles S. Peirce (see Peirce 1931-35, 1992, henceforth simply: *CP* (*Collected Papers*); Niiniluoto 1999b). In his early work in the 1860s, Peirce proposed a three-fold classification of inferences into deduction, induction, and hypothesis. Starting from Aristotelian syllogisms, he noted that a deductive argument can be inverted in two different ways. A typical *Deduction* is an inference of a result from a rule and a case:

In: R. Festa, A. Aliseda and J. Peijnenburg (eds.), *Confirmation, Empirical Progress, and Truth Approximation* (*Poznań Studies in the Philosophy of the Sciences and the Humanities,* vol. 83), pp. 255-275. Amsterdam/New York, NY: Rodopi, 2005.

(1)   *Rule.*   All *F* are *G*.
         Case.   This is an *F*.
         ∴ *Result.*   This is a *G*.

Induction is the inference of the rule from the case and result:

(2)   This is an *F*.
         This is a *G*.
         ∴ All *F* are *G*.

Hypothesis is the inference of the case from the rule and result:

(3)   All *F* are *G*.
         This is a *G*.
         ∴ This is an *F*.

(See *CP* 2.623.) By using modern notation, a typical example of hypothesis in the sense of (3) has the following logical form:

(4)   Given the law $(x)(Fx \rightarrow Gx)$,
         from Gb infer Fb.

Peirce described hypothetic inference as proceeding from effects to causes, while explanatory deductions proceed from causes to effects. This idea is related to the old Aristotelian tradition, advocated by many medieval and early modern scientists who called these two types of inferences *resolutio* and *compositio*, respectively (see Niiniluoto 1999c). These Latin terms were translations of the Greek terms *analysis* and *synthesis*, as used in geometry. Hypothesis is thus related to a propositional interpretation of the method of analysis or resolution.

From the deductive viewpoint, an inference of the form (4) commits the fallacy of affirming the consequent. Peirce of course knew that (4) is not valid in the sense of logical entailment: unlike deduction, induction and hypothesis are "ampliative" inferences. What was important in his account was the insistence that such an "inference to an explanation" has a significant role in science. Usually this role has been interpreted as the heuristic function of the discovery of new theories, or alternatively as the motive for suggesting or pursuing testworthy hypotheses (see Hanson 1961).

The main technical novelty in Peirce's treatment is the extension of (1) - (3) to *probabilistic inference*, where the universal premise 'All *F* are *G*' is replaced by the statistical statement 'Most *F* are *G*' or 'The proportion *r* of *F*s are *G*'. In his 1883 paper "A Theory of Probable Inference," Peirce formulated the simplest probabilistic variant of deduction (1) by the schema

(5)   The proportion *r* of the *F*s are *G*;
         b is an *F*;
         It follows, with probability *r*, that *b* is a *G*.

As Peirce noted, the conclusion here can be taken to be '*b* is a *G*', and the probability *r* indicates "the modality with which this conclusion is drawn and held to be true" (*CP* 2.720). According to Peirce, (5) is an explanatory statistical syllogism (*CP* 2.716). Here he anticipated C.G. Hempel's 1962 model of inductive-probabilistic explanation (see Niiniluoto 2000). Again, (5) has two inversions, one to be called induction, the other hypothesis.

Peirce gave an original analysis of the nature of probabilistic reasoning which covers deduction (in the sense of (5)), induction, and hypothesis. In his view, we should not try to assess the probability of the conclusion of such an inference. Rather, the probability of a mode of argument is its *truth-frequency*, i.e., its ability to yield true conclusions from true premises. For example, this applies to the probability *r* in (5). For ordinary deduction, which is necessarily truth-preserving, the truth-frequency is one. In the same way, it is possible to associate a truth-frequency with a general hypothetic inference like (4), or with the abductive inversion of (5). This truth-frequency is defined by the relative frequency of attribute *F* in the class of *G*s, and obviously depends on the context. In some cases, the probabilistic "validity" of such abductive inference may be high, but it is also possible that it is close to zero.

In his lectures 1898, Peirce called hypothesis also *abduction* and *retroduction* (see Peirce 1992). The latter term, in particular, refers to arguments which proceed backward in time. In his papers in 1901-03, Peirce defined induction in a new way as "the operation of testing a hypothesis by experiment" (*CP* 6.526). This is in harmony with the hypothetico-deductive (HD) model of scientific inference, but Peirce allowed also for cases where the test evidence is only a probabilistic consequence of the hypothesis. He characterized abduction as an "inferential step" which is "the first starting of a hypothesis and the entertaining of it, whether as a simple interrogation or with any degree of confidence" (*CP* 6.525). The general form of this "operation of adopting an explanatory hypothesis" is this (*CP* 5.189):

(6)   The surprising fact *C* is observed;
        But if *A* were true, *C* would be a matter of course,
        Hence, there is reason to suspect that *A* is true.

This schema shows how a hypothesis can be "abductively conjectured" if it accounts "for the facts or some of them".

Schema (6) is more general than the inference (4), since it is not any more restricted to any simple syllogistic form. Further, here *A* might be a general theory, rather than a singular premise expressing a cause of *C*. In this sense, we might call (4) *singular abduction*, and (6) *theoretical abduction*. Moreover, Peirce added, the conclusion is not *A* itself, but the assertion that "there is reason to suspect that *A* is true." Indeed, retroduction "only infers a *may-be*" from an actual fact (*CP* 8.238).

2. Abduction and Justification

It is easy to find examples of abductive inferences in our everyday life. Peirce himself regarded perceptual judgments as "extreme cases" of retroduction (*CP* 5.181). Suppose that I hear a sound in my room and immediately infer that there is a car outside my window. This is clearly an inference from an effect to its potential cause. Moreover, this is also an inference to the most plausible potential explanation, since the alternatives are quite artificial: my hearing might be a hallucination, or someone might have placed a radio outside my window.

Other examples of abduction, also mentioned by Peirce (*CP* 2.714), include retroductive historical inferences. A historian infers the existence of Napoleon Bonaparte from the documents available today. Again this is inference to the best explanation, since the alternative explanations of the present material would involve e.g. the phantastic assumption that someone has forged all the existing documents. A special class of retroductive inferences, which is also related to the tradition of geometrical analysis, is provided by detective stories (see Niiniluoto 1999c).

For Peirce, a paradigmatic example of abduction in science is Kepler's discovery of the elliptic orbit of the planet Mars (*CP* 1.71-74). Kepler had a large collection of observations of the apparent places of Mars at different times. At each stage of his long investigation Kepler had a theory which is "approximately true," since it "approximately satisfies the observations," but he carefully modified this theory "to render it more rational or closer to the observed fact." Here Peirce agrees with William Whewell against John Stuart Mill that Kepler's procedure essentially involved an inference which included the discovery of a new theoretical conception (i.e., the ellipse) not given in the mere description of the data.

The examples of abduction thus range from compelling everyday observations to the adoption of theoretical hypotheses in science by virtue of their explanatory power. In these cases, which include both singular and theoretical abduction, it appears that abductive arguments can sometimes serve as a *justification* of a hypothesis. This idea of justification by abduction can be understood in at least three different senses. All of them suggest that there is a probabilistic link between the premises and conclusion of abductive inference.

The first is Peirce's own account of *truth-frequency*, later followed by many frequentist theories of probability and statistics in the 20th century. The general reliability of the abductive inference may be relatively high in some kinds of circumstances. The problem with this approach is the question whether it warrants a tentative conclusion in any particular case.

The second approach is the theory of *confirmation*. Howard Smokler (1968) proposed that the following principles of Converse Entailment and Converse Consequence are satisfied by "abductive inference":

(CE)   If hypothesis H logically entails evidence E, then E confirms H.

(CC)   If K logically entails H and E confirms H, then E confirms K.

A variant of CE replaces deduction by explanation (see Niiniluoto and Tuomela 1973):

(CE*)   If hypothesis H deductively explains evidence E, then E confirms H.

The principles (CE) and (CE*) receive immediately a *Bayesian* justification, if an epistemic probability measure P is available for the language including H and E, and if confirmation is defined by the *Positive Relevance* criterion: E confirms H if and only if $P(H/E) > P(H)$. According to the Bayes's Theorem,

(7) $P(H/E) = P(H)P(E/H)/P(E)$.

If now H entails E, we have $P(E/H) = 1$. Hence,

(8)   If H logically entails E, and if $P(H) > 0$ and $P(E) < 1$, then $P(H/E) > P(H)$.

More generally, as positive relevance is a symmetric relation, it is sufficient for the confirmation of H by E that H is positively relevant to E. If inductive explanation is defined by the positive relevance condition, i.e., by requiring that $P(E/H) > P(E)$ (see Niiniluoto and Tuomela 1973; Festa 1999), then we have the general result:

(9)   If H deductively or inductively explains E, then E confirms H.

The notion of confirmation is weak in the sense that the same evidence may confirm many alternative rival hypotheses. A stronger notion of inference is obtained if one of the rival hypotheses is the *best* explanation of the facts. The strongest justification is obtained if the hypothesis is the *only* available explanation of the known facts. In such cases, abduction might be formulated as a rule of *acceptance*. But, as abduction is a form of ampliative inference, even in the strongest case abductive inferences are fallible or liable to error, and their conclusions may rejected or modified in the light of new information (such as new observations or the discovery of new rival hypothetical explanations).

Gilbert Harman (1965) formulated *inference to the best explanation* by the following rule:

(IBE)   A hypothesis H may be inferred from evidence E when H is a better explanation of E than any other rival hypothesis.

Comparison with Peirce's schema (6) suggests the following version of IBE:

(IBE′) If hypothesis H is the best explanation of evidence E, then conclude for the time being that H is true.

In analysing IBE′, it is again useful to distinguish between *deductive* and *inductive-probabilistic* explanations (see Niiniluoto 1999b). More precisely, when a hypothesis H explains evidence E relative to background assumptions B, the following conditions should be distinguished:

(a) E is deducible from H and B,

(b) E follows from H and B with the probability $P(E/H\&B)$ such that $P(E/B) < P(E/H\&B) < 1$.

Case (a) is typical in the HD model of science: theory H is a universal generalization, and $P(E/H\&B) = 1$. Case (b) is typical in situations where alternative causes are connected with their effects only with statistical probabilities. Bayes's Theorem was traditionally applied in such situations to calculate the "probabilities of causes."

The notions of "better explanation" in IBE and "the best explanation" in IBE′ can be analysed by the definitions of explanatory power. Let $\mathrm{expl}(H, E)$ be the *explanatory power* of theory H with respect to evidence E. Note that Hempel (1965) used the notion of *systematic power* to cover the explanatory and predictive power of a hypothesis relative to the total evidence E. The idea is that a part E_0 of the evidence E is known before considering the hypothesis H, and H is abductively introduced to explain E_0, but then H may be further confirmed by its ability to predict novel facts. Similarly, the degree of confirmation of a hypothesis H on the total evidence E usually depends on the explanatory and predictive power of H with respect to E. However, in typical cases, the novel facts that H has predicted become parts of the total evidence E that H explains. Therefore, I shall not systematically distinguish explanatory power and predictive power in this paper, and the issues with the so called "problem of old evidence" are ignored.

In the philosophical literature, there are various proposals for expl (see Hempel 1965; Hintikka 1968; Niiniluoto and Tuomela 1983). Two important examples are the following:

$$\mathrm{expl}_1(H, E) = (P(E/H) - P(E))/(1 - P(E))$$

$$\mathrm{expl}_2(H, E) = P(\sim H/\sim E) = (1 - P(H \vee E))/(1 - P(E)).$$

Let us say that H' is a *better$_i$ explanation* of E than H if and only if $\mathrm{expl}_i(H', E) > \mathrm{expl}_i(H, E)$, for $i = 1,2$.

For rival theories H, the maximal value of $\mathrm{expl}_1(H,E)$ is obtained by H with the highest likelihood $P(E/H)$. This leads to the following acceptance rule which formalizes IBE′:

(10)   Given evidence E, accept the theory H which maximizes the likelihood $P(E/H\&B)$.

This principle of *Maximum Likelihood* is a standard method in statistical point estimation. The likelihood criterion is also related to the following ratio measure of the *degree of confirmation* $\mathrm{conf}(H, E)$ of hypothesis H by evidence E:

$$\mathrm{conf}_1(H/E) = P(E/H)/P(E).$$

(See Festa 1999; Kuipers 2000.) It follows that

(11)   If H' is a better$_1$ explanation of E than H, then $\mathrm{conf}_1(H'/E) > \mathrm{conf}_1(H/E)$.

For the deductive case (a), the rule (10) does not help much, since all potential deductive explanations make the likelihood $P(E/H\&B)$ equal to one. Hence, the choice of the best explanation is usually taken to depend on other desiderata. Hempel's measure $\mathrm{expl}_2(H, E)$ implies that the best theory is the one which minimizes $P(H)(1 - P(E/H))$. It thus favors theories which have a high likelihood $P(E/H)$ and a high information content $\mathrm{cont}(H) = 1 - P(H)$. However, this proposal has the problem that even irrelevant additions to a theory improve it as an explanation: if A is any statement such that A is not entailed by H and $P(E/H\&A) = P(E/H)$, then $H\&A$ is better than H by the criterion of expl_2. Ultimately, $\mathrm{expl}_2(H, E)$ is maximized by choosing H to be a contradiction. (For the concept of minimal explanation, see Aliseda-Llera, 1997.)

On the other hand, if expl_2-measure is used as a truth-dependent expected utility (see Niiniluoto 1999a, p. 187), then its maximization recommends the acceptance of the hypothesis with the highest value of the *relevance measure* of confirmation:

$\mathrm{conf}_2(H/E) = P(H/E) - P(H).$

The rule expressing IBE' has now the form

(12)   Given evidence E, accept the explanation H of E such that H maximizes the difference $P(H/E\&B) - P(H/B)$.

Most Bayesian treatments of abduction as a method of belief change (see Douven 1999) are primarily interested in the behavior of posterior probabilities. The principle of *High Posterior Probability* recommends the acceptance of the most probable explanation:

(13)   Given evidence E, accept the explanation H of E such that H has the maximal posterior probability $P(H/E\&B)$ on E.

The rule (13) seems to be related to Peter Lipton's (1991) "likely" explanations, whereas his "lovely" ones might correspond to explanations with a high value of $\mathrm{expl}_2(H, E)$. For theories which deductively explain E, it follows from Bayes's Theorem (7) that rule (13) recommends the choice of the theory H that has the highest prior probability $P(H/B)$, i.e, H is initially the most plausible of the rival explanations. With the same assumptions, this H also maximizes the difference $P(H/E\&B) - P(H/B)$, so that (12) and (13) are equivalent in this case. However, if rule (13) is applied also to inductive cases (b), (11) and (13) may lead to different results: the theory with maximal likelihood need not be the one with the highest posterior probability, if it has a very low prior probability.

An important difference between the confirmation measures $conf_1$ and $conf_2$ is their behavior with respect to irrelevant additions to a hypothetical explanation:

(14)   Assume that H explains E but A is irrelevant to E with respect to H (i.e., $P(E/H\&A) = P(E/H)$ where A is not entailed by H). Then $conf_1(H/E) = conf_1(H\&A/E)$ and $conf_2(H/E) > conf_2(H\&A/E)$.

Measure $conf_2(H/E)$ thus favors a minimal explanation, and gives support only to the part of an explanatory hypothesis that is indispensable for the explanation of the evidence (cf. Niiniluoto 1999c, p. 190).

3. IBE and Truthlikeness

The idea of abduction as an acceptance rule has been discussed also in artificial intelligence. Atocha Aliseda-Llera (1997) connects IBE with theories of belief revision (see also Aliseda 2000). She formulates *abductive expansion* by the rule:

(AE)   Given a background theory T and a novel observation E (where T does not entail E), construct an explanation H such that T and H together entail E, and add E and H to T.

Abductive revision applies to cases where theory T and evidence E are in conflict:

(AR)   Given a background theory T and an anomaly E (where T entails the negation of E), revise the theory T to T' such that $T'\&E$ is consistent, and construct an explanation H such that T' and H together entail E, and add H and E to T'.

Kuipers (1999) adds a variant of AR where T' is obtained from T by concretization (cf. Niiniluoto 1999a, Kuipers 2000). Another variant would allow us to revise the observational data E into E' such that $T\&E'$ is consistent:

(AR')   Given a background theory and an anomaly E, revise the evidence E to E' such that $T\&E'$ is consistent, and construct an explanation H such that T and H together entail E', and add H and E' to T.

It is argued in Niiniluoto (1999b) that, in analyzing the notion of "the best explanation," we should also cover *approximate explanations*:

(c) E' which is close to E is deducible from H and B.

(See also Tuomela 1985.) This is related to Aliseda's abductive revision. It includes the problem of curve-fitting where the original observational data E is incompatible with the considered hypotheses H, so that $P(E/H\&B) = 0$. For this case, the probability $P(E/H\&B)$ has to be replaced by a measure of *similarity* or *fit* between E and H (see Niiniluoto 1994). It is suggested in Niiniluoto (1999b) that here the evidence may still indicate that the best hypothesis is *truthlike*. This principle might be called *inference to the best approximate explanation*:

(IBAE)   If the best available explanation H of evidence E is approximate, conclude for the time being that H is truthlike.

If degrees of truthlikeness are introduced (see Niiniluoto 1987, 1998), then there is a natural addition to IBAE: the greater the fit between H and E, the larger the degree of truthlikeness of H in the conclusion. A variant of IBAE could replace truthlikeness by the weaker notion of approximate truth:

(IBAE')   If the best available explanation H of evidence E is approximate, conclude for the time being that H is approximately true.

Theo Kuipers criticizes the original formulation of IBE for three reasons (see Kuipers 2000, p. 171). First, it does not include already falsified theories, i.e., theories incompatible with the evidence. Secondly, "it couples a non-comparative conclusion, being true, to a comparative premise, being the best unfalsified theory." Thirdly, it is difficult to see how it could be justified merely in terms of the true/false -distinction.

The first and third point of Kuipers are handled by the versions IBAE and IBAE' which allow falsified theories and introduce a link to the notions of truthlikeness and approximate truth. However, the second point can still be invoked against IBAE, as its premise is comparative but its conclusion is non-comparative.

Kuipers proposes an alternative to IBE which he calls *inference to the best theory*:

(IBT)   If a theory has so far proven to be the best one among the available theories, then conclude for the time being that it is the closest to the truth of the available theories.

The best theory is allowed to be inconsistent with the evidence. The phrase 'closest to the truth' can here be explicated on three levels: closest to observational truth, referential truth, or theoretical truth (see Kuipers 1999). For this purpose, Kuipers uses his own theory of truth approximation. In its "naive" version, the comparative notion of 'closer to the truth' involves a strong dominance condition: the better theory should have (set-theoretically) "more" correct models, and "less" incorrect models than the worse theory. In the "refined" version, the notion of betweenness between structures helps to make sense of the idea that a theory may be improved by replacing worse incorrect models by better incorrect models.

The phrase 'the best theory' in turn is defined in terms of empirical success. One theory is empirically more successful than another relative to the available data if it has "more" correct consequences and "less" counterexamples than the other theory. With these definitions, Kuipers is able to prove a *Success Theorem*: if theory Y is at least as similar to the truth as theory X, then Y will always be at least as successful as X relative to correct empirical data (Kuipers 2000, p. 160). Thus, higher truthlikeness explains greater empirical success. This means also that in our attempt to approximate the truth it is functional to use a method which is based on the *Rule of Success*: if theory Y has so far proven to be empirically more successful than theory X, accept the "comparative success hypothesis" that Y will remain to be more successful than X relative to all future data, and eliminate X in favor of Y (p. 114). In other words, it is rational to favor a theory which has so far proven to be empirically more successful than its rivals. This gives "a straightforward justification" of IBT in terms of truth approximation.

Kuipers' argument is very interesting. His methodological account of abduction resembles Peirce's defense of induction as a self-corrective inference: for Peirce, induction "pursues a method which, if duly persisted in, must, in the very nature of things, lead to a result indefinitely approximating to the truth in the long run" (*CP* 2.781).

The results of Kuipers, including his Success Theorem, depend essentially on his way of explicating truthlikeness. This approach is not without its problems (see Niiniluoto 1998). Here I will only make the reservation that the strong dominance condition implies that the comparative notions of empirical success and closeness to the truth define only partial orderings, so that many interesting theories are incomparable with each other. In particular, this means that if theory Y has been so far more successful than theory X, then X can never become more successful than Y in the future – the best prospect for X is to become incomparable with Y. Further, in many cases there will be no single theory which is better than all the available alternatives, so that a rule like IBT is inapplicable.

4. On the Justification of Abduction

It is interesting to note that many attempts to defend scientific realism by the famous "no miracle argument" appeal to forms of abduction which conclude that successful scientific theories are approximately true (see Putnam 1978; Psillos 1999). In other words, they involve something like the principle IBAE′ (but without making the notion of approximate truth precise). For example, Newton's mechanics is able to give approximately correct explanations and predictions of the behavior of ordinary macroscopic physical objects, and therefore it is a useful tool in engineering applications. This success can be explained by the hypothesis that Newton's mechanics is in fact truthlike or approximately true: there are in nature entities like forces whose relations to the movements of bodies are approximately correctly described by Newton's Laws. The no miracle argument adds that there is no alternative explanation of the same persistent facts (see Niiniluoto 1999a), so that realism is the only explanation of the continuing empirical success of scientific theories. To defend this argument against the charges of circularity (Fine 1986) and incoherence (van Fraassen 1980, 1989), one needs to defend abduction in the form of IBAE or IBAE'.

In fact, Larry Laudan (1984) in his well-known "confutation of scientific realism" demanded the realists to show that there is an "upward path" from the empirical success of science to the approximate truth of theories – and then a "downward path" from approximate truth to empirical success. In my own work, I have tried to reply to Laudan's challenge by using the concept of truthlikeness (see Niiniluoto 1984, Ch. 7), i.e., by appealing to something like IBAE and by making it precise with my own account of truthlikeness and its estimation (see Niiniluoto 1987). It is clear that Kuipers (2000) also gives a reply to Laudan by his "downward" Success Theorem and "upward" Rule of Success.

For these reasons, it is highly interesting to ask in what sense, in which way, to what extent, and under what conditions, abduction might be justified. Let us first repeat that the traditional formulations of IBE are clearly intended to be *fallible* rules of inference. It is too much to ask for a complete justification which would prove IBE to be necessarily truth-preserving. The same holds of the formulations of IBAE. But Kuipers' symmetric rule IBT, which concludes a comparative claim from a comparative premise, has more chances to be generally valid. But still it remains to be seen whether this is the case.

The second point to be made here is that the probabilistic account of IBE, given by the results (8) and (9), cannot be directly applied to our problem at hand. The results (8) and (9) can be understood so that they establish *a probabilistic link between explanatory power and truth*: posterior probability $P(H/E)$ is the rational degree of belief in the truth of H on the basis of E, and thereby confirmation, i.e., increase of probability by new evidence, means that we rationally become more certain of the truth of H than before. But a rule of the form IBAE needs a link between approximate explanation and truthlikeness. The notion of probability (at least alone) does not help us, since the approximate explanation of E by H allows that H is inconsistent with E, so that $P(E/H)$ and $P(H/E)$ are zero.

The same point can be made about Kuipers' principle IBT, where 'more successful' is now understood in terms of explanatory successes and failures. It is obviously a direct consequence of the following rule:

(15)   If Y is a better explanation of the available evidence E than X, then conclude that Y is more truthlike than X.

From (15) it would follow that, if Y is better than all of its rivals X, then Y is the most truthlike of the available theories. But again a direct analysis of (15) in terms of probability and confirmation is excluded.

One important approach to IBAE′ is to define the notion of *probable approximate truth* (see Niiniluoto 1987, p. 280). A theory (understood as a disjunction of constituents or complete states of affairs) is approximately true if it allows states of affairs that are close to the truth. If we fix a small distance δ to indicate what is to be counted as sufficient closeness to the truth, and an ordinary epistemic probability measure P is available, then it is possible to calculate the probability $PA(H/E)$ that a theory H is approximately true (within the degree δ) given evidence E. By this definition, a true theory is also approximately true. Hence, $PA(H/E)$ is always at least as large as $P(H/E)$. Thus, if H is probably true on E, then H is also probably approximately true on E. This means that probabilistic links between explanation and truth, like (8), induce probabilistic links between explanation and approximate truth as well.

But it is also possible that $PA(H) > 0$ even though $P(H) = 0$. This helps us to give a reply to van Fraassen (1989) who argues that the result (8) does not apply to hypotheses H with a zero prior probability (see Niiniluoto 1999a, p. 188). The ordinary notion of confirmation as positive probabilistic relevance can be modified by replacing P with PA: let us say that E *ap-confirms* H if $PA(H/E) > PA(H)$ (see Festa 1999). Let H^δ be the disjunction of H with states of affairs that are at most within the distance δ from some constituent in H. Hence, H is approximately true if and only if H^δ is true. Further, $PA(H) > 0$ if and only if $P(H^\delta) > 0$. Then (8) can be generalized to the following result:

(16)   If $PA(H) > 0$, and H^δ logically entails E, then E ap-confirms H.

(See Niiniluoto 1999a, p. 188.) Note, however, that this kind of result does not yet justify IBAE′, since here H is compatible with E.

Another challenge concerns the justification of IBAE. Given a definition of a truthlike theory (within the degree δ), it is again possible to calculate the probability that a theory is truthlike given some evidence (see Niiniluoto 1987, p. 278). This notion of *probable verisimilitude* may provide a probabilistic link between explanation and truthlikeness.

However, my own favorite method of connecting objective degrees of truthlikeness and epistemic matters is based on the idea of estimating verisimilitude by the expected degree of truthlikeness (see Niiniluoto 1987, p. 269). If $C_1, \ldots, C_m$ are the alternative complete states of affairs (constituents) expressible in some language, and the degree of truthlikeness of theory H would be $Tr(H,C_i)$ if C_i were the true state, then the *expected verisimilitude* $ver(H/E)$ of H given evidence E is defined by

(17) $ver(H/E) = \Sigma \, P(C_i/E) \, Tr(H,C_i),$

where the sum goes over all $i = 1, \ldots, m$. The value of $ver(H/E)$ generally differs from the probability $P(H/E)$. Again we may generalize the notion of probabilistic confimation: let us say that E *ver-confirms* H (relative to background assumption

B) if and only if ver($H/E\&B$) > ver(H/B). Then we have, for example, the following result:

(18)   If H and $\sim H$ are the rival hypotheses, and H entails E but $\sim H$ does not entail E, then E ver-confirms H.

(See Niiniluoto 1999a, p. 186.) (18) guarantees that E ver-confirms H if H is the only deductive explanation of E.

The measure of expected truthlikeness has the important feature that ver(H/E) may be non-zero, and even high, even though $P(H/E) = 0$. Therefore, (17) is an interesting methodological tool in analysing the relations between approximate explanation and truthlikeness. In order to reply to Laudan's challenge, we should investigate whether the following kinds of principles are valid:

(19)   If H' is a better approximate explanation of E than H, then ver(H'/E) > ver(H/E).

(20)   If H approximately explains E, and H may be inconsistent with E, then the expected verisimilitude of H given E is high.

If these principles hold, at least on some additional conditions, then (relative) explanatory success gives us a rational warrant for making claims about (relative) truthlikeness.

Laudan (1984, p.119) wonders whether my approach involves a confusion between "true verisimitude" and "estimated verisimitude." I plead not guilty, since this is a distinction which I have stressed and explicated. Laudan is right in stating that "estimated truthlikeness and genuine verisimilitude are not necessarily related," but this should not be the aim of the exercise: abduction is a form of ampliative inference, and its premises and conclusion are not "necessarily related." What the function ver hopefully does in a principle like (20) is to establish a sort of probability-based rational brigde between the approximate explanatory success of H and the truthlikeness of H. This is a generalization of the probabilistic bridge established between explanation and truth in (8). Indeed, ver(H/E) is a generalization of the concept of probability, since, by replacing degrees of truthlikeness with truth values (one and zero) in (17), it turns out that $P(H/E)$ is equal to the expected truth value of H on E (see Niiniluoto 1999a, p. 98).

It is important to add that the direct application of the ver-measure has to be modified in circumstances involving abductive revision. We shall see this at the end of the next section.

Instead of trying to study conditions (19) and (20) in the most general case, I shall illustrate the prospects and tricky problems of this endeavour in a special case in the next section. This treatment allows for a direct comparison with the approach of Kuipers as well.

5. Abduction in Monadic First-Order Language

In this section, I shall use the framework of monadic first-order logic L. The comparison to Kuipers would be even closer if we assumed that the language L has an operator of nomic necessity (see Niiniluoto 1987, Ch. 11, and Zwart 1998, pp. 59-70). Here it is sufficient to note that the results of this section can be formulated in the modal framework as well.

Let L be the language with Q-predicates $Q_1, ..., Q_K$, and constituents $C_1, ..., C_m$, where $m = 2^K$ (see Niiniluoto 1987). Each constituent C_i specifies a set CT_i of non-empty Q-predicates, and the Clifford-distance $d(C_i, C_j)$ between two constituents C_i and C_j is the relative cardinality of the symmetric difference $CT_i \Delta CT_j$. All the possible theories in L can be expressed as finite disjunctions of constituents. The most truthlike of these theories is the complete truth in L, expressed by the (unknown) true constituent C^* of L. The set of instantiated (non-empty) Q-predicates is denoted by CT^*. The degree of truthlikeness $\mathrm{Tr}(H, C^*)$ of a theory H in L is defined by the min-sum function as the weighted average of two factors: the minimum distance from the constituents allowed by H to C^*, and the normalized sum of all distances from the constituents allowed by H to C^*. The minimum distance helps to define the notion of approximate truth, while the additional sum-factor gives penalties to all mistakes made by a theory. For constituents, i.e., disjunctions with one member only, this distinction is not important: the degree $\mathrm{Tr}(C_i, C^*)$ is simply a function of the distance $d(C_i, C^*)$.

I shall also assume that a Hintikka-style inductive probability measure P is defined for the sentences of L. In particular, we have probabilities of the form $P(C_i / e_n)$, expressing a rational degree of belief in the truth of constituent C_i given evidence e_n, where e_n is a description in terms of the Q-predicates of a finite sample of n individuals. The set of Q-predicates exemplified in sample e_n is denoted by CT^e, and the corresponding constituent by C^e. In other words, C^e claims that the world is like the sample e_n in terms of exemplified Q-predicates. If we assume that evidence e_n is true, then $CT^e \subseteq CT^*$. Let c be the cardinality of CT^e, i.e., the number of different kinds of individuals exemplified in e_n. Some constituents of L are incompatible with e_n, so that their probability on e_n is zero. The basic result of Hintikka's system states that, for fixed c, the posterior probability $P(C^e / e_n)$ of constituent C^e on e_n approaches one in the limit, when the sample size n grows without limit, and the probabilities of all other constituents approach zero.

Let us start with what Laudan calls the "downward inference." The Success Theorem of Kuipers (2000) can now be formulated in our monadic framework by considering how the constituents are related to empirical data:

(21)   If $CT_i \Delta CT^* \subseteq CT_j \Delta CT^*$, then C_i will always be at least as successful as C_j relative to correct empirical observations.

The assumption implies that $CT_j \cap CT^* \subseteq CT_i \cap CT^*$. Hence, a correctly observed individual may be one of the following three cases: (i) a success to both C_i and C_j, (ii) a success to C_i but a counterexample to C_j, (iii) a counterexample to both C_i and C_j. Therefore, C_i must be at least as successful than C_j (i.e., at least as many successes and not more counterexamples relative to correct observations).

Result (21) entails immediately that the true (and most truthlike) constituent C^* of L is always at least as successful as any other constituent with respect to correct observations. However, only very few false constituents are comparable by the set-theoretic criterion given in (21). But if we apply the Clifford distance, which allows all constituents to be comparable, the corresponding general result fails:

(22)   If C_i is at least as truthlike as C_j, then C_i will always be at least as successful as C_j relative to correct empirical observations.

The Clifford distance does not guarantee that the more truthlike constituent C_i makes more correct positive existential claims than the less truthlike C_j, since this measure counts also the mistakes in the negative existential claims. Even when C_i and C_j make equally many true existential claims, it may happen that they have a different number of correct singular instances in the empirical evidence.

The result (21) is restricted to constituents (complete theories) only. If we wish to study its extension to arbitrary incomplete theories in L (i.e., disjunctions of constituents), the situation is more complicated. But, by applying the mini-sum measure of truthlikeness, at least it is possible to find extensions of (21) to particular types of theories under special kinds of conditions. For example, an existential-universal generalization g_i makes some positive existential claims PC_i and some negative existential claims NC_i with respect to Q-predicates, and leaves open some Q-predicates QC_i as question marks. Then the degree of truthlikeness of g_i decreases with the cardinalities of $PC_i - CT^*$ (the number of wrong existence claims), $NC_i \cap CT^*$ (the number of wrong non-existence claims) and QC_i (informative weakness of g_i) (see Niiniluoto 1987, p. 337). Assume that g_i and g_j are two existential-universal generalizations with the same number of question marks. If now g_i makes less mistaken existence claims and less mistaken non-existence claims than g_j, then g_i is more truthlike than g_j. Assume that 'less' is explicated by set-theoretical inclusion in the following way: $PC_i \Delta CT^* \subseteq PC_j \Delta CT^*$ and $NC_j \Delta CT^* \subseteq NC_i \Delta CT^*$. Then, for all correct observations, g_i is empirically at least as successful as g_j. This conclusion is a generalization of the result (21), since an existential-universal statement g_i with zero question marks is a constituent, and thus it is an instance of Kuipers' Success Theorem. However, again, this result cannot be generalized to the form which corresponds to (22).

Let us still illustrate that in general higher degree of truthlikeness is not sufficient for greater empirical success. Let g_i and g_j be universal generalizations which make negative existential claims NC_i and NC_j, respectively. (In this case, there are no positive existence claims.) Then the degree of truthlikeness of g_i increases with the cardinality of NC_i (the informative strength of g_i) and decreases with the cardinality of $NC_i \cap CT^*$ (the number of mistaken non-existence claims). (See Niiniluoto 1987, p. 336.) For illustration, assume that $k = 8$, $CT^* = \{Q_1, Q_2, Q_3, Q_4\}$, $NC_1 = \{Q_3, Q_4, Q_5, Q_6, Q_7, Q_8\}$, and $NC_2 = \{Q_4\}$. Then g_2 is a very weak false claim, while g_1 is a strong claim excluding correctly many Q-predicates. Even though g_1 also excludes incorrectly more Q-predicates than g_2, it is on the whole more truthlike than g_2. However, possible correct observations of individuals are either neutral between these hypotheses (the cases of Q_1, Q_2, Q_4), or in favor of g_2 over g_1 (the case of Q_3). Thus, in spite of its larger degree of verisimilitude, g_1 is not empirically more successful than g_2.

Similar remarks apply in situations where a new theoretical predicate M is added to the observational language L (see Niiniluoto and Tuomela 1983). Then, in the extended language L', each Q-predicate Q_i of L is split into two Q-predicates of L':

$$Q'_{i1}(x) = Q_i(x) \& M(x)$$

$$Q'_{i2}(x) = Q_i(x) \& {\sim}M(x).$$

Hence, the true constituent C^* of L is now entailed by several different theoretical constituents, but of course one of them, as the true constituent of L', is more truthlike than the others. More generally, it may happen that two theories in L' are empirically equivalent (in their deductive relations to statements in L), but one is more truthlike than the other. In particular, two theories may lack empirical content altogether, but still one of them is more truthlike than the other in L'. To avoid this, the relevant theories should be empirically testable in the sense that they include some assumptions about connections between theoretical and observational statements.

Even more dramatic problems arise in the following way. Assume that in fact Q'_{i1} is empty and Q'_{i2} is non-empty, so that Q_i is non-empty. Let C'_1 claim that Q'_{i1} is non-empty and Q'_{i2} is empty, and C'_2 claim that both Q'_{i1} and Q'_{i2} are empty Then (as far as these claims are concerned) C'_2 is more truthlike than C'_1 in L'. However, C'_2 mistakenly entails that Q_i is empty, while C'_1 correctly entails that Q_i is non-empty, so that the observational consequences of C'_1 are more truthlike than those of C'_2. Kuipers formulates an assumption of "relative correctness," which excludes constituents of the type C'_1, and with this additional condition he is able to prove a Projection Theorem which is a counterpart of theorem (21) in the language L' (see Kuipers 2000, p. 213).

Let us then return to the "upward" problem of abduction. Here we cannot assume (even hypothetically) that the truth values of statements in L are known, but rather we try to make inferences about truth or truthlikeness on the basis of empirical success. To express principles (19) and (20) in the monadic framework

L, let us first restrict our attention to the case where the rival explanations of the evidence e_n include all the constituents C_i of L (and nothing else). Constituents which are incompatible with e_n are allowed as well, even though their posterior probability on e_n is zero. The value of the expected verisimilitude $\mathrm{ver}(C_i/e_n)$ of C_i given e_n can be calculated by using the inductive probabilities of Hintikka's system. When the sample size n is suffiently large, $\mathrm{ver}(C_i/e_n)$ decreases with the distance $d(C_i, C^e)$ between C_i and C^e (see Niiniluoto 1987, p. 342). Hence,

(23)   If $d(C_i, C^e) < d(C_j, C^e)$, then, for large values of n, $\mathrm{ver}(C_i/e_n) > \mathrm{ver}(C_j/e_n)$.

Here it is natural to stipulate that C_i is a better approximate explanation of e_n than C_j if and only if $d(C_i, C^e) < d(C_j, C^e)$. Thereby (23) gives a proof of the comparative principle (19).

In Hintikka's system, $\mathrm{ver}(C^e/e_n)$ approaches the value one when c is fixed and n grows without limit. Here C^e is the boldest generalization compatible with the evidence e_n. Hence, asymptotically only the best explanation of e_n which perfectly fits e_n has the estimated verimilitude one. But, for a finite e_n, even when it is true, we cannot be completely certain that the evidence exhibits all the variety of the universe with respect to the Q-predicates of L. If we are almost certain that C^e is the true constituent, then $\mathrm{ver}(C_i/e_n)$ is approximately equal to $\mathrm{Tr}(C_i, C^e)$ (see Niiniluoto 1987, p. 275). If now C_i approximately explains e_n in the sense that $d(C_i, C^e)$ is small, then $\mathrm{ver}(C_i/e_n)$ is relatively high. Thereby we have given a proof of the principle (20).

We thus see that both the symmetric (19) and the non-symmetric principle (20) can be justified with the same framework. These results can be generalized to the case where the rival explanations include all the universal generalizations of language L (see Niiniluoto 1987, p. 344). The value $\mathrm{ver}(g/e_n)$ can be calculated even when generalization g is incompatible with e_n. Again, asymptotically the best explanation of e_n is the boldest generalization C^e compatible with e_n, but $\mathrm{ver}(g/e_n)$ can be relatively high when g is sufficiently close to C^e.

If these calculations are extended to the language L' with a new theoretical predicate, there will be in L' several constituents, and hence several universal generalizations, which asymptotically receive non-zero posterior probabilities on evidence e_n in L. Therefore, some generalizations in L', among them constituents that are incompatible with evidence e_n in L, may have relatively high degrees of estimated verisimilitude on e_n (see Niiniluoto 1987, pp. 275, 345). These degrees may depend also on additional theoretical background assumptions that are expressed in the vocabulary of L'.

The Bayesian framework within the context of a first-order language L has the consequences that all the alternative theories can be easily enumerated, and the complete truth C^* is always among the rival explanations. (The same remark can be made about Kuipers' framework.) Even though we have seen that the function ver allows us to assess also false and approximate explanations in an interesting way, ultimately the best explanatory theory has here a *perfect fit* with the observations. For these reasons, one may still doubt whether we have really succeeded in analyzing inference rules of the form IBAE and IBAE′ where the best available explanation is only approximate.

Therefore, to conclude this paper, we have to add some remarks about genuine abductive revision. Three different approaches are proposed in Niiniluoto (1987, pp. 285-288). Assume now that our rival hypotheses in L are all defined relative to a presupposition B which is false (even known to be false). For example, B may be a counterfactual idealizing assumption. The rival hypotheses are defined so that their disjunction follows from B, but they are all false. In this case, the role of the complete truth C^* is played by the most informative statement $C^*[B]$ in L which would be true if B were true. Suppose that we have evidence e which describes observations made in ordinary circumstances, not under the idealizing assumption B. Then the fit between e and the alternative hypothesis is not perfect.

One way of handling this situation is to transform the idealized statements $C_i[B]$ by concretization, i.e, by elimitating the false assumption B. Then our evidence e may be directly relevant to the concretized hypotheses via the functions P and ver. By modifying and applying the basic idea of the Converse Consequence principle CC, we have some reason to believe in the truthlikeness of $C_i[B]$, if e confirms its concretization.

The second way corresponds to the schema AR′ of abductive revision. Suppose that there is some way of transforming e into $e′$ which tells what the evidence would have been under the counterfactual assumption B. In other words, $e′$ is obtained from e by "substracting" the influence of those factors that are eliminated by B. Then we again apply the function ver to calculate the expected value $\mathrm{ver}(C_i[B]/e′\&B)$. If our transformation from e to $e′$ is reasonably reliable, then we have reason to claim on e that $C_i[B]$ is more truthlike than $C_j[B]$ if $\mathrm{ver}(C_i[B]/e′\&B) > \mathrm{ver}(C_j[B]/e′\&B)$. Applying this comparative idea to alternative hypotheses $g[B]$ relative to B, the following explication of IBAE is obtained:

(24)   If $\mathrm{ver}(g[B]/e′\&B)$ is maximal, conclude for the time being that $g[B]$ is truthlike.

The third approach is based on the idea that, under certain conditions, we have reason to believe that our evidence e is representative of the structure of the universe. (In Hintikka's system, this is the case with high probability, when the sample size is sufficiently large.) Suppose that we are able to define directly the distance $D(C_i[B], e)$ between a hypothesis $C_i[B]$ and evidence e. The method of least square difference in curve fitting problems is an example of such a distance measure D. (See also Zamora Bonilla 1996.) By our assumptions, even the shortest of these distances is larger than zero. Now $C_i[B]$ can be claimed to be more truthlike than $C_j[B]$ on e if $D(C_i[B], e) < D(C_j[B], e)$. Applying this comparative idea, the following explicate of the abductive rule IBAE is obtained:

(25)   If $D(C_i[B], e) > 0$ is minimal, conclude for the time being that $C_i[B]$ is truthlike.

In order to generalize (25) to all hypotheses $g[B]$, the distance function D has to be extended from constituents to their disjunctions (see Niiniluoto 1987, Ch. 6.7).

Another way of generalizing (24) allows that the hypotheses $g[B]$ are expressed in a language which contains theoretical terms and thereby is richer than the observational language of the evidence e.

6. Conclusion

Original forms of abduction reason from the explanatory success of a theory to its truth. Peircean and Bayesian treatments of such "upward" reasoning establish a probabilistic link between empirical success and truth. The main focus of this paper has been the generalization of abduction to cases where the conclusion states that the best theory is truthlike or approximately true. For this purpose, it is also important to study the "downward" inference from a theory to its empirical success. We have seen that the Success Theorem of Theo Kuipers can be formulated and justified within the framework of monadic first-order logic. But this presupposes his account of truthlikeness, where the dominance condition imposes strong restrictions to the comparability of theories. With my own notion of truthlikeness, which makes all theories comparable with respect to their closeness to the truth, it is not generally the case that higher truthlikeness guarantees greater empirical success. It is further shown that the notion of expected truthlikeness, explicated by the function ver which includes epistemic probabilities but is not identical with posterior probability, provides a fallible link from the approximate explanatory success of a theory to its truthlikeness. This idea can be applied also in cases where even the best potential theory is known to be false.

University of Helsinki
Department of Philosophy
P.O. Box 9
00014 University of Helsinki
Finland

REFERENCES

Aliseda, A. (1997). *Seeking Explanations: Abduction in Logic, Philosophy of Science and Artificial Intelligence*. Dissertation Stanford. Amsterdam: ILLC Dissertations Series (1997-04).

Aliseda, A. (2000). Abduction as Epistemic Change: A Peircean Model in Artificial Intelligence. In: P.A. Flach and A.C. Kakas (eds.), *Abduction and Induction: Essays on their Relation and Integration*, pp. 45-58. Dordrecht: Kluwer.

Douven, I. (1999). Inference to the Best Explanation Made Coherent. *Philosophy of Science (Proceedings)* **66**, S424-S435.

Festa, R. (1999). Bayesian Confirmation. In: M. Galavotti and A. Pagnini (eds.), *Experience, Reality, and Scientific Explanation*, pp. 55-87. Dordrecht: Kluwer.

Fine, A. (1986). *The Shaky Game: Einstein, Realism and the Quantum Theory*. Chicago: The University of Chicago Press.

Hanson, N.R. (1961). Is There a Logic of Discovery? In: H. Feigl and G. Maxwell (eds.), *Current Issues in the Philosophy of Science*, pp.20-35. New York: Holt, Rinehart, and Winston.

Harman, G. (1965). Inference to the Best Explanation. *The Philosophical Review* **74**, 88-95.

Hempel, C.G. (1965). *Aspects of Scientific Explanation*. New York: The Free Press.

Hintikka, J. (1968). The Varieties of Information and Scientific Explanation. In: B. van Rootselaar and J.F. Staal (eds.), *Logic, Methodology, and Philosophy of Science III*, pp.151-171. Amsterdam: North-Holland.

Kuipers, T. (1999). Abduction aiming at Empirical Progress or even Truth Approximation leading to a Challenge for Computational Modelling. *Foundations of Science* **4**, 307-323.

Kuipers, T. (ICR/2000). *From Instrumentalism to Constructive Realism: On Some Relations between Confirmation, Empirical Progress, and Truth Approximation*. Dordrecht: Kluwer.

Laudan, L. (1984). *Science and Values: The Aims of Science and Their Role in Scientific Debate*. Berkeley: University of California Press.

Lipton, P. (1991). *Inference to the Best Explanation*. London: Routledge.

Niiniluoto, I. (1984). *Is Science Progressive?* Dordrecht: D. Reidel.

Niiniluoto, I. (1987). *Truthlikeness*. Dordrecht: D. Reidel.

Niiniluoto, I. (1994). Descriptive and Inductive Simplicity. In: W. Salmon and G. Wolters (eds.), *Logic, Language, and the Structure of Scientific Theories*, pp. 147-170. Pittsburgh: University of Pittsburgh Press.

Niiniluoto, I. (1998). Verisimilitude: The Third Period. *The British Journal for the Philosophy of Science* **49**, 1-29.

Niiniluoto, I. (1999a). *Critical Scientific Realism*. Oxford: Oxford University Press.

Niiniluoto, I. (1999b). Defending Abduction. *Philosophy of Science (Proceedings)* **66**, S436-S451.

Niiniluoto, I. (1999c). Abduction and Geometrical Analysis. In: L. Magnani, N. Nersessian, and P. Thagard (eds.), *Model-Based Reasoning in Scientific Discovery*, pp.239-254. New York: Kluwer and Plenum.

Niiniluoto, I. (2000). Hempel's Theory of Statistical Explanation. In: J.H. Fetzer (ed.), *Science, Explanation, and Rationality: The Philosophy of Carl G. Hempel*, pp.138-163. Oxford: Oxford University Press.

Niiniluoto, I. and R. Tuomela. (1973). *Theoretical Concepts and Hypothetico-Inductive Inference*. Dordrecht: D. Reidel.

Peirce, C.S. (1931-35, 1958). *Collected Papers*. 1-6: edited by C. Hartshorne and P. Weiss. 7-8: edited by A. Burks. Cambridge, MA: Harvard University Press.

Peirce, C.S. (1992). *Reasoning and the Logic of Things: The Cambridge Conferences Lectures of 1898*. Edited by K.L. Ketner. Cambridge. MA: Harvard University Press.

Psillos, S. (1999). *Scientific Realism: How Science Tracks Truth*. London: Routledge.

Putnam, Hilary (1978). *Meaning and the Moral Sciences*. London: Routledge and Kegan Paul.

Smokler, H. (1968). Conflicting Conceptions of Confirmation. *The Journal of Philosophy* **65**, 300-312.

Tuomela, R.(1985). Truth and Best Explanation. *Erkenntnis* **22**, 271-299.

van Fraassen, B. (1980). *The Scientific Image*. Oxford: Oxford University Press.

van Fraassen, B. (1989). *Laws and Symmetry*. Oxford: Oxford University Press.

Zamora Bonilla, J.P. (1996). Verisimilitude, Structuralism, and Scientific Progress. *Erkenntnis* **44**, 25-47.

Zwart, S.D. (1998). *Approach to the Truth: Verisimilitude and Truthlikeness*. Amsterdam: ILLC Dissertation Series.

Theo A. F. Kuipers

QUALITATIVE AND QUANTITATIVE INFERENCE
TO THE BEST THEORY
REPLY TO ILKKA NIINILUOTO

Let me start with quoting from my Foreword to SiS:

> I like to mention Ilkka Niiniluoto's *Critical Scientific Realism* (1999) as, as far as I know, the most learned recent exposition of some of the main themes in the philosophy of science in the form of an advanced debate-book, that is, a critical exposition and assessment of the recent literature, including his own major contribution, viz. *Truthlikeness* of 1987. Despite our major differences regarding the topic of truth approximation, I like to express my affinity to, in particular, his rare type of constructive-critical attitude in the philosophy of science.

In the debate between realists and instrumentalists, I share with Niiniluoto a non-essentialist version of realism, taking truth approximation, and hence false theories, seriously. Our first major difference is his emphasis on what I call "actual truth approximation," whereas I focus on "nomic truth approximation." To be sure, he can deal with both, as far as first order languages are concerned, by adding modal operators for nomic truth approximation. Our second major difference is that I seek to remain "qualitative" for as long as possible, whereas Niiniluoto does not hesitate to go "quantitative," even though that makes arbitrary choices necessary. In this reply I will make some remarks on the first point, but focus on the second.

Niiniluoto starts with a clear survey of Peirce's main view on abduction, concluding with the important distinction between "singular" and "theoretical abduction" or between "individual" and "rule abduction" – to use Thagard's (1988) favorite terms – or simply between "individual" and "general abduction." Many expositions fail to make this distinction, but it is essential. In SiS (pp.75-6) I write

> [A]fter an explanation of an individual event by subsumption under a law the really important issue then is to explain this law. In my opinion the core of explanation lies in the explanation of observational laws by subsumption under a theory, in short, theoretical explanation of (observational) laws. After a successful theoretical explanation of a law, we get as an extra bonus a theoretical explanation of the individual events fitting into that law.

In: R. Festa, A. Aliseda and J. Peijnenburg (eds.), *Confirmation, Empirical Progress, and Truth Approximation* (*Poznań Studies in the Philosophy of the Sciences and the Humanities,* vol. 83), pp. 276-280. Amsterdam/Atlanta, GA: Rodopi, 2005.

Hence, theoretical or general abduction is the main kind of abduction in scientific research. This is not to say that individual abduction is nowhere important. On the contrary, in application contexts, e.g. when human experts or expert systems perform diagnostic reasoning, using a knowledge base, individual abduction is the primary form of abduction.

The Why and When of a Qualitative Approach

After surveying the three (probabilistic) ways in which abductive arguments can serve the justification of the relevant hypothesis, with (standard) "inference to the best explanation (as true)" (IBE) as the most far-reaching one, Niiniluoto turns his attention to IBE's functionality for truth approximation. He appreciates my turn to "inference to the best theory (as the closest to the truth)" (IBT), in particular for its symmetric character (unlike IBE, not only the premise but also the conclusion of IBT is comparative) and its basis in the Success Theorem, according to which "more truthlikeness" guarantees "being at least as successful." His main objection is that my qualitative notion of "more successfulness" and "more truthlikeness" are not frequently applicable. He concludes Section 3 with the correct observation: "in many cases there will be no single theory which is better than all the available alternatives, so that a rule like IBT is inapplicable" (p. 264). However, instead of forcing the existence of a best theory by going quantitative, as Niiniluoto favors, even when the subject matter gives no plausible distances between the structures (or sentences) to be compared, I am interested in the clues given by my analysis when there is no best theory or, more generally, when two theories are incomparable. In SiS I have been more explicit in this than in ICR:

> Finally, it is important to stress that the strict [qualitative] strategy does not lead to void or almost void methodological principles. If there is divided success between theories, the Principle of Improvement amounts, more specifically, to the recommendation that we should try to apply the Principle of Dialectics: "Aim at a success preserving synthesis of the two RS-escaping theories" [where RS refers to the Rule of Success, the purely methodological side of IBT].... Similarly, for truth approximation aims: if there is reason to suppose that two theories cannot be ordered in terms of 'more truthlikeness' in the strict sense, the challenge is to construe a theory which is more truthlike than both. In sum, the restricted applicability of strict notions of comparative success and truthlikeness does not exclude the possibility of clear challenges being formulated in cases where they do not apply, on the contrary." (SiS, p. 250).

Of course, when the subject matter suggests meaningful distances and probabilities, one may want to go further. Moreover, we should like to have as the quantitative variant of the (backward) Success Theorem, the expected success (ES) principle (ICR, p. 303, p. 310): the expected success increases with increasing closeness to the truth. Using Niiniluoto's plausible concept of "estimated truthlikeness" as the probabilistic specification of (quantitative) success we get (ICR, p. 314): the expected value of the estimated truthlikeness increases with increasing closeness to the truth. This would justify the quantitative use of IBT, but other (non-) probabilistic specifications of success might be defensible as well. Be that as it may, Niiniluoto focuses on the "forward" version of a kind of ES principle, his principle (19), according to which estimated truthlikeness increases with being a better approximate explanation, which, assuming a quantitative specification of the latter, amounts to: estimated truthlikeness increases with increasing success. To be sure, this sounds like a justification of the quantitative use of IBT, but it has a circular feature: success and estimated truthlikeness are based on the same evidence. Hence, for a genuine justification of such a use of IBT what we would need, besides meaningful distances and probabilities, is an additional link between estimated truthlikeness and "true" or "objective" truthlikeness, that is, something like the ES principle.

Monadic First-Order Languages

In Section 5 Niiniluoto compares several qualitative and quantitative principles in the special case of a (non-modal) monadic first-order language. Whereas I have to concede that the Clifford distance measure (roughly, counting the number of elementary differences) is rather arbitrary in this case and that inductive probabilities have no objective basis, I would also like to remark, in contrast to what Niiniluoto seems to think, that non-modal monadic constituents are directly interesting from the nomic perspective, for they allow straightforward nomic illustrations. For example, theories classifying chemical elements, such as the periodic table, may not only be read as claims about actually existing elements, but also, or even preferably, as claims of nomically possible existing elements, whether or not they have already been produced by nature or artificially. Hence, although a modal formalization is certainly possible, this is not necessary in this case.

Let me summarize the main results claimed by Niiniluoto. By (21) he specifies the Success Theorem for "constituent-theories" (and later informally for arbitrary theories and, in a restricted form, for theories with theoretical terms) and adds that only "very few" false constituents are comparable in my qualitative, set-theoretic way. However, apart from the general relativization of incomparability given above, the question is: what is "very few" in this context? Given a certain constituent, and hence a certain symmetric difference relative to the true constituent, the number of constituents closer to the truth, and hence, the number of possible qualitative improvements, is equal to the number of subsets of that symmetric difference. Hence, in absolute terms, this number may be very high.

The same holds for the number of possible worsenings. To be sure, *relative* to the number of qualitatively incomparable changes, both numbers are small, as a rule.

Niiniluoto concedes that (22), that is, his favorite quantitative, generalized, version of the success theorem (21) is invalid. This is so because the antecedence of (22) allows evidence in which relatively many of the correct Q-predicates of a less truthlike constituent are instantiated. Later Niiniluoto argues that the generalization of (22) to arbitrary theories and to theories with theoretical terms is invalid for similar reasons. In sum, the lesson is that the success theorem (21) is only valid for the qualitative case, which is only applicable in *relatively* few cases. More generally, I think that expecting (21) to be valid in any deterministic quantitative sense, however restricted, is too much. It seems more plausible to think in terms of the (probabilistic) ES principle mentioned above, and I am puzzled why Niiniluoto does not pay attention to it. In correspondence Niiniluoto refers to a theorem in this direction in (Niiniluoto 1984, (12), p. 92 and (16), p. 170) in terms of "true" and "fully informative" (or "non-misleading") evidence. However, this is still of a very limited kind, for the theorem's condition amounts to the claim that "the constituent corresponding to the evidence" is the true one. The theorem says that under that condition the estimated truthlikeness of that constituent, that is, the true one, approaches the maximum value. The remaining challenge is to generalize this result to the comparative case: increasing closeness to the truth should lead to an increasing expected value of the estimated truthlikeness.

Turning to the "upward" (or "forward") problem of abduction, and using Hintikka's system of inductive probabilities, Niiniluoto specifies (19) (estimated truthlikeness increases with increasing success) by (23) for monadic constituents and argues that a similar specification can be given of the non-symmetric variant of (19), viz. (20), and plausibly claims that both can be generalized to arbitrary theories and theories with theoretical terms.

Since all these claims presuppose a context, such as a monadic first-order language, in which all relevant theories are available, including the true one, this is not yet realistic. For this reason, Niiniluoto finally surveys three different ways of dealing with theories sharing a false idealizing presupposition as developed in his (1987). Here only the first one may have a qualitative counterpart. More specifically, it would be interesting to investigate its relation to my treatment of truth approximation by concretization, based on refined qualitative truthlikeness, in ICR (pp. 268-71), including Niiniluoto's interesting suggestion of "modifying .... the basic idea of the Converse Consequence principle," according to which confirmation of a concretization of an idealized assumption entails (some kind of) confirmation of that assumption.

                    Theo A. F. Kuipers

REFERENCES

Niiniluoto, I. (1984), *Is Science Progressive?* Dordrecht: Reidel.

Niiniluoto, I. (1987). *Truthlikeness*. Dordrecht: Reidel.

Niiniluoto, I. (1999). *Critical Scientific Realism*. Oxford: Oxford University Press.

Thagard, P. (1988) *Computational Philosophy of Science*. Cambridge MA: The MIT Press.

Igor Douven

EMPIRICAL EQUIVALENCE, EXPLANATORY FORCE, AND THE INFERENCE TO THE BEST THEORY

ABSTRACT. In this paper I discuss the rule of inference proposed by Kuipers under the name of Inference to the Best Theory. In particular, I argue that the rule needs to be strengthened if it is to serve realist purposes. I further describe a method for testing, and perhaps eventually justifying, a suitably strengthened version of it.

In his impressive work *From Instrumentalism to Constructive Realism* (Kuipers 2000; hereafter referred to as ICR) Theo Kuipers proposes a rule of inference under the name of Inference to the Best Theory (IBT) that is meant to be an ameliorated version of the Inference to the Best Explanation (IBE), a rule generally believed to be of crucial importance to the case for scientific realism. The present paper argues that, even though it does indeed greatly improve on IBE and eludes what many regard to be a fatal or near-fatal objection to the latter rule, IBT is, as it stands, too weak to serve realist purposes. However, it will also be seen that the rule can be strengthened so as to make it adequate to its purported task. The paper further considers the question whether there is any reason to trust the conclusions reached by means of IBT. It is argued that such reasons may well have to come from a test originally proposed in Douven (2002a) and summarized and subsequently further elaborated in the present paper. I start, however, by briefly discussing the argument that is at present the main source of antirealist sentiments, paying special attention to the thesis of empirical equivalence, which serves as one of the argument's premises and which, in my view, Kuipers dismisses too quickly. This will help to elucidate the role such rules as IBE and IBT play in the realism debate, and, later on, why from a realist perspective somewhat stricter criteria for the goodness of theories are required than the ones that accompany IBT in Kuipers' presentation.

In: R. Festa, A. Aliseda and J. Peijnenburg (eds.), *Confirmation, Empirical Progress, and Truth Approximation* (*Poznań Studies in the Philosophy of the Sciences and the Humanities,* vol. 83), pp. 281-309. Amsterdam/New York, NY: Rodopi, 2005.

1. Empirical Equivalence and Underdetermination

According to scientific antirealists, theory choice is radically underdetermined by the data. They conclude from this that we can never be in a position to warrantably attribute truth to theories beyond their observational consequences. The common antirealist argument for the underdetermination thesis starts from the premise that for each scientific theory there are empirically equivalent rivals, i.e., contraries that have the same observational consequences that it has (call this claim EE). If EE is correct, then no matter how many empirical tests a theory has already passed, this success cannot be taken as an indication that the theory is true, for each of its empirically equivalent rivals will or would pass the same tests just as successfully. Thus, unless the data refute a theory, no amount of them suffices to determine its truth-value. If we then further assume that if the data alone do not suffice to determine a theory's truth-value, then nothing does (call this Knowledge Empiricism, or KE for short), as antirealists typically do, it follows that the truth-value of any theory having non-observational consequences must forever remain beyond our ken.

Since the argument is deductively valid, scientific realists will have to rebut at least one of its premises. Traditionally, realists have believed the fault is to be sought in the second premise, KE. The present paper also is mainly concerned with the latter premise; for, as will become clear, if KE is correct, then rules such as IBE and IBT cannot be correct. In this section, however, I concentrate on the first premise, EE, and in particular on what Kuipers has to say about it.

For a long time, both realists and antirealists have taken the truth of EE for granted. However, in the past decade or so, philosophers have become more skeptical about it. It is no exaggeration to say that this change in attitude is mainly due to work done by Laudan and Leplin.[1] Kuipers seems to share the new skepticism regarding EE. In the context of a discussion of some intuitive arguments for the referentiality of theoretical terms, Kuipers notes that "it is difficult to make such intuitive arguments formally convincing" (ICR, p. 227). He then goes on as follows:

> However, there is no more reason to be pessimistic in this respect than about the presupposition of the referential skeptic, according to whom it will always be possible to invent 'empirically equivalent theories' ... which ... can explain the same variety of successes and success differences. It should be conceded that inventing such theories can not be excluded. In the same way, the skeptic can always tell a story that explains the seemingly veridical nature of our experimental procedures, without them really being veridical. We

---

[1] See, e.g., Laudan (1990), Laudan and Leplin (1991), and Leplin (1997).

> have to admit, however, that (natural) scientists, after a certain period of hesitation, make
> the inductive jump to the conclusion that a certain term refers ... . (p. 227)

As I understand this passage, Kuipers' point is that, although it may always be possible to come up with empirically equivalent rivals for any scientific theory, this possibility is typically not taken very seriously by scientists, and thus, I assume, we are to conclude that we (philosophers) should not take it very seriously either. Another way to put the same point may be to say that, just as Cartesian skepticism at most succeeds in raising philosophical, but definitely not real, doubts about the possibility of our gaining knowledge in general, so the argument from underdetermination at most succeeds in raising philosophical doubts about the possibility of our gaining scientific knowledge regarding the unobservable.

Let me first remark that at least the only well-developed antirealism to date, to wit van Fraassen's constructive empiricism, is not presented with the intention of simply rerunning the debate on skepticism within the philosophy of science. Skeptics are positively rare. For most (and perhaps even all) of us, skepticism is not a live option – we could not even *come* to hold it. We accept as a fact that we know quite a bit, and we regard any theory of knowledge that does not imply such as fundamentally defective. Nevertheless, we are sometimes willing to entertain skepticism; playing the skeptic's role can be a useful strategy for finding out whether a theory of knowledge is indeed defective for the reason just mentioned. But it will be clear to anyone familiar with van Fraassen's writings that his antirealism is not offered for such purely methodological reasons. Though van Fraassen agrees that we know quite a bit (cf. in particular 1989, p.178), according to him this bit is, and cannot but be, restricted to the observable part of the world. Antirealism thus definitely is a live option for him. More than that, he actually urges us to be (or become) antirealists.[2]

Leaving van Fraassen's intentions to one side, I also think Kuipers' rather dismissive remarks on EE, as well as the current skepticism about this thesis among many other authors, are not altogether well-founded. Even if the arguments the antirealist can advance in support of EE are perhaps not quite as convincing as she might wish (and as they were once believed to be), it seems to me that, especially if taken in conjunction, they are convincing enough to sustain a case for antirealism as a real contender (as opposed to a mere skeptical or methodological alternative) for scientific realism.

---

[2] At least this is what he does in his (1980). Later publications (in particular his 1985 and 1989) are more guarded on this point; some passages in these works suggest that van Fraassen has come to conceive of both scientific realism and scientific antirealism as (equally?) rational positions vis-à-vis science. See Kukla (1995, 1998) on the development of van Fraassen's thoughts on the (ir)rationality of scientific realism.

First, antirealists can point to some actual examples of empirically equivalent rivals. Special Relativity and the æther theory in the Lorentz/Fitzgerald/Poincaré version are demonstrably empirically equivalent, as are standard quantum mechanics and Bohm's mechanics. Admittedly, there are not many more such examples. But the antirealist seems perfectly able to explain why there are so few. As Fine (1984, p. 89), for instance, notes, in scientific practice it is typically quite hard to come up with even one theory that fits the data, let alone a number of such theories. By way of further explanation, we might add that it will in general not be a scientist's first ambition to find empirically equivalent rivals for extant theories (if only because success in this respect is very unlikely to be rewarded with a Nobel prize).

Secondly, there exist several proofs of EE; see Earman (1993), Douven (1995), and Douven and Horsten (1998). One might worry that the empirically equivalent rivals that these authors prove to exist for any scientific theory postulating unobservables are not genuine theories, but merely formal equivalents of the skeptic's Evil Demon or Brain in a Vat scenarios (cf. Laudan and Leplin 1991). However, although none of the proofs is constructive, they do give sufficient insight into the nature of the empirically equivalent rivals to justify the claim that they are not of that variety, or at least not all of them. And while it must be acknowledged that each of the currently available proofs of EE rests on non-trivial assumptions, these assumptions seem plausible enough for the proofs to lend considerable support to the thesis (even if the assumptions are not so obviously correct that the proofs can count as incontrovertible demonstrations of EE). And that seems to be all the antirealist needs. After all, we do not require the realist to *demonstrate* the correctness of her position, and so should not require this from her opponent. At any rate, in the face of these proofs it is at best misleading to assert, as Kuipers does, that it cannot be excluded that empirically equivalent rivals can be "invented."

In brief, I think it is fair to say that, although EE cannot be considered as being established once and for all, antirealists have, pending realist arguments to the contrary, bona fide grounds for holding that the existence of empirically equivalent rivals is to be seriously reckoned with (even if Kuipers is right that scientists typically do not do so).[3]

---

[3] One might suggest that there is a quick and easy way for the antirealist to end the recent skirmishes over EE, viz., by supplanting in the argument from underdetermination EE – according to which every scientific theory has empirically equivalent rivals – by the weaker premise that, for all we know, every scientific theory has empirically equivalent rivals. Logically speaking, this would suffice to make the argument from underdetermination go through. However, it should be noted that not any claim that will make the antirealist argument go through will also suffice to

Thus for the realist much hangs on whether she has an adequate response to the second premise of the argument from underdetermination, KE, according to which only the data can determine a theory's truth-value, if that can be determined at all. If this thesis is right, then it does indeed follow, given EE, that knowledge of the unobservable is unachievable. However, realists have objected that by endorsing KE, antirealists totally neglect the role played by explanatory considerations in theory validation. Scientists assess a theory not just on the basis of its conformity with the data; they also take into consideration the theory's explanatory force. And such considerations, realists claim, are truth conducive and not of merely pragmatic significance, as antirealists typically hold. If this realist claim is right, as also most scientists seem to believe, then of course it does not hold that we cannot possibly come to know the truth-value of a theory which makes claims that go beyond the observable. For although empirically equivalent theories necessarily conform (or fail to conform) to the same data, it may well be that one of them provides a better explanation of those data than the other(s). Under the current supposition, this would give reason to believe it is true.

The problems connected with this realist response to the argument from underdetermination are manifold. Chief among them is the fact that realists have so far been unable to answer the antirealist challenge to make plausible that explanation is a mark of truth (the mere fact that most scientists take it as such is, the current enthusiasm for naturalism notwithstanding, philosophically not a sufficiently good reason to believe it is). In section 4 I describe a strategy that may well provide the means to meet that challenge. However, I first want to consider a quite ingenious argument for scientific realism devised by Dorling (1992),[4] and argue that it fails. This intermezzo has a double motivation. First, if it were correct, Dorling's argument would show that any appeal to explanatoriness is dispensable in a defense of scientific realism, and thus that the whole project of justifying the confirmation-theoretical role realists assign to explanation (and thereby section 4 of this paper) is otiose. Second, the discussion of the argument allows me to introduce in a natural way van Fraassen's ideas about how antirealism is to be modelled in confirmation-theoretic terms. In section 4.4 I try to show how we can be maximally concessive to van Fraassen by presupposing his preferred confirmation theory and yet be in a position to find empirical support for (a version of) IBT.

---

make the argument a real threat to scientific realism. And indeed, the mere possibility that a scientific theory has empirically equivalent rivals is hardly sufficient to seriously challenge the realist claim that the theory constitutes (or may constitute) knowledge.

[4] The argument is also discussed in (ICR, p. 223*f*).

2. A Bayesian Defense of Scientific Realism

Dorling's (1992) argument focuses on local realism/antirealism disputes, i.e., disputes concerning the proper epistemic attitude towards a particular scientific theory (though as we shall see, it can, if successful at all, also be regarded as offering a defense of scientific realism *tout court*). His suggestion is that such disputes can be settled by simple Bayesian conditionalization on the available evidence. He tries to demonstrate this with the aid of an example. In his example, 'T_R' denotes some particular scientific theory not solely about observables, and 'T_P' denotes the set of observable consequences of T_R. Dorling then considers two philosophers (scientists?), one of whom is supposed to be a realist, the other an antirealist.[5] Their distinct attitudes are supposed to be reflected in the differences in the initial probabilities they assign to T_R: The realist assigns it a probability of .6, the antirealist a probability of .2. Since T_P is implied by T_R, both assign a conditional probability to the former given the latter of 1. The conditional probability of T_P given the negation of T_R is less straightforward, but Dorling assumes that the realist and antirealist agree that it is .2. We are now to suppose that we obtain enough evidence for T_P's correctness to make us virtually certain of it. What would that mean for the realist's and antirealist's confidence in T_R, respectively? Two easy calculations show that the realist now believes T_R to a degree of (approximately) .9 and that the antirealist believes it to a degree of (approximately) .6.[6] A first, rather unsurprising, observation Dorling makes about this result is that both the realist and the antirealist are now more confident in T_R than they were before they received the evidence for T_P. More surprising may be the observation that the antirealist's new confidence in T_R has increased to such an extent that she now has more confidence in T_R than in $\neg T_R$. She might thus be said to have been converted to realism with regard to T_R (Dorling 1992, p. 368f).

It seems that in this case the realism issue has been settled in favor of the realist, and – most relevant to our concerns – that this has been accomplished without any appeal to the explanatory power of the theory under consideration. The example is rather abstract, but according to Dorling many local realism debates in the history of science fit the example in all relevant respects.

---

[5] An assumption left implicit in Dorling's example is that both are Bayesian learners, i.e., both calculate new probabilities by means of Bayes' theorem. This assumption is far from innocent, of course, but we will go along with it here.

[6] A further, not entirely uncontroversial, assumption in Dorling's argument is that being virtually certain of a proposition allows one to conditionalize on that proposition. However, like Kuipers (ICR, p. 65ff), I approve of this practice and believe that the problems associated with it can be solved; see Douven (2002b). Also, as Howson (2000, pp. 198-201) shows, the argument can be modified so that conditionalizing on T_P is avoided.

Something that Dorling does not discuss but that is certainly noteworthy is that, if his argument is sound, it may even offer a defense of scientific realism in general. For it suffices that *some* local realist disputes have been, or can be, settled in favor of the realist in order to establish modern, relatively modest versions of realism such as Leplin's minimal epistemic realism, according to which "there are possible empirical conditions that would warrant attributing some measure of truth to theories" (Leplin 1997, p. 102). And although the same would not suffice as a defense of more ambitious versions of scientific realism, such as for instance Boyd's (1981; 1984; 1985), Devitt's (1991), or Niiniluoto's (1999), according to which scientific theories are typically approximately true,[7] a defense of these stronger versions along the lines indicated by Dorling is obviously possible as well.

Unfortunately I do not think Dorling's defense succeeds in the first place. It may be that in his argument he has managed to correctly represent *some* antirealists, but it would be a flagrant misrepresentation of the modern, sophisticated antirealist (such as van Fraassen's constructive empiricist), who bases her position on the argument from underdetermination, if we were to identify her with the antirealist Dorling puts on the scene. The latter's ontological claim "is simply the negation of [the realist's ontological claim]" (Dorling 1992, p. 363); for example, the antirealist "reject[s] the existence of atoms" (Dorling 1992, p. 367). That, however, is not at all what a sophisticated antirealist does. Remember that her point merely was that, since there are empirically equivalent rivals for every scientific theory, there is no way of *knowing* the truth-value of a theory which postulates unobservables, and thus also no way of knowing that the theory is *false*. According to this antirealist, what can at most be claimed is that a given theory is empirically adequate, meaning that that theory is a member of a class of empirically equivalent theories one of which is true; to claim that the theory is false would be no less justified than to claim that it is true or approximately true, as realists under certain circumstances consider justified. She therefore counsels *agnosticism* as the proper attitude with respect to what our theories tell us about the unobservable part of the world.

How does this affect Dorling's argument? This is made perfectly clear in the following passage from van Fraassen's (1989, p. 193*f*), which in fact anticipated Dorling's argument:

> Consider ... the hypothesis that there are quarks ... . The scientist has initially some degree
> of belief that this is true. As evidence comes in, that degree can be raised, to any higher

---

[7] The epistemological claim embodied in Kuipers' constructive realism, as presented in ICR, is certainly stronger than that made by Leplin but seems to be somewhat weaker than that of the versions of scientific realism just referred to. However, Kuipers is not very explicit on this point.

degree. That is a logical point: if some proposition X has positive probability, conditionalizing on other propositions can enhance the probability of X.

   The mistake in this argument is to assume that agnosticism is represented by a low probability. That confuses lack or suspension of opinion with opinion of a certain sort. To represent agnosticism, we must take seriously the vagueness of opinion … .

Van Fraassen then goes on to argue that a person's agnosticism concerning a proposition H is to be identified with her assigning a vague or interval-valued probability $[0, p]$ to H, where p is the probability of H's least probable consequence(s), and that conditionalizing on other propositions can at most have the effect of removing the upper limit on that interval, or, as one might also put it, it can only increase the vagueness of H's probability (I shall be more explicit on all this in §4.3). Thus, in Dorling's argument, T_P's becoming certain or nearly certain would for a sophisticated antirealist *à la* van Fraassen at most effect an increase of the upper bound on her degree of belief in T_R. But that would leave her as agnostic about T_R as she was before. In particular, she cannot be said to have converted to realism with regard to T_R.

Van Fraassen's way of modelling agnosticism is not the only one nor necessarily the best; see Hájek (1998) and Monton (1998) for recent criticisms. However, the point against Dorling's defense of scientific realism arguably stands on any reasonable construal of agnosticism (like, e.g., the one suggested in Earman's 1993, p. 35, which models agnosticism by assigning *no* probabilities to theoretical hypotheses). It thus appears that Dorling's defense is without any force against a sophisticated antirealist. Against such an antirealist, the realist has to make clear that reasons can be supplied for the claim that some particular theory is true which are not actually only reasons for the much weaker claim that it is empirically adequate, i.e., that the observable part of the world in every respect is and behaves *as if* the theory were true. In other words, the realist must provide reasons to believe that we can justifiably assign *sharp* high probabilities to particular scientific theories. Such reasons may well have to do with the explanatory force of those theories. That this is so, is the root intuition behind the rule of IBE. I shall now turn to this rule, and in particular to Kuipers' version of it.

3. The Inference to the Best Theory and Explanatory Force

The idea that explanatoriness is of confirmation-theoretical significance can be, and indeed has been, fleshed out in quite a variety of ways. Presumably the simplest of these, and certainly the one most frequently encountered in the literature, is the following:

IBE   Given evidence E and rival (potential) explanations H_1, ..., H_n of E, infer to the (probable/approximate) truth of the H_i that explains E best.

Even this is a rule schema rather than a rule, at least as long as it has not been supplemented by a precise account of explanation and by a set of criteria for the goodness of explanations. Here let me just note that realists agree that which theory of a collection of theories is to count as the best explanation, is to be determined on the basis of what are commonly called theoretical or non-empirical virtues, such as simplicity, elegance, inner beauty, fertility, co-herence with background theories and/or metaphysical suppositions.[8] Exactly how this is to be determined (how, for instance, these virtues are to be weighed against each other), is a matter of considerable controversy among realists, but we shall leave that discussion aside here.

It is also not important for van Fraassen's (1989, p.142*ff*) critique of IBE: That applies regardless of the precise understanding of the notion of explanation. The crucial point of that critique is that to make IBE a rationally compelling rule of inference it must be assumed that the truth generally is among the available potential explanations of the data to the truth of the best of which IBE allows us to infer. For, clearly, unless that is the case, IBE cannot be reliable. And since we will only rarely manage to consider all possible explanations of the evidence, that assumption seems to require some sort of privilege, viz., that we are somehow predisposed to come up with the truth when we contemplate what might explain the data. As van Fraassen (1989, pp. 143-149) convincingly argues, there is *a priori* scant reason to believe we are thus predisposed.

Numerous objections have been levelled against this so-called argument of the bad lot; see for instance Devitt (1991), Lipton (1991; 1993), Kitcher (1993), and Psillos (1996; 1999). However, for reasons given elsewhere, I believe that these objections fail (cf. Ladyman *et al.* 1997; Douven 2002a). In my view the argument of the bad lot is successful, at least to the extent that it shows IBE to rest on an unfounded assumption. But of course this is not to say that it succeeds in showing that there can be no rationally compelling rule of inference based on explanatory considerations. After all, it may well be that versions of such a rule other than IBE can do without the indeed not very plausible assumption of privilege that IBE requires. Kuipers has proposed just such a rule.

As a matter of fact, the feature of IBE the argument of the bad lot capitalizes on is one Kuipers had discovered as being undesirable independently of van Fraassen's critique (cf. Kuipers 1984; 1992; ICR). As

---

[8] Some authors count explanatory power itself among the theoretical virtues, but as McMullin (1996) points out, this is wrong.

Kuipers (ICR, p. 171) notes, the rule licenses a non-comparative conclusion – that a given theory is true – on the basis of a comparative premise, viz., that the particular theory is the best explanation of the evidence relative to the other theories available. That is to say, the rule displays a rather awkward asymmetry. Once the defect has thus been diagnosed, it is obvious how it can be repaired: One can either require a non-comparative premise for the rule to apply (for instance, that a given hypothesis is the absolutely best explanation, whatever other hypotheses have gone unnoticed) or one can have the rule license only a comparative conclusion when given a comparative premise as input. Kuipers opts for the second strategy, and proposes the following amended version of IBE, which he calls Inference to the Best Theory (p. 171):

IBT   If a theory has so far proven to be the best one among the available theories, then conclude, for the time being, that it is, of the available theories, the one closest to the truth.

For later purposes I should immediately note an interesting feature of this rule, namely, that it licenses an inference to the unqualified truth of the absolutely best theory, i.e., the theory that is better than any other theory, whether considered or not. For if a theory for a given domain is better than any other theory for that domain, then it must also be closer to the truth about the domain than any other theory. But no theory can be closer to the truth than the truth itself. Hence the absolutely best theory must be identical to the true theory of the domain. By consequence, if one is certain that a particular theory is the absolutely best, then applying IBT yields the same result as an application of IBE would have yielded.

IBT clearly is invulnerable to the argument of the bad lot; this rule could well be reliable without our being privileged in the way we must be privileged if IBE is to be a reliable rule. And if it can be shown to be compelling, then it is – in principle (see below) – sufficient for a defense of at least a convergent scientific realism such as Kuipers' constructive realism (and perhaps even for a defense of stronger versions of scientific realism – see Douven 2002a). After all, although the rule does not in general license belief in the unqualified truth of a theory, it does license believing that a theory is closer to the truth than any of its predecessors, provided it qualifies as better than those predecessors according to the criteria IBT assumes – and this is true even if the theory is about unobservables. This epistemic attitude contrasts sharply with the agnosticism we saw van Fraassen counsel in such cases.

In the previous paragraph the qualification 'in principle' was added because IBT requires a slight modification (or rather the criteria of goodness that accompany it need such modification) if it is to serve the (convergent) scientific realist's goal. It will be apparent from our presentation of the argument from underdetermination that for the realist it is crucial that the criteria for goodness IBT assumes are such that distinct empirically equivalent theories can satisfy them to *differing* degrees. And as Kuipers understands the notion of best theory, this is not the case. Another way to put the problem is that, given the criteria for goodness that Kuipers assumes, it follows from EE that there is never a unique best theory in cases in which the theory goes beyond the observable.

To see why, consider that Kuipers equates the best theory with the most successful theory (ICR, p. 170), where the latter notion is spelled out in purely empirical terms. Theoretical virtues, as described at the beginning of this section, have no part whatsoever in this, and thus the notion of best theory as supposed by IBT has little to do with the notion of best explanation that is typically invoked by realists in order to (or at least in the hope that it will enable us to) discriminate between empirically equivalent theories.[9]

To be more specific, according to Kuipers (ICR, p. 112) a theory T_1 is *at least* as successful as a second theory T_2 exactly if both of the following conditions hold:

(1) all individual problems of T_1 are also individual problems of T_2;

(2) all general successes of T_2 are also general successes of T_1.

T_1 is *more* successful than T_2 exactly if at least one of the following conditions holds:

(a) T_2 has some extra individual problems in addition to those it shares with T_1;

(b) T_1 has some extra general successes in addition to those it shares with T_2.

And, finally, T_1 and T_2 are *equally* successful exactly if T_1 is at least as successful as T_2 but not more successful than T_2.

Now an individual problem of a theory is a counterexample to what Kuipers calls a General Test Implication (GTI) of the theory; a general success of a theory is an established GTI of the theory. Without going into all the details, a GTI of a theory can be characterized as an empirical or observational law-like consequence of the theory (see ICR, p. 96 for a detailed account). Most significant here is, of course, the word 'observational'. For being an *observational* consequence of a theory, a GTI will be a consequence of any empirically equivalent rival theory as well; by the definition of empirical

---

[9] It will have been noted that IBT does not speak of best *explanation* but only of best *theory*. However, Kuipers (ICR, p. 170) seems to suggest that the only reason for this is that IBT is also meant to apply to theories that have already been falsified, and of course it would be odd to call a theory the best explanation of the data if the data refute it. My point is that as Kuipers understands the notion of best theory, it would be misleading to call it the best explanation even if the theory were unfalsified.

equivalence, empirically equivalent rivals have exactly the same observational consequences and thus also exactly the same GTIs. Consequently, empirically equivalent rivals are bound to have both the same individual problems (if any) and the same general successes (if any). They thus are bound to be equally successful in the sense just defined. Of course Kuipers is well aware of this, as witness, e.g., the remark in the quotation given in §1 that empirically equivalent theories "can explain the same variety of successes and success differences." The reason he does not seem to be too bothered by this is that, as we saw, he more or less refuses to take EE seriously, but not, as we also saw, for any good reason.

It should thus be clear that the theoretical virtues will have to be taken into account in determining the betterness and bestness of theories if we want to base our defense of (convergent) scientific realism on IBT. One obvious way to modify the definition of best theory is to let the best theory among the available theories be the one that is the most successful (in Kuipers' sense) of these theories *if there is a unique one;* else, let it be the one of the 'equally most successful' theories that does, on average, best with respect to the theoretical virtues (that Kuipers wants the notion of best theory also to apply to refuted theories is no impediment to this definition; refuted theories may be no less simple or beautiful than unrefuted theories).[10] Whether the foregoing is the optimal way to give theoretical virtues a role in determining the best of the available theories and, even more importantly, how these theoretical virtues are to be incorporated into the formal framework developed by Kuipers, are further and not readily answerable questions. Here let me just note that it is encouraging to know that in one of his most recent papers, Kuipers has made a start on the latter project (see Kuipers 2002; incidentally, in this paper he does seem to take EE seriously).

In whatever precise way theoretical virtues are going to play a role in comparing the goodness of theories, I shall henceforth assume that IBT operates on the basis of a definition of 'best theory' that takes these virtues into account in some formally and intuitively acceptable way. It is worth noting that this assumption does not in the least jeopardize the superiority of Kuipers' rule over IBE, as this solely depends on the distinction between the two rules with regards to input/output symmetry.

---

[10] Even given these more demanding criteria there is no guarantee that there will always a unique best theory; several "most successful" theories in Kuipers' sense may do equally well with respect to our additional criteria. But note that this does not bring the argument from underdetermination back. It follows from EE that *no* theory that qualifies as best, given Kuipers' criteria, can be unique. There is no plausible thesis, however, saying that there is never a unique best theory given the criteria just proposed.

4. What Justification is there for the Inference to the Best Theory?

In the previous section we saw that, in contrast to IBE, IBT does not rest on an unfounded and implausible assumption of privilege. Now it is one thing to show that a rule is invulnerable to certain objections, but it is quite another to justify the rule. And if IBT is to help us in blocking the argument from underdetermination we must, of course, make plausible that IBT is indeed justified.[11] It is evident that the rule has no analytical justification (at least not if we assume a traditional, correspondence notion of truth).[12] Thus it seems that, if it can be justified at all, its justification must rest on empirical grounds. Several authors in the realist camp have hinted at what at first sight is the evident way to proceed in order to obtain such an empirical justification of IBT.

I will start by describing this seemingly straightforward procedure and show that it is destined to beg important antirealist issues. It is also shown that, initial appearances to the contrary notwithstanding, Hacking's famous arguments from microscopy cannot *by themselves* save the procedure, for as they stand these arguments, too, beg the issue against the antirealist (§4.1). I will then argue that we can combine Hacking's arguments with the idea of the simple procedure for testing IBT in a way that yields a – slightly more complicated – testing procedure that is *not* question-begging (§4.2). In §4.4 it is shown that this new procedure can even be made to work within the confines of the confirmation theory that van Fraassen advocates (cf. §2). To that end this confirmation theory must first be presented in greater detail than has so far been done by van Fraassen or indeed anyone else; this I do in §4.3.

It should be emphasized right away that in this section I am only concerned to establish that there is no principled difficulty in empirically justifying IBT in a way that is acceptable to both the realist and the antirealist; I do not try to argue that IBT is in fact justified. One reason for this is that carrying out the latter task would require careful historical research, which is beyond my competence to undertake. Another is that the relation between empirical support and justification (*How much empirical support for IBT is needed in order for it to be justified?*) is a tangled issue that I want to sidestep here.

---

[11] Psillos (1999), in the course of defending IBE, argues that what is needed to justifiably apply some rule of inference is that there be no reason to doubt the rule; in the absence of such a reason, there is no requirement to actively seek to justify it. This is certainly an interesting proposal. However, for reasons given in Douven (2001) I do not endorse it.

[12] Kuipers (ICR, p. 171) seems to suggest that it does have such a justification, but that is definitely false for a version of the rule amended along the lines suggested at the end of the previous section.

4.1. *A Simple Tacking Argument for IBT (?)*

What was just referred to as the prima facie evident way to obtain an a posteriori justification of IBT is to check the success ratio of IBT, i.e., to investigate in what percentage of its applications IBT yields a correct result, where a correct result of course does not mean that the application of IBT led to the acceptance of a true theory, but only to the acceptance of a theory that is closer to the truth than the theories it once competed with for acceptance. To give a homely example of how such a check may be carried out: When in the morning we find a used plate on our kitchen table we conclude, by means of IBT, that one of our housemates made him- or herself a midnight snack (this is undoubtedly the very best explanation for the phenomena and thus IBT allows us to conclude to the unqualified truth of it – as was explained in the previous section). That might be wrong – surely there are other explanations for the phenomena. However, we can check whether our conclusion is correct simply by asking our housemates about it. If one of them did make a midnight snack, that would constitute a success for our rule of inference. Examples like this can be multiplied at will. And it may seem that, given enough of them, we obtain the required empirical justification of IBT. For would not the evidence in that case show the rule to be reliable?

The problem is that the example just given is an example of what, slightly adapting terminology used by Psillos (1996, 1999), we could call a horizontal inference to the best theory, i.e., an inference from phenomena to a conclusion strictly about observable entities or states of affairs (in contrast to vertical inferences to the best theory, in which we infer to a conclusion solely or partly about unobservables). So the antirealist might respond to a purported justification along the above lines, that all it shows is that IBT is reliable as long as what we come to accept by means of it is strictly about the observable realm. This does not in any way help the realist who is combating the argument from underdetermination. To block that argument by means of an appeal to IBT, the rule must also be shown to be reliable when it leads to the acceptance of hypotheses concerning the unobservable.

So now the all-important question is whether there exist examples of applications of IBT that could show that IBT also leads to correct conclusions when it is applied to hypotheses about unobservables. One is immediately inclined to answer this question positively. Were viruses not postulated for explicitly explanatory reasons at the end of the nineteenth century, and then discovered about fifty years later after the electron microscope had become available? This indeed appears to be an example of a successful application of

IBE or IBT[13] of the kind required to establish the reliability of vertical inferences to the best theory. And there seem to be many more of such examples (see Harré 1986, 1988, and Bird 1998).

This line of thought will not do to convince the antirealist, however. Recall that according to the antirealist our knowledge is restricted to the observable, where by 'observable' she means 'observable by the unaided senses'. She will thus certainly object to the assertion that the existence of viruses has ever been established. Perhaps they have been discovered, perhaps not. The tiny spots biologists identified (and identify) as viruses may in reality have been (and be) due to certain aberrations of the microscopes used. As with any theoretical hypothesis, the antirealist (van Fraassen type) will want to remain agnostic about whether or not viruses were discovered last century, so that to speak of the discovery of viruses would seem to beg a central antirealist question. Hence, a tacking argument for the justification of IBE or IBT such as the one proposed above seems bound to fail.

Is the antirealist not overly cautious in refusing to take data from microscopy as evidence for the existence of viruses and the like? No doubt the antirealist will respond, and rightly so to my mind, that the history of microscopy gives reason for caution at this point; more than once, entities were 'discovered' and then later shown (to use a realist terminology) to be artifacts, due to some aberration(s) of the instrument or the devices used to prepare the specimens for study (cf. Hacking 1981, p.138*ff*; Atlas 1986, p.52*f*).

But even though there is reason for caution when it comes to existence claims made on the basis of data obtained from microscopes, and even though realists may in general have been too quick to assume that, thanks to technological advances, erstwhile unobservable entities have become observable, it may be possible to argue for at least certain observation devices that they are veridical. Hacking (1981) has given two intuitively very appealing triangulation arguments for the conclusion that modern types of microscope are indeed veridical.[14] In one of these, Hacking notes that very different types of microscope give essentially the same output when given the same input. He argues that it is just too much to believe that this is due to some coincidence, and thus that it is reasonable to believe that the different types of microscope are all veridical. If this or the other triangulation argument is

---

[13] It is immaterial exactly what rule the scientists who postulated viruses were using. Even if it was not IBT that they were using, we may be sure that had that rule been applied to the theories available at the time, it would have led to the acceptance of viruses all the same – and that is what matters for present purposes.

[14] Hacking's argument only involves certain, and definitely not all, types of microscope. However, there is no reason to believe the argument cannot be extended to include other types of microscope. In fact, it seems possible to extend it to observation devices other than microscopes, such as X-ray crystallography, for instance.

correct, then it seems we may after all be able to hold in a non-question-begging way that entities once postulated on purely explanatory grounds were at a later stage actually seen by means of (a) certain type(s) of microscope. In other words, Hacking's arguments seem to provide exactly what is needed in order to make the tacking argument for the reliability of IBT go through.

But here another problem appears. Several authors have argued, quite convincingly to my mind, that Hacking's triangulation arguments are, implicitly, inference to the best explanation arguments[15]: That all the different microscopes figuring in Hacking's argument just summarized give a similar output, is reason to believe that they are veridical because that is the best explanation of the former fact – that, according to those authors, is what the argument must really be. Note that this does not mean the argument requires IBE. Given that its conclusion clearly seems to be the absolutely best explanation for the fact Hacking reports, and given the earlier noted fact that in case the absolutely best explanation is among the hypotheses considered, it makes no difference whether we apply IBT or IBE, Hacking's argument can make do with IBT. However, for the antirealist this will make little difference, for she accepts neither IBE nor IBT. It seems, then, that we were wrong to think Hacking's arguments can help us in empirically evaluating IBT.

So far we have considered two realist arguments – one for the conclusion that IBE/IBT is a reliable rule of inference, the other for the conclusion that our current microscopes are veridical – that at least individually are unsuccessful; each relies on an assumption that the antirealist cannot be expected to grant. What I have tried to show in Douven (2002a) is that, surprisingly, in combination they may well provide the means to test IBT in a way that does not beg any antirealist issues. The trick is to suitably link the two arguments, and the link needed is Glymour's (1980) account of bootstrap confirmation, or so I argue in the paper mentioned above. I summarize the procedure in the next subsection.

4.2. *A Bootstrap Test for IBT*

Since Duhem (at least), we have known that, in general, evidence bears on a hypothesis only relative to one or more auxiliary hypotheses. Some have taken this to imply that there can only be a relativized notion of confirmation (and even as opening the door to epistemological relativism). Now, it is Glymour's position that Duhem's insight does not entail that confirmation cannot be absolute. Though it is true that the basic confirmation relation is three-place rather than two-place, under certain specific circumstances we can go from relative confirmation to absolute confirmation, i.e., to the claim that certain

---

[15] Cf. van Fraassen (1985, p. 298), Devitt (1991, p.112), and Reiner and Pearson (1995, p. 64).

data (dis)confirm a certain theory, period. To make this more precise, let T be a finitely axiomatizable theory, consisting of the axioms H_1, ..., H_n and let D be our total evidence at this time. Now suppose that for each axiom H_i of T the following conditions hold:

(B1)   there are other axioms H_{j_1},..., H_{j_m} also of T such that (a) D confirms H_i when these latter axioms are taken as auxiliaries, and (b) there are possible (but non-actual) data D' that *disconfirm* H_i when H_{j_1}, ..., H_{j_m} are taken as auxiliaries, i.e., adopting these latter axioms as auxiliaries in testing H_i does not guard the latter hypothesis against disconfirmation whatever the data[16];

(B2)   there are no axioms H_{k_1}, ..., H_{k_l}, of T such that D disconfirms H_i when H_{k_1}, ..., H_{k_l} are taken as auxiliaries

Then, Glymour claims, we are allowed to conclude that the data confirm T, period, and not just that they confirm T with respect to T. In Glymour's (1980) presentation of it, this claim is backed up by a lot of sophisticated argumentation and is further buttressed by many examples from science, showing that the schema of bootstrap confirmation is an abstract but fair representation of the way theories are tested in actual scientific practice.

The following points out how the theory of bootstrap confirmation can link the two realist arguments considered in this section so as to yield a test that can help justify IBT. Since the theory I shall be concerned with consists of two axioms only, (B2) can further be ignored. After all, given any reasonable interpretation of the notion of (dis-)confirmation as used in (B1) and (B2) it will hold that, if evidence confirms a hypothesis H given another hypothesis H' as auxiliary, then it will not also disconfirm H given H' as auxiliary.

In the first, tacking argument we sought to show that IBT is a reliable rule; call the hypothesis that it is R. Hacking claimed to have provided empirical support for the hypothesis that modern types of microscope are veridical; call this hypothesis V. It was seen that the tacking argument for IBT has no force against the antirealist, because it assumes that thanks to technological advances, in particular the development of sophisticated types of microscope, shifts have occurred in the boundary between what can and cannot be observed (for only given that assumption can we claim, for instance, that we now have observational evidence for the existence of viruses). Hacking's argument was without force, too, because it relies on either IBE or IBT, neither of which the antirealist finds a compelling (or even acceptable) rule of inference.

---

[16] Some formulations in Glymour (1980) suggest that he actually intends, instead of (B1b), the slightly weaker condition that there exist possible data that do not confirm (rather than disconfirm) H_i relative to, H_{j_1}, ..., H_{j_m} The weaker condition may also suffice; cf. Douven (2002a).

But now consider the theory $T_{V\&R} = \{V, R\}$, and say that D, our total current evidence, comprises (among much else) the data Hacking adduces in his triangulation arguments as well as all the data available about events – alleged discoveries of erstwhile unobservable but postulated types of entities – that in our discussion of the tacking argument were said to be required in order to support the claim that vertical IBT is reliable, and suppose that, at least from a naive realist perspective, the latter data are favorable to R (i.e., from that perspective it seems that most applications of vertical IBT have been successful).[17] Clearly, if R is assumed as an auxiliary, then D is confirming evidence for V. Likewise, if V is assumed, then D is confirming evidence for R. But note that we now have already gone some way toward a bootstrap confirmation of $T_{V\&R}$. What we have called condition (B1a) is satisfied. So the only thing left to show is that (B1b) is satisfied as well.

Condition (B1b) requires that the assumption of R in testing V does not trivialize that test in the sense that it excludes a negative result, and that, similarly, the assumption of V in testing R does not trivialize this second test. However, it is obvious that (B1b) is satisfied. First, it is certainly conceivable that Hacking had obtained very different outputs from the different types of microscopes, even though they were given the same or similar inputs, the assumption of the reliability of IBT notwithstanding. And if he had, that would have cast considerable doubt on V. Secondly, making the assumption that contemporary types of microscope are veridical, be it in testing R or in testing some other hypothesis, cannot possibly make it the case that we will not find an unfavorable track record for IBT. For example, it might turn out that only very few of the putative unobservable entities once accepted on purely explanatory grounds "survived" the introduction of the electron microscope.

Thus our total current evidence (or rather, what we assume to be our evidence) confirms $T_{V\&R}$, i.e., it confirms in tandem the hypothesis stating that IBT is a reliable rule and the hypothesis stating that modern microscopes are veridical, and, using Glymour's idea of bootstrapping, it apparently does so without begging any antirealist issues. Now there are many objections that can be raised against these claims, both general objections to Glymour's account and more specific ones addressing the use made of that account in the test we have just outlined. The most pressing of these I have considered and, I believe, answered in Douven (2002a). Here let me only briefly repeat the answer to an objection that seems invariably to come first to people's minds when they are first confronted with the above construction. This is the objection that the proposed test of $T_{V\&R}$ is circular. This reaction is quite understandable. After

---

[17] If Harré and Bird, mentioned earlier, are right, then the data are indeed as here assumed. However, it seems to me that the historical evidence they cite to support this claim is rather meager.

all, the test combines two tests that are separately clearly question-begging; it may thus easily seem that any test combining them will have to be "doubly" question-begging. But there is in effect an easy way to see that this is not so: Just check whether it is possible to derive $T_{V\&R}$ from our test even before you have gathered any data (or, rather, assuming the relevant data are still unknown)! If $T_{V\&R}$ were really presupposed, that should be possible. As you will realize, however, condition (B1b) ensures that it is not. But then it is hard to see what substance there could be to the charge of question-begging.[18]

In order to render its logical structure more clearly, let me present the bootstrap test of $T_{V\&R}$ in a more formal fashion. To that end, we need some notation. Given a test of some hypothesis H_i, say that 'A_i' denotes the hypothesis or conjunction of hypotheses involved as auxiliaries in that test. (Note that, in the case of a bootstrap test, a theory $T = \{H_1, ..., H_n\}$ is tested by testing each of the $H_i \in T$ relative to other hypotheses belonging to the same theory, i.e., in such a test it holds that, for each H_i, either $A_i \in T$ or $A_i = \wedge_{j<m} H_j$ for some m such that $m < n$ and $H_j \in T$ for each $j \leq m$.) Furthermore, let '$C(D; H_i; A_i)$' mean that evidence D confirms hypothesis H_i relative to auxiliary A_i, and let '$D(D; H_i; A_i)$' mean that D disconfirms H_i relative to A_i. '$C(D; H_i; \varnothing)$' means, of course, that D confirms H_i absolutely.

To represent schematically our earlier test of $T_{V\&R}$ we need two rules of inference.[19] The first rule, a conjunction-introduction rule (&I), is utterly straightforward and reads as follows: If D confirms H_i relative to A_i and D confirms H_j relative to A_j, then D confirms $H_i \,\&\, H_j$ relative to $A_i \,\&\, A_j$, or, in natural deduction format:

$$\frac{C(D; H_i; A_i) \quad C(D; H_j; A_j)}{C(D; H_i \& H_j; A_i \& A_j)} \;\text{\&I}$$

The second rule captures what plainly is the most characteristic feature of bootstrap testing, namely, that it permits us to 'cancel' auxiliaries assumed in a test on the condition that the assumption did not trivialize the test. This cancellation mechanism is formalized by the following ('non-triviality') rule:

$$\frac{\Diamond \exists D_1 D(D_1; H_1; A_1) \wedge \cdots \wedge \Diamond \exists D_n D(D_n; H_n; A_n) \quad C(D; H_1 \& \cdots \& H_n; A_1 \& \cdots \& A_n)}{C(D; H_1 \& \cdots \& H_n; \varnothing)} \;\text{NT}$$

---

[18] Except for the fact that the argument might be rule-circular (cf. Psillos 1999). In Douven (2002a) I show that it is not, however.

[19] Remember that for our test clause (B2) of the definition of bootstrap confirmation above could be neglected. I am therefore not giving a rule of inference corresponding to it. However, it is obvious how the rule should read, so those who would like to have an inferential system entirely capturing the earlier definition can easily provide it themselves.

Recall that we assumed ourselves to be in the rather fortunate position of already having obtained data confirming R relative to V, that is to say, D, our total current evidence, is assumed to comprise both the data cited in Hacking's arguments as well as "sufficiently many" positive reports concerning the discovery of particular, earlier only hypothesized, unobservable types of entity (and only relatively few negative such reports). In formal clothing, then, the argument that our data D confirm $T_{V\&R}$ reads as follows:

$$\cfrac{\Diamond\exists D'\,\mathrm{D}(D';V;R) \wedge \Diamond\exists D''\,\mathrm{D}(D'';R;V) \qquad \cfrac{\mathrm{C}(D;V;R) \quad \mathrm{C}(D;R;V)}{\mathrm{C}(D;V\&R;V\&R)}\;{}^{\&\mathrm{I}}}{\mathrm{C}(D;V\&R;\emptyset)}\;{}_{\mathrm{NT}}$$

The application of NT in this derivation is justified by our respective observations that nothing in the way the test is constructed could have prevented Hacking from obtaining dissimilar outputs from the various microscopes he used and that there is no guarantee that historical research is not going to provide us with data disconfirming R relative to V.

Note that so far no specific assumptions have been made about the nature of the confirmation relation involved. Somewhat surprisingly, perhaps, it is not directly necessary to do so. As Glymour (1980) makes plain, the idea of bootstrapping is not tied to any particular notion of confirmation; it can be combined with a variety of confirmation theories. Still, some of the details of our test will depend on the particular theory to which we attach the bootstraps in order to test $T_{V\&R}$. For instance, the answer to the question whether bootstrap confirmation of $T_{V\&R}$ automatically yields confirmation of V and R separately, is clearly *yes* if the underlying confirmation theory is Hempel's (due to the Special Consequence Condition of that account), but is clearly *no* if the underlying theory is Bayesian confirmation theory.[20] Since we wish to justify IBT in a way that will also be acceptable to the antirealist, it would strategically be optimal if we can test $T_{V\&R}$ by means of the confirmation theory (plus bootstraps[21]) endorsed by the antirealist herself. It seems no exaggeration to say that van Fraassen is currently the realist's only serious opponent. Let us therefore attempt to formulate the test for $T_{V\&R}$ in the terms of the confirmation theory he advocates. We now immediately encounter a

---

[20] So in the former, but not in the latter case we could add to the rules of inference presented in the text a conjunction-elimination rule, i.e., a rule allowing us to infer both $\mathrm{C}(D; H_i; \emptyset)$ and $\mathrm{C}(D; H_j; \emptyset)$ from $\mathrm{C}(D; H_i \& H_j; \emptyset)$.

[21] As van Fraassen (1980, p. 222) rightly notes, bootstrapping itself is neutral with respect to the realism/antirealism controversy, so there is no reason to worry that the mere fact that we make use of the mechanism of bootstrapping will suffice to make the test unacceptable in the eyes of the antirealist.

problem, however, for much is unclear about this confirmation theory. So, before we can fill in further details of our test, we will first have to help ourselves to a number of assumptions concerning the confirmation theory van Fraassen endorses.

4.3. *A Bayesian Confirmation Theory Countenancing Vague Probabilities*

Van Fraassen is a Bayesian, but one who countenances vague probabilities. Vague probabilities have no place in standard Bayesian confirmation theory, and if we want to give them a place in it questions arise that are not readily answered (and that van Fraassen nowhere addresses). Yet to be able to cast our bootstrap test in terms of the confirmation theory van Fraassen seems to advocate, these questions need to be answered. In this subsection I make certain assumptions about what the answers should be. However, I should note that it is not my objective to develop a Bayesian confirmation theory countenancing vague probabilities; among other things, this means that I do not try to give full-fledged justifications for the assumptions I make.

We want to combine Bayesian confirmation theory with the idea that hypotheses concerning the unobservable should be assigned probabilities that are vague over an interval including 0, whereas hypotheses strictly about observables should have sharp probabilities.[22] One question that arises then, concerns the fact that under certain circumstances even an antirealist may want to change a vague probability assigned to a proposition into a sharp (positive) probability. Suppose A, a convinced antirealist who assigns vague probabilities to any hypothesis not strictly about observables (but assigns sharp probabilities to hypotheses about observables), believes hypothesis H to be about unobservables and assigns it a probability interval $[0, x]$ for some $x : 0 \leq x \leq 1$. Now A discovers that her belief concerning H was erroneous and that the entities H is about are not unobservable after all. She may have heard you describe those entities as unobservables, whereas in fact you said they were observable. Or consider this: According to van Fraassen (1980, p. 57*ff*) it is up to science to delimit the observable from the unobservable; and of course scientists may find out – and report to A – that certain entities or processes once classified as being unobservable are visible to the unaided eye after all. Also, A may find out, or at any rate come to believe, that, although the entities H is about are unobservable, they are, for some reason, to be treated epistemologically on a par with observables. The question now is this: How

---

[22] Some antirealists may want to remain agnostic even about propositions strictly about observables. However, it is difficult to see how such an epistemic attitude could be motivated by the argument from underdetermination as presented in §1.

302 *Igor Douven*

should *A* change her degree of belief in *H* under these and similar circumstances?

It seems that upon discovering that *H* is not really about unobservables, *A* should no longer assign a vague probability to it, as this would conflict with her policy of assigning vague probabilities only to hypotheses about unobservables. However, it should be remarked that, if all belief change is to proceed via conditionalization, as it is according to standard Bayesian confirmation theory, then the only kind of case in which there can be a transition from a vague probability assigned to some hypothesis to a sharp probability assigned to it, is the one in which one comes to accept evidence that refutes the hypothesis so that its probability becomes 0 sharp.[23] Yet one would suppose that a confirmation theory countenancing both vague and sharp probabilities would contain a rule telling us what to do in the indicated kind of cases.

To have such a rule, I will make the following assumption:

Assumption I If a person *A* believes hypothesis *H* to be about unobservables and therefore assigns it a probability interval $[0, x]$, for some *x*: $0 \leq x \leq 1$, then, if *A* comes to learn that *H* is strictly about observables (or strictly about entities that are epistemologically on a par with observables), the rational thing to do for her is to change her probability for *H* from $[0, x]$ to *x*.

As intimated above, I will not attempt to fully justify the assumptions made in this subsection. However, I do want to note that this assumption is at least *prima facie* plausible, for it seems that any reason one could have for assigning a probability to *H* that is either lower or higher than the upper bound of the old, vague probability would have been a reason to have a lower respectively higher upper bound on the probability interval assigned to *H* when one still believed the hypothesis was about unobservable entities.

Further questions about van Fraassen's confirmation theory that are relevant to our project concern the notions of confirmation and disconfirmation. There are well-known definitions of confirmation and disconfirmation, respectively, for standard Bayesian confirmation theory – *E* confirms *H* exactly if $p(H \mid E) > p(H)$, and disconfirms *H* exactly if $p(H \mid E) < p(H)$ – but

---

[23] For those familiar with van Fraassen's theory of vague probabilities, the reason for this is that if one is agnostic about *H*, then one's representor, i.e., roughly, the set of probability functions compatible with one's opinion, will include a probability function that assigns 0 to *H*. Since conditionalizing one's representor on new evidence comes down to conditionalizing each probability function in one's representor, and since conditionalizing a function that assigns 0 to *H* will result in another function that also assigns 0 to *H*, the only sort of occasion on which someone agnostic about *H* can come to have a sharp opinion about *H* purely by conditionalization is when all the probability functions in her representor come to assign *H* a probability of 0.

it is not straightforward to derive from these definitions, definitions of confirmation and disconfirmation for the theory we are assuming here. I will not try to supply such definitions for the theory, but will only make one assumption about a sufficient condition for confirmation and another about a sufficient condition for disconfirmation:

Assumption II   If $p(H) = [0, x]$ and $p(H \mid E) = y$, with $y > x$, then E confirms H.

Assumption III   If $p(H) = [0, x]$ and $p(H \mid E) = [y, z]$, with $z < x$ and $0 \leq y \leq z$, then E disconfirms H.

Assumption II concerning confirmation should be unproblematic. As regards what counts as disconfirmation of a hypothesis that is assigned a vague probability, it seems to me that the only cases that are entirely unproblematic are those in which the evidence refutes the hypothesis (so that $p(H \mid E) = 0$). However, although the matter is not quite so uncontroversial as in this latter kind of case, I do think it is sensible to say that disconfirmation occurs in the kind of case satisfying the condition of Assumption III.

4.4. *The Bootstrap Test Reconstructed*

I already said that our bootstrap test is sensitive to the underlying confirmation theory that is used. For reasons that will soon become clear, the effect of using the above confirmation theory is that we can obtain a positive test result only for a theory $T_{V\&R'} = \{V, R'\}$ in which R' is a slightly restricted version of R,[24] namely the thesis stating that IBT is reliable if it is used to derive conclusions about observables and/or unobservables of a specific type. For simplicity I assume that this type can be specified in terms of size. Let us say that, *if V is correct*, then entities of size S, but not of any smaller size, can be seen by means of our current microscopes (no doubt other features of an entity than just its size will be relevant to the question whether it can be seen by means of a microscope – provided the microscope is veridical – but as I said, I am simplifying here). Then R' is the hypothesis stating that IBT is reliable if it is used to derive conclusions about entities of size S or larger. Clearly, since R' is a weaker hypothesis than R, replacing $T_{V\&R}$ by $T_{V\&R'}$ diminishes the significance of the test.[25]

---

[24] Or, more carefully, I can only see how to obtain a positive test result for $T_{V\&R'}$; it is not excluded that, in a manner I am presently unaware of, a bootstrap test for $T_{V\&R}$ can be constructed on the basis of the confirmation theory here assumed.

[25] The diminution may in the end not be very considerable, though, given that, as was said in note 14, the test Hacking proposes for microscopes may well be extensible to other observation devices, so that we could have a bootstrap test with V replaced by a hypothesis V' stating that these other devices plus our current microscopes are veridical. R' could then in turn be replaced by a stronger hypothesis (even though it may not be possible ever to replace it by R).

To begin our test of $T_{V\&R'}$ within what we assume to be the confirmation theory accepted by the antirealist (or by van Fraassen, in any event), first note that both V and R are about unobservables: V says that certain unobservables (things too small to be seen by the naked eye) *can* be seen by means of our microscopes; R' says that IBT is reliable when it licenses inferences about observables *and also when it licenses inferences about unobservables that can be seen by means of our microscopes in case V is true.* So, the antirealist will want to remain agnostic about these hypotheses and accordingly assign vague probabilities to both of them. Suppose that $p(V) = p(R') = [0, .5]$ represents her prior degrees of belief in the two hypotheses (nothing much hinges on the exact values of the upper bounds, and nothing at all on the assumption that $p(V) = p(R')$).

Let us now see whether the data support V relative to R' and R' relative to V, in that order.

Put briefly, the argument that the data support V relative to R' is as follows. Clearly V is about unobservables of size S or larger; it says that entities of at least size S can be seen by means of certain types of microscope. Thus R' is sufficient to draw from the data reported in Hacking's triangulation arguments, and from the fact that V is the very best explanation for those data, the conclusion that V is very likely correct. So, assuming R' as an auxiliary, we may conclude that the probability of V is 1 or close to 1, i.e., relative to R' the data bestow a sharp probability on V and a probability that is higher than what we assumed to be the upper bound of the antirealist's prior degree of belief in V. Hence, by Assumption II the data confirm V relative to R'.

The latter part of this argument is evident. However, I am sure it can do no harm to state in rather more detail the sub-argument for the conclusion that the data make V very likely given R' as an auxiliary. In this sub-argument we start from the premise that V, the hypothesis stating that entities of a size no smaller than S can be seen by means of modern microscopes, is the very best explanation of the data. By IBT, it follows from this that V is true (recall that in case a hypothesis constitutes the very best explanation for the data, IBT allows us to infer to the unqualified truth of that hypothesis). Now our auxiliary, R', states that conclusions concerning hypotheses about observables and/or unobservables of size S and larger, reached on the basis of IBT, are very likely true. Thus the conclusion that V is true is very likely true, i.e., V is very likely true, or, put differently again, V is very likely. That is, given R', we have a deductively valid argument for the conclusion that our current microscopes are very likely veridical.

It may be noted that we do not reach the conclusion that V is very likely by updating the antirealist's initial degree of belief in V via conditionalization.[26] Now this may seem strange, for we clearly are assuming a non-standard Bayesian confirmation theory that, besides the rule described in Assumption I, normally only regards conditionalization as a legitimate way of updating one's degrees of belief. However, because in the argument R' is assumed as an auxiliary, a third rule becomes (at least temporarily) available, namely IBT, and it is this rule that allows us to conclude that V is very probable given the data.

As a next step in our test, we must see whether the data confirm R relative to V. At this point Assumption I becomes relevant. For one problem we seem to encounter is that the realist and antirealist may disagree over what the data are. Assume that a realist would report the data relevant to this part of the test as follows: There exist data about discoveries by means of modern types of microscope of entities that once were postulated on explanatory grounds, and these data are in fact favorable to IBT. Evidently the antirealist will demur at this description. She will only be willing to grant that all observable phenomena are *as if* such entities were discovered. More specifically she will want to remain agnostic about the question whether the apparent discoveries were real discoveries, and thus will assign a probability to sentences such as

Viruses were discovered fifty years after they were postulated,   (*)

that is vague over the whole interval [0, 1] (here the assumption that the antirealist's upper bound is 1 is appropriate because the antirealist will certainly acknowledge that (*) is empirically adequate: all observable phenomena are in any case *as if* viruses were discovered fifty years after they were postulated).

However, if V is assumed as an auxiliary, then (*) and kindred sentences are about entities to which we have epistemic access no less than we have epistemic access to objects that can be seen by the unaided eye. Hence, given V, Assumption I applies to such sentences, and the antirealist's probability for them should "collapse" to 1. This collapse should also occur in the case of her probability for R', which we assumed to be [0, .5]. After all, given V, R' is a hypothesis about observables and unobservables that are to be treated epistemologically on a par with observables. We may now assume that conditionalization on the data (provided they are as favorable to IBT as the realist hopes they are) will raise the antirealist's degree of belief in R' to some value $x > .5$. It then follows from Assumption II that the data confirm R' relative to V.

---

[26] As explained earlier, that initial degree of belief *could* not be updated by conditionalization and result in a sharp, non-zero probability.

                                    Igor Douven

To complete the test, we must check whether bootstrap condition (B1b) is satisfied, i.e., whether in both cases disconfirmation would have been possible or whether this was prevented by the specific choice of the respective auxiliaries. Here we can refer to what was said in our general exposition of the bootstrap test. Had the outputs from the various types of microscopes Hacking obtained been very dissimilar despite the fact that the inputs used were the same or at least similar to each other (clearly the occurrence of such data could not be prohibited by assuming R'), the upper bound of the antirealist's degree of belief in V would certainly have been lowered. Thus in that case the data would, by Assumption III, have disconfirmed V. Furthermore, we have only assumed that the track record of IBT was favorable on the assumption of V; in actuality we could certainly find an unfavorable track record for it on the same assumption. If we do, then, we may plausibly suppose, this will lead the antirealist to have a degree of belief in R' that is sharp (because the rule of Assumption I applies) but that also is lower than the upper bound on her initial degree of belief. Given Assumption III, the data would thereby disconfirm R'. Hence condition (B1b) is satisfied.

5. Conclusion

We have seen that the realist seems to be in need of some rule like IBE if she is to successfully block the argument from underdetermination. It was argued that the realist is much better off with Kuipers' rule of inference, IBT, than with the more popular IBE, provided we adopt somewhat stricter criteria for the goodness of theories than the ones Kuipers proposes. As regards the question of the rule's legitimacy, in the main part of this paper I have argued that, initial appearances to the contrary notwithstanding, it is possible to test IBT, or in any event – depending on what confirmation theory we employ – it is possible to test it in combination with the hypothesis that the types of microscope currently in use are veridical, and that this can be done in a way that does not beg any of the antirealist's central tenets. So at least in principle it seems possible for the realist to justify IBT in a non-question-begging fashion. It has even been seen that we can bootstrap test (again in combination with the hypothesis concerning our microscopes) the reliability of IBT in a restricted, but for the realist still interesting, domain even if we employ as the underlying confirmation theory of the test the one preferred by van Fraassen.

ACKNOWLEDGMENTS

I am greatly indebted to Roberto Festa, Theo Kuipers, and Jan-Willem Romeyn for valuable comments on an earlier version of this paper.

Erasmus University Rotterdam
Department of Philosophy
P.O. Box 1738
3000 DR Rotterdam
The Netherlands
e-mail: douven@ fwb.eur.nl

REFERENCES

Atlas, R. (1986). *Basic and Practical Microbiology*. New York: Macmillan.

Bird, A. (1998). *Philosophy of Science*. London: UCL Press.

Boyd, R. (1981). Scientific Realism and Naturalistic Epistemology. In: P. Asquith and R. Giere (eds.), *PSA 1980*, vol. II, pp. 613-662. East Lansing MI: Philosophy of Science Association.

Boyd, R. (1984).The Current Status of Scientific Realism. In: Leplin (1984), pp. 41-82.

Boyd, R. (1985). Lex Orandi est Lex Credendi. In: Churchland and Hooker (1985), pp. 3-34.

Churchland, P. and C. Hooker eds. (1985). *Images of Science.* Chicago: University of Chicago Press.

Devitt, M. (1991). *Realism and Truth.* Oxford: Blackwell.

Dorling, J. (1992). Bayesian Conditionalization Resolves Positivist/Realist Disputes. *Journal of Philosophy* **89**, 362-382.

Douven, I. (1995). Boyd's Miraculous No Miracle Argument. In: P. Cortois (ed.), *The Many Problems of Realism*, pp. 89-116. Tilburg: Tilburg University Press.

Douven, I. (1999). Inference to the Best Explanation Made Coherent. *Philosophy of Science* (Proceedings) **66**, S424-S435.

Douven, I. (2000). The Antirealist Argument for Underdetermination. *Philosophical Quarterly* **50**, 371-375.

Douven, I. (2001). Quests of a Realist. *Metascience* **10**, 354-359.

Douven, I. (2002a). Testing Inference to the Best Explanation. *Synthese* **130**, 355-377.

Douven, I. (2002b). A New Solution to the Paradoxes of Rational Acceptability. *British Journal for the Philosophy of Science* **53**, 391-410.

Douven, I. and L. Horsten (1998). Earman on Underdetermination and Empirical Indistinguishability. *Erkenntnis* **49**, 303-320.

Earman, J. (1993). Underdetermination, Realism, and Reason. In: P. French, T. Uehling, Jr., and H. Wettstein (eds.), *Midwest Studies in Philosophy,* vol. XVIII, pp. 19-38. Notre Dame: University of Notre Dame Press.

Glymour, C. (1980). *Theory and Evidence.* Princeton: Princeton University Press.

Hacking, I. (1981). Do We See Through a Microscope? *Pacific Philosophical Quarterly* **62**, 305-322. Reprinted in Churchland and Hooker (1985), pp. 132-152; page reference in text to the reprint.

Hájek, A. (1998). Agnosticism Meets Bayesianism. *Analysis* **58**, 199-206.

Harré, R. (1986). *Varieties of Realism.* Oxford: Blackwell.

Harré, R. (1988). Realism and Ontology. *Philosophia Naturalis* **25**, 386-398.

Howson, C. (2000). *Hume's Problem: Induction and the Justification of Belief.* Oxford: Clarendon Press.

Kitcher, P. (1993). *The Advancement of Science.* Oxford: Oxford University Press.

Kuipers, T. (1984). Approaching the Truth with the Rule of Success. *Philosophia Naturalis* **21**, 244-253.

Kuipers, T. (1992). Naive and Refined Truth Approximation. *Synthese* **93**, 299-341.

Kuipers, T. (2000/ICR). *From Instrumentalism to Constructive Realism.* Dordrecht: Kluwer.

Kuipers, T. (2002). Beauty, a Road to the Truth? *Synthese* **131**, 291-328.

Kukla, A. (1995). The Two Antirealisms of Bas van Fraassen. *Studies of History and Philosophy of Science* **26**, 431-454.

Kukla, A. (1998). *Studies in Scientific Realism.* Oxford: Oxford University Press.

Ladyman, J., I. Douven, L. Horsten and B. van Fraassen (1997). A Defence of van Fraassen's Critique of Abductive Inference. *Philosophical Quarterly* **47**, 305-321.

Laudan, L. (1990). Demystifying Underdetermination. In: C. Savage (ed.), *Scientific Theories,* pp. 267-297. Minneapolis: University of Minnesota Press.

Laudan, L. and J. Leplin (1991). Empirical Equivalence and Underdetermination. *Journal of Philosophy* **88**, 449-472.

Leplin, J. (1984). *Scientific Realism.* Berkeley: University of California Press.

Leplin, J. (1997). *A Novel Defense of Scientific Realism.* Oxford: Oxford University Press.

Leplin, J. (2000). The Epistemic Status of Auxiliary Hypotheses. *Philosophical Quarterly* **50**, 376-379.

Lipton, P. (1991). *Inference to the Best Explanation.* London: Routledge.

Lipton, P. (1993). Is the Best Good Enough? *Proceedings of the Aristotelian Society* **93**, 89-104.

McMullin, E. (1996). Epistemic Virtue and Theory Appraisal. In: I. Douven and L. Horsten (eds.), *Realism in the Sciences*, pp. 13-34. Leuven: Leuven University Press.

Monton, B. (1998). Bayesian Agnosticism and Constructive Empiricism. *Analysis* **58**, 207-212.

Niiniluoto, I. (1999). *Critical Scientific Realism*. Oxford: Clarendon Press.

Psillos, S. (1996). On van Fraassen's Critique of Abductive Reasoning. *Philosophical Quarterly* **46**, 31-47.

Psillos, S. (1999). *Scientific Realism: How Science Tracks Truth*. London: Routledge.

Reiner, R. and R. Pierson (1995). Hacking's Experimental Realism: An Untenable Middle Ground. *Philosophy of Science* **62**, 60-69.

van Fraassen, B. (1980). *The Scientific Image*. Oxford: Clarendon Press.

van Fraassen, B. (1985). Empiricism in the Philosophy of Science. In: Churchland and Hooker (1985), pp. 245-308.

van Fraassen, B. (1989). *Laws and Symmetry*. Oxford: Clarendon Press.

Theo A. F. Kuipers

WHAT IS THE BEST EMPIRICALLY EQUIVALENT THEORY?

REPLY TO IGOR DOUVEN

Douven's paper certainly is a very constructive one relative to ICR. It supports my turn to "inference to the best theory" (IBT) as a critically revised version of the standard rule of "inference to the best explanation" (IBE), it argues for an important refinement of IBT, and it shows ways of empirically testing it. I have nothing to criticize in the main argument, but I would just like to make some local remarks. But let me start by remarking that Igor Douven (1996) introduced the important notion of an OUD shift, that is, a shift in the Observable/Unobservable Distinction. It helped me a lot in clarifying the long-term dynamics of scientific research as described in ICR. Specifically, in Section 9.2.2., I first discuss in general what arguments can be given for specific and general, separate and comparative reference claims. I then deal with theoretical arguments, followed by experimental and then combined ones. Finally, I deal with the consequences of the acceptance of specific reference claims together with experimental and/or theoretical criteria for applying them, that is, when an OUD-shift has taken place. In this reply I first indicate how theoretical virtues of theories can be taken into account within my general approach. I then indicate what I like to call the "referential Douven test."

Theoretical Virtues

At the end of Section 3 Douven pleads for taking "theoretical virtues" into account in evaluating the merits of theories, in particular when the theories to be compared are empirically equivalent. He rightly remarks that I have studied this problem more recently, see (Kuipers 2002), with special emphasis on aesthetic virtues or, more precisely, "nonempirical features which (certain) scientists (have come to) find beautiful, that is, to which they ascribe aesthetic value" (Kuipers 2002, p. 299). Let me quote the abstract of that paper:

> In this article I give a naturalistic-cum-formal analysis of the relation between beauty, empirical success, and truth. The analysis is based on the one hand on a hypothetical

In: R. Festa, A. Aliseda and J. Peijnenburg (eds.), *Confirmation, Empirical Progress, and Truth Approximation* (*Poznań Studies in the Philosophy of the Sciences and the Humanities,* vol. 83), pp. 310-312. Amsterdam/New York, NY: Rodopi, 2005.

variant of the so-called 'mere-exposure effect' which has been more or less established in experimental psychology regarding exposure-affect relationships in general and aesthetic appreciation in particular (Zajonc 1968, Temme 1983, Bornstein 1989, Ye 2000). On the other hand it is based on the formal theory of truthlikeness and truth approximation as presented in my *From instrumentalism to constructive realism* (2000).

The analysis supports the findings of James McAllister in his beautiful *Beauty and revolution in science* (1996), by explaining and justifying them. First, scientists are essentially right in regarding aesthetic criteria useful for empirical progress and even for truth approximation, provided they conceive of them as less hard than empirical criteria. Second, the aesthetic criteria of the time, the "aesthetic canon," may well be based on "aesthetic induction" regarding nonempirical features of paradigms of successful theories which scientists have come to appreciate as beautiful. Third, aesthetic criteria can play a crucial, schismatic role in scientific revolutions. Since they may well be wrong, they may, in the hands of aesthetic conservatives, retard empirical progress and hence truth approximation, but this does not happen in the hands of aesthetically flexible, "revolutionary" scientists.

For critical commentaries on this paper I refer the reader to the contributions by Miller (this volume) and Thagard (the companion volume). Here I shall merely focus on the formal point of the paper, according to which "more truthlikeness," besides being empirically at least as successful, entails sharing at least as many nonempirical features with the true theory, as far as "distributed" features are concerned, that is, features that hold for all relevant models. Hence, if we have reasons to assume that the (strongest) true theory has certain nonempirical features, such features may guide theory choice aiming at truth approximation. Quoting the introduction of the paper, with some insertions, I claim:

> … an aesthetic success [or, more generally, a theoretical success] can be just as good a signpost to the truth as an extra case of explanatory success, albeit in a more modest degree. The relevant difference is that the justified desirability of such an explanatory success can be more reliably established than that of an aesthetic [theoretical] feature, which is why the latter should be approached with more care.

Hence, I would like to claim that the paper on beauty essentially answers Douven's implicit question, as far as "distributed" virtues are concerned, when he writes: "In whatever precise way theoretical virtues are going to play a role in comparing the goodness of theories, I shall henceforth assume IBT to operate on the basis of a definition of 'best theory' that takes these virtues into account in some formally and intuitively acceptable way." (p. 292) Unfortunately, I have not presented that paper in terms of IBT. However, in these terms, the amended IBT would at least include the subrule that of two empirically equivalent theories the theoretically more successful one should be chosen, for the time being, as the closest to the truth. As an aside, in response to Note 9, this implies that Douven's argument against speaking of "the best explanation," instead of "the best theory," would disappear. However, my own argument, according to which it sounds problematic to leave (very much) room for the possibility that the best explanation is already known to be falsified, remains valid.

The Referential Douven Test

The above suggested (partial) emendation of IBT immediately implies that I do not agree with Douven's claim in Note 12, in which he suggests that such a subrule has no justification in terms of truth (approximation), for which reason he focuses in Section 4 on the empirical justification of the amended IBT. To be sure, I would like to agree with Douven's point at the beginning of Section 4.1 that realists have to justify more than, to use his term borrowed and adapted from Psillos (1999), horizontal inference to the best theory. The latter rule corresponds to the observational version of the three versions of IBT that I distinguish (ICR, p. 228, see also Kuipers 2004 for a more detailed analysis): *Inference to the best theory on the observational/referential/ theoretical level (as the closest to the relevant truth)*. More specifically, entity realists have to justify in addition the referential version and theory realists the theoretical one.

However, empirical justifications, rightly advocated by Douven, must also have some relation to the truth approximation account. More specifically, whereas explanatory successes are based on (low-level) inductive generalizations or "object-induction," that is, induction of a regularity about (the behavior of) a certain kind of objects, theoretical successes are based on "meta-induction," that is, induction of a recurring nonempirical feature correlating with empirical success. Object-inductions are not very trustworthy, but they are certainly more trustworthy than meta-inductions. In a way, the bootstrap tests described by Douven must give an empirical justification of both types of induction, with the relevant differences, of course. However, it may well be that his tests are essentially tests of the methodological substrate of IBT, that is, the (also to be amended) rule of success (RS) (ICR, p. 114). In ICR (p. 227) I already suggested one particular form of such a test in relation o OUD and OUD shifts mentioned above: "Here it should be mentioned that an interesting test can be attached to the rationality of RS, which is a version of a test suggested by Douven (1996) and which we would like to call the Douven test. If most of our RS-choices on the basis of the old OUD, remain in tact after the OUD-shift, it suggests that RS is not only in theory, but also in practice, very fruitful for truth approximation." In particular, this outcome of the (referential) Douven test would give empirical support to the referential version of IBT attached to RS, in addition to its theoretical support.

REFERENCES

Bornstein, R. (1989). Exposure and Affect: Overview and Meta-Analysis of Research, 1968-1987. *Psychological Bulletin* **106** (2), 265-289.

Douven, I. (1996). *In Defence of Scientific Realism*. Dissertation. Leuven: University of Leuven.

Kuipers, T. (2002). Beauty, a Road to the Truth? *Synthese* **131**, 291-328.

Kuipers, T. (2004). Inference to the Best Theory, Rather Than Inference to the Best Explanation. Kinds of Induction and Abduction. In: F. Stadler (ed.), *Induction and Deduction in the Sciences*, pp. 25-51. Dordrecht: Kluwer Academic Publishers.

Psillos, S. (1999). *Scientific Realism. How Science Tracks Truth*. London: Routledge.

McAllister, J. (1996). *Beauty and Revolution in Science*. Ithaca: Cornell University Press.

Temme, J. (1983). *Over Smaak Valt te Twisten. Sociaal-Psychologische Beïnvloedingsprocessen van Esthetische Waardering*. With a summary in English. Dissertation. Utrecht: University of Utrecht.

Ye, G. (2000). *Modeling the Unconscious Components of Marketing Communication: Familiarity, Decision Criteria, Primary Affect, and Mere-Exposure Effect*. Dissertation Rotterdam. Tilburg: University of Tilburg.

Zajonc, R. (1968). *Attitudinal Effects of Mere Exposure*. Monograph supplement 9 of *The Journal of Personality and Social Psychology*.

TRUTH APPROXIMATION BY EMPIRICAL AND NONEMPIRICAL MEANS

Bert Hamminga

CONSTRUCTIVE REALISM AND SCIENTIFIC PROGRESS

ABSTRACT. This paper exploits the language of structuralism, as it has recently been developed with stunning effectiveness in defining the relations between confirmation, empirical progress and truth approximation, to concisely clarify the fundamental problem of the classical Lakatos concept of scientific progress, and to compare its way of evaluation to the real problems of scientists facing the far from perfect theories they wish to improve and defend against competitors.

I opt basically for the structuralist terminology adopted in Kuipers (2000), because that is balanced with care to deal with a range of issues far wider than the one dealt with in this contribution. It should be added that this does not commit me to any position on any subject, because structuralism is not a (meta-) theory, it is a language, able to express anything that can be said in other concept systems created to describe knowledge and its dynamics.

Introduction

Section 1 below is an informal explanation of the language used in the paper to discuss theory structure and theory development. It is a customary merger of two developments, that of structuralism as introduced by Sneed (1971), and the verisimilitude definitions as found in Kuipers (2000), hereafter referred to as ICR. This should enable the reader to understand the claims in later sections about scientific progress, where Lakatos' "novel fact" idea is analyzed, and other remarks made by Lakatos that he did not formalize but that seem to be (more) promising in dealing with scientific progress, in the section on "new things." The requirements for formalizing new things in a scientific theory lead naturally to some of the hardest problems in comparing theories and establishing scientific progress, mainly that basic theory structures define their own standard of comparison, as a result of which rival standards of comparison may block the road to progress.

1. The Truth Seeker's Pocket Maps in 61 Kb

The ingenuous thing about structuralism is that it does not start with what a theory asserts in some language (Japanese, mathematics, or, in Inca culture,

In: R. Festa, A. Aliseda and J. Peijnenburg (eds.), *Confirmation, Empirical Progress, and Truth Approximation* (*Poznań Studies in the Philosophy of the Sciences and the Humanities*, vol. 83), pp. 317-336. Amsterdam/New York, NY: Rodopi, 2005.

ropes with knots). It aims directly at delineating the set of all states and events relevant to the theory X, called Mp. Mp consists of all the kinds of things about which the theory says they can, or cannot happen in the world we live in. Mp is the set of logical possibilities, and the Truth T is defined as the *complete* subset of *real* (also called *nomic*) possibilities in the whole set Mp of *logical* possibilities. The rest of the logical possibilities, states and events in Mp – T, are those that, though logically possible, are impossible in reality.

As an example of Mp – T consider logical, but not real possibilities, in a theory of mechanics, like a pendulum that increases in amplitude, or any orbital deviation of a planet. In a theory of light, an example would be rays suddenly projecting your name on a wall. In economics, a sharp drop of tariffs followed by decreasing import volumes. In general: events you can *describe*, so they are members of the set Mp, but they will not happen because they are *impossible in reality*, and if you have a good theory, your theory will tell you so.

So a theory X, defining its own set of possibilities in Mp may consist of hits (gray) and misses (black) as follows:

	(T)-possibilities	**(T)-impossibilities**
X hits	internal matches	external matches
X misses	internal mistakes	external mistakes

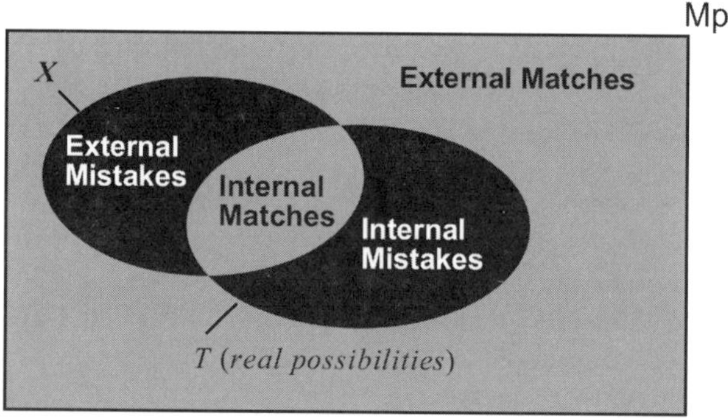

Fig. 1. Hits and misses

If you knew T, you would know all hits (gray) and misses (black) of theory X as depicted in Fig.1. But science stems from the sorry fact that we don't, and so both scientists and philosophers of science have some hard work to do.

Mp

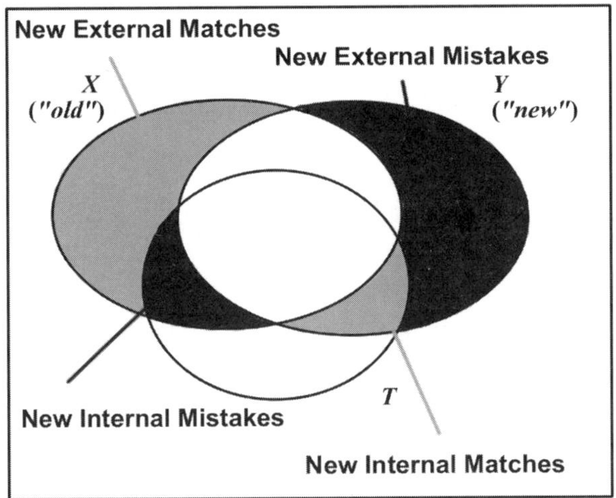

Fig. 2. Comparative hits and misses

If you knew *T*, and have an *old* theory *X*, you could tell about every *new* theory *Y* (see Fig. 2.) whether it has *new* hits (gray) and/or *new* misses (black) of any of the two respective kinds, and for the "superior" label you could require: "*no new misses*," and, if you are exigent, "*some new hits*." Or, briefly, *no black, some gray*.

But you don't know *T*.

It is nevertheless possible to establish whether or not we have moved closer to the Unknown *T* in the sense that some new *Y* is closer to that limit than some predecessor *X*.

In real science, only a limited set $R(t)$ of the real possibilities in *T* has been *established* (by observation) at time *t*. Since at a given time *t* scientists do not dispose of *T* but only a limited stock of established data $R(t)$, real *im*possibilities are, by definition, never established, so there is asymmetry inside and outside $R(t)$: what is inside $R(t)$ has occurred and is hence established as being possible in reality. What is outside $R(t)$ has, alas, *not* been established as being *im*possible.

In the absence of *T*, $R(t)$ provides the consolation of yielding an identification of *new established internal matches*, and the *new established internal mistakes*, both only part of the full sets of possibility hits and misses that you can only know by knowing *T*. If you are not equipped with Truth and only with $R(t)$, the only evaluable success criterion is: "*no new established mistakes*," and, if you are exigent, "*some new established matches*," so unlike

God, we humans are stuck with the filled gray (new *established* hits) and black (new *established* misses) areas in Fig. 3. The rest of the Truth paradise, the areas with black and gray *borders* only, seems lost. But not completely. There is a second consolation:

At t, scientists have a set of laws related to the subject Mp. Those laws are "established," in the sense of being, like $R(t)$, accepted by the scientific community. Since every individual law is a local restriction containing just a few variables only, it classes as impossible only a few elements of Mp, and leaves the larger part of the elements of Mp "free," as possibilities. Hence sets S of possibilities allowed by an "established" law Mp cover a relatively large part of Mp.

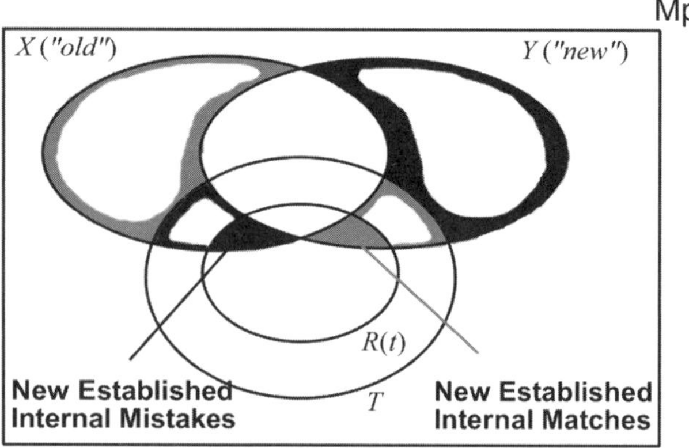

Fig. 3. Success in possibility hits

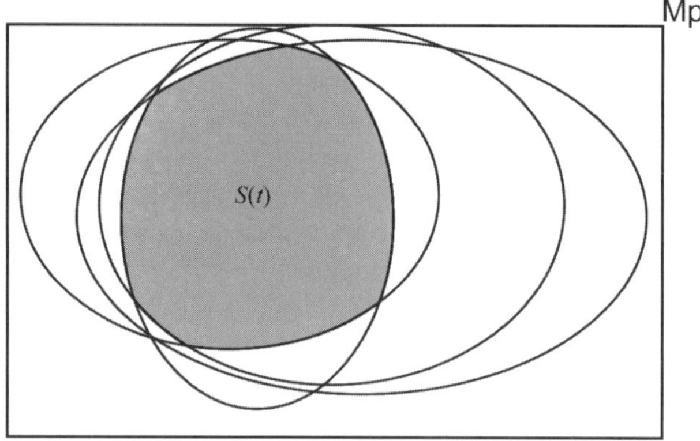

Fig. 4. Laws

As a second consolation in the absence of Truth, theories could be required to simultaneously cover all established laws on the subject. Unlike $R(t)$, $S(t)$, the possibilities in the intersection of all established laws, allows symmetrical evaluation: inside $S(t)$ are law-established possibilities (possible according to all laws), outside $S(t)$ are the *law-established im*possibilities (impossible according to at least one law).

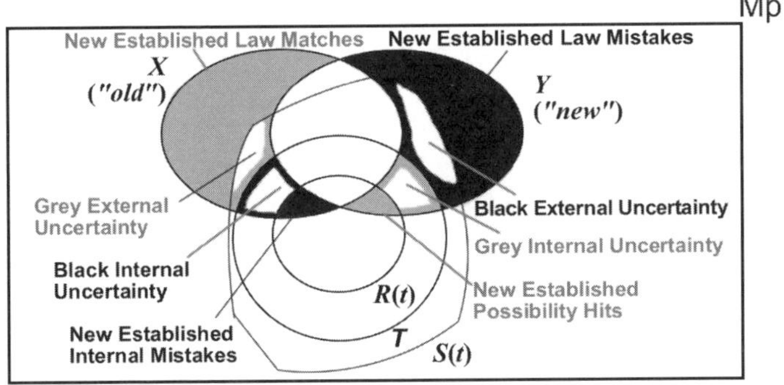

Fig. 5. Overall assessable success

Hence theories should be judged according to their *established law misses* (*im*possibilities according to at least one established law). And hits too, because with laws (unlike individual observed members of R(t)) every non-miss is a *hit*. Hence, in Fig. 5, the upper left gray surface in X can be labeled "new established law matches". In Fig. 5 the success criterion is: "*No black*," and, if you are exigent: "*Some gray*."

The "identifiability" loss resulting from not knowing the Truth consists of the four "uncertainty" surfaces indicated by white areas bordered with black or gray: we know the total area of black and gray, internal and external uncertainty (ignorance). But our present established observations $R(t)$ and established laws $S(t)$ do not allow us to classify any of these uncertain possibilities in any single one of these four uncertainty sets. We can perfectly identify this four-part area as a whole, we know that from the vantage point of Truth it is completely black and gray, but we do not know which of the two is where. One of the great merits of the verisimilitude apparatus is that it allows you so clearly to say what it means *not to know the Truth*.

Fig. 5 is the illustration of the Success Theorem (Kuipers ICR, p. 160): a pictorial way to state this theorem is:

if, cast in the picture Fig. 5, your X and Y satisfy Kuipers' closer-to-the-truth criterion, that is *if* they show no black and some gray, *then* they would certainly show also no black and some gray at a future time t when you would (as you never can) have reached the state of Truth-bliss where established observations $R(t) = T =$ established laws $S(t)$, that is, you would have known the Truth and thus could have cast your X and Y in Fig. 2.

Or:

if you adopt the Kuipers definition of "closer-to-the-truth" (ICR, p. 150-1), it can be proved that no Truth possibility for the four types of uncertainty as indicated in Fig. 5 can affect "closer-to-the-truth" inferences made with the help of $R(t)$ and $S(t)$.

2. Lakatos on Novel Facts

In one of the crucial paragraphs of his classical paper "Falsification and the Methodology of Scientific Programs," Lakatos writes: "For the sophisticated falsificationist, a theory is acceptable or scientific only if it has corroborated excess empirical content over its predecessor (or rival)." (Lakatos 1970, p. 116) The falsificationist definition of "content" of X is: what is *forbidden by* (impossible according to) X. In truth approach terms it is Mp $- X$. The *excess*, or *new* content of Y over X is hence what was allowed by X and is forbidden by Y: in Fig. 6 it is the horizontally hatched area. It consists of *all new external matches* and *new internal mistakes* together. A part of this horizontally hatched excess content surface is called "empirical" by Lakatos. That is the part that is, in principle, testable by observation. The resulting "testable" *empirical excess content*, subset of *excess content* is crosshatched in Fig. 6. Now, for being "acceptable" or "scientific" in the eyes of the sophisticated falsificationist, some of that cross hatched area (that is – some empirical excess content) should be *corroborated*. What does that mean in truth-approach terms?

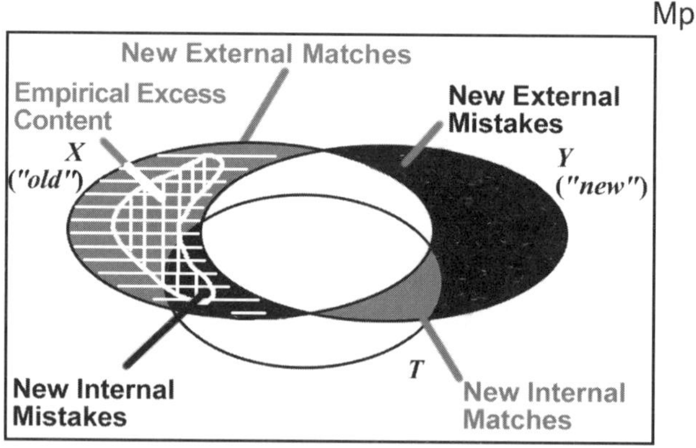

Fig. 6. Excess content

It means something logically inconceivable: as far as the cross hatched area (that is the empirical excess content of Y over X) overlaps T, we are in the "black area" and observation of one of its elements would *by definition* imply falsification of Y, which is hardly a rational demand for a new theory. The rest of the cross hatched area is in the gray, which means they are new matches, but *external* matches, hence *impossibilities*, and nothing can be *observed* to be impossible.

Not knowing the Truth of course does not alter this purely logical conclusion.

It is exactly at this juncture that Lakatos coins a new term, which may indicate his dim awareness of the problem rising here. To explain the notion of corroborated excess empirical content, the sentence quoted above continues: "that is, only if it leads to the discovery of novel facts." In the light of Fig. 6, it is illuminating to follow Lakatos on novel facts in his paper. Novel facts are explicated as "facts improbable in the light of, or even forbidden by" the preceding theory (Lakatos 1970, p. 116), which clearly makes one think of a weaker new theory Y allowing more possibilities than X, hence less falsifiable than X. But is not weakness the exact reverse of the falsificationist aim of science? A novel fact is called a "hitherto unexpected fact" (Lakatos 1970, p. 118), in a footnote, "similar facts" are disqualified and we again read the expression "improbable or even impossible in the light of previous knowledge" (Lakatos 1970, p. 118, footnote 2). Then, on p. 120 the "rather rare – corroborating instances of excess information" are called the "crucial ones" in falsification... "these receive all attention." Again on p. 121 they are denoted by "the few crucial *excess-verifying instances*" that are "decisive." "*Exemplum docet, exempla obscurant*" (Lakatos 1970, p. 121: meaning "One example teaches, examples obscure"). In a footnote the term 'verification' (single quotation marks in original) is coined as shorthand for *corroboration of excess content* (Lakatos 1970, p. 137, footnote 2) and it reads "it is the 'verifications' which keep the program going, recalcitrant instances notwithstanding" (Lakatos 1970, p. 137). These words make one think of a new Y allowing for new possibilities ruled out by a stronger, more falsifiable old X. The novel fact is the occurrence of one of those possibilities ruled out by X but allowed by the weaker, hence less falsifiable Y. *But then Lakatosian excess content means new possibilities, where Popper's excess content meant new impossibilities*. This seems not to be an acceptable reading of Lakatos.

The best way to find out what may have been on Lakatos' mind is to pin down the examples that Lakatos gives of "novel facts" on the Truth seeker's maps introduced above:

- Einstein's theory explained "*to some extent* some known anomalies" of the latest version of the Newtonian theory (Lakatos 1970, p. 124)

                    Bert Hamminga

Explaining known anomalies of a predecessor would clearly at least mean to turn an internal mistake of *X* into a possibility of *Y*, so that amounts to a *new internal match*, albeit a special type of match, because deeming something possible is supposedly not *sufficient* to deserve the title "explanation." We shall deal with that below. In any case, it is not in the hatched area of Fig. 6.

- Einstein's theory "forbade events like transmission of light along straight lines near large masses about which Newton's theory had said nothing but which had been permitted by other well-corroborated scientific theories of the day; moreover, at least some of the unexpected excess Einsteinian content was in fact corroborated (for instance, by the eclipse experiment)" (Lakatos 1970, p. 124).

At first sight this seems a relevant example in the sense of being in the crosshatched area of Fig. 6: a new external match (old external miss). Your new theory stamps "impossible" a straight line near a large mass: it predicts a curve, so the theory says that, near a large mass, there will *always* be a curve and *never* be a straight line. A novel fact predicted! And you find a curve (eclipse experiment)! The actual observation you add to $R(t)$ is a gray new internal match (a light curve near a large mass, *not* in the hatched area), but what counts to Lakatos is that, not knowing the Truth, the test *could* have yielded a new internal mistake (a straight line), which *would* have been in the black-bordered part of the horizontally hatched area of Fig. 6. That is the Popperian spirit: we ran the risk of falsification and got away with it. But "this time a curve" is not enough to conclude "never a straight line," and only the latter claim refers to cases in the set where Lakatos wants to be, that of the new external matches. This "forbidding" of straight lines by Einstein's theory might give us a taste for some more measurements after observing a first curve, just to make sure that we were not just lucky enough to make a measurement error causing the deviation from a strictly linear trajectory. If we would stick to Lakatos: "*Exemplum docet, exempla obscurant*," that is: more measurements may yield more curves, but those are not novel facts. Multiple measurements would not even in this case be rational. This casts doubt on whether Lakatos fully apprehended the theoretical problem situation of the kinds of empirical research involved in this example.

- The example, originally adduced by Popper, of the Bohr-Kramers-Slater theory, "refuted in *all* its new predictions" is given by Lakatos as "a beautiful example of a theory which satisfied only the first part of Popper's criterion of progress (excess content), but not the second part (corroborated excess content)" (Lakatos 1970, p. 124).

That fits the excess (falsified) content definition, because it is in the black part of the crosshatched area: a new internal mistake. Intuitively few scientists would call it progress, but in the Popperian mood one could consider regular internal mistakes to be the "collateral damage" of the basically sound method of trial and error. Another point is whether the issue on which Y made the error is new in the sense of being (partly) outside the *vocabulary* of X. That would mean the "falsity knowledge" is new. We have to redraw the Mp map and mark the corresponding impossibility set, as well as all other sets we already had, on that new map. But this "redrawing" is a most complicated procedure, the *pièce de résistance* of philosophy of science, associated with the "incommensurability" and the "reduction" literature. Some more about it will be presented below.

- Then there is the story of radiation in quantum theory, where Lakatos is, as elsewhere, keen to deny the cruciality of the classical experiment, here the Lummer-Pringsheim experiments, and crowns a theory shift, here the Bose-Einstein statistics as the prime mover towards progress in terms of novel facts: "in Planck's version it predicted correctly the value of the Boltzman-Planck constant and in Einstein's version it predicted a stunning series of novel facts" (Lakatos 1970, p. 167).

It is doubtful here whether Lakatos considers the correct prediction by a theory of an already known law parameter to be a novel fact of that theory. It is clearly desirable for a new theory to entail established laws (on pp. 157, 159 of ICR, Kuipers formulates as a criterion that there should be no – and above all no new – established law mistakes, see Fig. 5). In the Mp maps it results in the situation where the new theory, the set Y, is a subset of the established law set $S(t)$. The subset-condition assures that Y entails all established laws.

- Dealing with the Bohr program it is mentioned that there were novel facts "in the field of spectrum lines" (Lakatos 1970, p. 144), later specified: "Bohr's celebrated first paper of 1913 contained the initial step ... predicted facts hitherto not predicted by any previous theory: the wavelengths of hydrogen's line emission spectrum. Though some of these wavelengths were known before 1913 – the Balmer series (1885) and the Paschen series (1908) – Bohr's theory predicted much more than these two known series. And tests soon corroborated its novel content: one additional Bohr series was discovered by Lyman in 1914, another by Brackett in 1922 and yet another by Pfund in 1924."

These line emission spectrum results are all *internal matches*, some of them *new* internal matches (only the latter in the solid gray over Fig. 6). Neither the old nor the new internal matches are in the hatched area, the excess content, which consists of all new external matches and all new internal mistakes together. One could even be question whether these examples meet the MRSP progressivity criterion of not being similar, and being improbable, or even impossible in the light of existing knowledge.

In conclusion: where Lakatos adduces examples of *false* novel facts, this notion of novel fact always unproblematically identifies with *falsified excess empirical content*, but "verified" (quotation marks after Lakatos) novel facts seem consistently to fall in the class of new internal matches, and if so do not meet the logical requirements of (*corroborated*) *excess empirical content*. They are not in the hatched area of picture Fig. 6. Lakatos consistently requires the theory to state about a "'verified' novel fact" that it is *possible* and so we should identify it as a *new internal match* of the theory.

But that is not all that is required by Lakatos to pass as a novel fact. The new theory Y should not only deem it possible, but should *predict* it. And it should be *not similar* to known facts, *improbable* or even *impossible* in the light of old theory X. This is the subject of the next section.

3. Lakatos on New Things

Lakatos (1970, p. 100) introduces a planet orbit anomaly that prompts a hunt for hitherto undiscovered forces that might explain the anomaly. Such as 1) another planet (this is inspired by the discovery, on Sept. 23, 1846, of Neptune by astronomer-observer Johann G. Galle on instructions derived from Newton-orbit deviations of Uranus from astronomer-mathematician Urbain Jean Joseph Le Verrier). Lakatos adds to this, for didactic purposes of showing the untenability of dogmatic falsificationism, a second and third option that might have been explored if the first option had led to failure: 2) a cloud of cosmic dust, 3) a magnetic field. But he clearly considers the list of *a priori* possible explanations to be *unlimited*.

In the truth-approach terms, the problem situation is a falsification, that is, a theory X, a planet p and planet orbit observation data o such that:

$$<p, o> \in R(t) \cap \mathrm{Mp} - X$$

In other words: $<p,o>$ is an established internal mistake of old X. Other planets, clouds and magnetic fields may, in Lakatos' terms, feature in some *auxiliary hypothesis H*, yielding a new theory Y such that:

$$<p, o> \in R(t) \cap Y$$

that is $<p, o>$ is a new established internal match of new Y. How can such an old impossibility of old X become a new possibility of new Y? The requirement is:

$$<p, o> \in Y - X$$

If in the Lakatosian spirit that inspires his orbit anomaly example, Y succeeds X in a theory series of a research program that generates auxiliary hypothesis H in order to make progressive problem shifts, there must have been an auxiliary hypothesis H_{XY} that "turned X into Y." H_{XY} "cuts off" $X - Y$ and "adds" Y-X (where logical inspection of a particular theory should reveal whether $X - Y$ and $Y - X$ are logically non-empty, and empirical research (as opposed to logical inspection) should establish which parts violate established laws and observations). It is efficient to see H_{XY} as a function turning X into Y:

$$Y = H_{XY}(X)$$

In words: H_{XY} turns the observed (X-) impossible planet orbit data $<p, o>$ into a (Y-) possibility. And it would surely be nice to actually observe in space something that the theory allows to attribute responsibility for the force that would explain the impossibility at first (X-) sight.

What does *explanation* mean here in truth-approach terms? This depends on quantity and precision of the data you got about p and o. In the worst case (not Le Verrier's) you have observed only some locally concentrated points with high error margins, and you have no idea of the mass of p, leaving you, given X, with a *large class of possible, all anomalous* orbits. The observation of some similarly deficient data on another planet $<p', o'>$ might only allow you to conclude that many combinations of hypothetical extensions of data of the two planets yield Y-matches (and many other equally hypothetical extensions will yield Y-misses). In short: given the extended $R(t)$, $Y = H_{XY}(X)$ is possible (leaves you without established internal mistakes, at least not with respect to the anomaly under scrutiny). X is false, because the observation $<p, o>$, not by identifying one possibility, but by identifying a huge error range the elements of which might all be real possibilities, *but certainly all would be anomalies*, has yielded an *established internal mistake*. You do not know which of the many small shotgun pellets triggered by the vague observation killed X, you are only sure one of them did, forcing you to go for an H_{XY}, in other words for an alternative theory Y.

A simple illustration of this predicament is a theory claiming some types of series of observations always to be on straight lines, and a highly imprecise series of three observations shimmering in the gray areas of Fig. 7 respectively: you have not been able to obtain single elements of $R(t)$, but whatever precisely are the respective true points, given the borders of the areas in which they are shimmering, the three of them together will certainly refute your linear theory.

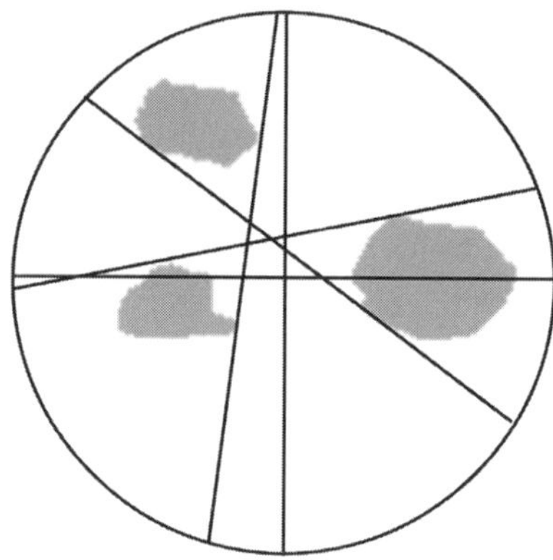

*Fig.*7. Linear theory

If you then assume by auxiliary hypothesis another *yet unknown* thing curving your lines, the degree of curvature depending, according to your theory, on some specifications of the new thing that you will need to observe, if you even manage to find such a new thing and its observed specifications have a vagueness similar to the terrible black blots you obtained from the anomalous *known* thing, then what are you left with? Instead of the set of straight lines some of which are drawn in Fig. 7, you now have a set of curves, one of which should hit the three gray blots simultaneously. But since the degree of curvature depends on a vague observation, there will be lots of versions of every single curve (you can still shape a given curve by choosing some specific point in the observed blot of your new thing as the true value). As far as your old thing is concerned, there is a very wide set of possible true point triples, such that each element of them is in one of your three black blots. *If one of these many possibilities yields three points on one version of one of your curves, you will have saved your theory,* that is, the wide set of possible true points in the range of your vague observations will not *all* be impossible according to your theory. It is not falsified. But how happy will you be, facing the fact that you are very near to the idea of drawing a freehand curve through the three black blots in Fig. 7, something most people from very early age are able to do?

At the other extreme, you may have been so lucky as to have obtained observations of the known planet $<p, o>$ and the new planet $<p', o'>$ such that *any other results of these observations* would have been $Y = H_{XY} (X)$-*impossibilities*. That would be a *powerful ace, giving you game and set*, and, of course, any scientist's dream. But that dream never comes true in practice: as soon as there is a situation where observed $<p, o>$ and $< p', o'>$ are compatible with $Y = H_{XY} (X)$, a set (even if very small indeed) of alternative values $<p, o>$ and $< p', o'>$ will remain in some error margin, to which your enemy camp will cling as long as it can, if necessary by increasing the margin of conceivable errors by conceiving ever wilder assumptions on what could have caused you to make mistakes in your chain of explanatory assumptions. As a scientist, you can powerfully stun your colleagues, but never in the logically Absolute (Big A) sense which many reputable scientists have not been shy to claim. That is not to say such claims are irrational (ineffective) in such rhetoric struggles. After all, real world scientific theories undeniably serve as propaganda instruments of those who back them. A surrender of the enemy camp to avoid *falsity* (as opposed to *disgrace* and its consequences) has probability zero.

In the vicinity of the lucky case extreme ("given H_{XY}, any other outcome of observation than the one obtained would have been a falsification") it is appropriate to defend H_{XY} as *the* "explanation." The new thing discovered can justly be defended as being *relevant*. Near the worst case ("given H_{XY}, the outcome of the observation is only one of very many possibilities") H_{XY} can only be treated as an invitation to try and find an "explanation" in that direction. Then H_{XY} is only a delineation of a set of *possible* explanations, even without excluding other sets. That worst case means this: although you found a novel fact, you cannot even say you have good reasons to claim it as *necessary* to the solution of the problem of the theory you scrutinize. Between these extreme neighborhoods of bliss and despair, you will simply be struggling without even knowing how to call your H-speculations.

And this, roughly, is how Imre Lakatos would no doubt have tackled dogmatic falsificationism had he been in the happy possession of truth approach terms.

As already stated, on Sept. 23, 1846 Galle found among the host of absolutely minuscule new light sources observable by the latest telescope something small, shiny and moving that could do the job a certain "Neptune" was supposed do to Uranus. This did not *logically entail* progress, but it was too close to the happy extreme for astronomers on inconsistent tracks to start torturing their brains to find assumptions of possible errors in the chain of explanations and summon up guts to put their reputation at risk by advertising and testing such assumptions. Surrender. In this case, at that time.

And that resignation of the opposition is, curiously enough, what scientists call "progress," "truth," "explanation," "proof" and "fact."

4. False Observation Reports and False Laws

Constructive Realism is meant to show how, after the expulsion from Truth's Paradise, facing the Suffering of Observation $R(t)$ and the Evil of Laws $S(t)$ we may stick to Truth as our Epistemological Savior. The toil and trouble resulting from the expulsion would be bearable (ICR, p. 157, 159) if we could be sure that T is a superset of $R(t)$: *no false observation reports*, and a subset of $S(t)$: *no false laws*. This would leave us with some loss of the gray and black we long for (the four types gray/black, internal/external uncertainty depicted in Fig. 5, but we could Absolutely rely on the gray and black areas that remain (the four solidly gray/black filled areas of new established law/possibility hits/misses in Fig. 5).

But the expulsion has obviously been so ruthless that falsity in observation reports and in accepted laws are, in real science, even more frequently admitted than traffic offenses on real roads. And, what is worse, in science the rules for such offenses, apart from conscious fraud and the most naive parts of wishful thinking, are badly canonized. In many cases, observation and law errors are not thought to be blameworthy and on removal there will be praise only.

At no time is the correction of observations and laws done more vigorously than at times where a new theory Y is accused of inconsistency with old facts explained by old X. Adherents of new Y are usually young. That means they have energy and want to be right. The old guys still in favor of old X are happy to make the agenda of Y-inconsistencies with X-successes seem big and insurmountable and the young go for it. It is of utmost curiosity that this extraordinarily neat cooperation between adherents of old and new theories has been stamped by philosophers as something less than the heyday of rationality. It is a perfect example of how, in all times and all cultures, and even among all non-human social mammals, the old-wise and the young-vigorous relate to the best benefit of survival of the group. Do not all fathers say "you can't" while their sons go for it? This competitive instinct serves the preparation for coping with external danger: in danger, if you're a young adult in a monkey group, you are the one who goes out to throw the branches and stones. The old and wise sit high and signal.

But how could the competitive agenda that the oldies are so happy to add to and the young are trying to work through be drawn on a truth approach Mp-map? This is the subject of the next section.

5. Inadequate Mp's

How should one draw a new planet in the Mp of X and $Y = H_{XY}(X)$? One could draw an Mp where "new planets" are logical possibilities, a new Y containing some of those planets and an "old X" not containing any of them. But this is not a fair description of what Sneed (1971, p. 117 *passim*), calls "existing expositions". If old X said nothing about new planets, it said neither that they were possible nor that they were impossible. Old X said nothing about it, so such new planets are not in the vocabulary, not in old X's Mp. Some tend to consider the absence of new planets a "hidden" assumption of X. But Lakatos, in his treatment of *ceteris paribus* makes clear that that would not be fair either. He makes it clear that the falsification (established internal mistake) of X by $<p, o>$ in his planet example above should best be written:

Ceteris Paribus $\Rightarrow <p, o> \in R(t) \cap \mathrm{Mp} - X$

And *ceteris* refers to an unknown and infinite set of blanks in our knowledge that we promise ourselves to go for if necessary. So once a falsification prompts us to go for an auxiliary H_{XY}, a new planet, where our new Y, the only knowledge we gained about the ceteris paribus clause is negative: "no new planets" is *out* in new Y's clause.

Ceteris (**save H_{XY}**) *Paribus* $\Rightarrow <p, o> \in R(t) \cap Y$

So, it is fairest to say that the Mp of Y contains new planets as possibilities, whereas the Mp of X did not. In such a case we need the *new* Mp of Y to depict the comparison between X and Y, starting by drawing the Mp of old X as a subset. This is called a *field extension case* (Hamminga 1983, p. 67). The vocabulary specifies the *kinds* of things you have, but for the Mp we have to specify their numbers. Planets were already in the vocabulary before Galle and Le Verrier, Neptune was a field extension: adding one specimen of a certain *kind* of thing already featuring in old X (apparently not similar enough for Lakatos to deprive it of the stamp "novel fact"!). In economics many theories start to be articulated in simple fields (like two factors of production, two goods, two trading countries), after which the field is extended to see whether the established theorems remain derivable.

More than with field extension, set theoretical trouble is caused by *vocabulary extension*: the introduction of new *kinds* of things ("black holes," or, in economics "technological change"). There often is no way to represent

                    Bert Hamminga

to the satisfaction of its adherents an old theory in a new Mp containing new *kinds* of things.

These are all situations in which a new Y brings with it a larger Mp(Y), that is Mp(Y) $\supset$ Mp(X). In the Mp of Y, we have a straw-X that is a very wide theory due to its lack of specification and hence infinite permissibility about new Y-variables. No adherents of old X would subscribe to that, facing the new Mp(Y).

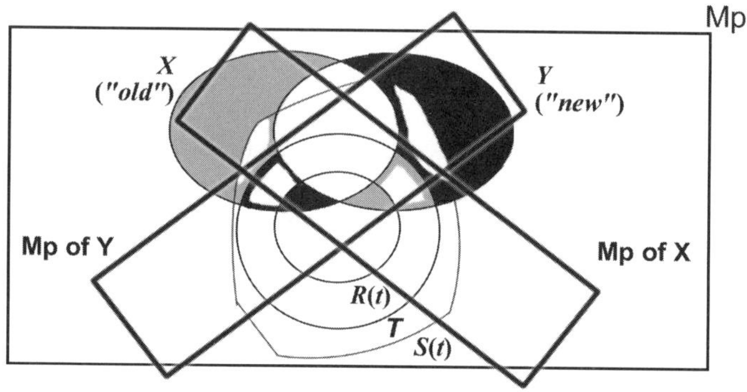

*Fig.*8. Rival Mp's

The real trouble starts once the adherents of an old X claim that the vocabulary extension of Y yields *losses*. Mp of X is partly outside the Mp of Y too. In an example like that in Fig. 8 (a picture that is identical to Fig. 5 except for the overlay of **Mp of X** and **Mp of Y**) neither party will easily stop thinking it is ahead in the competition: old X-supporters see new Y mainly in the black, new Y-supporters see their own theory mainly in the gray. There is a need to redefine the theory battlefield Mp accordingly, in order to define reduction relations between a new Y and an old X. In the confusion of such times of change, the structuralist's research objects, that is the contending scientists themselves, will fight among themselves about what structuralists would call the "proper" super-Mp.

Here we find, not a sharp demarcation but indeed a typical characteristic of scientists as opposed to others in society who defend and attack views: as we know from general public discussion in the media, defenders of vested interests are used to safely staging in the area of their own Mp, there to shoot down as much enemy-black as they can. As a politician, your media adviser trains you how *not* to be drawn out of your Mp by an interviewer. That is what makes public political debates so boring to the educated public. In science discussions, this behavior in discussion is frowned upon decisively more than elsewhere. You just don't usually "score" with it in scientific discussions of physical, economic, biological or other theories (where one does, this reveals a weakness of the relevant subsection of the scientific community). Popper did voice a cultural trait of scientific circles when he wrote "... we are prisoners in a Pickwickian sense: if we try, we can break out of our framework at any time. Admittedly we shall find ourselves again in a framework, but it will be a better and a roomier one; and we can at any moment break out of it again." (Popper 1970, p. 56)

During the heydays of breakthroughs, when scientists firmly believe they came dramatically closer to the truth (sometimes so much closer that they even judge some cheating to be justified), their strategy usually involves drastic vocabulary changes. As Lakatos writes: "Physicists rarely articulate their theories sufficiently to be pinned down and caught by the critic" (Lakatos 1970, p. 128). In later stages, once Y chased X off the scene, a straw-X, thoroughly bleached in its enemy camp, will appear, partly as flat error, partly as a special case of Y, allegedly killed with a single bullet by a crucial experiment. That straw-X continues to be contested only in esoteric journals of historians of science, for as a rule few are interested in studying how much less stupid, say, pre-Copernicans, pre-Adam Smithians, pre-Einsteinians or pre-Keynesians were than their overthrowers later made us believe. Trying to make some generalizations of such "rehabilitating-the-predecessor" studies, however, is extremely relevant to the verisimilitude program, because we should be aware that, after something has gone down, we should not start considering it to have run as badly as those who always argued for its replacement want us to believe. That awareness is of utmost importance in choosing the version of a rejected older theory X to use as basis for reconstruction and comparison with a new Y. Philosophico-logical life is easy when for the illustration of the inter-theory relations you use the presentation of the old theory made by adherents of the successor, and not the presentation by those who believed and defended it. No doubt victorious rivals did some nice work in harmonizing the concepts. But that work is often done, let us say by way of understatement, *far too nicely* and is too much directed to make sure the reader will not fall in any doubt concerning the progress from old to new – which is exactly what philosophers would like to form their own independent opinions about. It is certainly too discouraging to take on the aim of strict determination of the *criteria for correct super-Mp models* in terms of which the comparative truth-analysis of two vocabulary-different theories X and Y should be executed. But any small progress in that direction clearly makes the whole analytical apparatus of verisimilitude significantly more realistically applicable to existing and past theory-controversies in science. And the least one can expect is some nice knowledge of the widely played game in science of getting that super-Mp accepted which created the best picture of your favorite theory in comparison to the rivals.

In economics there is even a well known example where the rehabilitation of the predecessor is done by the scientists themselves, where part of the predecessor's vocabulary has been reintroduced, and where the new *Y*, starting as an outright rival of the old *X*, gradually became considered to be a special case of a covering theory closer to the old *X* than to the new *Y*. This is the macro-economic program of the "Neo-Classical Synthesis," in which the Keynesian theory became embedded in a modified version of the preceding Neo-Classical theory. Die-hards considering themselves "real" Keynesians, like Davidson (1991) and Leijonhufvud (1968), found themselves forced into the role of protesters against the Neo-Classical mainstream of a "new old" Neo-Classical theory that, under special conditions, allows for what were thought to be the main Keynesian implications. The plausibility of these special conditions in a variety of cases became the center of economic policy discussions. In these kinds of real science discussions, the pushing and pulling at Mp's is a most interesting spectacle. Crucial paragraphs in the rhetoric of the main papers in this history answer questions like "what are we really talking about?", "what are we really interested in?". And the answer often implies an Mp-shift. The main thing a reader learns to handle while studying such papers is the technical integration (or a different way of technical integration) of some generally used common sense concept in existing theory ("price," "consumption," "investment," "money," "technical change," "tariff"). Such a concept turns your attention to some new aspect of markets, and often causes your attention to some other aspects fade. Here, vocabulary shifts and meaning shifts seem to be the chief engine of change, and the exploration of what can be done to work out the theory in the new vocabulary encounters more enthusiasm than the precise, let alone neutral scrutiny of possible losses associated with it.

Whatever may come out of further research on how competing scientists handle Mp's, the determination of a "Big V" Vantage point yielding a True View of the battle between such an *X* and *Y* seems to me necessarily – I am sorry, even after all words I've spent, not to have a more precise expression – an *act of philosophy*. But this undoubtedly is the Achilles heel of any general conception of the historical dynamics of anything, whether it is about society, technology, science, art or music, and certainly for those who are on a university's payroll, it is flatly irrational to claim one should stop trying.

6. The Net and the Spider

In Victorian times, it had become a matter of civilized hygiene to take the scientific nets as the object of thought and to abstract from the spiders (that is, the scientists). Not everybody thought this would lead to progress of knowledge. In economics, one of the fields suffering the most from Victorianism, a number of mostly non-British scholars prefered refuge in an underworld (Heilbroner 1986) to abstract tidiness.

How does one bring Victorians down to earth? Keynes tried to exploit the opportunity provided by the collapse of the economy in the 1930s to stress that clean hands obeying rules that are considered proper are unsuitable to promote economic progress, and our hope should be based on the the greed of those not shunning dirty hands to enrich themselves: "For at least another hundred years we must pretend to ourselves and to everyone that fair is foul and foul is fair; for foul is useful and fair is not" (Keynes, 1930).

Could the the same hold for scientific progress? In Keynes' time, no philosophers of science dared to claim *that*. Popper stressed the (theory-)net as an objective thing to study apart from the (scientist-)spider, but he did have an eye for him: he set out to teach the spider *rational* behavior (Popper 1937). The training method Popper chose was clean Victorianism: *writing*, assuming the spiders would learn and improve themselves by reading. That Victorian strategy, so timelessly depicted in that cartoon where an English butler (wearing, if I recall correctly, white gloves), defeating the Indians with the right hand, holding the Encyclopedia Britannica on "Fighting" in the left, turned into a disappointment in Popper's case: normal scientists (in Kuhn's sense) turned out to be, as Popper put it, "badly taught" (Popper 1970, pp. 52-3). In European philosophy a firm methodological wall had been built with vigor to keep spiders out of the analysis of nets, the erection of which has often been hailed as progress towards "objectivity." The "sociologist" tearing his clothes to approach the truth about science could not easily count on much attention.

Of course, structuralism is Victorian in the sense of being a completely clean-hands, white gloves, "net-oriented" type of research. But it is at least capable of describing *real* nets and such descriptions are to some extent – not to be exaggerated – testable. In verisimilitude analysis spiders come in *at t*, dropping $R(t)$ and $S(t)$. The success theorem might be thought of as providing an operational definition for comparative web quality that can impossibly point in the direction opposite to that of the result we could have obtained by evaluating its distance to the Truth, had we known it. And its tastiest fruit is an agenda of new and answerable empirical questions about real science.

PHiLES Institute
Jan Evertsenstraat 18
5021 RE Tilburg
The Netherlands

REFERENCES

Balzer,W. and J.D. Sneed. (1977) Generalized Net Structures of Empirical Theories, Part I. *Studia Logica* **36-3**, 195-211.

Cools, K., B. Hamminga, and T.A.F. Kuipers. (1994). Truth Approximation by Concretization in Capital Structure Theory. In: B. Hamminga and N.B. de Marchi (eds.), *Idealization IV: Idealization in Economics*. (*Poznań Studies in the Philosophy of the Sciences and the Humanities*, vol. **38**.), pp. 205-28. Amsterdam: Rodopi.

Davidson, P. (1991). *Money and Employment*. Collected Writings, Vol 1. New York: New York University Press.

de Marchi, N.B. and M. Blaug, eds. (1991). *Appraising Economic Theories. Studies in the Methodology of Research Programmes*. Hants: Elgar.

Hamminga, B. (1983). *Neoclassical Theory Structure and Theory Development*. Berlin: Springer.

Hamminga, B. (1991). Comment on Hands: Meaning and Measurement of Excess Content. In: de Marchi and Blaug (1991), pp. 76-84.

Hamminga, B. and N.B. de Marchi, eds. (1994). *Idealization VI: Idealization in Economics. Poznań Studies in the Philosophy of the Sciences and the Humanities*, vol **38**. Amsterdam, Atlanta: Rodopi.

Hands, D.W. (1991). The Problem of Excess Content. In: de Marchi and Blaug (1991), p. 58-75.

Heilbroner, R.L. (1986). *The Worldly Philosophers*. New York: Simon&Schuster.

Keynes, J.M. (1930). *The Economic Possibilities for our Grandchildren. Essays in Persuasion*. Cambridge: Cambridge University Press.

Kuipers, T.A.F. (2000/ICR). *From Instrumentalism to Constructive Realism, On Some Relations Between Confirmation, Empirical Progress, and Truth Approximation. Synthese Library*, vol. 287. Dordrecht: Kluwer.

Lakatos, I. (1970). Falsification and the Methodology of Scientific Research Programmes. In: I. Lakatos and A. Musgrave (eds.), *Criticism and the Growth of Knowledge*, pp. 58-75. Cambridge: Cambridge University Press.

Lakatos, I. (1971). History of Science and Its Rational Reconstructions. In: R.C. Buck and R.S. Cohen (eds.), *Boston Studies in the Philosophy of Science*, vol. 8, pp. 91-136. Dordrecht: Reidel.

Lakatos, I. (1976). *Proofs and Refutations*. Cambridge: Cambridge University Press.

Leijonhufvud, A. (1968). *On Keynesian Economics and the Economics of Keynes*. Oxford: Oxford University Press.

Popper, K.R. ([1936] 1972). *The Logic of Scientific Discovery*. London.

Popper, K.R. (1970). Normal Science and Its Dangers. In: Lakatos, I. and A. Musgrave (eds.), *Criticism and the Growth of Knowledge*. Cambridge: Cambridge University Press.

Sneed J.P. (1971). *The Logical Structure of Mathematical Physics*. Dordrecht: Reidel

Theo A. F. Kuipers

DOMAIN AND VOCABULARY EXTENSION
REPLY TO BERT HAMMINGA

Over the years, Bert Hamminga, a philosopher of economics, has demonstrated an interest in empirical progress and truth approximation. Thanks to him, I have been able to develop an alternative way of truth approximation (ICR, pp. 271-2, pp. 288-98; SiS, pp. 33-4) especially for economic research or, more generally, research driven by an "interesting theorem," as analyzed by Hamminga (1983). Hamminga's present contribution concludes by considering economic theorizing once again. Before doing so, he presents a set of very adequate "pocket maps" for truth seekers and a provocative diagnosis of Lakatos' notion of "novel facts" and requests further attention to two important topics: domain extension and vocabulary extension.

I will come back to these extensions in some detail, but let me first remark that I like the pocket maps very much. I see them as a fresh representation of my basic ideas. I have just one critical and a (related) constructive remark. When dealing with the "second consolation," in Section 1, Hamminga suggests a difference between $R(t)$ and $S(t)$ which seems to me somewhat exaggerated. The suggestion is that $R(t)$ is asymmetric in that its members have been established as real or nomic possibilities, whereas its non-members have not (yet) been established as nomic impossibilities. However, I would say that $S(t)$ is similarly asymmetric in that the non-members of $S(t)$ have been established-in-a-certain-sense as nomically impossible, viz. by accepting a law that excludes them, whereas its members, as far as non-members of $R(t)$ are concerned, have not (yet) been established as nomically possible. In both cases, only the first types of (non-)members are used for comparisons, viz. established members of $R(t)$ (and hence of $S(t)$) and established non-members of $S(t)$ (and hence of $R(t)$), respectively.

Maybe the intuitive asymmetry Hamminga has in mind has to do with the different status of being established, which inspired me to make the following terminological proposal.

Recall that $R(t)$ contains all established possibilities, that they represent (new) established internal matches if they belong to Y (and not to X), that they

In: R. Festa, A. Aliseda and J. Peijnenburg (eds.), *Confirmation, Empirical Progress, and Truth Approximation* (*Poznań Studies in the Philosophy of the Sciences and the Humanities,* vol. 83), pp. 337-340. Amsterdam/New York, NY: Rodopi, 2005.

represent (new) established internal mistakes if they do not belong to Y (but do belong to X). Now laws are established on the basis of $R(t)$. One can even say that $R(t)$ results in large measure from testing hypothetical laws. These laws are of course a kind of inductive generalizations. Hence, instead of speaking of established laws, one might speak of "induced laws." More importantly, the non-members of $S(t)$ might well be called "induced impossibilities," instead of "law-established impossibilities" and the like (established law mistakes and matches). This would emphasize the different nature of establishment of nomic *im*possibilities as opposed to possibilities. Similarly, non-members of $S(t)$ not in Y (but in X) would become "(new) induced impossibility matches" and members of $S(t)$ in Y (but not in X) would become "(new) induced impossibility mistakes."

Hamminga's argumentation for the diagnosis that the notion of novel fact of Lakatos is not so much related to Popper's notion of "empirical content" of a theory, that is, to its forbidden possibilities, but to its allowed possibilities, seems quite convincing to me. In addition, following Lakatos, such a possibility should be excluded by the old theory and experimental evidence should exemplify it beyond reasonable doubt. However, Hamminga goes on to relate the particular case, a thusfar not discovered planet, Neptune, to the general topic of domain extension. Here, I have some doubt about his treatment. Moreover, I have some problems with his treatment of vocabulary extension. These kinds of extension are the subject of the rest of this reply.

Domain Extension

Under the heading "inadequate Mp's," Hamminga introduces the possibility that the old theory did not say anything about new planets and hence that the domain of the theory has to be modified. However, I think that this need not be the case for the introduction of a new planet, depending on the type of conceptual possibilities Mp is made of. Either the intended applications, constituting the domain of the theory, deal with two-object (planet-sun) systems or with "planetary systems" in general, systems with at least two objects, one of which has considerably more mass than all the others. In both cases, there are lots of conceptual possibilities that can take a new planet into account, among other objects. However, if the new theory concerns the transition from the first (two-object) to the second (general) type of theory, the new Mp is of a really different nature than the old one. It has to take the so-called many-objects (more than 2) problem into account; as is well known this has to be done in an approximate way. This would correspond to Hamminga's (1983) notion of "field extension" for international trade when going from two countries to three or more. However, in the case of Neptune, taking more than one planet into account (by approximation) had already been done before 1983.

To be sure, substantial domain extension is an important activity in scientific research and the ICR theory of truth approximation does not yet deal with this. It assumes a fixed domain. That is, I start from a fixed primitive set of applications D which amounts to the subset T of $Mp(V)$ when conceptualized by a (richer) vocabulary V. Hence we may write: $T = Mp(D)$. (For some refinement in terms of a domain vocabulary, see ICR, pp. 326-7 Section: 13.5). This leads directly to two types of strengthening of a theory, as suggested by Zwart (1998/2001), that is, by reducing the set of allowed possibilities or by enlarging the domain (see the last paragraph of Ch. 8, pp. 206-7, in particular the last fives lines):

> Finally, variable domains can also be taken into account, where the main changes concern extensions and restrictions. We will not study this issue, but see (Zwart 1998 [/2001], Ch. 2-4) for some illuminating elaborations in this connection, among other things, the way in which strengthening / weakening of a theory and extending / reducing its domain interact.

Hence, at least some new pocket maps will have to be designed in order to take empirical progress and truth approximation by domain extension – or domain reduction – into account, leaving the theory fixed. In my reply to Zwart I make a first attempt to formally design such maps.

Vocabulary Extension

Hamminga also turns his attention to vocabulary extension. However, here his formulations frequently suggest an essentialist kind of realism that I would not wish to endorse. Talk about 'proper' or 'correct super-Mp's', is more or less explicitly declined in ICR (p. 231) by rejecting the ideal language assumption and replacing it by the Popperian refinement principle. This only allows the possibility of comparing various kinds of success of vocabularies (ICR, p. 234).

To be sure, change of vocabulary is not an easy task to deal with, and my trivial kind of fusion of languages (ICR, pp. 230-5) is such that the number of theory comparisons that can be made will be very small. However, simple extension of language to deal with new kinds of objects and attributes is relatively easy. As a matter of fact, the treatment of stratified theories in observational and theoretical terms is formally a matter of vocabulary extension.

Let me finally note that Hamminga's very intriguing last section, about "The Net and the Spider", refers to Popper's net metaphor as a metaphor for

 Theo A. F. Kuipers

theories. In the very last section of ICR I argue that that metaphor is much better suited for vocabularies (see also my reply to Mooij). Be this as it may, Hamminga is right in suggesting that the spiders, that is, the researchers, only figure implicitly in the picture developed in ICR.

REFERENCES

Hamminga, B. (1983). *Neoclassical Theory Structure and Theory Development*. Berlin: Springer.

Zwart, S. (1998/2001). *Approach to The Truth. Verisimilitude and Truthlikeness*. Dissertation Groningen. Amsterdam: ILLC Dissertation Series 1998-02. Revised version: *Refined Verisimilitude. Synthese Library*, vol. 307. Dordrecht: Kluwer Academic Publishers.

David Miller

BEAUTY, A ROAD TO THE TRUTH?[1]

ABSTRACT. Calling into service the theory of truth approximation of his (1997) and (2000), Kuipers defends the view that "beauty can be a road to the truth" and endorses the general conclusions of McAllister (1996) that aesthetic criteria reasonably play a role in theory selection in science. My comments pertain first to the general adequacy of Kuipers's theory of truth approximation; secondly to its methodological aspects; thirdly to the aetiolated role that aesthetic factors turn out to play in his account; and fourthly to the question before us, with a remark on McAllister's doctrine that scientific revolutions are characterized above all by novelty of aesthetic judgement.

0. Introduction

As Borges wryly observes, although the world is under no obligation to be interesting, our hypotheses about it cannot easily escape that obligation.[2] Kuipers and I too have a duty to provide an interesting discussion of the question before us, *Can beauty be, or is beauty, a road to the truth in science?*, even though it seems, to me at least, to be a question to which the correct answer is rather obvious and rather uninteresting. The answer is that, like anything else, aesthetic considerations may point us in the right direction, but again they may deceive us. The truth may not be beautiful, or it may not be beautiful in a way that can be captured by beautiful theories. That is pretty much my conclusion. I shall escort you back to it by a scenic route rather than by the direct one.

---

[1] These remarks were prepared originally in response to a paper under the same title given by Theo Kuipers at the Annual Conference of the British Society for the Philosophy of Science, held at the University of Sheffield on July 6th-7th, 2000. They have been revised to take account of the revisions introduced into Kuipers (2002), to which all the unattributed quotations refer. I should like to express my profound thanks to Kuipers and to Roberto Festa for all that they have done to make the complex task of revision as simple as it could be. Table 1 and Appendix 1 have been added to this revision.

[2] J.L. Borges, "La muerte y la brújula", *La Muerte y la Brújula,* Emecé Editores, Buenos Aires, 1951, p. 139: "Usted replicará que la realidad no tiene la menor obligatión de ser interesante. Yo le replicaré que la realidad puede prescindir de esa obligatión, pero no las hipótesis."

In: R. Festa, A. Aliseda and J. Peijnenburg (eds.), *Confirmation, Empirical Progress, and Truth Approximation* (*Poznań Studies in the Philosophy of the Sciences and the Humanities,* vol. 83), pp. 341-355. Amsterdam/New York, NY: Rodopi, 2005.

By appealing to his well known theory of approximation to the truth, Kuipers undertakes to defend what sounds like an equally modest conclusion, the conclusion that "beauty can be a road to the truth, namely as far as the truth is beautiful in the specific sense" (and here I paraphrase the explanation of "distributed" in Kuipers 2002, p. 295) that all the conceptual possibilities allowed by the truth have "features that we have come to experience as beautiful" (p. 296); and to defend the non-triviality of this conclusion. There is indeed some sense in bringing ideas about approximate truth to bear on the question. For beauty is evidently a quality that comes only in degrees, with no obvious maximum; nothing is perfect. We cannot hope to identify beauty and truth. But perhaps we might identify degrees of beauty with degrees of truth, or with degrees of truthlikeness. Specialist aesthetes may even suggest that if beauty is linked to degrees of (approximation to) truth, then what should be linked to degrees of truthlikeness (that is, approximation to the whole truth) is the sublime. But I doubt that anything that Kuipers or I has to say deserves such subtlety. In the present discussion beauty represents almost any aesthetic quality that can sensibly be attributed to scientific hypotheses.

My remarks, which I hope are not too ungracious, are arranged in the following way. First I shall give a brief presentation of the theory of truth approximation that Kuipers endorses. Secondly I explain why it gives no succour to the doctrine that in the empirical evaluation of scientific theories success is what counts. Thirdly I shall say why, in my opinion, the theory provides also almost no help in answering, in any manner except the obvious one, the question before us. Finally I shall consider briefly the question before us, with reference to the controversial ideas of McAllister (1996), and compare these ideas with some of those in the Metaphysical Epilogue of Popper's *Postscript.*

1. Approximation to the Truth

Popper's original proposal about verisimilitude or approximation to the truth was that the content of a theory be divided into a desirable part, its truth content, and an undesirable part, its falsity content (1963, Chapter 10.3). A theory is the better the greater is the former and the smaller the latter. As Tichý and I showed, amongst false theories truth content and falsity content increase together (Tichý 1974, Miller 1974). This implies that no false theory can be a better approximation to the truth than any other false theory is.

Almost at once Harris (1974) ventured a more liberal construe of content; he substituted for the set of all truths an incomplete theory $\mathbf{T}$, not much further specified, and for the truth and falsity contents of $\mathbf{Y}$ the sets $\mathbf{Y} \cap \mathbf{T}$ and $\mathbf{Y} \cap \mathbf{T}^*$, where $\mathbf{T}^* = \{\neg z \mid z \in \mathbf{T}\}$. This enabled him to establish a parallel negative result: **Z** better approximates **T** than **X** does only if it has no falsity content. Had Harris defined the falsity content of **Z** as $\mathbf{Z} \cap \overline{\mathbf{T}}$, where $\overline{\mathbf{T}} = \{z \mid z \notin \mathbf{T}\}$ is the complement of **T** in the class of all propositions, thus retaining the idea that everything in a theory's content is either desirable or undesirable, he would have discovered something rather more positive: that false theories are then comparable, but only if they have the same truth content; the false **Z** is a better approximation to **T** than the false **X** is only if **X** is logically stronger than **Z**. This is proved in Appendix 0.

My (1977) implicitly understood the content of a theory in another way; not as the class of its consequences but as the class of the models that it excludes. There is only one model that it is undesirable to exclude, the world we live in (suitably trimmed); we wish to exclude as many of the rest as we can. A theory approximates the truth better the more it excludes models that it is desirable to exclude, and the less it excludes models that it is undesirable to exclude. Since all false theories exclude the one model that should not be excluded, a false theory can here approximate the truth better than another does, but only if it excludes more of the models that should be excluded; the false **Z** better approximates the truth than the false **X** does only if **Z** is logically stronger than **X**.

If **X** *and* **Z** *are false theories, then*

	POPPER 1963	**HARRIS 1974** (amended)
Contents	sets of propositions	sets of propositions
Desirable	complete theory **T**	incomplete theory **T**
Undesirable	propositions outside **T**	propositions outside **T**
X p **Z** *only if*	**X** = **Z**	**X** ⊢ **Z**

	MILLER 1977	**KUIPERS 1982**
Contents	sets of excluded worlds	sets of excluded worlds
Undesirable	the actual world **T**	physically possible worlds
Desirable	worlds outside **T**	worlds outside **T**
X ⪯ **Z** *only if*	**Z** ⊢ **X**	—

Table 0. Four comparative approaches to approximation to the truth

Kuipers (1982), working independently, combined this representation of content by excluded models with the incompleteness of truth proposed by Harris. Now any elements of some (closed) set **T** of models may count as undesirably excluded and the others as desirably excluded. Kuipers understands **T** as the set of physical (or natural) possibilities; the class of models not excluded by the laws of nature. Here at last it is possible for two false theories, neither stronger than the other, to be comparable in the degree to which they approximate the truth. For all that it overemphasizes what theories permit, rather than what they exclude, this theory of Kuipers's is a decided improvement on most of its predecessors (and on most of its successors too).

Table 0 summarizes the extent to which three of the four comparative (or qualitative) theories sketched above are degenerate. (Harris's proposal could also be accommodated in the north-west corner.) X and Z here are false theories (theories with undesirable elements), and $X \preceq Z$ means that Z is at least as good an approximation to the truth as X is. Note that contents in the lower row are sets of excluded worlds; this is why the actual world is undesirable (= undesirably excluded). In none of the above four theories is it possible for a false theory to be a better approximation to the truth than a true theory is. It is of some interest that although "only if " could be replaced by "if and only if " in the north-west and south-west corners, this is not permissible in the north-east (Harris) corner. For $X \vdash Z$ does not imply that $X \cap T = Z \cap T$ (that is, that X and Z have the same truth content in the sense of Harris), and hence does not imply that $X \preceq Z$. It is not obvious that the artificial symmetry of Table 0 reflects anything very deep. If we note how = is weakened to $\vdash$ as we move east and south, Table 1 may seem to do considerably better. But remember that T does not have a constant reference: in the westerly column T is complete, whereas in the easterly column it is incomplete. The proof that Table 1 is correct is sketched in Appendix 1.

If X and Z are false theories, then $X \preceq Z$ if & only if

POPPER 1963	**HARRIS 1974** (amended)
$X \wedge T = Z \wedge T \vdash Z \vee T = X \vee T$	$X \wedge T \vdash Z \wedge T \vdash Z \vee T = X \vee T$
MILLER 1977	**KUIPERS 1982**
$X \wedge T = Z \wedge T \vdash Z \vee T \vdash X \vee T$	$X \wedge T \vdash Z \wedge T \vdash Z \vee T \vdash X \vee T$

Table 1. The same four comparative approaches to approximation to the truth

Kuipers's theory, as here presented, has two components: it represents the achievement of the theory Y by the models of T that it admits, together with those of the remaining models that it excludes. This is the class of models that is shaded in Figure 0. (The theories under discussion are represented throughout by rectangles.) Given this representation, Z is said to be at least as good as X if it has at least as great an achievement. This relation is illustrated in Figure 1, where the areas within convex broken borders are understood to be empty: the rectangle Z has all the models of T that the rectangle X has, and X has all the remaining models that Z has. Note that the first inclusion, concerning the models of T, is a descendant of Popper's original requirement that the better theory should have the less falsity content; whilst the second inclusion is identical with his requirement that the better theory have the greater truth content. In Figure 0, that is to say, the shaded area on the right represents the richness of the truth content of the theory **Y**, whereas the shaded area on the left represents the poverty of its falsity content. The theory **Y** is true [physically possible] if & only if it wholly includes **T**; that is, if & only if there are no physical possibilities that it excludes.

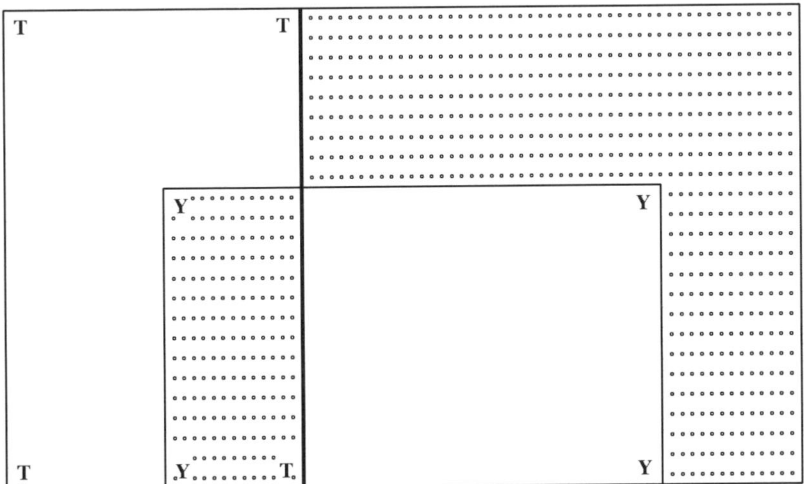

Fig. 0. The achievement of the theory **Y**

2. Empirical Decidability

Kuipers has since suggested that this theory of approximation to the truth is most attractively presented on what he calls a dual foundation: **Z** is at least as good an approximation to the truth as **X** is if and only if it has all the models of **T** that **X** has, and also has all the consequences of **T** that **X** has (1997, §2.3; 2000, §8.1.5). This is the same as saying that **Z** has greater instantial success than **X** does, in the sense that every refutation of (or counterexample to) **Z** is also a refutation of **X**; and that **Z** has greater explanatory success too, in the sense of having greater truth content. Although at the level of logic, there can be no objection to this, we should not be unduly impressed. In particular, we must avoid the conclusion that in the empirical evaluation of scientific theories success is more important than failure, the pipe-dream of justificationists of all ages. Kuipers himself has indeed written recently: "the selection of theories should exclusively be guided by more empirical success, even if the better theory has already been falsified" (2000, §1.3). Of course, no one who works in the truth approximation industry thinks that falsified theories are uniformly valueless. But a consideration of how hypotheses of the form $\mathbf{X} \preceq \mathbf{Z}$ and $X \mathrm{p}\, \mathbf{Z}$ ("$\mathbf{Z}$ is a better approximation to the truth than $\mathbf{X}$ is") are themselves evaluated makes it incontrovertible that falsification remains the key to empirical evaluation.

Before its technical inadequacy became apparent, Popper's theory of verisimilitude was ritually criticized for introducing a concept, the concept of being a better approximation to the truth, for which there is no criterion of application. *How,* he was often asked, *can you know that* $\mathbf{Z}$ *is a better approximation to the truth than* $\mathbf{X}$ *is if you do not know what the truth is? And if you do know what the truth is, why bother with theories that are false?* Popper's original rejoinder that "I do *not* know – I can only guess ... [b]ut I can examine my guess critically, and if it withstands severe criticism, then this fact may be taken as a good critical reason in favour of it" (p. 234), was indeed sufficient to squash this particular line of criticism, even if, in the eyes of some of us, the idea of "a good critical reason" must be handled cautiously (though preferably not at all). Anyway, it became clear in time that for a falsificationist it quite suffices if statements of relative proximity to the truth are, like other empirical statements, falsifiable (Miller 1975, §VII). Popper's own theory meets this condition without difficulty, though not very adventurously, since on his account "$\mathbf{Z}$ is a better approximation to the truth than $\mathbf{X}$ is" is falsified whenever $\mathbf{Z}$ is falsified.

With regard to its openness to empirical scrutiny, Kuipers's theory of approximation to the truth is similarly placed, though there are more interesting possibilities. The opportunities for falsifying the hypothesis $\mathbf{X} \preceq \mathbf{Z}$ ("$\mathbf{Z}$ is as good an approximation to the truth as $\mathbf{X}$ is") are shown by the rectangles with two broken borders in Figure 1: they consist (on the left) of counterexamples to $\mathbf{Z}$ that are allowed by $\mathbf{X}$ (I-differences in Kuipers's terminology), and (on the right) of genuine phenomena explained by $\mathbf{X}$ but left unexplained by $\mathbf{Z}$ (E-differences). The I-differences, being (it is assumed) observable instances that are sufficiently repeatable, may indeed furnish refutations of the hypothesis $\mathbf{X} \preceq \mathbf{Z}$, though Kuipers allows that the situation may be complicated by pseudo-counterexamples. The E-differences are understood by Kuipers as low-level universal hypotheses (observational laws); although they are of course not empirically verifiable, let us for present purposes admit them to the class of statements on which empirical agreement can be obtained. In other words, let us accept that we can determine empirically that a genuine phenomenon is being explained. The empirical decidability of judgements of relative approximation to the truth remains unremittingly negative. If either E-differences or honest I-differences are discovered, the hypothesis $\mathbf{X} \preceq \mathbf{Z}$ is falsified. But no amount of differential success for **Z**, compared with **X**, will permit us to conclude that the comparative hypothesis is true.

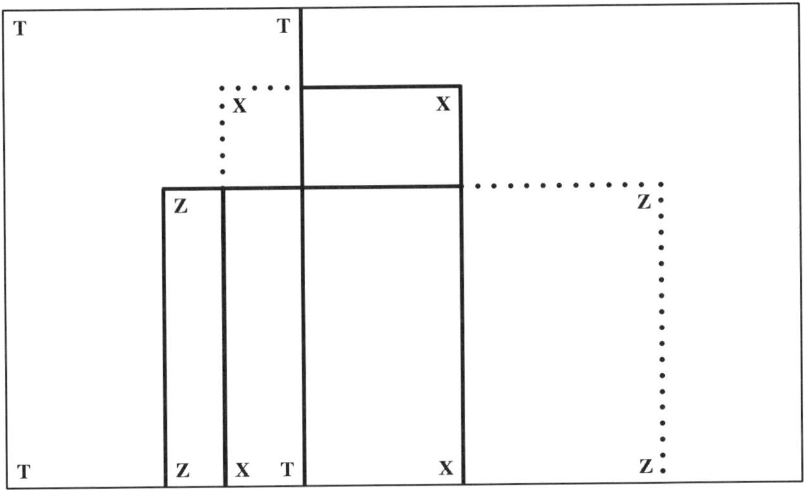

Fig. 1. **Z** is at least as good an approximation to **T** as **X** is (Kuipers)

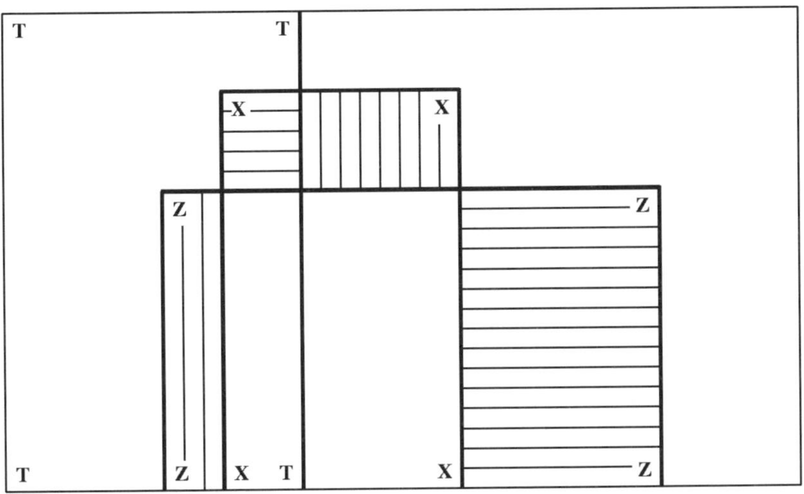

Fig. 2. When **Z** $\preceq$ **X** and **X** $\preceq$ **Z** are both refuted

Through the smokescreen of justificationist special pleading with which Kuipers envelops his discussion (see on p. 310 how glibly the innocent phrase "easier to determine" is transformed to mean "harder" – that is, "more justified") it is possible to discern that this point is fully admitted. It is conceded that neither type of discriminating counterexample (I-differences and E-differences) can offer more than "a modest indicator ... a signpost that tells you: do not take this direction" (p. 311). This is what falsificationism has been saying for 65 years. Experience can tell us only what not to endorse. It does not tell us what to endorse. In a real case, of course, it is not unlikely that both comparative hypotheses, $\mathbf{Z} \preceq \mathbf{X}$ and $\mathbf{X} \preceq \mathbf{Z}$, will suffer refutation. That is, neither the horizontally shaded portion of Figure 2, nor the vertically shaded portion, will be empty. It is disappointing to find that Kuipers's theory can provide no further help in such a case (or in any other case where the areas that the two theories assert to be empty are disjoint). If they are both falsified, neither of these comparative hypotheses is a better approximation to the truth than the other one is. For them falsification is an irreversible death sentence.

3. Kuipers on Beauty

Explanatory success and instantial success are the most important of what Kuipers calls "desirable features" of scientific theories. But there are others. If, for example, we have "the expectation, based on inductive grounds, that all successful theories and hence the (relevant) truth will have a certain aesthetic feature" then that feature will be "held to be desired" (pp. 309*f.*). If $\mathbf{Z}$ is shown to have the feature and $\mathbf{X}$ is shown not to have it (an A-difference) then this, according to Kuipers, "as far as it is correctly determined... is a reliable, modest, signpost... in favour of" the hypothesis $\mathbf{X}$ p $\mathbf{Z}$ (p. 312). A few words of explanation and debunking are required.

In the structuralist iconography favoured by Kuipers (for example, in 1997, §2.4; 2000, §8.1.6), a theory is simply a set of models or (conceptual) possibilities. In addition (p. 304)

> a feature of a theory will be understood as a "distributed" feature, that is, a property of all the possibilities that the theory admits. ... A theory is frequently called symmetric because all its possibilities show a definite symmetry. According to this definition a feature of a theory can be represented as a set of possibilities, namely as the set of all possibilities that have the relevant property.

To the penultimate sentence is appended a note: "However, it is easy to check that 'being true' of a theory $\mathbf{X}$ in the weak sense that $\mathbf{T}$ is a subset of $\mathbf{X}$ is not a distributed feature, let alone 'being true' in the strong sense of the claim '$\mathbf{T}=\mathbf{X}$'." In any event, a feature of a theory is itself a theory (though perhaps not one with a serviceable linguistic formulation), and the theory $\mathbf{Z}$ possesses the feature $\mathbf{Y}$ if and only if $\mathbf{Z} \subseteq \mathbf{Y}$; which is to say that $\mathbf{Z}$ implies $\mathbf{Y}$. Kuipers goes on to say that "desired features are features that include all desired possibilities...; undesired features are features that include all undesired possibilities" (*loc.cit.*). Recall that T is the set of desired possibilities. It follows that the features possessed by the truth T are all and only the desired features.

Kuipers remarks that the relation p of better approximation to the truth can be cast in terms of features: $X \preceq Z$ holds if & only if Z possesses all the desired features that X possesses, and X possesses all the undesired features that Z possesses ((UF) and (DF) on p.305). It would therefore not be wrong to conclude that if Y is a desired aesthetic feature, and Z is correctly shown to possess Y while X is correctly shown not to possess Y, then X cannot be as good an approximation to the truth as Z is; for $Z \preceq X$ cannot hold if Z has true consequences that X does not have. In §6 Kuipers presents a series of such judgements in the language of confirmation, support, and reasons. He says, for instance, that an appropriate "A-difference ... gives a specific reason for" the hypothesis X p Z, and that it "nondeductively and theoretically ... confirms" it. This has all the persuasiveness and logical legitimacy of the claim that the discovery that Cain is the child of Adam, but not of Eve, confirms the hypothesis that all children of Eve are children of Adam. Kuipers adds rightly that "the A-difference makes the reversed [comparison] ... impossible"; that is to say, that $Z \preceq X$ is refuted. Unless one is a gullibilist, this is all that there is to it.

In my opinion some imagination is needed to convert these truisms into the conclusion that a meditation on the aesthetic features of the world, or of scientific theories, can set a scientist on the road to the truth.

Kuipers summarizes his investigation as follows (pp. 322*f.*):

> ... the answer to the title question "Beauty, a road to the truth?" hence is: yes, beauty can be a road to the truth as far as the truth is beautiful, in the specific sense that it has distributed features that we have come to experience as beautiful. This is a nontrivial answer because it is not immediately obvious that a common feature of a theory and the truth can be considered a (modest) signpost to the truth.

For Y is a common feature of Z and the truth if and only if both Z and T imply Y; that is, if Y is a true subtheory of Z, part of Z's truth content. Kuipers recognises that this is a formal matter, not particularly concerned with aesthetics: "every nonempirical feature can indicate the road to the truth if there are inductive grounds for assuming that the truth probably has this formal feature" (p.323). Translated, this becomes: if there are inductive grounds for assuming that Y is probably true, then Y can indicate the road to the truth.

All this may or may not be edifying. But if Kuipers's way of tackling the question is correct, his answer not only is not particularly concerned with aesthetics, it is not in any way concerned with aesthetics. For according to what he says, the truth is beautiful (or sensuous, or has any other feature you like) if and only if all conceptual possibilities, and presumably all physical possibilities too (for example a world packed with buildings resembling Consolação Church in São Paulo), are beautiful (or sensuous). Were this so, which of course it is not, aesthetics would be trivialized. The fact is that the ascription of aesthetic (or other) features to a theory rarely has much to do with whether its models possess those features. I am unable to convince myself that many aesthetic features are distributed in Kuipers's sense. When we say that a physical theory is beautiful, or elegant, or symmetric, or languid, we do not mean that all the states of affairs that the theory permits are beautiful or elegant, or symmetric or languid (this is well known with regard to symmetry, Kuipers's example). The implication from theory to instances may perhaps hold if we replace "beautiful" or "elegant" or "symmetric" or "languid" by a fierce metaphysical feature such as "mechanical" or "spiritual" or something of that kind. That may indicate how far apart metaphysics and aesthetics are in this area. Sometimes, too, a theory may have a feature (in the everyday sense) if and only if all its logical consequences have that feature. (Indeed, a theory is true if and only if its consequences are true. But falsity fails this test without difficulty.) That may indicate that scientific theories are not always well represented by the structures in which they hold.

4. McAllister on Beauty

In a brief exploration of possible "a priori reasons" for expecting the truth in some way to be beautiful (p. 301), Kuipers commits the old mistake of thinking that since "[p]hysics presupposes, to a large extent, the so-called principle of the uniformity of nature" (p. 301), the thesis that the truth has some aesthetic qualities possesses some degree of a priori validity. This is wrong. Science does not and need not presuppose order, since what it does is to propose order (Miller 1994, p. 27). Likewise science does not presuppose that any non-empirical features (proto-aesthetic features that we might come to find beautiful) are concomitants of empirical success. It may proceed by proposing theories possessing these features (and sometimes theories lacking them, despite the recommendations of Maxwell 1998), and seeing what the world makes of them. Nothing of course can prevent us from conjecturing that some non-empirical (or proto-aesthetic) features are associated with empirical success (which is indeed the only sense I can make of the idea that we can know any such thing a priori, or that science presupposes it). But if we make the conjecture seriously then we have the obligation, as Kuipers recognises (2002, §6), to indicate circumstances in which we would abandon the conjecture. Artless it may be, but it is not entirely inert. McAllister (1996). makes it evident that some scientists have taken it rather seriously.

We should distinguish positive and negative uses of the doctrine that the truth is beautiful. A positive use assumes that all beautiful theories are true, and licenses the inference from beauty to truth: Franklin thought that the Crick/Watson model of DNA was "too pretty not to be true" (McAllister 1996, p. 91). A negative use assumes that all true theories are beautiful, and licenses the inference from ugliness to falsehood: Dirac thought that quantum electrodynamics was so ugly that it could not be true *(op. cit.*, p. 95). Given the richness of artistic creation, I do not know how anyone can defend inferences of the first kind, quite apart from their shameless circularity. But perhaps there could be something to be said for the critical or negative use within science of aesthetic judgements. Anyway, we shall henceforth restrict the thesis that beauty is a road to the truth to the doctrine that aesthetic considerations, in Kuipers's words, provide "a signpost that tells you: do not take this direction" (p. 311).

It is interesting to compare the thesis that beauty is a road to the truth with the almost equally vague thesis, which we may call simply empiricism, that experience is a road to (theoretical) truth. This too, as we know, is defensible only if understood in a negative sense. Does the practice of science presuppose empiricism? I think not, though every individual appeal to experience does assume a special case of it: that such-and-such an empirical report is true. Indeed it is by assuming that some deliverances of experience are true that we have learnt that experience often delivers the wrong answer, and that the appeal to experience must be carefully overseen. It seems that there are some limited but well-tested generalizations about experience that we judge to be true, and others that we judge to be false. That is one of the lessons to be learnt from optical illusions. It is simply incorrect to suppose that true theories are compatible with every kind of experience. To refute a scientific theory ("Steel rods do not bend when placed in water") it is not enough to claim airily that "it conflicts with experience" (which is no doubt true). This is saloon-bar science. A refutation sticks only if it cites a much more specific conflict with experience.

Similarly a bare accusation of ugliness in a theory carries little weight. (Dirac's objection to QED was not mere revulsion. Whether it was genuinely aesthetic is a point that I cannot go into.) The objection is not that the under-lying thesis, that all true theories are beautiful, is a piece of metaphysics, for we knew that already. The objection is that this thesis is false (presumably it is inconsistent) if it means that all true theories are beautiful in all respects; hence the non-specific ugliness of a theory is not a sufficient excuse for branding it as false. But that does not imply that there might not be a much more detailed thesis that is correct. For example, it might be the case, for all I know, that all true theories (or all true theories at a certain level) show certain particular symmetries. Violation of these symmetries in a theory would therefore point to the conclusion that the theory is false; and if no empirical tests are possible, that may be the best pointer that we have. In this sense aesthetic considerations may provide a road, an utterly conjectural road of course, as to where the truth lies.

The above remarks may be seen as an attempt to patch up the naive justificationism and inductivism of McAllister (1996), with material from Chapter IV ("A Metaphysical Epilogue") of Popper (1982) (briefly reported in his 1976, §33). Here I can give only a summary, which is not, I think, in conflict with anything said by Kuipers on the subject (at the end of 2002, §6). According to Popper a metaphysical research programme guides scientific development by setting standards for satisfactory explanations. Its role is largely negative. According to McAllister, specific aesthetic judgements are often crucial to what he calls criteria of theory choice in science. Many of his examples of aesthetic factors (mechanicism, determinism, and so on) are in reality metaphysical theories. In other words, metaphysical and aesthetic ideas may assist our judgement when empirical considerations fail to make the wanted discriminations. McAllister's central thesis is that since "a scientific revolution constitutes a radical transformation in a community's criteria for theory choice" (p. 126), and empirical criteria are not greatly revised, in revolutions it is the aesthetic standards that suffer the greatest upheaval. Unfortunately the premise here is an awesome non sequitur, and is false. The significant change in a scientific revolution is in the field of theories competing for supremacy. It need not be concluded that the rules of the competition have to change too. Sometimes they do. Sometimes they do not.

Appendix 0

Harris's proposal was that, if $\mathbf{T}$ is any theory, the truth content $\mathbf{Y}_T$ of the theory $\mathbf{Y}$ is $\mathbf{Y} \cap \mathbf{T}$ and the falsity content $\mathbf{Y}_F$ is $\mathbf{Y} \cap \mathbf{T}^*$, where $\mathbf{T}^* = \{\neg y \mid y \in \mathbf{T}\}$. The suggested amendment is to define $\mathbf{Y}_F$ as $\mathbf{Y} \cap \bar{\mathbf{T}}$, where $\bar{\mathbf{T}} = \{y \mid y \notin \mathbf{T}\}$. So suppose that $\mathbf{X}_T \subseteq \mathbf{Z}_T$ and $\varnothing \neq \mathbf{Z}_F \subseteq \mathbf{X}_F$. Choose $z \in \mathbf{Z}_F = \mathbf{Z} \cap \bar{\mathbf{T}} \subseteq \mathbf{X} \cap \bar{\mathbf{T}}$, and any $x \in \mathbf{Z}_T = \mathbf{Z} \cap \mathbf{T}$. It is clear that the biconditional $x \leftrightarrow z \in \mathbf{Z}$. But if $x \leftrightarrow z \in \mathbf{T}$, then $z \in \mathbf{T}$. Hence $x \leftrightarrow z \in \mathbf{Z} \cap \bar{\mathbf{T}} \subseteq \mathbf{X} \cap \bar{\mathbf{T}}$, and so $x \leftrightarrow z \in \mathbf{X}$. Since $z \in \mathbf{X}$, we may conclude that $x \in \mathbf{X}$, and hence that $x \in \mathbf{X} \cap \mathbf{T}$. We have shown that $\mathbf{Z}_T \subseteq \mathbf{X}_T$, and hence $\mathbf{Z}_T = \mathbf{X}_T$. It is evident that if $\mathbf{Z}$ is more truthlike than $\mathbf{X}$ then $\mathbf{X} \vdash \mathbf{Z}$. For similar results see Kuipers (1997, §2).

Figure 3 shows that the possibility allowed in the above proof may be realized. **T** is the theory axiomatized by the literal P, and contains the four propositions represented by solid balls. The theories $\neg P \wedge R$ and R have the same truth content, $\{P \vee R, \top\}$, but $\neg P \wedge R$ has a greater falsity content. Hence R is more truthlike than $\neg P \wedge R$.

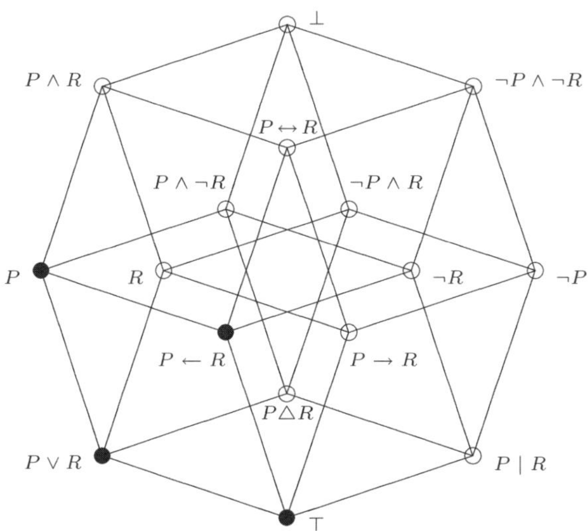

Fig. 3. R is more truthlike than $\neg P \wedge R$

Appendix 1

Theorems 1-3 of §3 of Miller (1974), according to which the truth and falsity contents of false theories vary directly with their contents, suffice to show that the condition given under POPPER 1963 is necessary and sufficient for $\mathbf{X} \underline{p} \mathbf{Z}$ in his theory. And since for false $\mathbf{X}$ and $\mathbf{Z}$ both $\mathbf{X} \wedge \mathbf{T}$ and $\mathbf{Z} \wedge \mathbf{T}$ are contradictory, the condition given under MILLER 1977 is necessary and sufficient for $\mathbf{X} \underline{p} \mathbf{Z}$ in his theory. The condition given under KUIPERS 1982, which is the same as in Table 0, is his standard definition of $\mathbf{X} \underline{p} \mathbf{Z}$ expressed in terms of theories (sets of sentences) rather than in terms of structures (2000, §7.2.2).

As before, the trickiest case is the condition given under HARRIS 1974 (amended). The necessity of the condition follows from the result of Appendix 0. For its sufficiency, we shall make use of a number of straightforward, though not very well known, results from the general theory of deductive theories. Suppose that $\mathbf{X} \wedge \mathbf{T} \vdash \mathbf{Z} \wedge \mathbf{T} \vdash \mathbf{Z} \vee \mathbf{T} = \mathbf{X} \vee \mathbf{T}$. Choose some sentence u not in $\mathbf{T}$ that belongs to the falsity content (undesirable part of the content) of $\mathbf{Z}$. Then $\mathbf{Z} \vee \mathbf{T} \vdash u \vee t$ for each $t \in \mathbf{T}$. Since $\mathbf{Z} \vee \mathbf{T} = \mathbf{X} \vee \mathbf{T}$, it follows that $\mathbf{X} \vee \mathbf{T} \vdash u \vee t$ for each $t \in \mathbf{T}$. In other words $(\mathbf{X} \vee \mathbf{T}) \wedge \neg u \vdash t$ for each $t \in \mathbf{T}$, and therefore $\mathbf{X} \wedge \neg u \vdash t$ for each $t \in \mathbf{T}$. That is, $\mathbf{X} \wedge \neg u \vdash \mathbf{T}$. It follows that $\mathbf{X} \wedge \neg u \vdash \mathbf{X} \wedge \mathbf{T} \vdash \mathbf{Z} \wedge \mathbf{T}$, and hence $\mathbf{X} \wedge \neg u \vdash \mathbf{Z}$, which by assumption implies u; and hence $\mathbf{X} \wedge \neg u \vdash u$. In other words, $\mathbf{X} \vdash u$. We have shown that for every u in $\overline{\mathbf{T}}$, if $\mathbf{Z}$ implies u then $\mathbf{X}$ implies u. In other words, the falsity content of $\mathbf{Z}$ is included within the falsity content of $\mathbf{X}$. But the second half of the assumption, $\mathbf{Z} \vee \mathbf{T} = \mathbf{X} \vee \mathbf{T}$, says that the truth content of $\mathbf{Z}$ is identical with the truth content of $\mathbf{X}$. The assumption (which is the condition given under HARRIS 1974) is therefore sufficient for $\mathbf{X} \preceq \mathbf{Z}$.

University of Warwick
Department of Philosophy
Coventry CV4 7AL UK

REFERENCES

Harris, J.H. (1974). Popper's Definitions of "Verisimilitude". *The British Journal for the Philosophy of Science* **25**, 160-166.

Kuipers, T.A.F. (1982). Approaching Descriptive and Theoretical Truth. *Erkenntnis* **18**, 343-378.

Kuipers, T.A.F. (1997). The Dual Foundation of Qualitative Truth Approximation. *Erkenntnis* **47**, 145-179.

Kuipers, T.A.F. (2000). *From Instrumentalism to Constructive Realism. On Some Relations between Confirmation, Empirical Progress, and Truth Approximation.* Dordrecht: Kluwer Academic Publishers.

Kuipers, T.A.F. (2002). Beauty, a Road to The Truth. *Synthese* **131** (3), 291-328.

Maxwell, A.N. (1998). *The Comprehensibility of the Universe. A New Conception of Science.* Oxford: Clarendon Press.

McAllister, J.W. (1996). *Beauty and Revolution in Science.* Ithaca & London: Cornell University Press.

Miller, D.W. (1974). Popper's Qualitative Theory of Verisimilitude. *The British Journal for the Philosophy of Science* **25**, 166-177.

Miller, D.W. (1975). The Accuracy of Predictions. *Synthese* **30**, 159-191.

Miller, D.W. (1977). On Distance from the Truth as a True Distance (short version). *Bulletin of the Section of Logic* (Wrocław) **6**, 15-26.

Miller, D.W. (1994). *Critical Rationalism. A Restatement & Defence.* Chicago and La Salle: Open Court Publishing Company.

Niiniluoto, I. (1987). *Truthlikeness.* Dordrecht: D. Reidel Publishing Company.

Popper, K.R. (1963). *Conjectures and Refutations. The Growth of Scientific Knowledge.* London: Routledge & Kegan Paul.

Popper, K.R. (1974). Intellectual Autobiography. In: P.A. Schilpp (ed.), *The Philosophy of Karl Popper*, pp. 3-181. La Salle: Open Court Publishing Company.

Popper, K.R. (1976). *Unended Quest. An Intellectual Autobiography.* Glasgow: Fontana/Collins. Now published by Routledge, London.

Popper, K.R. (1963). *Quantum Theory and the Schism in Physics.* London: Hutchinson. Now published by Routledge, London.

Tichý, P. (1974). On Popper's Definitions of Verisimilitude. *The British Journal for the Philosophy of Science* **25**, 155-160.

Theo A. F. Kuipers

TRUTH APPROXIMATION BY EMPIRICAL AND AESTHETIC CRITERIA

REPLY TO DAVID MILLER

As he explains in the first note, Miller wrote the first version of his contribution on the basis of the version of my paper "Beauty, a road to the truth(?)", as it was in July 2000. The final version (Kuipers 2002)[1], is not only revised to some extent in the light of Miller's comments but it is also substantially enlarged, in particular Section 2 ("Aesthetic induction and exposure effects"). This explains why Miller in his final version does not touch upon this specimen of "naturalized philosophy," based on results in experimental psychology.

My reply will deal with our remaining disagreements regarding the nature and role of empirical and aesthetic criteria in the evaluation of scientific theories. But I would first like to express my appreciation for his exposition of four theories of truth approximation in his Section 1. Tables 0 and 1 are of course attractive from my point of view. They suggest that my approach is a kind of improvement not only of Popper's original approach but also of the two parallel improvements of that approach by Harris and Miller. Technically, they represent a very informative structuring of the possibilities for "content theories" of verisimilitude, as Zwart (1998/2001) calls them. However, although Miller apparently can appreciate my approach to truth approximation in general, he is rather reserved in two respects: the role I assign to empirical successes and my analysis of aesthetic considerations.

Empirical Success

Several paragraphs of Section 2 (Empirical Decidability) deserve comment. In the first paragraph, Miller suggests that I find successes more important than

---

[1] Contrary to the recurrent reference in the text to a title with question mark and despite my proof corrections, the paper was published without. This provided Miller the opportunity to keep using "my title" for his contribution.

In: R. Festa, A. Aliseda and J. Peijnenburg (eds.), *Confirmation, Empirical Progress, and Truth Approximation* (*Poznań Studies in the Philosophy of the Sciences and the Humanities,* vol. 83), pp. 356-360. Amsterdam/New York, NY: Rodopi, 2005.

failures, but this seems to me inadequate. In his quote of mine, 'more empirical success' is clearly meant to be neutral in this respect, as is clear from my crucial definition of 'more successful' (ICR, p. 112), and my notion of "divided (empirical) success" (see below). There is one exception to this. What I started to call "lucky hits" in the paper on beauty and truth has no counterpart in something like "unlucky failures." In this respect, my analysis even has a bias in favor of "failures."

Regarding the second paragraph, I regret having hitherto overlooked Miller's (1975) remark about falsifiability of truth approximation claims. When I wrote it, I believed I was the first to have stressed the (empirical) falsifiability of such claims. It is a crucial point and, as Miller illustrates in the third paragraph, my theory of truth approximation gives rise to more interesting possibilities for falsification of the comparative claim than Popper's original theory. On second thoughts, I like the term 'pseudo-counterexample' as an equivalent to 'lucky hit', viz. if used with respect to a comparative claim, e.g. Z is closer to the truth than X. However, it may be a confusing term since, on the level of theories, a lucky hit of X relative to Z is a genuine counterexample of Z but pseudo-example of X.

Regarding the fourth and last paragraph, I essentially agree that the conclusion "do not take this direction," attached to a "discriminating counterexample," is very much in the Popperian spirit. However, I nevertheless have no serious problems with phrases like 'easier to determine', 'harder' and 'more justified'. They not only suggest that every judgment is conjectural, but also that they may differ in the degree, informally conceived, to which they are conjectural. More importantly, I would like to claim to have explicated for the first time a clear notion of comparative success, with a straightforward relation to the HD method on the one hand and to truth approximation on the other. In a way, these are the core claims of ICR (see p. 162 for a summary).

Miller's regret, in the same paragraph, for the case of what I called 'divided (empirical) success' seems too strong to me. Note first that this notion is essentially symmetric between successes and failures for it amounts to: some successes of the one theory are failures of the other, and *vice versa*. However, although there remain in my approach only two negative claims[2] (Z is neither closer to the truth than X, nor X to Z), a clear heuristic task is suggested in addition: try to find a third theory which improves upon both (which is by implication also clearly defined), that is what I call (ICR, p. 115) the heuristic principle of dialectics. This relativizes the, strictly speaking, correct verdict of a death sentence for the two falsified comparative claims.

---

[2] This in contrast to Niiniluoto's quantitative approach, which also in this case gives an estimate of truthlikeness of both theories.

Aesthetic Criteria

Miller is essentially right in the concluding sentence of the penultimate paragraph of Section 3: "if there are inductive grounds for assuming that [the aesthetically appreciated (nonempirical) hypothesis] Y is probably true, then Y can indicate the road to the truth." However, I cannot see that as debunking, at most demystifying and disenchanting. Compare it with: "if there are inductive grounds for assuming that hypothesis Y is probably true, then Y can indicate the road to *the strongest true hypothesis* (i.e. the (relevant) truth)."

I should stress that I agree with Miller's claim in the last paragraph of Section 3 that my analysis essentially deals with all kinds of nonempirical features, provided they have become desirable on (meta-)inductive grounds. As I explain in the expanded Section 2 of my paper on beauty and truth, this frequently happens to go together with "affective induction."

However, regarding Miller's claim that standard aesthetic features are usually not of the distributed kind, and his mentioning of real world instances, such as certain churches, of beautiful physical possibilities, I would like to quote a long passage from my 2002 paper that apparently did not convince Miller, but it may convince others:

> *Third*, our truth approximation claims regarding aesthetic features are, at least in this paper, restricted to a certain formal type of aesthetic features. More precisely, the 'underlying' objective nonempirical features of aesthetic features, and objective (nonempirical and empirical) features of theories in general, will be restricted to a certain formal type. A feature of a theory is called 'distributed' when it corresponds to an objective property of all (formal representations of) the conceptual possibilities admitted by the theory. Note first that aesthetic features of theories are not supposed to be associated with (the set of) its real world instances, but with the corresponding (set of) conceptual possibilities. However, it may well be that the aesthetic appreciation concerns a non-formal type of representation of certain conceptual possibilities. The famous Feynman diagrams in quantum electrodynamics provide an example. But also in such a case, it is assumed that there is in addition a formal, i.e. logico-mathematical, representation of the conceptual possibilities, such that the aesthetic feature is co-extensional with an objective property of the relevant formal conceptual possibilities. The corresponding distributed feature is called the objective feature underlying the aesthetic feature. Aesthetic features of which the objective nature cannot be explicated in the suggested distributed way fall outside the scope of my truth approximation claims, and demand further investigation. However, it should be stressed that some standard aesthetic features are of the distributed type. Regarding simplicity, for example, it is important to note that the members of the set of conceptual possibilities satisfying a simple formula all share the property to 'fit' in this simple formula. Regarding symmetry, representing a kind of order, we may note that a theory is frequently called symmetric because all its possibilities show a definite symmetry. For example, all admitted orbits may have a certain symmetrical shape. Regarding inevitability and its opposite contingency ..., it is also plausible to assume that at least certain types of both properties can be localized within conceptual possibilities. (Kuipers 2002, p. 295)

To be sure, in the suggested cases of simplicity and symmetry the claim that the relevant features are distributed is rather trivial, but this merely illustrates that the restriction to distributed features is not as restrictive as one might think at first sight.

However, I also would like to quote the most important passage on non-distributed features:

> Of course, there may well be aesthetic features that can neither be represented as a set of conceptual possibilities nor as a set of such sets. They may be of a more holistic kind. For example, a theory may be called symmetric not only because of its symmetric possibilities, but also because it is closed under a certain operation: given a model, applying the operation leads again to a model of the theory. Other examples of holistic, at least non-distributed, features of theories are diversity (of admitted/desired possibilities) and convexity. In general, all formal features that postulate membership claims in response to given members cannot be distributed. For such non-distributed features an alternative formal analysis will have to be found to complete the naturalistic analysis to a full-fledged naturalistic-cum-formal analysis of such features. (Kuipers 2002, p. 319)

Finally, I would like to refer to my reply to Paul Thagard in the companion volume, and of course to the paper itself, a paper that I would of course have summarized in a section of Ch. 8 (Intuitions of Scientists and Philosophers) of ICR if I had completed it earlier.

At the beginning of Section 4, Miller rejects the claim that physics presupposes order: scientists propose order instead. However, proposing order would not make sense if we did not believe that nature could be ordered. Presupposing order is only meant in this modest sense (at least by me). However, I do claim on this basis that certain non-empirical features may be expected to accompany successful theories on a priori grounds, with the consequence that we will come to appreciate them as beautiful if that turns out to be the case. But such a priori considerations need not be possible for all nonempirical features that we come to appreciate as beautiful, that is, features that become the subject of aesthetic induction. However, my refined claim in this respect can be tested:

> Finally, my refined claim about aesthetic induction can be falsified: determine a nonempirical, (not necessarily) distributed feature which happens to accompany all increasingly successful theories in a certain area from a certain stage on and which is not generally considered beautiful, and increasingly so, by the relevant scientists. (Kuipers 2002, pp. 318-9)

I should stress that I very much agree with the fourth paragraph of Section 4, starting with 'Similarly ...', in particular from "But that does not imply ..." onwards. I would claim to have explicated in my (2002) paper why there is a grain of truth in what Miller writes there.

Moreover, I completely agree with Miller's last point that McAllister goes much too far in suggesting that a fundamental change in aesthetic standards is not only a sufficient condition but also a necessary condition for a genuine revolution. In particular, his example that Einstein's theory is not a revolutionary transition seems to me too far from scientific common sense. See (Kuipers 2002, pp. 317-8).

But all this does not devalue McAllister's notion of aesthetic induction as a very illuminating and stimulating one. Let me finish with a general remark made by Jeffrey Koperski after the BSPS session (see Miller's first note). "You might have encountered less opposition to your basic ideas if you had called it a 'naturalization of aesthetic intuitions'". In the final version I have used the phrase 'naturalistic(-cum-formal) analysis' a couple of times. As indicated already, in Section 2 of the expanded final paper, I present and analyze indirect evidence for aesthetic induction in science, viz. "exposure effects," established in experimental psychology and dealing with aesthetic appreciation in the arts.

REFERENCES

Miller, D.W. (1975). The Accuracy of Predictions. *Synthese* **30**, 159-191.

Kuipers, T.A.F. (2002). Beauty, a Road to The Truth. *Synthese* **131** (3), 291-328.

Zwart, S.D. (1998/2001). *Approach to The Truth. Verisimilitude and Truthlikeness.* Dissertation Groningen. Amsterdam: ILLC Dissertation Series 1998-02. Revised version: *Refined Verisimilitude, Synthese Library*, vol. 307. Dordrecht: Kluwer Academic Publishers.

Jesús P. Zamora Bonilla

TRUTHLIKENESS WITH A HUMAN FACE
ON SOME CONNECTIONS BETWEEN THE THEORY OF VERISIMILITUDE AND THE SOCIOLOGY OF SCIENTIFIC KNOWLEDGE

ABSTRACT. Verisimilitude theorists (and many scientific realists) assume that science attempts to provide hypotheses with an increasing degree of closeness to the full truth; on the other hand, radical sociologists of science assert that flesh and bone scientists struggle to attain much more mundane goals (such as income, power, fame, and so on). This paper argues that both points of view can be made compatible, for (1) rational individuals only would be interested in engaging in a strong competition (such as that described by radical sociologists) if they knew in advance the rules under which their outcomes are to be assessed, and (2), if these rules have to be chosen "under a veil of ignorance" (i.e., before knowing what specific theory each scientist is going to devise), then rules favoring highly verisimilar theories can be prefered by researchers to other methodological rules.

The theory of verisimilitude is a theory about the aim of science. In a well known paper (Popper 1972), written before developing his own approach to the topic of verisimilitude, Popper described that aim as the production of testable explanations of whatever facts we thought to be interesting to explain, though he also recognised that it was rather improper to talk about the aims *of science*, since only *scientists* have goals, properly speaking, and these may look for a wide variety of things. Most discussions about the concept of truthlikeness have obviously been concerned with the first of these questions – say, what is the cognitive goal of science, assuming that one such goal exists – but they have largely ignored the second one, i.e., what the connection may be between that epistemic goal and the actual motivations and behavior of scientists. In this brief paper I would like to make a contribution to the second topic, though the ideas I am going to suggest will perhaps illuminate some aspects of the first question. To cut a long story short, I defend here three hypotheses. The first is that, besides other interests, scientists have *epistemic* ones that can be reconstructed as the pursuit of a kind of "truthlikeness" (for example, the notion of truthlikeness proposed by Kuipers; see note 4 below). My second hypothesis is that scientists can engage in the negotiation of a

In: R. Festa, A. Aliseda and J. Peijnenburg (eds.), *Confirmation, Empirical Progress, and Truth Approximation* (*Poznań Studies in the Philosophy of the Sciences and the Humanities,* vol. 83), pp. 361-369. Amsterdam/New York, NY: Rodopi, 2005.

"methodological social contract," i.e., a set of norms indicating the circumstances under which a theory must be taken as better than its rivals; these norms act as the "rules of the game" of the research process, and tell scientists who must be deemed the winner of the game; some norms of this type are needed, because each scientist needs to know what 'winning' amounts to, if they are to become interested in playing the game at all. The last hypothesis will be that the choice of these norms is made by scientists "under the veil of ignorance," i.e., without having enough information about how each possible norm will affect the success of the theories each researcher will be proposing in the future. The main conclusion is that, under these circumstances, researchers will tend to prefer methodological norms which promote the truthlikeness of the theories which must be accepted according to those norms. This conclusion could be tested by studying whether the actual methodological norms used by scientists through history have been consistent, so to speak, with the maximization of verisimilitude.

Traditionally, philosophical explanations of science were developed under the tacit assumption that scientists disinterestedly pursued epistemic values, such as truth, certainty, generality, and so on. Even though sociologists in the school of Merton had put forward the fact that scientists were mainly motivated by other kinds of interests, this sociological school proposed the hypothesis that science was governed by an unwritten rule which forced scientists to disregard their personal or social motivations. This hypothesis presupposed that scientists were able to judge in an objective way which theory was the best solution to a given problem, and agreed in unanimously declaring that theory "the most appropriate one," even while many of them might have proposed alternative solutions. This utopian vision of the mechanism of scientific consensus has been challenged since the seventies by several new schools in the sociology of science, particularly by the two called "Strong Program" and "Ethnomethodology." According to these new radical schools, the role of non-epistemic interests – either social or personal – in the process of scientific research was absolutely determining. Through an overwhelming amount of empirical work, both in the history of science and in "laboratory studies," these scholars claimed to have shown that scientists tended to take their decisions motivated almost exclusively by this other kind of interests, and, as a conclusion, they advanced the thesis that "established" scientific knowledge did not as a matter of fact mirror the hidden structure of the world, but only the "all-too-human" struggles between scientists. Scientific consensus would thus not be the outcome of a match between theoretical hypotheses and empirical evidence, but the result of quasi-economic negotiations for the control of power in society and within scientific disciplines.

Of the two kinds of motivations these sociologists have assumed to explain scientists' behavior (i.e., interests rooted in roles and social classes, and interests related to status within the scientific profession), I think the second one better reflects the actual interests of individual scientists. In the first place, "wide" social interests are less apt to explain researchers' ordinary choices when the problems they try to solve have weak or uncertain social implications. In the second place, the relevant social groups are usually fewer than the number of competing solutions, so that it is impossible to match them one by one in order to explain a researcher's choice of a solution as the consequence of his belonging to a certain social group.

"Recognition" or "authority" can thus be the main goal of individual scientists, in the sense that, when they face a choice among several options, if one of them clearly leads to a higher degree of recognition, they will always choose this one. But it seems difficult to accept that this can be the *only* motivation for devoting one's life to scientific research: after all, "recognition" could be gained through many other activities, from politics to sports or to the arts, all of them more rewarding than science in terms of fame and income, and perhaps less demanding in terms of intellectual effort. If somebody has chosen to spend his youth among incomprehensible formulae and boring third-rate laboratory work, we may assume that he will at least find some pleasure in the acquisition of knowledge. So I propose to make the benevolent assumption that, in those cases where no option clearly entails an advantage in terms of recognition, a researcher will tend to base his choices of theory (or hypothesis, or description of facts) on the epistemic worth of the available options. This presupposes that each scientist is capable of establishing an (at least partial) ordering of the available options according to some set of "cognitive values," "epistemic preferences" or "epistemic utilities," but I do not go so far as to suppose that different researchers make necessarily the same ordering. Stated somewhat differently, I assume that scientists have some *informed beliefs* about the correctness or incorrectness of the propositions they handle during their work. After all, if researchers are able to obtain knowledge about which actions will cause them to reach a higher level of recognition, as radical sociologists of science easily assume, it would seem absurd to deny that they may also gain information about which theories are probably more correct, which empirical data are more relevant and which strategies of reasoning are logically sound. My assumption, hence, is simply that in those cases when choosing the most valuable option from the epistemic point of view does not diminish the expected level of recognition, this option will be chosen.

The question, hence, is what are the epistemic values of scientists, those which make them prefer *ceteris paribus* some theories, hypotheses or models to others. The sociology of science literature is not very helpful here, since it has either usually ignored this question, or it has just tried to show that epistemic preferences did not play any relevant role at all. Perhaps this question might be answered with the help of an opinion poll among scientists, but it would be difficult to decide exactly what should be asked; a particularly problematic issue about this poll would be to design the questions in a neutral way with respect to rival methodological or epistemological theories. Still more problematic would be the fact that scientists' cognitive preferences will probably be *tacit*, and it may be difficult for them to articulate in a coherent and illuminating way those factors which lead them to value a particular theory, experiment or model. On the other hand, I think we should avoid the arrogant stance of those philosophers who think that real scientists are not a reliable direct source of criteria in epistemological matters. Perhaps the average scientist is not very good at deriving *philosophical* implications from his own work, and perhaps he is as ignorant of the formal and conceptual complexities of scholastic philosophy of science as we often are about *his own discipline's* complexities. But, since scientists are our paradigmatic experts in the *production* of knowledge, it can hardly be denied that their practices will embody, so to speak, the best available criteria for determining what should count as "knowledge."

One common theme in the so-called "deconstructionist" approaches to the sociology of science is that "knowledge is negotiated." I do not deny it is. As a social institution, science is a "persuasion game," and in order to get recognition you have to make your colleagues accept that the hypotheses advanced by you are better than the rest... even better than those hypotheses which they themselves proposed! Agreement, hence, will hardly be an immediate outcome of a direct confrontation with an intersubjective experience; only after many rounds of "I will only accept that if you also accept this" moves will an almost unanimous agreement be reached. But it is also difficult to believe that the *full* negotiation process of a scientific fact is reducible to that kind of exchange, for in science it is compulsory to offer a *justification* of whatever proposition a researcher accepts or presents as being plausible; many readers of the sociological literature about the "rhetoric" of science may legitimately ask why that "rhetoric" has any force at all, why does each scientist not simply ignore *all* his colleagues' arguments and stubbornly reject any hypotheses proposed by a "rival." I guess that a more verisimilar account of the process will show instead, that what each scientist tries to "negotiate" is the coherence of a proposition with the *criteria of acceptance* shared by his colleagues. This entails that if, during the negotiation of a fact, you have employed a type of argument, or other propositions as premises, so as to "force" your colleagues to accept a hypothesis, you will be constrained to accept in the future the validity of that type of argument, or the rightness of those propositions. Otherwise, those colleagues you happened to persuade by means of that strategy could reverse their former decisions and reject your hypotheses, if they realize that you do not honor the very arguments you had used to persuade them.

As long as the decision of accepting a given fact or model (probably under the pressure of negotiation with colleagues, both rivals and collaborators) is constrained by the necessity of supporting that decision with *reasons* which are coherent with the kinds of reasons one has employed in other arguments, a fundamental point in the analysis of "scientific negotiations" must be why certain *types of reasons* are accepted by scientists, especially the types of reasons which serve as justifications of the use of other (lower-level) reasons. If we call these "higher-level" reasons *methodological norms*, then our question is just why certain methodological norms are used within a scientific community, instead of other alternative sets of criteria. I plainly accept that even these norms can be a matter of "negotiation," but we must not forget that the requirement of coherence entails that, if a scientist has acquired through former "negotiations" the compromise of abiding by his decisions to accept certain criteria, it is possible that the future application of those criteria will force him, for example, to reject a hypothesis he himself had proposed. So, as long as *reasons* are used in negotiation processes – reasons whose domain of application is necessarily wider than the negotiation of a particular fact or theory – it will be *uncertain* for a scientist whether in other cases it will be still favorable for him (i.e., for the acceptability of his own hypotheses) that those reasons are widely accepted. Or, stated somewhat differently, when a researcher decides to accept or to contest a given methodological norm, it is very difficult for him to make an informed estimate of how much support his own theories will receive from that norm "in the long run."

If my argument of the last two paragraphs has some plausibility, it follows that the decision of accepting or rejecting a given methodological norm must be made *under a "veil of ignorance,"* in the sense that the personal interests of scientists will hardly serve them as a guide in their choice. This does not entail that those interests actually play no role in the negotiation about "proper methods," but we can guess that their influence will tend to be rather blind, if it happens to exists at all; i.e., even if a group of scientists accepts a methodological norm *because they believe* that it will help them to fulfil their professional aspirations, it is equally likely that the actual effect of the norm will be to undermine those aspirations. Under these circumstances, the only reasonable option for scientists is to base their choice of methodological norms on their epistemic preferences referred to above, since these preferences will allow then to make a much easier, direct evaluation of those norms. So, the wider the applicability of a norm, the more exclusively based on epistemic reasons alone it is likely to be.

My suggestion is, then, to look for those methodological norms which are more prevalent in the history of science, and to use them as data to be explained by a hypothesis about the nature of the scientists' epistemic preferences. This strategy would allow us to take the theories about the aim of science not – or not only – as metaphysical exercises, but also as *empirical hypotheses*, which could be tested against the history of science. The question is, hence, *what the epistemic preferences of scientists can be if they have led to the choice of the methodological norms observed in the practice of science?* Nevertheless, before arguing in favor of a particular hypothesis about those epistemic preferences, it is important to clarify some delicate points:

a) In the first place, there is probably no such thing as "the" epistemic preferences of scientists, for different scientists can have different preferences, and even one and the same scientist can change his preferences from time to time, or from context to context. The very concept of a "negotiation" applied to the choice of a set of methodological norms entails that the rules chosen will not necessarily correspond to the optimum choice of every scientist; instead, it can resemble a process of bargaining in which everyone agrees to accept something less than their optimum, in exchange for concessions made by the other parties. In particular, the chosen methodological norms may be different in different scientific communities or disciplines, as well as they may vary in time. So, we might end up with the conclusion that the best explanation of actual methodological practices is a *combination* of different epistemic preferences. This does not preclude that simpler explanations will be prefered *ceteris paribus*.

b) In the second place, the hypotheses taken under consideration should not refer to epistemic utilities which are too complicated from the formal point of view. The strategy defended here is, to repeat, that actual scientific practices are our best criteria to look for what constitutes "objective knowledge," and that these practices tend to reflect the cognitive intuitions of the "experts" in scientific research. It is hardly believable that these intuitions need to be reconstructed by means of excessively intricate epistemic utility functions, particularly when the functions are so complicated that no relevant, easily applicable methodological norms can be derived from them.

c) In the third place, our strategy suggests we should inspect scientific practice in order to look for "negotiations" about methodological norms, rather than about facts or theories. Most empirical reports from historians and sociologists of science refer to the second kind of negotiation, where conflicting methodological criteria are *used*, rather than *discussed*; the conclusion of many case studies is that a researcher or group of researchers managed to "impose" a new method, but usually it is left unexplained why the *other* scientists accept that "imposition" at all, if it goes against their own interests. So, either negotiations on method are not frequent, or they have tended to be neglected in the study of historical cases, or probably both things are partially true. In fact, methodological norms can often be "negotiated" in an absolutely tacit way: as long as they constitute the "rules of the game" of a scientific discipline or subdiscipline, they determine the rewards a researcher could expect to have if he decided to become a member of it; so researchers can negotiate the rules by "voting with their feet," i.e., by going to those fields of research in which, among other things, the norms are most favorable from their own point of view.

d) In the fourth place, the empirical character of this strategy does not preclude a *normative* use of it, in a "Lakatosian" sense: once a certain hypothesis about the epistemic preferences of scientists has been sufficiently "corroborated" (which entails, among other things, that there are no better available explanations of the methodological practice of scientists), we could use that hypothesis to show, e.g., that in some historical episodes the "optimum" methodological norms have not been followed, due to the influence of some factors; these factors might include the inability to reach an agreement about the norms or the presence of non-epistemic interests which suggested to researchers that other methodological norms could have been more favorable to them. On the other hand, even if an agreement about norms exists, some choices of models or facts made by some researchers may be contrary to the recommendations of the norms, especially if strong enough enforcement mechanisms fail to become established by the scientific community.

e) In the fifth and last place, and again from an evaluative point of view, a hypothesis about the epistemic preferences of scientists must not be identified with the thesis that those are the preferences they *should* have. Some philosophers may find logical, epistemological or ontological arguments to criticize the cognitive goals revealed by scientists in their methodological choices. But it is difficult to understand, in principle, how the generality of scientists – i.e., of society's experts in the production of knowledge – could be "mistaken" about the validity of their own goals. In this sense I think we should not dismiss an argument such as Alvin Goldman's against some critics of scientific realism, when he asserts that we should put into brackets any philosophical theory which, on the basis of complicated lucubrations on the concept of meaning, "proved" that the statement "there are unknown facts" has no logical or practical sense (Goldman 1986). On the other hand, knowledge advances thanks to the invention of new ideas that may look strange at first sight, and this is also true in the case of philosophy; so new epistemological points of view should nevertheless be welcomed and discussed, even if their discussion does not end up in their acceptance.

Going back to the question of what the epistemic preferences of scientists can be, I have argued elsewhere that some notions of verisimilitude can help to explain some of the most general methodological practices observed in science.[1] In particular, the approach developed by Theo Kuipers is very successful at giving a justification of the hypothetico-deductive method as an efficient mechanism for selecting theories closer and closer to the truth. Hence, under the perspective offered in this short paper, we can understand Kuipers' contribution to the theory of verisimilitude as an *explanation* of the *fact* that that method is as widely employed as it is: if the epistemic preferences of scientists are such that they consider a theory to be better than another just if the former is closer to the truth than the latter,[2] then they will tend to prefer the hypothetico-deductive method to any other system of rules, *were they given the chance of collectively choosing a norm about the method for comparison of theories they were going to use within their discipline.*

One possible challenge for those epistemologists who defend other kinds of cognitive utilities would be to justify that these other preferences explain, just as well as the theory of verisimilitude, the methodological norms *actually* adopted by scientists. Sociologists of science should also try to offer alternative explanations of the extreme popularity of the hypothetico-deductive method. Even if these better explanations were actually provided (which I do not discard *a priori*), the methodological approach to the theory of verisimilitude would have had the beneficial effect of promoting the kind of research which had led to those empirical and conceptual results.

---

[1] See, for example Zamora Bonilla (2000, pp. 321-35)

[2] One of the simplest definitions of truthlikeness proposed by Kuipers is the following: if X is the set of physically possible systems, theory A is closer to the truth than theory B if and only if $\text{Mod}(B) \cap X \subseteq \text{Mod}(A) \cap X$ (i.e., all "successes" of B are "successes" of A) and $\text{Mod}(A) \cap \text{Comp}(X) \subseteq \text{Mod}(A) \cap \text{Comp}(X)$ (i.e., all "mistakes" of A are "mistakes" of B). Here, a "success" is taken as a physical system rightly described, and a "mistake" as a physical system wrongly described. Kuipers shows that the hypothetico-deductive method is "efficient" for truth approximation by proving that, if A is *actually* more truthlike than B, then for any possible set of empirical data, A will always have more empirical success than B, and hence, if scientists follow the hypothetico-deductive method (which commands then to prefer those theories with more confirmed predictions), then it cannot be the case that a theory that is less verisimilar is prefered to a more verisimilar theory. See, for example, Kuipers (1992, pp. 299-341).

ACKNOWLEDGMENTS

Research for this paper has been made possible under Spanish Goverment's research projects PB98-0495-C08-01 and BFF2002-03656.

U.N.E.D.
Depto. Logica y Filosofia de la Ciencia
Ciudad Universitaria
28040 Madrid
Spain

REFERENCES

Goldman, A. (1986). *Epistemology and Cognition*. Cambridge, MA: Harvard University Press.

Kuipers, T.A.F. (1992). Naive and Refined Truth Approximation. *Synthese* **93**, 299-341.

Popper, K.R. (1972). The Aim of Science. In: *Objective Knowledge*, pp. 191-205. Oxford: Clarendon Press.

Zamora Bonilla, J.P. (2000). Truthlikeness, Rationality and Scientific Method. *Synthese* **122**, 321-335.

Theo A. F. Kuipers

ON BRIDGING PHILOSOPHY AND SOCIOLOGY OF SCIENCE
REPLY TO JESÚS ZAMORA BONILLA

There is a difficult relationship between present-day sociologists of science and social epistemologists, on the one hand, and "neo-classical" philosophers of science, on the other. Both parties have difficulty in taking each other seriously. Hope should be derived from those scholars who seriously try to build bridges. Of course, bridge builders have to start from somewhere and the most promising constructors with a philosophy of science background are in my view Alvin Goldman (1999), Ilkka Niiniluoto (1999), and, last but not least, Jesús Zamora Bonilla (2000). In the latter's contribution to this volume Zamora Bonilla continues his very specific project of clearly specifying a kind of research agenda for studying bridge issues, in critical response to Ilkka Kieseppä's reservations about a methodological role of the theory of verisimilitude and David Resnik's arguments against the explanation of scientific method by appeal to scientific aims. Some of his main points are the following. (1) Gaining "recognition" is the dominant personal motivation of scientists, followed by trying to serve epistemic values. (2) Epistemic values can be served by methodological norms. (3) The norms have to be chosen under a "veil of ignorance" regarding the fate of the theories that will be proposed by certain scientists and hence the recognition they will get from them. (4) Hence, the most common norms in practice will best serve the relevant epistemic values. (5) Conversely, an adequate epistemic theory should enable us to justify these norms. (6) The HD method is very popular among scientists and is favorable for truth approximation, at least when both are explicated along the lines of ICR or along related lines, as presented by Zamora Bonilla. (7) The theory of truth approximation even justifies the popularity of the HD method.

Zamora Bonilla concludes with:

> One possible challenge for those epistemologists who defend other kinds of cognitive utilities would be to justify that these other preferences just as well explain as the theory of verisimilitude the methodological norms *actually* adopted by scientists. Sociologists of

In: R. Festa, A. Aliseda and J. Peijnenburg (eds.), *Confirmation, Empirical Progress, and Truth Approximation* (*Poznań Studies in the Philosophy of the Sciences and the Humanities,* vol. 83), pp. 370-372. Amsterdam/New York, NY: Rodopi, 2005.

science should also try to offer alternative explanations of the extreme popularity of the hypothetico-deductive method (p. 366).

I would like to accept the first challenge and make the second somewhat more precise. Before doing so, though, I quote a statement from SiS (pp. 349-50).

> To be sure, scientists not only aim at cognitive goals like empirical success or even the truth of their theories, but they also have social aims like recognition and power, and hence means to reach such aims. And although these goals frequently strengthen each other, [the existence of] such convergences by no means implies that the conscious pursuit of these social goals is good for science.

By arguing that epistemic values are subordinate to recognition and methodological norms subordinate to epistemic values, the latter on the basis of a veil of ignorance regarding the ultimately resultant recognition, Zamora Bonilla greatly relativized the possible negative effects of the conscious pursuit of recognition for the pursuit of epistemic values such as empirical success and truth.

To What Extent Are Instrumentalist Epistemic Values Sufficient?

A dominant line of argumentation in ICR is that the instrumentalist methodology, that is, HD evaluation of theories, is functional for truth approximation. Hence, that methodology serves the sophisticated realist cognitive values, and hence, conversely, these values can explain and justify the popularity of this methodology, to whit comparative HD evaluation. So far I agree with Zamora Bonilla. However, I would also claim that this methodology serves instrumentalist epistemic values, notably empirical success, at least as well. At first sight, Zamora Bonilla seems to disagree, but this might be mere appearance. The reason is that his own explication of truth approximation (see Zamora Bonilla 2000, and references therein) is essentially of an epistemic nature. Like Niiniluoto's (1987) notion of "estimated truthlikeness," it is not an objective notion. However, unlike Niiniluoto's notion, that of Zamora Bonilla is not based on an objective one. On the other hand, like my objective explication, and in contrast to Niiniluoto's explication, Bonilla's explication straightforwardly supports HD evaluation. Hence, the question is whether Bonilla's explication goes further than instrumentalist epistemic purposes. If so, my claim would be that even his explication is more than strictly necessary for explaining and justifying HD evaluation. However, this is not the occasion to investigate this in detail.

For the moment the interesting question remains whether there are other reasons to favor the (constructive) realist epistemology relative to the instrumentalist one. In ICR I have given two such reasons, one of a long-term and one of a short-term nature. Only the realist can make sense of the long-term dynamics in science, practiced by instrumentalists and realists, in which theoretical terms become observation terms, viz., by accepting the relevant theories as the (strongest relevant) truth. This general outlook enables the realist to relativize for the short term a counterexample to a new theory that is an observational success of a competing theory by pointing out the possibility that the latter may be accidental (ICR, p. 237, p. 318) or, to use my favorite new term, that it may be "a lucky hit." In sum, although both epistemologies can explain and justify the popularity of the instrumentalist method, only the realist can make sense of the regular occurrence of long-term extensions of the observational language and the occasional short-term phenomenon of downplaying successes of old theories.

The Proper Challenge to Sociologists Regarding Non-Falsificationist Behavior

None of this alters the fact that the suggested explanations-cum-justifications of HD evaluation provide an invitation to sociologists of science to offer alternative explanations of the popularity of HD evaluation. To be more precise, sociologists of science have shown convincingly that scientists frequently demonstrate non-falsificationist behavior. However, they have been inclined to look for "social" explanations for that type of behavior, whereas in the view of Zamora Bonilla and myself, straightforward cognitive explanations can be given. Certainly the most important reason is the relativization of the cognitive role of falsification in the process of (even purely observational) truth approximation. This amounts to the difference between HD testing and HD evaluation of theories. Moreover, both methods leave room for many sensible ways in which a *prima facie* counterexample of a favorite theory can be questioned as such. For both claims, see ICR, Section 5.2.3, or SiS, Section 7.3.3. Finally, there is the possibility of the lucky hit nature of successes of a competing theory, referred to above. Hence, in all these cases there are cognitive reasons to expect that non-falsificationist behavior may serve epistemic purposes. To be sure, and this is a major point made by Zamora Bonilla, on average this may well be useful for gaining recognition. Hence, in regard to non-falsificationist behavior, the proper challenge to sociologists is to look for cases that cannot be explained in this convergent way.

REFERENCES

Goldman, A. (1999). *Knowledge in a Social World.* Oxford: Oxford University Press.

Niiniluoto, I. (1987). *Truthlikeness.* Dordrecht: Reidel.

Niiniluoto, I. (1999). *Critical Scientific Realism.* Oxford: Oxford University Press.

Zamora Bonilla, J.P. (2000). Truthlikeness, Rationality and Scientific Method. *Synthese* **122**, 321-335.

TRUTHLIKENESS AND UPDATING

Sjoerd D. Zwart

UPDATING THEORIES

ABSTRACT. Kuipers' choice to let logical models of a theory represent the applications or evidence of that theory leads to various problems in ICR. In this paper I elaborate on four of them. 1. In contrast to applications of a theory, logical models are mutually incompatible. 2. An *increase* and a *decrease* of a set of models both represent an increase of logical strength; I call this the ICR *paradox of logical strength*. 3. The evidence *logically implies* the strongest empirical law. 4. A hypothesis *and its negation* can *both be false*. My conclusion therefore reads that we should not identify (newly invented) applications of a theory with its logical models, but with *partial* models that can be extended to the logical model(s) of the language used to formulate the theory. As an illustration I give a model theoretical account, based on partial models, of the HD-method and crucial experiments.

1. Introduction[1]

It was in 1606 that Willem Jansz was probably the first European actually to set feet on the Australian coast; without wanting to be Eurocentric, let me baptize this event the discovery of Australia. The funny thing is that, epistemically, Jansz's discovery may be interpreted as an *increase* and as a *decrease* of our geographical knowledge. How can this be? To start with, we may consider the discovery of Australia as an *increase* of our geographical knowledge. After all, our knowledge of the number of continents becomes more accurate, and on a more sophisticated level, the discovery increases our knowledge about the patterns of ridges and trenches in the earth crust. Doing so, the discovery adds to our geographical theories. Considered as an increase of knowledge, the (logical) models of the theories represent *sets* of continents. The discovery of Australia means we must dismiss all models consisting of four continents; only those models survive that consist of five or more continents. Since the extension of the set of admissible models decreases, the logical strength of the new geographical theory increases, since logical weakness varies with the extension of model sets.

How, then, are we to underpin the claim that the discovery of Australia may be interpreted as a *decrease* of our geographical knowledge? The answer to this question becomes clear if we realize that we may also consider the single

---

[1] Thanks are due to J. van Benthem who drew my attention to this example.

In: R. Festa, A. Aliseda and J. Peijnenburg (eds.), *Confirmation, Empirical Progress, and Truth Approximation* (*Poznań Studies in the Philosophy of the Sciences and the Humanities*, vol. 83), pp. 375-395. Amsterdam/New York, NY: Rodopi, 2005.

individual continents to be the models of our geographical theories. This is not as strange as it might appear at first sight. All our geographical theories and hypotheses have to be true for *all* continents. The discovery of Australia will probably be accompanied by the falsification of some previously accepted (geographical) generalization that holds for the four known continents – one such generalization we shall encounter in due course. Consequently, as logical weakness varies with the extension of the set of models, we must conclude that our geographical knowledge has *decreased* since the set of models has increased. The lesson to be drawn from this example is that we have to be cautious when modeling the dynamics of updating theories. In the remainder of this paper I call structural entities *physical* models when added to our body of knowledge, they lead to an *in*crease of logical strength, and *logical* models if they lead to a *de*crease of strength. More specifically, sections 3.3 and 4.1 elaborate the details of the discovery-of-Australia paradox using the distinction between physical and logical models.

The claim of this paper is that in ICR, Theo Kuipers fails to distinguish adequately between logical and physical models in the sense just explained, which leads to conceptual problems. I have chosen the following course to substantiate my claim. In the next section I introduce the Δ-definition of verisimilitude and Kuipers' interpretation of it, which have led him to his $R(t)$ and $S(t)$ account of updating theories. The difficulties to which this interpretation is bound to lead are the topic of the third section. At the end of the third section I show that the problems mentioned are due to the identification of logical and physical models. The confusion is avoided if physical structures are identified with partial (logical) models. In the fourth section I put forward the model theoretical formulation of the Hypothetical Deductive method – from now on the HD-method – using partial models. The last section is dedicated to the conclusions.

In the present paper I want to achieve the largest possible conceptual clarity at the cost of technical complexities, therefore I mainly consider finite propositional languages with the usual Boolean connectives such that all theories are axiomatizable. Additionally, I do not differentiate between theories and hypotheses, both being paraphrased as deductively closed sets of sentences.[2] The restrictions on the languages allow me to refer to a theory by one formal sentence φ, being the axiom of the theory. Kuipers often refers to theories as sets of models, such as X, Y or T, where T is reserved for the empirical truth. Our restricted context enables us to identify these capitals with $\mathrm{Mod}(\varphi)$, $\mathrm{Mod}(\psi)$ and $\mathrm{Mod}(\tau)$. When I analyze the HD-method I have to consider monadic predicate

---

[2] Kuipers suggests that one might distinguish between hypotheses and theories by differentiating their claims. He continues "However, we will not do so in general ..." with the truth as the only exception. (ICR, 7.2.1)

languages, and again, I suppose them to have finite vocabularies. "The truth", below, refers to the set of all empirically true L-sentences, which, for finite languages, is finitely axiomatizable. Of course, most scientific theories build their quantitative observational statements on arithmetic, which is not axiomatizable. It is therefore a nontrivial simplification to assume that the truth of a scientific language L is axiomatizable. My goal, however, is to achieve the greatest possible conceptual clarity, and I therefore omit the well-known intricacies of the axiomatization of mathematics.

2. The ICR Model of Theory Updates

Kuipers' account of the dynamics of scientific theories and his formulation of the HD-method is conceptually related to his basic Δ-definition of verisimilitude. A clear understanding the Δ-definition is therefore a prerequisite for understanding Kuipers' account of the dynamics of theories (ICR 7.2.1).

2.1. *The Δ-definition and Kuipers' Interpretation*

The notion of verisimilitude originates (via C. S. Peirce) with Karl Popper, one of the most prominent philosophers of science who emphasized the importance of falsification and the HD-method. According to Popper, one falsified theory is to be preferred to another falsified theory if the first bears more similarity to the true theory – the set of all true sentences of the language of both theories – than the other. Within this point of view two questions need to be answered. First, the question of the *definition* of verisimilitude has to be addressed. It reads: How do we define the idea of φ being more similar or closer to the truth τ than its adversary ψ? Second, we have to answer the *epistemic* question, since, as we are ignorant of the truth, how do we *know* that a theory is closer to the truth than another? It is well known that Popper's answer to the first question failed (Miller 1974; Tichý 1974), and that there have been many alternative proposals. In Zwart (2001), I introduced and evaluated eight verisimilitude definitions. One very basic way to put a measure on the proposition of a language is proposed by David Miller (1978); being unfamiliar with this proposal, Kuipers came up in 1982 with the same formalism but gave it a different interpretation. This Δ-definition is still foundational to the ICR framework.

Let us consider the Δ-definition for a finite propositional language L. Suppose that all and only those sentences of L are empirically true that are a consequence of sentence τ. Then, according to the Δ-definition of verisimilitude, the theory axiomatized by ψ is at least as close to the truth as the theory axiomatized by φ, $\psi <_\tau \varphi$, iff

(1) $\mathrm{Mod}(\psi) \, \Delta \, \mathrm{Mod}(\tau) \subseteq \mathrm{Mod}(\varphi) \, \Delta \, \mathrm{Mod}(\tau)$

The Δ in (1) is the *symmetric difference* between sets; thus (1) is equivalent to

$$(2) \ a. \ \mathrm{Mod}(\psi) - \mathrm{Mod}(\tau) \subseteq \mathrm{Mod}(\varphi)\text{-}\mathrm{Mod}(\tau) \text{ and}$$
$$b. \ \mathrm{Mod}(\tau) - \mathrm{Mod}(\psi) \subseteq \mathrm{Mod}(\tau)\text{-}\mathrm{Mod}(\varphi). \qquad \text{(ICR, p. 151)}$$

What do the two clauses of the Δ-definition mean? Let us first consider the clause $2a$. Using some basic set theory, we see that it is equivalent to $\mathrm{Mod}(\psi) \subseteq \mathrm{Mod}(\varphi) \cup \mathrm{Mod}(\tau)$, i.e., all models of ψ are models of φ or τ. Some basic model theory then teaches that this is equivalent to the claim that $\psi \vDash \varphi \vee \tau$. Applying some logic again, we see that this final constraint comes down to $\mathrm{Cn}(\varphi) \cap \mathrm{Cn}(\tau) \subseteq \mathrm{Cn}(\psi)$, where $\mathrm{Cn}(\psi)$ is equal to $\{\alpha \in \mathrm{Sent}(L) | \psi \vdash \alpha\}$. If we want to put this last characterization of $2a$ into words, we will say that all true consequences of φ are also consequences of ψ.[3]

The interpretation of the second clause, however, is less uncontroversial, although its model theory is also straightforward. This theory teaches that $2b$ is equivalent to $\mathrm{Mod}(\varphi) \cap \mathrm{Mod}(\tau) \subseteq \mathrm{Mod}(\psi)$, stating that all L-structures being models for φ and τ are also ψ-models. According to basic model theory, then, the last formula is equivalent to $\varphi \wedge \tau \vDash \psi$, and this claim is unchallenged. How are we to interpret the last expression using L-sentences? To answer this question we continue in the same way as we did in the previous paragraph. Formally, it is undisputed that $2b$ is equivalent to $\mathrm{An}(\varphi) \cap \mathrm{An}(\tau) \subseteq \mathrm{An}(\psi)$, where $\mathrm{An}(\psi)$ equals $\{\alpha \in \mathrm{Sent}(L) | \alpha \vdash \psi\}$, which is the set of all *antecedences* of ψ (note the similarity to the Cn formulation of $2a$). To put the second clause into words, therefore, I take it to state that all truth-antecedences of φ are truth-antecedences of ψ. Since the notion of antecedence is not very well known, I will rephrase the last characterization using the term explanation.[4] The second clause, then, comes down to the expression that all truth-explanations of φ explain ψ, where truth-explanations are simply L-sentences implying the truth τ. This interpretation does not fit to the ICR-interpretation of $2b$ as will be revealed later. Without further qualifications, we may say that clauses $2a$ and $2b$ are similar, and if the truth is axiomatizable, as it is in finite languages, the most concise formulation of the Δ-definition reads: $\psi <_\tau \varphi$ iff $\varphi \wedge \tau \vDash \psi \vDash \varphi \vee \tau$.

Besides the question whether the truth is axiomatizable, the completeness of the true theory also plays an important role in the interpretation of the Δ-definition. If the truth τ is complete, such that every L-sentence either implies $\neg\tau$ or is implied by τ, the only demand of clause $2b$ is that no false theory can

---

[3] Detailed proofs of these contentions can be found in Zwart (2001).

[4] This stipulation of the term explanation disregards the descriptive adequacy condition that explanations need to be true. In my opinion the gain in understanding counterbalances this minor drawback. Also Kuipers uses this "liberal realist sense" of explanation (ICR, p. 148).

improve any true theory, even if the latter is almost the tautology.[5] Under the assumption that the true theory is complete, the Δ-definition reads: ψ is at least as close to the truth iff all true consequences of φ are consequences of ψ, while ψ is true or φ is false.

Of course, it remains to be seen whether $2b$ is a desirable constraint on verisimilitude definitions, especially when the truth is complete. Generally, strong but false theories seem preferable to true but weak theories, and this is the reason I dismissed clause $2b$ in my refined version of the verisimilitude definition. Whatever one thinks about this question of adequacy, clearly the Δ-definition is a perfectly consistent and interesting definition of verisimilitude. If one favors the Δ-definition, the following question arises: how does standard falsificationist methodology fare with the Δ-explanation of verisimilitude? Is there any chance that the HD-method leads us Δ-closer to the truth? The next subsection sketches Kuipers' answer to the epistemic question of verisimilitude, depicting the way he relates the Δ-definition to the dynamics of theories using confirming or falsifying instances and accepted laws.

Let us return to the introduction of the basic truthlikeness approach of ICR in regard to finite propositional languages. As Kuipers wants to put forward a structuralist and linguistic definition of verisimilitude,[6] he identifies the set of all L-structures with the structuralist set Mp of all possible models, and he interprets the latter as *conceptual possibilities* (ICR, p. 143). Furthermore, domain D is the set of intended applications of the theory under consideration, and Kuipers assumes that the combination of D and Mp results in a "unique, time-independent subset $Mp(D)$" (ICR, p. 147), which he equates with the models of the true theory T. As explained, for propositional languages, T equals $\text{Mod}(\tau)$, and any theory φ corresponds with a set of models, $\text{Mod}(\varphi)$, as the deductively closed sets of L-sentences are one-one mapped into the power set of the L-structures. Furthermore, according to Kuipers' terminology a theory φ is empirically true if its claim that $\text{Mod}(\tau)$ is a subset of $\text{Mod}(\varphi)$ is true, and φ is false if this subset relation fails. In the next section we shall see how this definition deviates from the usual definition of truth-values.

As mentioned before, the completeness of the empirical truth of L is an important issue. Although I do not want to run the risk of committing the fallacy of the bandwagon, it is interesting to see that, here, Kuipers takes a minority stance. All authors on verisimilitude but Kuipers assume the truth to be complete. In our toy language this means there is only one L-structure that is a model of τ. According to Kuipers, a verisimilitude proposal defines *actual truthlikeness* if it

---

[5] Clause $2b$ is equivalent to $(\text{Mod}(\tau) \cap \text{Mod}(\varphi) - \text{Mod}(\psi)) = \varnothing$. If τ is complete, $2b$ claims that $\text{Mod}(\tau) \subseteq \text{Mod}(\varphi \wedge \psi)$ or $\text{Mod}(\tau) \subseteq \text{Mod}(\neg\varphi \wedge \psi)$ or $\text{Mod}(\tau) \subseteq \text{Mod}(\neg\varphi \wedge \neg\psi)$. Since τ is complete, this is equivalent to $\text{Mod}(\tau) \not\subseteq \text{Mod}(\varphi \wedge \neg\psi)$.

[6] In the sense of e.g. Suppes (1957) and Sneed (1979).

assumes the truth to be complete (ICR, p. 142), and *nomic* truthlikeness if the truth admits more than one model. He illustrates this idea with electrical switching circuits. As different combinations of open and closed switches in one switching circuit can cause a light bulb to be switched on, a physical theory may be true on different situations in the physical world. In this sense a physical theory picks out the physically possible worlds from the set of all logically possible ones. Thus in the present context of finite propositional languages, according to Kuipers, the paradigmatic case is the one in which more than one *L*-structure is a model of the truth τ, and the truth is therefore *in*complete. Kuipers calls Mod(τ) the set of all physical or *nomic possibilities* and its complement the set of the nomic *impossibilities*. Clearly, in this setting, if a theory φ is false, i.e., Mod(τ) is not a subset of Mod(φ), the intersection of Mod(τ) and Mod(φ) may still be non-empty; Kuipers calls structures in this intersection the *correct* models (ICR, p. 150) or the examples of φ, whereas the *L*-structures of the symmetric difference of Mod(φ) and Mod(τ) are the incorrect models or the counterexample of φ (ICR, p. 137, p. 150).

This idea of correct or incorrect models is closely related to Kuipers' interpretation of 2*b*. Since the second clause is equivalent to Mod(φ)∩Mod(τ) ⊆ Mod(ψ), according to Kuipers the intuitively preferable interpretation of the second clause reads *all correct models of φ are (correct) models of ψ* (ICR, p. 151, 178). Combining the sentence interpretation of the first clause and the model interpretation of the second, Kuipers come to his preferred "dual foundation" (ICR, p. 173) of the Δ-definition, i.e., "'more truthlike' amounts to 'more true consequences and more correct models'" (ICR, p. 137). It turns out that this formulation is misleading and has led to the four problems that are the subject of section 3.1.

2.2. *R(t) and S(t)*

Kuipers and Niiniluoto are two of the few philosophers who, in their endeavor to define the verisimilitude notion, seriously took up the challenge of answering the epistemic question.[7] This question is difficult since we only have very limited knowledge of the whereabouts of the truth. Kuipers formulates his epistemic rule in terms of a *strongest accepted law*, S(t), and the *realized instances* of that law, R(t). He uses both sets to formulate the correct data hypothesis.

> The data at a certain moment *t* (not to be confused with the actual truth!) can be represented as follows. Let *R(t)* indicate the set of realized possibilities up to *t*, i.e., *the accepted instances*, which have to be admitted. ... Up to *t* there will also be some ... accepted laws, which have to be accounted for. On the basis of them, *the strongest accepted law* to be accounted for is the general hypothesis *S(t)* associated with the intersection of the sets constituting the accepted laws. ... In the following ... we shall need the ... *correct data*

> (*CD-)hypothesis R(t)* $\subseteq T \subseteq S(t)$, guaranteeing that $R(t)$ only contains nomic possibilities, and that hypothesis $S(t)$ only excludes nomic impossibilities. (ICR, p. 157)

The nomic possibilities are all conceptual possibilities that are physically possible. Thus, applying this quotation to a traditional toy example, we take the data record reporting that object1, object3 and object7 are made of iron and expand when heated, then they are instances of the accepted law saying that all iron objects expand when heated. In Kuipers' terminology object1, object3 and object7 are elements of $R(t)$, which is a subset of $Mod(\tau)$ ($:= T$) since it is supposed to be true that the objects observed are made of iron and expand when heated. $S(t)$ is the law stating that iron expands when heated, which excludes all nomic impossibilities being all iron objects that do not expand when heated.

After having introduced $R(t)$, $S(t)$ and the CD-hypothesis, Kuipers defines the notion of *successfulness*. Theory ψ is at least as successful as φ regarding $R(t)$ and $S(t)$ iff

(3)   *a.* $Mod(\psi) - S(t) \subseteq Mod(\varphi)\text{-}S(t)$ and
     b. $R(t) - Mod(\psi) \subseteq R(t) - Mod(\varphi)$ (ICR, p. 160)

Note the similarity between (3) and (2). In fact, the conjunction of the CD-hypothesis and the TA-hypothesis (2), stating that ψ is at least as close to the truth as φ, implies, and therefore explains (3). Kuipers calls this implication the *Success Theorem* (ICR, p. 158).

How does the updating of theories evolve through time? Note 16 of chapter 7 in ICR gives a clue to the answer. It extends the success theorem toward two successive points in time. It reads:

> The TA-hypothesis of course also implies that theory Y will always remain at least as successful as theory X in the face of supplementary correct data (for which $R(t) \subseteq R(t') \subseteq T$ and $T \subseteq S(t') \subseteq S(t)$ holds, for t' later than t).

Intuitively, we learn from note 16 that if time advances and new instances of the strongest laws extend the existing set of instances toward $R(t')$, then $R(t)$ is a subset of $R(t')$. If we, in addition, succeed in formulating a (logically) stronger accepted law $S(t')$, the latter is a subset of $S(t)$, since all nomic impossibilities excluded by $S(t)$ also have to be excluded by $S(t')$. In the words of Kuipers "... we assume the idealization that $R(t)$ can only grow in time and that $S(t)$ can only shrink" (ICR, p. 205). These and similar quotations clearly show that Kuipers does not draw any formal distinction between the nomic possibilities, the elements of $Mod(\tau)$, and nomic impossibilities, the elements of $Mod(\neg\tau)$.

As an illustration, I apply the previous ideas about $R(t)$ and $S(t)$ to the case of Kepler's Laws. Let $S(1609)$ refer to the conjunction of Kepler's first and the second law (ellipses of the planets and equal areas), and let $S(1619)$ refer to the conjunction of $S(1609)$ and Kepler's harmonic law of 1619. If the laws are paraphrased in a formal language L, $S(1619)$ implies $S(1609)$. Additionally, consider two intended applications $R(1609)$ and $R(1619)$. The first is the system of the planets, without their moons, and the sun in our solar system, and the second system is the union of $R(1609)$ and the system Jupiter and its moons, and the earth and its moon. Let us finally, for the sake of the argument, agree that $S(1619)$ is not only true on $R(1619)$ – and therefore on $R(1609)$ – but is true *tout court*. It is an easy exercise, then, to put this information in the $R(t)/S(t)$ format of the previous quotation; the answer reads $R(1609) \subseteq R(1619) \subseteq T \subseteq S(1619) \subseteq S(1609)$. Kuipers formulates the interpretation of this formula as follows: "The three mentioned systems belong to the relevant set of intended applications of Kepler's theory and Kepler's three laws are (approximately) true for all three systems. The systems happen to be (partially overlapping) parts of our entire solar system."

The first of these two sentences refers to $R(1619) \subseteq T \subseteq S(1619)$; and the second to $R(1609) \subseteq R(1619)$. Apparently, the interpretation of the two inclusion symbols in the first part differs from the meaning of the inclusion symbol in the second part. The first indicates that $S(1619)$ is true for applications in $R(1619)$, the second shows that $R(1609)$ is a subsystem of $R(1619)$. In addition, $S(1619) \subseteq S(1609)$ means that law $S(1619)$ logically implies the law $S(1609)$. This multiple interpretation of the '$\subseteq$' symbol illustrates the underlying failure in ICR, viz. letting sets of logical models indiscriminately represent laws (S) and their instantiations (R). This issue is the second of the problems addressed in the next section.

3. Applications and Logical Models

In this section I want to elaborate the consequences of the diachronic version of the CD-hypothesis (4) and the way Kuipers' combines this hypothesis with the Δ-definition.

$(4)\ R(t) \subseteq R(t') \subseteq T \subseteq S(t') \subseteq S(t)$, (note 16, chapter 7)

First, I point out four difficulties arising from the formal ICR framework as it stands, then I consider Kuipers' notion of strongly false sentences, and finally I will get to the cause of the troubles.

3.1. *Four Problems*

The first problem with Kuipers' interpretation of the CD-hypothesis concerns a straightforward logical point. In standard logic or model theory is it commonly assumed that various possible worlds cannot hold simultaneously since logical models are *mutually incompatible*. Thus if one possible world holds, no other possible world applies.[8] This, however, is not a feature of the R-models of the structuralist definition. The different elements of $R(t)$ may very well apply simultaneously. For instance, Jupiter and its moons, *and* the earth-moon system, *and* our solar system all validate Kepler's laws; they are not at all mutually incompatible. In this sense, the structures validating a theory are more similar to *consequences* of a theory than to *logical* models of the theory. In fact, the elliptical orbit of Mercury is rather a logical consequence of the first law of Kepler than one of its logical models.

I would like to call the second problem with ICR-interpretation of (4) the ICR-*paradox of logical strength*. Suppose scientists have made theoretical and experimental progress between the points in time t and t', where t precedes t'. Intuitively, then, the strongest accepted law at time t', $S(t')$ is logically stronger than $S(t)$, and $R(t')$ is stronger than $R(t)$. The former is in accordance with the ICR framework; indeed, since $S(t') \subseteq S(t)$ and $S1$ is logically stronger than $S2$, i.e. $S1$ logically implies $S2$, iff "the models of $S1$ form a subset of those of $S2$" (ICR, p. 148). This part of (4) agrees with model theory and common intuitions, since an increase of the logical strength of the strongest law paraphrases a *de*crease of the set of models representing this law. Regarding $R(t)$ and $R(t')$ things are different. Under the same assumption that t precedes t', the empirical evidence on time $R(t)$ is logically at least as weak or even weaker than the evidence available on time t', $R(t')$. Yet, in ICR, an increase of the set of models, $R(t) \subseteq R(t')$, represents an increase of the logical strength of the empirical evidence. Thus, regarding the evidence, an *increase* of the sets of models represents an *increase of the logical strength*. In brief, regarding the $R(t)/S(t)$ methodology of ICR, *increase and decrease* of a set of models both correspond to the increase of logical strength. This is a paradoxical situation, since the models used to represent S and R are the same.

The third problem related to the $R(t)/S(t)$ representation of experiments and laws, is the fact that the linguistic representation of the experiments *logically implies* (!) the strongest accepted universal empirical laws, since (4) implies $R(t) \subseteq S(t)$. As we saw, Kuipers paraphrases $R(t) \subseteq S(t)$ as "$R(t)$ is not in conflict with $S(t)$" (ICR, p. 157). If we take the ICR formalities seriously, and in accordance with the $S1/S2$ quote of the previous paragraph, $R(t) \subseteq S(t)$ not only means that $R(t)$ is not in conflict with $S(t)$, but that $R(t)$ logically implies $S(t)$.[9] Consequently, Tycho Brahe's observations would *logically imply* the laws of Kepler, and if Brahe's observations were true, Kepler's laws, being logically implied by these observations, *cannot* be false. Clearly, this consequence of the ICR framework is

---

[8] Actually, this point was also made by Niiniluoto (1987, pp. 380-382).

[9] Note that in addition, the correct data hypothesis also implies that the *claim* of $R(t)$, viz. $R(t) \subseteq T$, implies the *claim* of $S(t)$, $T \subseteq S(t)$.

in sharp contrast with everything philosophers of science have learned about induction from Hume until the present day.

The paradoxical taxonomy of truth-values in the ICR framework is the fourth problem triggered by the choice to identify experimental outcomes with logical models of the true theory. In standard model theory a sentence is (empirically) true iff it is a logical consequence of the true theory, which is the strongest empirically true sentence of the language. In addition, a theory is empirically false iff it implies the negation of the truth. If the true theory is complete, these definitions render all sentences of L true or false. If the truth is incomplete, any sentence receives the truth-value *indeterminate,* unless it is a consequence of the truth or it implies the negation of the truth. Consequently, a sentence is true iff its negation is false, and a sentence is indeterminate iff its negation is also indeterminate. This is not in line with the ICR framework. The identification of a counterexample with a *logical* model of the language to formulate law S (cf. ICR sect 7.2) makes the standard truth-value taxonomy inappropriate for ICR. Ideally speaking, one counterexample suffices to bestow the truth-value false upon laws and theories. Consequently, if (5) holds for theory φ, the identification mentioned implies that φ is *false*, since φ has a counterexample and $\mathrm{Mod}(\tau) - \mathrm{Mod}(\varphi) \neq \varnothing$.

(5) $\varphi \nvDash \tau$ and $\varphi \vDash \tau$,

In standard model theory under these circumstances theory φ receives the truth-value indeterminate since τ does not imply φ, and φ does not imply $\neg\tau$ and the same holds for $\neg\varphi$. In short, in ICR theory φ is true iff all models of τ are also models of φ, else it is false; consequently, in the ICR framework, besides all model theoretically false sentences, all the indeterminate sentences are also false.

Of course, it is not wrong per se to propose an alternative to Tarski's definition of truth and falsehood. The choice made in ICR, however, leads to a taxonomy of truth-values that I call the *paradox of false hypotheses.* First, let us devote ourselves to some close reading. Consider the following passage on page 148:

> [A] hypothesis is defined as the combination of a subset X of Mp and the (weak) claim that T is a subset of X (i.e., all nomic possibilities satisfy the conditions of X). Hypothesis X is true or false when its claim '$T \subseteq X$' is true or false, respectively.

If we substitute $\mathrm{Mod}(\varphi)$, $\mathrm{Mod}(\neg\varphi)$ and $\mathrm{Mod}(\tau)$ for X, X^c (the complement of X in Mp) and T, respectively, the last quotation clearly leads to paradoxical situations. Let us suppose that hypothesis $\mathrm{Mod}(\varphi)$ (and its claim that $\mathrm{Mod}(\tau) \subseteq \mathrm{Mod}(\varphi)$) is false in the sense that both $\mathrm{Mod}(\varphi) \cap \mathrm{Mod}(\tau)$, and $\mathrm{Mod}(\varphi) \cap \mathrm{Mod}(\tau)^c$ are non-empty (where $\mathrm{Mod}(\tau)^c$ refers to the complement of $\mathrm{Mod}(\varphi)$). Additionally, suppose that $\mathrm{Mod}(\varphi)^c \cap \mathrm{Mod}(\tau)$ and $\mathrm{Mod}(\varphi)^c \cap \mathrm{Mod}(\tau)^c$ are also non-empty such that hypothesis $\mathrm{Mod}(\varphi)^c$ and its claim that $\mathrm{Mod}(\tau) \subseteq \mathrm{Mod}(\varphi)^c$ are false. The claim of hypothesis $\mathrm{Mod}(\varphi)$, then, and the claim of its negation *are both false* (since the complement of a model set Mod(φ) equals Mod($\neg\varphi$)). Thus, according to the last sentence of the quotation, hypothesis Mod(φ) and its negation Mod($\neg\varphi$) are false, and the same holds for φ and $\neg\varphi$.[10]

The conclusion of the present subsection reads as follows: Kuipers' decision to identify relevant observations and experiments with elements of some subset $R(t)$ of Mod(τ) introduces serious conceptual problems into the ICR framework. In the sequel, the notion "dual foundation" will not only refer to the consequence/model interpretation of the Δ-definition but also to the identification of logical models and (new) applications, because of their close relation.

3.2. *Strong Falsehoods*

In this subsection I want to consider Kuipers' notion of a "strongly false" sentence since it helps us to understand Kuipers' interpretation of the Mod(τ) elements. As we have seen in section 2.1, Kuipers favors the dual foundation of the Δ-definition. Nevertheless, he also endeavors to give an overall consequence interpretation of this definition. To interpret clause $2b$ using logical consequences, Kuipers claims about the consequences of the negation of the truth, Cn($\neg\tau$), that "all its non-tautological members are false" and

> all its members transmit their falsehood to their non-tautological consequences, in the same way as members of Cn(τ) transmit their truth to their consequences. It is therefore plausible to call the members of Cn($\neg\tau$) 'strongly false'... (ICR, p. 183)

Consequently, the members of Cn(ψ) $\cap$ Cn($\neg\tau$) are characterized as "the strongly false consequences of ψ," and since clause $2b$ is equivalent to Cn(ψ)$\cap$Cn($\neg\tau$) $\subseteq$ Cn(φ), constraint $2b$ is translated into "all the strongly false consequences of ψ are (strongly false) consequences of φ." This formulation sounds as a reasonable counterpart of the consequence interpretation of $2a$, which read "all true consequences of φ are also consequences of ψ." The strong falsehood paraphrase, however, also leads to a less attractive reading of the Δ-definition. If φ and ψ are both false, the false consequences of φ form a subset of the false consequences of ψ if and only if the same holds for the true consequences of φ and ψ.[11] Therefore, the Δ-definition may also be paraphrased as stating that: $2a$: all the false consequences of φ are false consequences of ψ, and $2b$: all the strongly false consequences of ψ are (strongly false) consequences of φ. The exchange of φ and ψ in the previous reformulation shows the peculiarity of Kuipers' notion of "strongly false" sentences.

---

[10] It is difficult to find examples where this taxonomy of "false sentences" is standard. In group theory, for example, it would mean that φ := "there are exactly two different identity elements" and $\neg\varphi$:= "it is not the case that there are exactly two different identity elements" both are false.

[11] The Tichý/Miller result also holds if the truth is incomplete (Harris 1974).

Obviously, the notion of a "strongly false" sentence deviates drastically from Tarski's truth definition.[12] The term "strongly false" suggests that all "strongly false" sentences are at least false. It turns out, however, that not only some strong falsehoods fail to be false in the Tarskian sense, but *none of them are false*. If we read falsehood and truth in the standard Tarskian sense, the claim that all members of $Cn(\neg\tau)$, "transmit their falsehood to their non-tautological consequences, in the same way as members of $Cn(\tau)$ transmit their truth to their consequences" (p. 183) is flatly wrong. It is the truth preserving character of the deduction rules that guarantees the truth of the consequences of a true statement. These rules of deduction, however, do not preserve falsity as is suggested in the previous quotation as many consequences of a false statement are true (in the sense of Tarski and Kuipers). The non-tautological consequences of $\neg\tau$ are (only) false in Kuipers' sense and fail to have a truth-value in Tarski's sense.

How, then, are we to make sense of this idea of a strong falsehood? Here, again, we must consult the structuralist interpretation of the so-called "correct and incorrect models" of φ, being the φ-models that are, or fail to be models of τ, respectively. From Kuipers' structuralist viewpoint, a statement is called a strong falsehood iff it is true on *all* physical situations (structures) on which $\neg\tau$ is (Tarskian) true. For instance, a sentence proclaiming the rectangularity of Mars' orbit, and that of Jupiter and of Venus etc. is not only false in respect of Mars but also in respect of Jupiter and Venus and all other planets, and is therefore strongly false. It is this interpretation of models as physical structures that explains why consequences of the negation of the truth may be called "strong falsehoods," something that is very difficult to explain if the elements of *Mp* are considered to be models of the language of the theories and laws in question.

3.3. *The Diagnosis*

Let us go back to the start of the seventeenth century and consider the discovery of Australia again, and let us assume that (6) was considered the geographical law $S(1)$.

(6) "Europe is one of the two smallest continents on earth."

All models of (6) are finite linear orderings and the elements of these orderings represent the continents, the first or the second one of which carries the name Europe. After the discovery of Australia, however, all models in which Europe is the smallest continent are dismissed and those models remain in which Australia is the smallest and Europe is the second smallest continent. Scientists were left with the law $S(2)$.

---

[12] Moreover, Fig 6.1 of ICR suggests that, even if the truth is *incomplete*, all sentences that are not true are false. Consequently, all empirical sentences implying the truth *and their negations* are supposed to be false.

(7) "Australia is the smallest, and Europe is the smallest but one continent
    on earth."

Clearly, (7) logically implies (6) and $S(2)$ is a subset of $S(1)$. In accordance with
the ICR framework, the models dismissed are taken to be conceptual possibilities
that are rightly excluded from the set of nomic possibilities. In another sense,
Australia is an application of the geographical laws such as (6) and (7). It is a
member of the empirical evidence of geography and therefore, according to the
ICR framework, needs to be a member of $R(2)$. Consequently, if $R(2)$ is a subset
of $\mathrm{Mod}(\tau) \subseteq S(2)$, Australia has to be of the same kind as the logical models of
$S(2)$, and this last step epitomizes the basic failure of the ICR framework.
Australia as a physical model clearly should not be identified with the logical
models of the established geographical laws or theories, since the models of the
latter are finite orderings of elements and not the (descriptions of) continents. The
conclusion therefore reads that $R(t)$ should not be framed as a subset of T.

Identifying the applications of a theory with its logical models, the
structuralist verisimilitude definition seems to have fallen victim to the notorious
ambiguity of the term 'model'. In the way the 'dual foundation' sounds plausible,
the term 'model' refers to physical models to which the relevant theory is applied.
In this way the moons of Jupiter are models for the laws of Kepler, or Australia is
a model for the geographical theories. The term 'model' in clause $2b$, however, is
unrelated to the fact that Australia validates some geographical laws. Clause $2b$
models are *logical models*, being linear orderings, and not *physical models*, which
are structures in the real world falsifying or confirming theories. The four
problems of section 3.1 can be avoided if the applications of the theories are not
identified with the logical models of the language used to formulate the laws, $S(t)$,
and the 'strongly false' notion may be dismissed.

Kuipers' ideas about theories picking out the physically possible models from
all logically possible ones cannot consistently be paraphrased in the basic ICR
framework using standard model theory. They can, however, be reformulated in
various different ways. One of them is to use a richer formal framework. In my
(2001), I showed how to reformulate Kuipers' ideas about nomic possibilities
using modal logic. In the next section, I show how the method of examples and
counterexamples can also be formulated extensionally using partial models.[13]

---

[13] The two-set view of theories is not the source of my misgivings, witness Zwart (2001, section 2.6); it
is the identification of the logical models of the laws and their instantiations which causes the trouble.

4. The Partial Models Account of the HD-Method

In ICR, Kuipers' account of the HD-method is intimately connected with his formulation of the comparative success of theories in terms of $R(t)/S(t)$, and the identification of applications and logical models (or the dual foundation interpretation). On page 187 we read:

> In sum, $R(t)$-X and $Q(S(t))\cap Q(X)$ summarize the results of HD-evaluation of theory X and reflect the intuitively appealing dual foundation of HD-results. These HD-results can be compared with the results of another theory, which may lead to theory choice on the basis of the appropriate rule of success.

Since Kuipers defines $Q(X)$ to be the set of all supersets of X, $Q(S(t))\cap Q(X)$ is uniquely determined by the set X-$S(t)$ (ICR, p.159). Thus, the HD-evaluation report of a theory is summarized by the same sets $R(t)$-$\text{Mod}(\varphi)$ and $\text{Mod}(\varphi)$-$S(t)$ used to evaluate the comparative success of theories. Somewhat above the previous quote on page 187 we read:

> The evaluation report of X can now be summarized by $R(t)$-X, indicating the set of all problems of X, reinterpreted as 'established wrongly missing models', and by $Q(S(t))\cap Q(X)$, indicating the set of all successes of X, reinterpreted as 'established laws' or, more neutrally, 'established true consequences'.

The formal ICR account of the HD-method is inadequate as it is fundamentally connected to the dual foundation – i.e., with the identification of the intended applications with the logical models of the laws. In the present section I will put forward a model theoretic account of the HD-method in terms of *partial models*. To keep things as simple as possible, my account of the HD-method is based on finite monadic predicate languages used to formulate the theories and empirical data, and the difference between laws and theories is kept out of consideration.

4.1. *Partial Models*

Before introducing my model theoretic account of the HD-method, I show how the discovery-of-Australia paradox can be solved using partial models. The key to understanding this paradox is the fundamental *open-endedness* of the logical models of the language used to express the theory. These models may permit or prohibit extensions needed to express the increase of our knowledge. Take, for example, the discovery of Australia. Some logical models of geographical theories before 1606 excluded the possibility of more than four continents. After 1606, they had to be rejected since they could not serve as a model for the true sentence stating that there were at least five continents. Other logical models before 1606 left open the possibility of more than five continents, and had to be extended in the sense that the referent of the constant name Australia had to be added. Consequently, new developments go hand in hand with an *increase* as well as *decrease* of the set of models. The decrease consists in the rejection of those former models excluding the possible extension needed to formulate the new theories, and the increase consists in the appropriate extension of the remaining models.

Kepler's example also shows how rejection of old models and extension of the remaining ones combine in a logical reconstruction of scientific developments. Let us go back to 1609, and assume that astronomy covered Kepler's law of the elliptical planetary trajectories and the equal area law, applying them to all planets in our solar system. Furthermore, let us suppose that, ten years later, the harmonic law was added to astronomical theory and the three laws of Kepler were applied to all objects in our solar system, the moons of Jupiter and the moon of the earth included. In a reconstruction of these events, many logical models do not survive these developments. On the level of the laws, the logical models that do not allow for Kepler's harmonic law are dropped, and additionally, those models are dismissed that do not allow for the appropriate referents of the moons of Jupiter and the earth. Besides this decrease in the number of models, however, the remaining models have to be extended with the referents of the moons of the Jupiter and the earth, since these were the new applications of the astronomical theory. Once again, the identification of the applications with the logical models of the language used to formulate the theory would create paradoxical situations.[14]

4.2. *Model Theory of the HD-Method*

To examine the HD-method let us consider a monadic predicate language L with nonlogical (and observational) vocabulary $\{Ax, Sx, Wx, a, b, c, d\}$. The four constants refer to four individual birds living in the last decade of the seventeenth century, and the three predicates refer to "being Australian," "being a Swan" and "being white," respectively. Let us suppose that at the start of the decade just mentioned, the following truthful empirical evidence is available. Bird a is a white swan spotted in France, and b is a white swan found in America. Furthermore, assume X is a group of rather daring biologists who, on the basis of this evidence, jump to the conclusion that all swans are white:

$$(8)\ \varphi := \forall x(Sx \rightarrow Wx).^{15}$$

Model theoretically this means that according to the X-biologists all logical models verifying all true sentences are φ-models. In other words, by claiming that

---

[14] In my opinion, using partial logic to define $R(t)$ and $S(t)$ – see the contribution by J. van Benthem – does not solve the problems either. If $R(t) := A^+$, $S(t) := -A^-$ and an increase of model sets comes down to an increase of logical strength, then updating causes the decrease of the logical strength of $S(t)$ ($-A^-$ is the complement of A^-). The same problem holds for $R(t)$ if an increase of a set of models means a decrease of logical strength.

[15] In this example I consider being white not to be an essential property of being a swan.

φ is true, the *X*-biologists say there is no way to empirically falsify φ; thus it is physically impossible to find an environment where there are nonwhite swans. Consequently, in all φ-models, all individuals in the extension of swans belong to the extension of white things.

Imagine, then, some group *Y* of more cautious biologists. Realizing an entire new continent has been discovered, they believe only that all non-Australian swans are white:

$$(9) \quad \psi := \forall x((Sx \wedge \neg Ax) \to Wx).$$

Clearly, φ implies ψ and all models of φ are models of ψ. Notice that all *L*-structures $\mathfrak{M}$ are models of ψ iff all swan individuals are white or Australian (or both). Suppose, additionally, that the strongest truth of the matter in *L* reads as follows:

$$(10) \quad \tau := \forall x((Sx \wedge \neg Ax) \to Wx) \wedge \exists x((Sx \wedge Ax \wedge Wx) \wedge \exists x((Sx \wedge Ax \wedge \neg Wx)$$

According to τ there are white and nonwhite swans in Australia, and all non-Australian swans are white. As ψ is a consequence of τ, it is an empirically true consequence of φ. If we disregard our present knowledge, how are we to picture the 1695 state of affairs, where *a* and *b* are white swans, and the difference of opinion between the daring and the cautious biologists (assuming that the observational evidence is true and leaving the true strongest empirical law out of the consideration)?

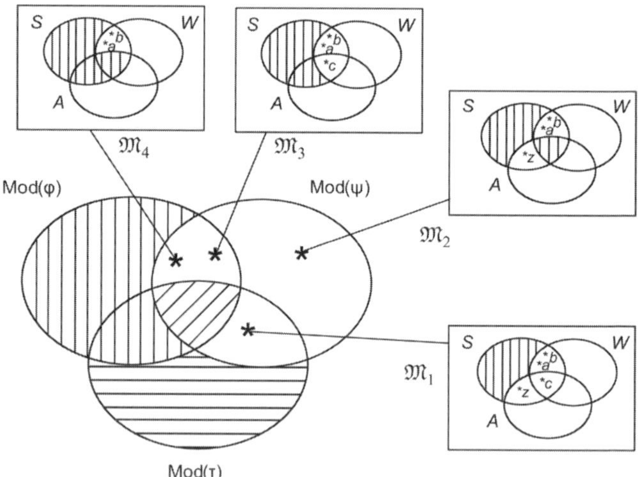

*Fig.*1. The four "model-schemes" of φ, ψ and τ

Figure 1 is the model theoretic image of the information just described. It pictures the model sets of φ, ψ and τ such that $\text{Mod}(\varphi)$ and $\text{Mod}(\tau)$ are subsets of $\text{Mod}(\psi)$. Furthermore, the models of φ are open-ended regarding the swans in Australia but they exclude the existence of swans that are not white. The models of the truth are open-ended regarding white non-swans inside or outside Australia, and nonwhite things in or outside Australia. Theory ψ leaves all those conceptual possibilities open. It does not even exclude the situation in which all Australian swans are nonwhite (model $\mathfrak{M}_2$). Let us return to the differences of opinion between the two imaginary groups of biologists at the end of the seventeenth century who were, to be sure, ignorant of the true theory.

Since the cautious and the daring biologists had different opinions about theory φ, as good scientists they started to search in all possible regions, well known for their overwhelming abundance of various species of birds, trying to falsify φ. Suppose, notwithstanding their efforts, they failed to falsify φ since all swans they encountered, in and outside Australia, turned out to be white and many biologists became convinced of the truth of φ. How are we to present these developments model theoretically? All newly discovered white swans are represented by objects in the intersection of S and W as in $\mathfrak{M}_3$; the Australian ones in the a, b-region, the non-Australian ones in the c-part. Note that, in contrast to the representation in the ICR framework, the new individuals are *not models themselves*, and $\mathfrak{M}_4$ does not have the required open-endedness to survive the new observations, although φ is still alive. In addition, the findings strengthened some biologists their belief that in all $\mathfrak{M}$ of $\text{Mod}(\tau)$, the extension of Sx is a subset of the extension of Wx as is the case in $\mathfrak{M}_3$ and $\mathfrak{M}_4$.

In 1697 things changed drastically because Willem de Vlamingh found a black swan in Australia. How are we to present this counterexample of φ in Figure 1? Since φ turned out to be false, its negation is true and $\text{Mod}(\tau)\text{-Mod}(\varphi)$ is not empty. An interesting question, then, is whether there may still be models in the intersection of $\text{Mod}(\varphi)$ and $\text{Mod}(\tau)$. If $\text{Mod}(\varphi)\cap\text{Mod}(\tau)$ is nonempty, then not all models of τ are models of all true sentences since the structures that are model for φ and τ fail to be a model of the negation of φ which is true. This contradicts the assumption that all elements of $\text{Mod}(\tau)$ must verify all empirically true sentences, and therefore $\text{Mod}(\varphi) \cap \text{Mod}(\tau)$ has to be empty, and $\text{Mod}(\tau)$ is a subset of $\text{Mod}(\neg\varphi)$. The black swan found by de Vlamingh has to be presented as a swan-individual that is not white and lives in Australia. The element z exemplifies such an individual.

The foregoing example and its model theoretical picture clearly show that falsification of a deterministic law is an *all-or-nothing affair*. If a deterministic law φ is true, *all* the models of the possibly incomplete truth are in $\text{Mod}(\varphi)$, and if it is false *none* are. By no means can we present this situation of one falsifying instance as one nonwhite swan flapping about in the intersection of $\text{Mod}(\tau)$ and

Mod($\neg\varphi$) region whereas all the confirming instances reside in the adjacent Mod(φ) $\cap$ Mod(τ) region as is done in the ICR account of falsification. If φ is true, Mod(τ) is a subset of Mod(φ), and if φ has one counterexample it is false, and Mod(τ) is a subset of Mod($\neg\varphi$) and Mod(τ) $\cap$ Mod(φ) is empty. It is impossible, therefore, to identify a counterexample of a theory φ with an entire model of the true theory that fails to be a model of the theory φ. Model theoretically a φ-counterexample is framed as the extension of previously existing $\neg\varphi$-models that did not preclude this required addition.

4.3. *Crucial Experiments*

Although Kuipers explains some methodological aspects of crucial experiments, he does not go into the formal details (see, e.g., ICR, Subsection 7.5.2). In the present section, I show that regarding those experiments, too, the ICR-identification of applications and logical models leads to trouble. Let me describe the standard situation of a crucial experiment. A crucial experiment is an experiment used to decide between two rival theories. The theories agree on at least one experimental outcome e; e.g., regarding Newton's corpuscular and Huygens' wave theory of light, e might be the refraction of light. A crucial experiment is a prediction about some (pseudo) observable phenomenon e' about which the two rival theories disagree; one theory predicts e' and the other $\neg(e')$. In the theories-of-light example, e' is the prediction of the velocity of light in water being greater than in air. In this specific example, Newton's theory of light turned out to be false as Fizeau showed $\neg(e')$ to be true.

What would be the outline of an HD-evaluation report in $R(t)/S(t)$ terms? First, the reliable experiment e on which the consistent theories φ and ψ agree is an individual success of both theories and has to be an element of $R(t) \subseteq$ Mod(τ) and Mod(φ) $\cap$ Mod(ψ). Next, if e' is the situation in which φ is true, and $\neg(e')$ is the situation in which ψ is true, e' is an element of Mod(φ)-Mod(ψ), and $\neg(e')$ is an element of Mod(ψ)-Mod(φ).[16] These outlines of the HD-evaluation report are already problematic since two theories agreeing on some prediction e and disagreeing on another prediction e' cannot have any (logical) model in common. Since e' and $\neg(e')$ are consequences of the conjunction of φ and ψ, this conjunction is inconsistent and therefore the intersection of Mod(φ) and Mod(ψ) is empty. Consequently, if one identifies the applicational situations of theories with their logical models, and theories agree on some situation (e.g., the refraction of light), then the conjunction of the theories needs to be consistent. The theories cannot disagree on any prediction. This is just another undesirable consequence of the identification mentioned.

---

[16] Again, in the context of finite monadic predicate languages X, Y, and T are equivalent to Mod(φ), Mod(ψ) and Mod(τ), respectively.

The foregoing analysis leaves us with the next question: How do we describe, in standard model theoretical terms, the situation in which theories agree on true evidence e but disagree on some prediction e'? This is where the notion of partial models enters again. If two theories φ and ψ agree about e, e.g., the refraction of light, then e is not a logical model of φ or ψ, but e can at most be seen as related to a *part* of such a model, that part which is necessary for the verification of e.[17] Thus, as prediction e turns out to be true, it can be taken to be a common part of the models of the mutually exclusive theories φ, ψ, and τ. If the theories φ and ψ disagree about another prediction e', then, for instance, the φ-models are extended so as to verify prediction e', and therefore *cannot be extended* to verify $\neg(e')$. On the contrary, the ψ-models are extended such that they verify $\neg(e')$, prohibiting an e'-verifying extension. Theories disagreeing on observational predictions cannot therefore have any logical model in common. To consider the most simple illustration of a crucial experiment situation, let $\tau := p \wedge q \wedge r$, $\varphi := p \wedge \neg q$, and $\psi := p \wedge q$. Theories $Cn(\varphi)$ and $Cn(\psi)$ agree to explain p, where p is already accepted evidence and it is the "partial model" common to the models of φ, ψ and τ; theories $Cn(\varphi)$ and $Cn(\psi)$, however, do not have any model in common as they disagree about the prediction q. At the very moment that prediction q turns out to be true, it is known that $Mod(\varphi) \cap Mod(\tau) = \varnothing$ and that $Mod(\tau)$ may still be a subset of $Mod(\psi)$.

5. Summary and Prospects

In this paper we have seen that the Δ-definition of Miller and Kuipers is a sound verisimilitude definition. In answering the epistemic question of verisimilitude, however, Kuipers proposes to treat the elements of *Mp*, on the one hand, as the *logical models* used to formulate the strongest established laws, $S(t)$, and, on the other hand, as *physical models* of theories, which are situations in which the theories may be true or false. I have mentioned four problems connected with the use of models in the ICR framework. First, in contrast to applications of a theory, logical models are mutually incompatible; second, an increase of logical strength is represented as an *increase and as a decrease* of a set of models. This is what I have called the ICR *paradox of logical strength*; third, the combination of the correct data hypotheses with the formal rules of ICR leads to the conclusion of the individual evidence *logically implying* the strongest established empirical law; fourth, it becomes conceptually possible for a hypothesis *and its negation* both to be false. I used the ICR notion of "strongly false" sentences as a lever to get to

---

[17] Note that when updating theories we lack knowledge about sufficient conditions since they are logically stronger than, and would replace the theories considered. Moreover, in regard to crucial experiments all sufficient conditions are inconsistent!

grips with the source of the problems mentioned. According to my diagnosis, the source of the difficulties in ICR is the identification of the logical models and the physical models. The four problems mentioned and my diagnosis underpin the following conclusion. In the sense of the example of the discovery of Australia, Kuipers should not identify the newly invented applications of the theory with the logical models of the language used to formulate the laws, $S(t)$, or theories; a correct representation uses partial models to paraphrase the applications of the theories. In addition, I have shown that the increase and decrease of our knowledge paradox of the discovery of Australia also can be disentangled using partial models. By identifying the applications with partial models, while formulating the theories using logical models, we can give a model theoretically sound representation of the HD-method and crucial experiments.

This leaves us with the question of how we are to answer the epistemological question of verisimilitude if the ICR endeavor does not seem viable. Within the context of my refined verisimilitude proposal (Zwart 2001), to answer the epistemic question I have used insights in the theory of belief revision, especially the orderings of epistemic entrenchments. I shall publish my finding in this area elsewhere.

ACKNOWLEDGMENTS

The present paper is partly subsidized by the Grotius Project of the University of Amsterdam; its support is hereby gratefully acknowledged.

Delft University of Technology
Faculty of Technology, Policy and Management
Postbox 5015, 2600 GA Delft
The Netherlands
e-mail: s.d.zwart@tbm.tudelft.nl

REFERENCES

Balzer, W., C. Ulises Moulines and J. Sneed (1987). *An Architectonic for Science*. Dordrecht: Reidel.

Chang, C.C. and H. J. Keisler (1973). *Model Theory*. Amsterdam, London: North-Holland Publishing Company.

Harris, J.H. (1974). Popper's Definition of "Verisimilitude". *The British Journal for the Philosophy of Science* **25**, 160-166.

Kuipers, T.A.F. (1982). Approaching Descriptive and Theoretical Truth. *Erkenntnis* **18**, 343-387.

Kuipers, T.A.F. (2001). *From Instrumentalism to Constructive Realism*. Dordrecht, Boston, London: Kluwer Academic Publishers.

Miller, D. (1978). On Distance from the Truth as a True Distance. In: *Essays on Mathematical and Philosophical Logic*, pp. 415-435. Dordrecht: Reidel.

Sneed, J. D. (1979). *The Logical Structure of Mathematical Physics*. Revised second edition. Dordrecht: Reidel.

Suppes, P. (1957). *Introduction to Logic*. Princeton, New York, Toronto, London: D. van Nostrand Company.

Zwart, S.D. (2001). *Refined Verisimilitude*. Dordrecht, Boston, London: Kluwer Academic Publishers.

Theo A. F. Kuipers

ONE VERSUS MANY INTENDED APPLICATIONS
REPLY TO SJOERD ZWART

I am grateful to Sjoerd Zwart for the fact that he elaborated for the present purposes a point of view already present in Zwart (1998/2001), that may be representative for many logicians dealing with problems in the philosophy of science. His main claim is that I am mixing up what he calls "logical models" and "physical models." From my point of view, however, the main divergence is that between the (dominant) model theoretic and the structuralist perspective, according to which the target of theorizing is one particular "intended application" and a set of "intended applications," respectively. Although the suggested model theoretic perspective may be dominant, I would like to leave room for the possibility that an alternative model theory will be further developed and become respected, not to replace the present dominant one but as an alternative that is more suitable for certain purposes. However, contrary to Zwart's suggestion, I do not see this alternative as a non-Tarskian move in some deep sense. Starting from Tarski's basic definition, which is that of "truth in a structure" (Hodges 1986), there are at least two coherent ways of defining that a theory is true or false, depending on whether one has one or more than one intended application in mind. In this reply I will refrain from trying to give a response to every detail of Zwart's account. Instead, by focusing on some of his main points and by starting the elaboration of the ICR approach to domain revision, I wish to show that the suggested alternative model theory makes sense. Moreover, readers will be able to form their own opinion about whether Zwart's exposition of the dominant approach applied to a couple of examples, in particular in Section 4, is illuminating or, which is my opinion, functions as a Procrustean bed, even for examples that do not seem to be very representative of scientific theories and domains, let alone for examples that are representative, such as theories about planetary systems.

In: R. Festa, A. Aliseda and J. Peijnenburg (eds.), *Confirmation, Empirical Progress, and Truth Approximation* (*Poznań Studies in the Philosophy of the Sciences and the Humanities,* vol. 83), pp. 396-402. Amsterdam/New York, NY: Rodopi, 2005.

Two Perspectives on Four Problems and "Strong Falsehoods"

Since Zwart in his Section 2 gives an adequate summary of some crucial elements of my approach in ICR, I start immediately by responding to the four problems set out in his Section 3.1 and his questioning of my definition of "strong falsehoods" in Section 3.2.

Like Niiniluoto, Zwart objects to my earlier talking about "possible worlds" that are compatible in some sense, which is indeed problematic as long as one has one target "possible world" in mind. To avoid this connotation, I systematically talk in ICR about 'conceptual possibilities' instead. This is a general term, which may in specific contexts refer to possible kinds, states, systems, etc.

By calling it the "ICR-paradox of logical strength," the second problem may seem more serious. Representing the "realized possibilities" at t and t', later than t, by $R(t)$ and $R(t')$, and assuming that some (in a certain sense) new experiments have been performed between t and t', that is, some new possibilities have been realized, it follows that $R(t)$ is a proper subset of $R(t')$. The paradoxical air arises from the fact that, although the evidence has increased in some sense, the linguistic representation of $R(t')$ is weaker than that of $R(t)$. The point is, of course, that these linguistic representations merely represent the theories of $R(t)$ and $R(t')$, respectively, in the logical sense of the set of sentences that are true on all their respective members. On the other hand, the evidence has increased in the sense that the relevant claim "$R(t')$ is a subset of T" is stronger than "$R(t)$ is a subset of T" (where T represents the target set of nomic possibilities). I fail to see this as a paradox, let alone an interesting one.

Zwart's third problem amounts to the fact that "the linguistic representation of the experiments $[R(t)]$ *logically implies* (!) the strongest accepted universal empirical law $[S(t)]$" (p. 383), due to the fact that $R(t)$ is a subset of $S(t)$. Again there is no problem if we look carefully. Indeed, the linguistic representation of $R(t)$ logically entails the linguistic representation of $S(t)$. However, the claim associated with $R(t)$, viz. "$R(t)$ is a subset of T", does not at all entail the claim associated with $S(t)$, i.e., that it represents a law, viz. "T is a subset of $S(t)$". Following Ruttkamp (2002, and her contribution to the companion volume), I like to call the first observation "the problem of *over*determination" of theories by data. As soon as we have performed some experiments, represented by $R(t)$, the linguistic representation of the latter entails an enormous number of theories. Only some of them will be true, in the sense of true of all members of T, and, under certain conditions, just one will be the strongest. The task of further experimenting and theorizing is to zero in on true ones or the strongest true one.

Although my distinction between the sets of conceptual possibilities $R(t)$ and $S(t)$ and the *claims associated with them* seems perfectly clear in ICR, a symbolic distinction may be useful for some specific purposes. With thanks to Roberto

Festa, I mention the following plausible symbolization of the claims. The claims associated with $R(t)$ and $S(t)$ may be denoted, in a transparent way, by $r(t)$ and $s(t)$. Hence, $r(t)$ and $s(t)$ amount to, respectively:

$$r(t) \equiv R(t) \subseteq T \qquad s(t) \equiv T \subseteq S(t)$$

If one likes one can also express the data available at t by: $d(t) \equiv r(t)\ \&\ s(t)$.

From the above explicit definitions and from the following "temporal assumption" about the monotonic nature of data:

$$R(t) \subset R(t')\ \&\ S(t') \subset S(t)$$

it immediately follows that $r(t')$ implies $r(t)$ (but not vice versa) and that $s(t')$ implies $s(t)$ (but not vice versa). Hence $d(t')$ implies $d(t)$ (but not vice versa). Of course, as Zwart will not dispute, it also immediately follows from the definitions that $r(t)$ does not imply $s(t)$.

Although essentially terminological, I find the fourth problem mentioned by Zwart, that is, my non-standard definition of a theory being false, the most interesting one. It stimulated me to explicitly talk about an "alternative model theory" in the introduction of this reply (and in the introduction of Part III of the ICR synopsis in this volume), for it enables me to indicate clearly where the dominant and the alternative model theory deviate. For further motivation of the following, see also the section "Some reflections on the definition of a 'false theory'" in my reply to Burger and Heidema, which was written before the present reply.

I assume Tarski's definition of a sentence, e.g. a (finitely) axiomatized theory, being either true or false in (or on) a structure of the relevant language. From the point of view of one target intended application, that is, one target structure, say t, it is plausible to call a theory true or false depending on whether it is true or false on that structure. Since every sentence gets a truth-value on a structure and assuming that the linguistic representation of t, that is, the strongest true theory, is axiomatizable, it will be a complete theory. As long as the target structure is, for some reason or other, not yet clearly determined, but only known to belong to a certain set of structures, say Z, it is plausible to call a theory true when it is true on all Z-structures, false when it is false on all Z-structures, and indeterminate otherwise. By zeroing in from Z to t, all true/false-judgements remain the same, and the judgements 'indeterminate' can be replaced by either 'true' or 'false'.

However, when the target is a fixed set of structures, T, these definitions are no longer plausible. Then it becomes plausible to call a theory true ("as a hypothesis," ICR, p. 184) when it is true on all T-structures, and false otherwise. The reason for the latter is that there apparently exists a counterexample to the claim that the theory is true on all T-structures. Note that this definition not only makes sense for theories in the empirical sciences, e.g. about planetary systems, but also for certain kinds of mathematical claims. For example, in group theory one says that a purported theorem is true when it has been proved on the basis of the definition of a group and false when an appropriate (type) of counterexample has been provided. Of course, there is nothing mysterious about the fact that, *according to this definition*, a theory and its negation can both be false. However, a theory and its negation cannot both be true.

This brings me immediately to the notion of "strong falsehoods." The rationale of this terminology also presupposes the alternative perspective. As a matter of fact there are two plausible definitions. Although the first is even more plausible than the second, I restricted attention in ICR to the second because that permitted an adapted "consequence reading" of the relevant truthlikeness clause. But let me start with the first definition. A theory is strongly false in the first sense (sf1) if it is not merely false in the sense of being false on *some T*-structure, but being false on all *T*-structures. A theory is strongly false in the second sense (sf2) if it is non-tautological and true on all *non-T*-structures or, equivalently, not merely false (in the sense of being false on some *T*-structure), but in addition true on all *non-T*-structures. From the dominant perspective both definitions make little (new) sense. For the first holds that, whether or not the target structure is already fully fixed, all and only false theories, that is, false in the dominant sense, would become sf1. In other words, sf1 coincides with false in the dominant sense. For the second holds that the definition is empty for a fixed target structure; and only indeterminate theories become sf2, viz. those theories that are only false on some Z-structures. Of course, as Zwart clearly illustrates, if you mix up two sets of definitions it is plausible that you may get queer results. For example, sf2-theories are not false in the dominant sense, but indeterminate; however, as indicated, they are not merely indeterminate.

Truth Approximation by Revision of the Domain of Intended Applications

I would like to take the opportunity to elaborate a point that was put forward by Sjoerd Zwart earlier (1998/2001). Let me start by quoting from ICR (p. 207).

> Finally, variable domains can also be taken into account, where the main changes concern extensions and restrictions. We will not study this issue, but see (Zwart 1998[/2001], Ch. 2-4) for some illuminating elaborations in this connection, among other things, the way in which strengthening/weakening of a theory and extending/reducing its domain interact.

More specifically, I propose a coherent set of definitions of 'more truthlikeness', 'empirical progress' and 'truth approximation' due to a revision of the domain of intended applications. This set of definitions seems to be the natural counterpart of the basic definitions of similar notions in ICR as far as theory revision is concerned. Regarding theory revision, there will be some overlap with Zwart's contribution.

Assuming the distinction between a vocabulary V and a *sub*vocabulary Vd as "domain vocabulary," a distinction that was only made explicit in the last chapter of ICR, we may also assume that the domain of intended applications can be represented as a well-defined subset D of the set of conceptual possibilities (potential models) $Mp(Vd)$ generated by Vd. Leaving out the distinction between theoretical and observational terms, the target set of structures $T(D)$ is the set of nomic possibilities corresponding to D within the set of conceptual possibilities $Mp(V)$ generated by V.

For subsets X and Y of $Mp(V)$ representing theories the basic definition for "Y is at least as close to T as X" in ICR (cf. (2) in Zwart's contribution; see also Miller's evaluation of the basic definition, his own definition and some other explications of verisimilitude) amounts to

(a) $Y-T \subseteq X-T$
(b) $T-Y \subseteq T-X$

With a plausible strengthening to capture 'closer to', viz. by requiring a *proper* subset relation at least once, this results in 'more truthlikeness due to theory revision', keeping (the vocabularies and) the domain fixed. Note that this definition already makes formal sense without the specific interpretation of X, Y and T.

However, we may also consider what happens when we change the domain and fix the theory. The following definition is then plausible. For domains $D1$ and $D2$, and $T(D1)=T1$ and $T(D2)=T2$, "X is at least as close to $T2$ as to $T1$" iff "$T2$ is at least as close to X as $T1$"

(a') $T2-X \subseteq T1-X$
(b') $X-T2 \subseteq X-T1$

again with a plausible strengthening to capture 'closer to', now resulting in 'more truthlikeness due to domain revision'.

When drawing plausible pictures, with X fixed, $T1$ replacing T and $T2$ replacing Y, it is interesting to see that the revisions go in the opposite direction: whereas Y approaches T, starting from X, $T2$ approaches X, starting from $T1$. The divergent graphic representation of the proper subset requirements reflects this in particular. The same switch of direction is relevant for design research. Whereas a new drug may be better for a certain disease than an old one, a certain drug may be better for another disease than the original target disease, a phenomenon which was nicely captured by the title of a study by Rein Vos (1991): "Drugs looking for diseases."

Let us turn to the notion of empirical progress. In ICR the basic definition of 'empirical progress by theory revision' is based on the subsets R and S of $Mp(V)$, representing the set of realized possibilities of D (at a certain moment) and the strongest law about D (inductively) based on R, respectively. The minimal condition for empirical progress, viz. "*Y* is at least as successful wrt *R/S* as *X*", amounts (in)formally to the following two conditions":

(a-*R*) $R \cap X \subseteq R \cap Y$: all established examples of *X* are examples of *Y*
(b-*S*) $S \cup X \supseteq S \cup Y$: all established successes of *X* are successes of *Y*

Whereas the paraphrase of the first formal clause will be clear enough, that of the second may need some explication. A superset of $S \cup X$ represents a law, for it is entailed by *S*, which is also entailed by *X*, hence a success of *X*. The formal clause guarantees that such a success is also a success of *Y*.

Crucial in ICR is the Success Theorem, according to which "being closer to the truth" entails "being at least as successful," assuming the correct data hypotheses (CD): $R \subseteq T \subseteq S$. This theorem provides several reasons (ICR, p. 62) for claiming that empirical progress by theory revision is functional for truth approximation (by theory revision).

Turning to domain revision, let *R*1 and *R*2 represent the relevant sets of realized possibilities of the corresponding domains *D*1 and *D*2, and *S*1 and *S*2 the corresponding strongest laws. Now it is plausible to define "*X* is at least as successful wrt *R*2/*S*2 than wrt *R*1/*S*1" or "*X* is at least as successful wrt *D*2 than wrt *D*1 (as far as the available data are concerned)" iff

(a'-*R*) $R1 \cap X \subseteq R2 \cap X$: all established *R*1-examples of *X* are *R*2-examples
(b'-*S*) $S1 \cup X \supseteq S2 \cup X$: all established *S*1-successes of *X* are *S*2-successes

As a counterpart to the Success Theorem, regarding theory revision, we would now like to prove the following

Success Theorem, regarding domain revision:
"*X* is closer to *T*2 than to *T*1" entails
"*X* is at least as successful wrt *R*2/*S*2 than wrt *R*1/*S*1"

To prove this theorem we again need, of course, the relevant CD-hypotheses, viz. $R1 \subseteq T1 \subseteq S1$ and $R2 \subseteq T2 \subseteq S2$. But we need some more, again plausible, relational assumptions. If a realized possibility of *D*1 also is one of *D*2, it is recognized as such. Formally: $R1 \cap T2 \subseteq R2$, and of course vice versa: $R2 \cap T1 \subseteq R1$. Similarly, a law entailed by *S*1 that also holds for *D*2, is recognized as such, that is, it is also entailed by *S*2. Formally: $S1 \cup T2 \supseteq S2$, and vice versa $S2 \cup T1 \supseteq S1$.

It is easy to check that the latter assumptions are implied when we start from sets *R* and *S* relating to a domain *D* covering *D*1 and *D*2, and define $R1 = R \cap T1$, $R2 = R \cap T2$ and $S1 = S \cup T1$, $S2 = S \cup T2$. These definitions amount to the assumptions that, for example, *R*1 captures all realized possibilities of *D*1, and *S*1 the strongest established law about *D*1.

The consequence of the theorem is that 'empirical progress by domain revision' is functional for 'truth approximation by domain revision' in a similar way as the corresponding type of theory revision.

Turning to the quote from ICR at the beginning of this subsection, the result of all this is that a theory may indeed come closer to the (relevant) truth by either theory revision in the form of strengthening or weakening a theory and by domain revision in the form of extending or reducing the domain. Moreover, a theory may come closer to the (relevant) truth by a combination of such revisions. Similar conclusions apply for the possibility for a theory to become more successful. Due to the two success theorems the latter will be functional for truth approximation, but there is, of course, no guarantee.

REFERENCES

Hodges, W. (1986). Truth in a Structure. *Proceedings of the Aristotelian Society, New Series* **86**, 135-151.

Ruttkamp, E. (2002). *A Model-Theoretic Realist Interpretation of Science*. Dordrecht: Kluwer Academic Publishers.

Vos, R. (1991). *Drugs Looking for Diseases*. Dordrecht: Kluwer Academic Publishers.

Zwart, S. (1998/2001). *Approach to The Truth. Verisimilitude and Truthlikeness*. Dissertation Groningen. Amsterdam: ILLC Dissertation Series 1998-02. Revised version: *Refined Verisimilitude, Synthese Library*, vol. 307. Dordrecht: Kluwer Academic Publishers.

Johan van Benthem

A NOTE ON MODELING THEORIES

ABSTRACT. We discuss formats for formal theories, from sets of models to more complex constructs with an epistemic slant, clarifying the issue of what it means to update a theory. Using properties of verisimilitude as a lead, we also provide some connections between formal calculus of theories in the philosophy of science and modal-epistemic logics. Throughout, we use this case study as a platform for discussing more general connections between logic and general methodology.

1. Logic and Methodology of Science

The border-line between logic and methodology of science is often hard to detect. Topics straddling it include the proper representation of information and patterns of reasoning. Theo Kuipers' ongoing work on theory structure, confirmation, truth-likeness, and theory change lies at this interface. It uses methods from logic, but also, logicians would be well-advised studying the general issues which it addresses! Instead of pursuing this interface in its proper depth, in this brief note, I merely address one theme raised by Sjoerd Zwart in his piece for this volume addressing Kuipers' view of theories. What is the appropriate format for formal theories, and for the ways they may change – if all goes well – toward some target theory: the Truth, if you like? Theories are basic building blocks in scientific reasoning, but just as well in the logic of daily practice. This leads to obvious formal analogies, as shown below, but it also reflects a material one. How people reason on "gala occasions" in the pursuit of science is just an extension – and in some cases even a simplification – of our reasoning about the tasks of domestic daily life. Cognitive rationality is the same throughout.

2. Theories in Logic

In logic, theories serve as information structures across a wide range: mathematical theories like Peano Arithmetic, knowledge bases in computer

In: R. Festa, A. Aliseda and J. Peijnenburg (eds.), *Confirmation, Empirical Progress, and Truth Approximation* (*Poznań Studies in the Philosophy of the Sciences and the Humanities,* vol. 83), pp. 403-419. Amsterdam/New York, NY: Rodopi, 2005.

science, or the informal background knowledge and presuppositions that ordinary language users live by. Hence, computational demands of expressiveness and complexity in both scientific and everyday tasks affect the optimal definition of theories, when viewed as the longer- and shorter-term data structures involved in human reasoning. Accordingly, theory structure in modern logic is a live topic, as new definitions keep appearing – even though this issue of "logical architecture" is not yet an established agenda item, the way it has been in the philosophy of science. A real chronicle of this development goes beyond this note, and what follows is just a sketch.

2.1. *Theory Formats*

The basic view of theories in logic is quite austere. Theories T in a logical language are plain sets of syntactical sentences, or alternatively, they can be viewed semantically as the associated set of models

$$\mathrm{MOD}(T) \quad = \quad \{\, m \mid m \vDash T\}$$

The two viewpoints are inversely related. The bigger the syntactic theory (more sentences), the smaller the set of models (fewer semantic situations qualify). Either way, there are various types of motivation. A theory may be a stable set of axioms, as in the foundations of mathematics, but it may also be a more ephemeral set of premises in ordinary reasoning. Moreover, in the former setting, some axiomatic theories intend to describe one unique model, say the standard natural numbers, while others are meant to describe a class of mathematical systems, the more the better, as in group theory. In ordinary reasoning, these aspects may play together, witness the common distinction into a long-term "background theory" constraining the class of situations one deals with, and a short-term "local theory" of assertions describing one situation out of these, viz. the current topic of conversation.

For some reasoning tasks, this format is too poor, and more fine-structure is needed. The above two viewpoints already differ qua information. The model set approach abstracts away from "details of syntax" – but in doing so, it loses the *packaging* into the chunks in which the information was presented. Each sentence ϕ in T corresponds to a set of models $\mathrm{MOD}(\phi)$, and it is only their intersection which is recorded in $\mathrm{MOD}(T)$. Some modern theory definitions therefore work with

$$\mathrm{MOD}^{*}(T) \quad = \quad \{\, \mathrm{MOD}(\phi) \mid \phi \in \mathrm{T}\}$$

E.g., $\{A \vee B, \neg A\}$ is a different packaging for the theory $\{\neg(A\&B), B\}$, though the over-all set of models is the same, being just the valuation with $V(B)=1$, $V(A)=0$. Such details are crucial to counterfactual reasoning and theory revision, when seeking maximal sets of chunks consistent with new information that contradicts the theory as a whole (witness authors like Kratzer, Veltman, Segerberg; cf. also Gärdenfors 1987). And even this richer format suppresses information about presentation, as logically equivalent formulas defining chunks are identified. But logically equivalent assertions can have very different properties qua ease of understanding, or suggested inferences. In actual argumentation in everyday life, or science, presentation is all-important – but we will disregard such more proof-theoretic perspectives here, despite their relevance to general methodology.

Information chunks may also be *ordered*, making theories consist of layers of less or more important parts, more or less prone to revision. Relative importance has many determinants: such as credibility, ease of computation, or elegance. Thus, preference orderings of components of theories, syntactic or semantic, come into play – as "entrenchment relations" between assertions in the study of belief revision, or in accounts of data bases as "ordered theories." These modern ideas echo earlier ones in the philosophy of science about different importance of assertions involved in hypothetico-deductive explanation: core laws, facts, and auxiliary assumptions. Modern logics of default reasoning use refined entrenchment orderings, showing how these orderings may be updated through incoming information.

This completes our sketch, which is by no means complete. E.g., another major topic in the philosophy of science has been the *vocabulary* of theories, such as the distinction between observational and theoretical terms. Issues of language design and vocabulary update, too, are highly relevant to modern logical theory formats, even though they have been neglected so far – but we let such refinements rest here.

2.2. *Information Update*

A major interest in my own logical work is information update. Here, theories serve as information states about the relevant reality, which get updated as new information comes in. Changes range from straight addition to more drastic revision. I will stick with the former, for comparison with congenial themes in Kuipers' work. Typically, information update decreases the set of options for the actual situation.

Ann, John, or Mary may or may not have been at the party.

We want to find out, using the following information:

    (i) John comes if Mary or Ann comes
    (ii) Ann comes if Mary does not come.
    (iii) If Ann comes, then John stays away.

Initially, our information state consists of all possible party compositions

$\{MAJ,\ MA\text{-}J,\ M\text{-}AJ,\ M\text{-}A\text{-}J,\ \text{-}MAJ,\ \text{-}MA\text{-}J,\ \text{-}M\text{-}AJ,\ \text{-}M\text{-}A\text{-}J\}$

The first assertion $(M \vee A) \to J$ removes three of these, and updates to

$\{MAJ,\ M\text{-}AJ,\ \text{-}MAJ,\ \text{-}M\text{-}AJ,\ \text{-}M\text{-}A\text{-}J\}$

The remaining updates are as follows:

$\neg M \to A \qquad \{MAJ,\ M\text{-}AJ,\ \text{-}MAJ\}$

$A \to \neg J \qquad \{M\text{-}AJ\}$

The final information state has just one situation $M\text{-}AJ$, representing the actual state of affairs with Mary and John present, while Ann is not. This scenario shows how, semantically,

information update proceeds by elimination of possibilities.

Dually, information grows syntactically by adding more and more assertions to our current stock.

Now, the main thrust in logics of information update appears to differ from that in the formal philosophy of science. Update logics do not emphasize the statics of knowledge representation, but rather the *many-agent* dynamics of update in communication. Language use revolves around what people know about each other's information. For instance, I will normally try to say something informative, of which I believe that you do not know it yet. Thus, modern update logics will typically handle compound action–knowledge assertions such as the following

$[C!]\ K_i\ \phi$ after public announcement of C, agent i knows that ϕ

E.g., before you answer my question, I do not know the answer; while afterwards, I do. This makes cognitive actions like announcements and agent-relative knowledge an explicit part of the logical system. Also, knowledge is not the only yardstick. Language use involves a rich repertoire of propositional attitudes toward assertions. Some are known, others believed, suspected, entertained, desired, or hoped for... The dynamics of communication also involves more private communication, gossip, white lies, and many other subtle cues on which we act in our daily lives. These skills may be more subtle than those in science, which is public and "above board." This may be the rationale behind Keith Devlin's recent provocative point (cf. http://www-csli.stanford.edu/~devlin/) that someone with the intellectual powers to follow the epistemic complexities of a soap opera on TV should have no problem whatsoever with something as simple as pure mathematics. But despite this potential disparity, it makes sense to compare epistemic perspectives across logic and philosophy of science.

3. Theories in Kuipers' Work

Theo Kuipers' formal analysis of scientific theories has ranged widely through his work. (Here and henceforth, we refer to the relevant chapters in the monograph [Kuipers 2000, 2001], and occasionally also the forthcoming paper Kuipers (forthcoming). These go back to many earlier papers, not listed separately here.) At one extreme lies his interest in structuralist theories in Sneed's "structuralist" tradition, with their sophisticated account of families of models, and expansions of vocabularies. In other settings, he sticks with much simpler formats to make his points. A noticeable example is his definition of relative truthlikeness.

3.1. *Verisimilitude and Flat Theories*

Here is what it means for theory A to be at least as good a fit for theory T as B:

$$A \geq_T B \quad \text{if} \quad \text{(a)} \ B \cap T \subseteq A \quad \text{and}$$
$$\text{(b)} \ A - T \subseteq B$$

Thus, A does at least as well as B on recognizing T-instances, while doing no worse than B on non-T-instances. This so-called Miller-Kuipers definition employs the bare logical view of theories, though with an original take on an important relation between them. One interesting issue here is the nature of T. If we think of approaching the Truth, in line with the above update story, we might expect T to be a singleton set consisting of just the actual world. But in Kuipers' work, T is often a larger set of models, standing for the physically possible situations. This is a structuralist way of thinking: physical theories describe whole families of possible situations, such as "mechanical systems" obeying the laws of Newtonian mechanics. This is more like the above case of group theory, with many models for a theory being an advantage, as they stand for broad applicability. A similar point is made in Kuipers' recurring electrical circuit example with light bulbs and switches. Among all possible propositional combinations of atomic assertions "switch i is open," "bulb j is burning," only a certain subset corresponds to physically possible states of the circuit – reflecting the proper dependencies between the propositional variables. Scientific theorizing must then lead us toward this true set of physically possible systems.

From the perspective of the earlier update logics, having a set of worlds, rather than a single world as a target of an elimination sequence is quite reasonable. E.g., when a group of agents pools what they know, they need not expect one single world to emerge, but rather the smallest set of worlds representing their joint information. More generally, instead of zooming in on one object, the update process then becomes a way of learning the extension of a certain predicate, say, the *actual situations*. This is completely in line with many learning tasks, and with the corresponding epistemic logic of knowing which objects satisfy some given predicate.

3.2. *Theories in Progress*

But there may be more structure. Instead of identifying a theory with just the set of models where it holds, Kuipers also considers a more partial view involving a pair of sets

$$R(t), S(t) \quad \text{where} \quad R(t) \subseteq S(t),$$

with $R(t)$ the models shown to satisfy the theory so far at time t, and $S(t)$ those satisfying all laws recognized so far as following from the theory. One can view this as what we know so far about the True theory, with R the result of experimental verifications (or analyses of empirical systems), while the information encoded in S might be the result of general deductions out of T, or even inductive leaps. More generally, this move replaces any theory T viewed as a set of models by pairs of an "inner" and an "outer approximation":

$$T_{\text{down}}, T^{\text{up}} \quad \text{with} \quad T_{\text{down}} \subseteq T^{\text{up}}$$

This two-set view of theories seems the main source of Sjoerd Zwart's misgivings. The reason has to do with update. Suppose that we learn more about a partial theory $T_{\text{down}}, T^{\text{up}}$. Intuitively, this should mean that we get a new partial theory T', i.e., a situation $T'_{\text{down}}, T'^{\text{up}}$ tightening the bounds to

$$T_{\text{down}} \subseteq T'_{\text{down}} \subseteq T'^{\text{up}} \subseteq T^{\text{up}}$$

This update from T to T' involves both decrease of a set of models (viz. the upper bound) and increase of another: the lower bound. Therefore, update is not quite a matter of eliminating models by adding more laws to be true: we also eliminate potential laws by adding models. Now, is this a flaw – or worse, an inconsistency? The heterogeneity seems just a feature of the more sophisticated notion of "theory," though one must take care to use consistent terminology appropriate to the new setting. Indeed, the heterogeneity reflects a natural distinction. Sometimes we learn necessary conditions for a certain predicate, say the predicate 'T' defining the right models for our theory. These are constraints of the form

$$\forall x \, (Tx \rightarrow \phi(x))$$

where ϕ is some property of models – a "law" in Kuipers' sense. This sharpens the upper bound. The counterpart of these are sufficient conditions of the form

$$\forall x \, (\psi(x) \rightarrow Tx)$$

Now, the prime examples for this direction were assertions "model m is an instance of T." These would rather have the atomic form Tm. But often, one does not verify that a single situation falls under a physical theory, but rather a whole class – say, all pendulums in a certain range. Then the sufficient condition format is more appropriate. Conversely, the law-format of necessary conditions also subsumes the special case of one negative instance, represented by an atomic assertion $\neg Tm$. For the latter is equivalent to the universal formula $\forall x\,(Tx \to x \neq m)$.

3.3. *Theories in Partial Logic*

More radically, we can even reverse the update direction altogether! Here is another view of partial theories. Instead of the lower and upper bound, we follow practice in three-valued logical semantics, when defining a partial predicate T of objects (in this case, "actuality" of models). For this purpose, one specifies two disjoint sets: the positive extension consisting of the known instances and the negative extension consisting of the known counter-examples. This is encoded in an ordered pair

$$T^{+}, T^{-}$$

Here the two given sets are disjoint, leaving in general a third "grey zone" of those objects which are neither definitely inside, nor outside of T. Of course, the two representations in 3.3 and 3.2 are equivalent, as we have the identities

$$A^{+} = A_{\text{down}}, \ A^{-} = -\,A^{\text{up}} \qquad \text{with} - \text{for set-theoretic complement}$$

Update now amounts to increase in both sets: adding more positive instances and also adding more negative instances. This is indeed the way we learn the extension of many basic predicates of our ordinary language in daily life.

Partial logic poses interesting challenges when extending classical notions to such a new setting. A well-known case is implication between assertions. Classically, A logically implies B if B is true whenever A is. By contraposition, then also: A is false whenever B is. In partial logic, however, the latter clause does not follow from the former, which only implies that A is not true whenever B is not true. Typically, in a three-valued perspective, 'not true' is not the same as 'false'. Indeed, if we want both implications to hold, we need the stronger so-called "double-barreled consequence" (cf. the survey of partial logic in Blamey 1986):

(a) whenever A is true, B is true,   and
(b) whenever B is false, A is false.

Similarly, it is not entirely clear how to extend the above definition of verisimilitude to partial theories. Such an extension makes sense, e.g., when comparing two partial theories versus the complete true theory, or two complete theories versus the best known partial approximation of the true theory. The question is now how to extend the definition given in 3.1. Our proposal would be to read the earlier implications in the double-barreled sense, making sure that all positive successes for B are also successes for A, while clear blunders for A are also blunders for B:

$$A \geq_T B \quad \text{if} \quad \text{(a)} \ T^+ \cap B^+ \subseteq A^+ \quad \text{and} \quad T^+ \cap A^- \subseteq B^-$$
$$\text{(b)} \ T^- \cap A^+ \subseteq B^+ \quad \text{and} \quad T^- \cap B^- \subseteq A^-$$

In some ways, this is even clearer and more symmetric than what we had before. Moreover, when the partial theories are total (i.e., the given two parts exhaust the whole set of models), this stipulation reduces to the original one. Nevertheless, the three-valued partial logic perspective also allows other definitions of verisimilitude, and the general conclusions below do not depend on this particular choice.

4. Lifting One Level

Does the partial view of theories contradict the original elimination view of update? The answer is negative, as we can lift the setting.

4.1. *From Partial Models to Sets of Models with Complete Denotations*

Let M be a model with a domain of objects satisfying a bunch of totally defined predicates P, plus some "unknown" partial predicate T of objects. Now take all standard models $M(T)$ over the same objects as M which leave the interpretation of all predicates in P unchanged, while providing complete two-valued interpretations for T extending the definite facts about membership of T already true in M. These models $M(T)$ represent the definite ways the predicate T can turn out to lie on the basis of M. In particular, then, we can associate the pair T_{down}, T^{up} with the set of all those models $M(T)$ whose T lies in between these two bounds. Thus, theories now become sets of sets, each representing a different option for the extension of T that is still open. In particular, an old standard theory T corresponds to a singleton set, as there is no variation possible. Closely related to this perspective, an information state T in update semantics, which can still shrink arbitrarily, will correspond to the set of all subsets of its current T.

Clearly, the above "increasing updates" adding new positive or negative instances now become eliminative ones after all, removing those complete extensions for the predicate T either lacking, or having the object mentioned in the update. Increase becomes decrease at the price of one extra level of sets. It just depends on how we want to model things. We can think of this more generally as shifting from updating in terms of objects and their properties to updating in terms of propositions about predicate extensions. Just as earlier, we eliminate all models $M(T)$ that fail to satisfy the relevant assertion. But is this just a trick, or is there more to the new setting? I think there is.

4.2. *Updating with Arbitrary Assertions*

The richer setting allows for further updates, not covered by the format of learning a law, or a positive example. To see this, consider a scenario with four people $\{1, 2, 3, 4\}$, two of which $(2, 3)$ are women. Some people are criminals, but we do not know which. Now, successively, we learn the following about the gang T involved:

>   (i) 1 belongs to the gang
>   (ii) the gang has 2 members
>   (iii) the gang contains a woman

The first update might just lead to the partial theory $\{1\}$, $\varnothing$. But the second one is of a different character: it gives a global property of the gang, without telling us about precise members. It rather puts a constraint on membership patterns. In terms of updates, the original information state has all 16 subsets of $\{1, 2, 3, 4\}$ as possible extensions of T. The update for assertion (i) reduces this to the eight subsets containing person 1. The update for assertion (ii) reduces this drastically to

$$\{1, 2\}, \{1, 3\}, \{1, 4\},$$

even though it does not add to our direct knowledge about individual gang members. Finally, assertion (iii) is again of a global type, reducing this information state to

$$\{1, 2\}, \{1, 3\}.$$

Speaking logically, these more general updates correspond to further assertions one might make about the predicate T. Thinking of a more definite language, one might think of a monadic predicate logic with predicates for existing structure in the model M plus a unary predicate T for the special objects we are interested in. In particular, updates on partial theories of the earlier special forms are so-called Horn clauses, whereas general assertions may involve disjunctions: either these two objects, or these, either this woman or that, etc. That is, assertions might have such forms as

$$(Tm \ \& \ \forall x \ (Tx \to \phi(x))) \lor (\neg Tn \ \& \ \forall x \ (\psi(x) \to Tx))$$

In ordinary reasoning, as we just saw, such disjunctive updates are not far-fetched. I think they also occur in science. It is quite possible to find that the models for our scientific theory must be either this class or that, without being able to determine which one. Think of a theory given by differential equations allowing two types of solution, where the choice must come from further principles, yet unknown.

4.3. *Recovering the Old Inside the New*

The question which does come up then is what makes the original partial theory format special in this broader world. For the information states corresponding to partial theories T_{down}, T^{up}, this is easy to see. We formulate two simple technical results comparing the two levels. Their only purpose here is to show that, instead of hunts for 'one right viewpoint' or inconsistencies, there are interesting connections to be observed and proved between different levels of looking at a "theory."

> *Fact* A set U of models $M(T)$, or the associated family of T-denotations, can be represented in the format $T_{\text{down}} \subseteq T^{\text{up}}$ iff the following two conditions hold:
>
> (a) The set U is closed under arbitrary unions and intersections
>
> (b) U is convex: any set in between two of its members is in it.

We can think of these sets U as especially "coherent" information states in the larger setting. Moreover, techniques from update logics (van Benthem 1996) characterize the above special updates. Take the case of adding a law, i.e., a necessary condition.

> *Fact* The updates corresponding to satisfying a law are those maps Φ on families of sets U which satisfy the following three conditions:
>
> (a) Introversion: $\Phi(U) \subseteq U$
>
> (b) Continuity: $\Phi(\bigcup_{i \in I} U_i) = \bigcup_{i \in I} \Phi(U_i)$
>
> (c) Tightness: $\{ X \mid \Phi(\{X\}) = \{X\} \}$ is a set-theoretic ideal.

Further questions of this sort arise concerning verisimilitude. We have some first technical results on relating the notion $A \geq_C B$ for partial theories to one for the corresponding families of complete models $M(T)$ – but we will leave this tangential issue here.

4.4. *Other Instances of "the Shift" in Logical Modeling*

The above is not an isolated discussion. Similar shifts in levels of representation occur in other areas. First, consider natural language, which involves many different updates. There is informational update, as we learn more facts – but e.g., there is also enrichment of the relevant context, as more individuals are introduced into the current narrative. Major systems of dynamic semantics treat this via a mix of representations. An assertion like "He came in" will enrich the current context with a link between the pronoun 'he' and some object d, while at the same time, eliminating pairs <world, context> in which the assertion CAME-IN(d) is false. The current semantic literature calls this approach mixing of eliminative and "constructive" update. The balance between the two will be dictated by considerations of parsimonious representation, and perhaps focus.

The contrast also shows in AI, with non-monotonic versus monotonic reasoning. E.g., in *predicate circumscription*, one takes only models of the premises so far with minimal extensions for the predicates, and valid conclusions are all assertions true in such models. Thus, saying that a and b are P implies that *only* a and b are P – but learning a new fact that c is P may subsequently invalidate this inference. The usual motivation for non-monotonic reasoning thinks of the given premises as *all we know*, using the small set of predicate-minimal models to draw "eager conclusions." Now, many critics find this unconvincing, as the additional fact does not contradict the earlier two – while we should only call a conclusion from premises valid if we are not going to be bothered by additional information. Again, this has to do with choosing a larger model set for the premises, encoding in advance what further information we might receive. As in the above discussion of theories, there seems to be no "truth of the matter" between non-monotonic and monotonic approaches. It is up to us to choose if we want a small model set, supporting many inferences, but jolted by every new update into a drastic revision – or a larger model set allowing for lots of updates, through the smooth elimination of scenarios that will no longer occur.

Finally, the move to sets of sets occurs frequently with "higher constraints" on possibilities. Suppose you are pondering the future course of a game played against me. For many purposes, this is adequately modelled by the well-known picture of a branching tree of all future worldlines from now on which you consider possible. Each worldline has moves by you and countermoves by me, and the total tree imposes restrictions on this interaction. But now, suppose you believe a "uniformity" about my strategic behavior. Either, I always chose the left-most move available to me in the game tree, or always the right-most move. Then one set of future world-lines no longer suffices, as you need to consider two separate future trees, one for each strategy of mine. As in the above, such a move up in sets also suggests a richer logical language with assertions exploiting the additional structure.

414   *Johan van Benthem*

5. Logics and Calculi

Typical aspects of a logic-style analysis are the use of formal languages, and the calculisation of valid inferences for notions of interest. Now, some key notions in Kuipers' methodology of science have the same flavor. Hence a precise logic connection would be of interest. Here is an example.

5.1. *The Logic of Verisimilitude*

Verisimilitude as defined in Section 3.1 satisfies a number of formal axiomatic properties, making it function a bit like a logical conditional (van Benthem 1987). We list a few:

$$A \geq_T B \,,\, B \geq_T C \qquad \text{imply} \qquad A \geq_T C$$

$$A \geq_T B \,,\, A \geq_{T'} B \qquad \text{imply} \qquad A \geq_{T \cup T'} B$$

$$A \geq_T B \,,\, A \geq_{T'} B \qquad \text{imply} \qquad A \geq_{T \cap T'} B$$

$$A \geq_T B \qquad \text{implies} \qquad A \geq_B T$$

Tarski and the Polish logicians proposed a logical "calculus of theories" in the 1930s – notions like this provide interesting extensions of its repertoire. Now, the above way of looking at theories provide a method for embedding issues of verisimilitude into standard logical systems. This accounts for the above formal inference patterns. Let the working language be monadic first-order logic with unary predicates A for sets, or equivalently, a basic modal language with operators $\square$, $\Diamond$, over an S5–style semantics, with proposition letters a corresponding to sets A.

> *Fact*   Reasoning with verisimilitude can be faithfully translated into modal S5.

Here is the reason. The above has inferences from some set of premises of the form $A \geq_T B$ to some such conclusion. More generally, these are Boolean inferences over atoms stating a relation of relative truthlikeness. Clearly, it suffices to translate these atoms into a modal setting. But this is easy, as an assertion $A \geq_T B$ interpreted as in 3.1 can be translated as a modal statement of the form

$$\square\, (((t\&b) \to a) \,\&\, ((\neg t\&a) \to b)))$$

As a consequence of this logical translation, the logic of verisimilitude is completely axiomatizable, and decidable at the same low complexity as S5, taking non-deterministic polynomial time (NP).

The translation even extends to cover truthlikeness of partial theories. For this purpose, a well-known trick suffices. For each partial theory A^+, A^-, take a pair of proposition letters a^+, a^-, and add a modal premise $\Box \neg(a^+\& a^-)$. On top of this, reasonable definitions for verisimilitude between partial theories will translate straightforwardly into the modal language.

5.2. *Epistemic Logic*

From a logical point of view, there are really two languages involved in the above. The modal language with the universal box serves as a classification device: it provides a description of one particular model $M(T)$, explaining the available sorts of objects (worlds, if you wish). But there is also the larger setting of a family of such models, which determines what we know about the extension of various predicates. For that we need an additional epistemic vocabulary with operators

$$K_i\, \phi$$

stating that ϕ is true in all *i*–accessible members $M(T)$ of the family, as seen from the actual model. A typical epistemic formula might be

$$K_i\Box(a \to p) \;\&\; \neg K_i \Diamond(q \,\&\, a),$$

expressing that agent *i* knows that all A-situations satisfy law P, while she does not know that A-situations can co-occur with the property Q. And of course, the K-language will also allow for "social" knowledge about other people's knowledge. In this semantics, over an information state for a partial theory, the objects in A^+ are precisely those for which people know that they belong to A, and those in A^- the ones they know not to belong to A. Moreover, one can show that the epistemic model supports no further "hidden knowledge" on constraints governing the eventual filling of the grey zone. Thus, a partial theory A^+, A^- corresponds to an epistemic model concerning the real form of A which validates exactly the right assertions. (Cf. Fagin *et al.* 1995, and Van der Hoek and Meijer 1995 on the state of the art in modern epistemic logic, plus connections with partial logic.)

But we can also make finer distinctions now, such as one between $A \geq_T B$ and the stronger $K_i A \geq_T B$. In addition, knowledge assertions may be specified for individuals, or even for common knowledge $C_G\, \phi$ in groups G. Thus, we plug in to the existing, and fast-growing body of research on epistemic logic. This includes dynamic logics for information update, telling us, e.g., how relations of truthlikeness can change as new information comes in, through assertions of the form

$[P!]\, C_G\, \phi$ after a public announcement of P (say, in some journal publication), community G has common knowledge of ϕ

Even more expressively than this, one might consider the size of the total domain of objects as subject to ignorance. In that case, one might add dynamic operators referring to addition or removal of objects from the current domain – as studied by Theo's Groningen colleague Gerard Renardel in his dynamic logics with 'creation and destruction' of objects in the current domain.

As stated before, this setting imports special concerns from dynamic-epistemic logic with procedural details of communication and other forms of information passing, and that between different agents forming different groups. This may be too fine-grained for much methodology of science, but it might be an interesting aspect of method in scientific communities as well – especially, if one could show that their communicative conventions differ systematically from those found in daily life. E.g., many puzzles of actual communication involve learning useful factual things from knowledge of *ignorance* by others – with card games as a clear example. Is there any counterpart to this in the methodology of science, unless our aim is to describe the mechanics of its social competition?

6. Logic and Methodology of Science

This note has taken rather long to make some simple points about representation of theories, update, and knowledge. Nevertheless, it may also serve more broadly to show how methodology of science and logic meet in significant themes. The above story is only a beginning. One might take ideas across in either direction. E.g., the methodologists' idea of one theory being a better approximation of the truth than another makes good sense in epistemic logic, too – as do even more general "quality comparisons" between model-theoretic constructs vis–à–vis some given standard. Likewise, scientific practice has speech acts like "proposing" or "entertaining" a theory which might enrich the repertoire of communicative actions in logical semantics. Vice versa, powerful research trends in modern logic such as belief revision, or many-agent perspectives, seem of immediate relevance to understanding the workings of scientific methodology.

At the editors' request, I conclude with a few general reflections on logic and the methodology of science. In the years around 1980, I did some work at this interface – with van Benthem (1982) as a fair example. My expectation then was that the model theory of formal theories would blossom into a flourishing new field in between the philosophy of science and mathematical logic. These hopes have not really been fulfilled. The two research communities seem rather disjoint, despite occasional journal issues or conference sessions where practitioners are thrown together in hopes of sudden love and understanding. In particular, logicians have looked elsewhere for affection, taking AI, computer science, and linguistics as sources of inspiration for flourishing new interface communities. Despite all this, and even in this wider modern landscape, I still feel that logic and methodology of science share many common causes.

First, qua subject matter, I see hardly any difference at all. Pioneering 19th century logicians like Bolzano, Mill, or Peirce, were also innovative methodologists of reasoning in the broadest sense. The fact that general methodology moved out of logic textbooks is largely a historical accident of the agenda-contracting turn toward the foundations of mathematics in the early 20th century. But a blend of logic and general methodology still persists with influential later authors like Lewis, Hintikka, van Fraassen, or Gabbay – to name just a few. It would be tedious to draw a sharp boundary in their work between logic proper and general methodological concerns. But also less anecdotically, it is a matter of simple observation in the literature that key topics in general methodology such as confirmation, explanation, theory structure, or causality, also occur in logic, AI, or linguistics. A good example of a happy marriage of this sort is the theory of belief revision in Gärdenfors (1987), mixing sources from both logic and the philosophy of science in the most natural fashion. A similar blend may be found in the formal account of events and causal reasoning in Shoham (1985). More recent examples are modern studies in the logic of time and space (surveyed in van Benthem 1995), or investigations of abduction and other forms of non-monotonic inference (Flach and Kakas 2000, as well as Aliseda-Llera's contribution to this volume), where logic runs into philosophy of science, computer science, and general argumentation theory. And of course, as said before, Theo Kuipers' own work provides lots of pointers for fruitful confluence.

Moreover, these concerns are shared for a reason. As stated at the start of this Note, I believe in the continuity of human cognition, from daily life to more exalted intellectual discipline, when practicing, law, theology, or science. This does not mean that no differences exist at all between science and common sense – but the specialized intellectual activities of science are probably best understood as special "parameter settings" of one general cognitive functioning. As a student, I was made to memorize Ernest Nagel's list of supposedly crucial differences between "science" and "common sense" at the beginning of his classic book *The Structure of Science*. Nowadays, I find that all points mentioned there are matters of degree – and that one cannot even understand the success of science without seeing it as an extrapolation of human common sense behavior. Instead of arguing for this view here, let me just stick out my neck for it. I would be happy to defend the position that virtually every issue in understanding science has its "domestic" counterparts in common sense cognition. Of course, philosophers of science have much to say about larger aggregation levels of knowledge, and longer-term historical phenomena across generations of scientists, which do not normally surface in logic textbooks. But eventually, even these broad-band perspectives seem equally crucial to the functioning of our natural language and common sense reasoning in daily life.

I do not pretend that I forecast all the above commonality around 1980. To the contrary, I find many things I wrote in those days remarkably un-visionary. Let me list a few predictive failures. First, the more fruitful logic interface for methodology has not been model theory and mathematical logic, but rather philosophical logic and logic systems of a more computational slant. Also, the computational turn which has shaken up research agendas in both logic and philosophy of science is conspicuously absent. And finally, I now find much of what I wrote about philosophers "just stating definitions" patronizing and misleading. There may indeed be a historical division of labor between mainstream logicians and philosophers of science, in that the former take definitions of formal systems and proving technical theorems as their standard, while the latter look for interesting formal models explicating phenomena in the (empirical) sciences. But difficult issues of modeling and concept explication are also crucial to logic itself – and indeed, some of the breakthrough achievements in logical semantics of natural language, or recent dynamic logics of information update, have been precisely of the latter sort. Delicate modeling skills may even be a rarer talent among students than hard-hitting theorem-proving powers.

It would take a much longer story to substantiate the above claims. But even with the broad outline I have painted, one clear message of my piece is this. Despite the distance between our communities, logicians should find trips to the work of Theo Kuipers and his colleagues in the philosophy of science highly rewarding!

University of Amsterdam & Stanford University
Institute for Logic, Language and Computation
Plantage Muidergracht 24
1018 TV Amsterdam, The Netherlands
e-mail: johan@science.uva.nl
http://staff.science.uva.nl/~johan/

REFERENCES

van Benthem, J. (1982). The Logical Study of Science. *Synthese* **51**, 431-472.

van Benthem, J. (1987). Verisimilitude and Conditionals. In: T. Kuipers (ed.), *What is Closer-to-the-Truth?*, pp.103-128. Amsterdam: Rodopi.

van Benthem, J. (1995). Temporal Logic. In: D. Gabbay, C. Hoggar, J. Robinson (eds.), *Handbook of Logic in Artificial Intelligence and Logic Programming*, vol. 4, pp. 241-350. Oxford: Oxford University Press.

van Benthem, J. (1996). *Exploring Logical Dynamics*. Stanford: CSLI Publications.

van Benthem, J. (2001). Language, Logic, and Communication. In: *Logic in Action*. Amsterdam: ILLC Spinoza project, ILLC.

Blamey, S. (1986). Partial Logic. In: D. Gabbay, F. Guenthner (eds.), *Handbook of Philosophical Logic*, vol. 3, pp. 1-70. Dordrecht: Kluwer.

Fagin, R., J. Halpern, Y. Moses and M. Vardi (1995). *Reasoning about Knowledge*. Cambridge MA: The MIT Press.

Flach, P. and A. Kakas, eds. (2000). *Abduction and Induction, their Relation and Integration*. Dordrecht: Kluwer.

Gärdenfors, P. (1987). *Knowledge in Flux*. Cambridge, MA: The MIT Press.

Kuipers, T. (2000/ICR). *From Instrumentalism to Constructive Realism*. Dordrecht: Kluwer.

Kuipers, T. (2001/SiS). *Structures in Science. Heuristic Patterns Based on Cognitive Structures. Synthese Library*, vol. 301. Dordrecht: Kluwer.

Kuipers, T. (*forthcoming*). Inductive Aspects of Confirmation, Information, and Content. To appear in the Schilpp-volume *The Philosophy of Jaakko Hintikka*.

Shoham, Y. (1985). *Reasoning about Change: Time and Causation from the Standpoint of AI*. Cambridge, MA: The MIT Press.

Theo A. F. Kuipers

LOGICS OF SCIENTIFIC COGNITION

REPLY TO JOHAN VAN BENTHEM

In brainstorming about a suitable title for the present book the editors pondered over "Logics of Scientific Cognition." For some reason this title did not make it, but Johan van Benthem's contribution would have fitted that description perfectly. It breathes the same flexible spirit that was already present in his review-like programmatic paper "The Logical Study of Science" (Van Benthem 1982) and that seems to me essential for the very idea of confronting logic and philosophy of science. As also mentioned in my reply to Batens, unfortunately for philosophy of science, after 1982, with a few exceptions (e.g. Van Benthem 1987), Van Benthem directed his logical skills mainly to a number of other areas of the application of logic, notably informatics, linguistics and the philosophy of nature, in particular the logic of time. It is to be hoped that his present contribution is a sign of a partial return to philosophy of science.

In this reply to Van Benthem I start with a number of miscellaneous remarks that cropped up when reading his Sections 2 and 3.2. I shall then indicate that the notion of "partial theory" propagated by him in Section 3.3 not only seems to be directly useful for a couple of interesting topics in the philosophy of science. It has also some resemblance to an interesting special type of theory that I have called "partition theory" (Kuipers 1982). One reason for not dealing with Van Benthem's Sections 4 and 5 is to challenge him, his colleagues and his students to really cross the threshold and to reside for a substantial time in the philosophy of science. There is really much interesting work to do for the mutual benefit of philosophy of science and logic. For this general theme, see also a few other replies in this volume, notably to Aliseda, Meheus, Batens, Zwart, Mormann, and Burger and Heidema, and in the companion volume those to Kamps and Ruttkamp.

In: R. Festa, A. Aliseda and J. Peijnenburg (eds.), *Confirmation, Empirical Progress, and Truth Approximation* (*Poznań Studies in the Philosophy of the Sciences and the Humanities,* vol. 83), pp. 420-427. Amsterdam/New York, NY: Rodopi, 2005.

Miscellaneous Remarks

On Section 2.1: *Theory Formats*

(1) Van Benthem states that in ordinary reasoning theorizing about (using the terms of my reply to Sjoerd Zwart) *one* intended application and *a class* of intended applications may play together. I would like to emphasize that precisely the same applies in the theory-experiment interplay in science (ICR, Section 7.3, briefly indicated in Van Benthem's Section 3.2). In experimental contexts we try to establish by "actual truth approximation," given a vocabulary, the true description (or theory, if you like) of one experiment, that is, its set-up, initial conditions, and the outcome. By "nomic truth approximation" we try to approach the strongest true theory of what is physically possible, that is, the set of physical or nomic possibilities, using the results of (attempts at) actual truth approximation with regard to experiments, including inductive leaps based on them.

(2) Van Benthem rightly notes that by defining theories as sets of models, and, I should add, hence also by conceiving theories as deductively closed sets of statements, one "loses the *packaging* into chunks in which the information was presented." The modern definitions of theories he is hinting at by $MOD^+(T)$ may be very useful for localizing possible causes for counterexamples to theories, and hence for theory revision in favor of closer-to-the-truth ones in the basic sense of ICR. This is *a fortiori* the case when information chunks are ordered. By quoting from SiS, I would like to stress that in the philosophy of science very nice, partially computational, work in this direction has already been done:

> Finally, *Darden* [1990] considers the solution of an anomaly of a theory as a task for diagnostic reasoning, which has been developed for expert systems, guided by the tracing of a defect in a technical system. She shows, using Mendelian examples (see also Darden (1991)), that the decomposition of all presuppositions of a theory may provide the points of departure for solving the anomaly, and that its solution, as in Lakatos' famous analysis (1976), may or may not lead to fundamental theory revision, that is, a revision staying within a research program or breaking through the barriers of the relevant program. (SiS, p. 301).

On Section 2.2: *Information Update*

(3) I shall return later to the claim that "information update proceeds by elimination of possibilities." Here I just want to note that the party illustration of "elimination of possibilities," in the sense of possible models, typically amounts to actual truth approximation by applying *given* general facts, where the latter happen to be, in conjunction, so strong that they leave only room for one final possibility. That is, relative to the given language, an incomplete true

description is extended to the complete true one by eliminating all the other ones. By leaving out at least one such general fact and adding an initial condition, we obtain formal analogues of actual truth approximation as it occurs in the HD prediction and DN explanation of individual facts. These types of actual truth approximation are essentially contained in the general approach to propositional languages presented in (Kuipers 1982, Section 4). This general approach also covers another special case: actual truth approximation by successive corrections of a provisional description of a certain situation, whether that situation is realized by nature or by experiment. Only this special case is also presented in ICR (Section 7.3.1).

Moreover, they differ from "nomic truth approximation" by the interplay of realized possibilities and established empirical laws. The crucial reason for the latter is that in all variants of actual truth approximation there is, by definition, one target model, whereas a set of models is the target in the case of nomic truth approximation. More generally, the target is a set in all cases in which we want to establish a certain constellation of possibilities, which happens to be interesting for one reason or another. To mention just one social science example, consider an anthropologist who wants to establish in some way or other "the true moral theory" holding in some society and who has as his target the largest set of "morally allowed actions," of which we may hope that it is far greater than one. Note that all mentioned types of truth approximation can formally be illustrated by playing with the party example as well as by my favorite electric circuit example.

(4) Regarding *"many-agent* dynamics of update in communication" I just quote a passage from SiS:

> Moreover, Thagard [1988] suggests that we consider and simulate a scientific community as a collection of interacting parallel computational systems. Here the idea of social adequacy comes into the picture. The suggested analogy raises the question of the possibility and the desirability of a further division of labor than the well-known one between theoreticians and experimenters. In this connection he thinks in particular of the division of labor between 'constructive dogmatists' and 'skeptical critics', to use our favorite terms for what is essentially the same idea. (SiS, p. 299).

On Section 3.2: *Theories in Progress*

(5) By calling R "the result of experimental verifications" van Benthem suggested to me a new explication of the fact that scientists frequently talk about "verification," where philosophers of science prefer to be more cautious and merely talk about "confirmation." However, in view of the logical relation between 'the strongest true theory' T and one of its models, it is on closer inspection perfectly reasonable to say that the model m that truly characterizes a realized possibility, verifies T, by its definition in terms of nomic

possibilities. Similarly, it verifies any theory X that has m as one of its models. In both cases, just because T and X, respectively, are true on that model. Of course, it would be totally wrong to conclude, for example, that theory X has then been verified in the relevant general sense.

Hence, it might make sense to speak of verification of a theory by a model in contrast to general or overall verification of a theory. Note that this point is directly related to Emma Ruttkamp's observation that theories are frequently overdetermined by the data, instead of underdetermined (see Ruttkamp 2000, her contribution to the companion volume, and my reply).

According to this sense of verification of a theory by a model, an "actualized model" (in the sense of an actualized conceptual possibility) compatible with the theory verifies it. More roughly: a "positive instance" and even a "neutral instance" (see below) of the theory verify it. Hence, a theory can be – at the same time – verified by some models (instances) and falsified by others. This new sense of verification perfectly shows the non-standard way in which I use the term 'model'. See the synopsis' introduction to Part III and my reply to Zwart. The main interest of the new sense of verification may even be restricted to just revealing my non-standard use of the term 'model'.

Note too that the verification of a theory by a model need not imply that we have a confirming instance or model of the theory in the sense of deductive confirmation, discussed in ICR Ch. 2, for the relevant model may be a neutral instance. For example, although both a black raven and non-black non-raven are "verifying and confirming" models of "all ravens are black," a black non-raven is a "verifying but not confirming" model of that hypothesis. This amounts to similar reasons to those in the case of verification in the general sense (as treated in ICR, Ch. 2, pp. 21-2) for not conceiving "verification" simply as an extreme, viz. maximal, form of "confirmation," as is usually done by philosophers of science.

(6) As announced, I come back to van Benthem's claim in 2.2 that "information update proceeds by elimination of possibilities." In 3.2 this is rightly qualified by "we also eliminate potential laws by adding models." Indeed, in addition to my favorite dual formulations of 'more successful' and 'more truthlike' in terms of consequences and models (ICR, p. 182, p. 189), we get in this way also a "dual formulation" of updating the data $R(t)$ and $S(t)$. They are updated to a superset $R(t')$ and subset $S(t')$, respectively, for t' later than t, by eliminating potential laws and possibilities, respectively. Of course, conversely, there is also a dual formulation in terms of "addition" (or "introduction"): adding new possibilities and laws, respectively.

(7) Van Benthem plausibly distinguishes between necessary and sufficient conditions for satisfying, for example, a certain theory. Moreover, he suggests

that the latter form is more appropriate when one concludes that "a whole class of systems – say, all pendulums in a certain range" falls under a physical theory. Although this way of speaking is not false, it may be still more appropriate to conceptualize this as a case of domain change, in particular domain narrowing, and hence change, in particular strengthening, of the relevant truth. For the topic of truth approximation by domain change, see my reply to Sjoerd Zwart.

Partial Theories and Partition Theories

From my point of view two intuitions come together in the notion of a "partial theory." First, the *three-valued division* of objects, for example of models, into a positive, a negative and a, so far, grey or neutral extension. Second, (*epistemological*) *symmetry* in the sense that "negative instances" are known in the same way as "positive instances." Both aspects are very important, but they need not go together.

Partition Theories

Let me first turn to "*two*-valued" epistemically symmetric theories. I have (only) paid special attention to them in my first paper on truth approximation (Kuipers 1982, Section 7.2, pp. 368-371), where I called them "partition theories," for reasons that will soon become clear. Take an electric circuit with just one bulb and some network of switches. We can immediately assume, for good reasons, that the true theory about the circuit is of the following type: the bulb will light if and only if a certain complex condition for the on and off state of the switches holds. Hence, whereas we may suppose that all conceptually possible switch states are physically possible, they are *partitioned* into two sets, viz. switch states with and without a lit bulb. When we propose a theory for the circuit, we can restrict ourselves to theories of the same form, which hence also partition the set of switch states into two subsets. As a consequence, such a theory can have two types of *real* counterexamples. A lit bulb and a switch state that does not satisfy the theory's condition and a non-lit bulb and a switch state that does satisfy the condition. The same feature occurs in many other, more interesting, cases. For example, equilibrium theories in physics and economics. The true theory about a balance and hence candidate theories are of the form: the balance is in equilibrium if and only if …. A quite different example is a grammatical theory. Such a theory is also of the form: a sentence is grammatical, as judged by native speakers, if and only if it satisfies such and such conditions. Hence, a grammatical sentence *not satisfying* the theory's condition is a counter-example, but an *un*grammatical sentence *satisfying* the condition is one, too.

In the rest of Section 7.2 of (Kuipers 1982) 'more truthlike' and 'more successful' of partition theories is elaborated, where useful, deviating from the way this is done for theories in general. Taking two types of real counterexamples into account turns out to be less complex than one might expect, even less complex than for theories in general, in which case only one type has to be taken into account. The reason is of course that for theories in general we have also to take a kind of *virtual* counterexamples into account. Recall that $R(t)$ represents the realized, or established *real or nomic*, possibilities at time t and $S(t)$ the, on their basis, strongest established law at t. The complement of $S(t)$ can be said to represent "virtual impossibilities." Hence, whereas $R(t) - X$ represents the established *real* counterexamples of theory X, $X - S(t)$ represents the established *virtual* counterexamples. The asymmetric role attributed to $R(t)$ and $S(t)$ in ICR has everything to do with the special nature of virtual impossibilities and virtual counterexamples. Of course, in some sense, partition theories also have virtual counterexamples. The point is that they are now implicitly represented by real counterexamples. For, to each real counterexample, e.g. a lit bulb with improper switch condition, corresponds a virtual counterexample, e.g. a (non-defective) non-lit bulb in the same switch state.

To be sure, although partition theories frequently occur in the empirical sciences, it is by no means the case that other theories can be missed. On the contrary, in my opinion, the great variety of "empirical laws" implies that as a rule we have to take virtual impossibilities explicitly into account. Only in rather special cases may we have good reasons to suppress them.

Partial Theories

The idea of partial theories, whether or not with a partitioning character in the sense above, does indeed seem to be very useful for philosophy of science purposes. Let me start by indicating how it could refine the general (meta-) theories of confirmation, empirical progress, and truth approximation as presented in ICR. As a matter of fact, ICR and remark (5) above, already contain several indications in this direction. Realized examples of a theory X, that is, members of $R(t) \cap X$, may be of two kinds. X may or may not have some conditional deductive consequence for a member, that is, there may or may not be some partial description of that member for which X entails the rest of the (complete) true description. For confirmation theory, this leads to the distinction between "neutral evidence" (recall the black non-raven for "all ravens are black") and (genuine) "confirming evidence" (a black raven as well as a non-black non-raven), see ICR, e.g. pp. 27-8. Moreover, for the theory of empirical progress it leads to the symmetric evaluation matrix, according to which, besides negative and positive results, neutral results have to be taken into account when comparing theories (ICR, p. 117). Finally, for the theory of truth approximation and in particular for the definition of 'more truthlike' it would lead to a distinction between types of "correct models," e.g. something like "positive models" and "neutral models," besides "negative models" of the theory.

As suggested, ICR contains hints with respect to the implementation of the partial character of empirical theories in the first two meta-theories, that is, about confirmation and empirical progress. However, I must confess that I am not sure to what extent I in fact assume in these hints that the empirical theories also are of a partitioning nature. Be this as it may, ICR does not contain hints for implementing "partiality" in the theory of truth approximation. Van Benthem's proposal at the end of Section 3.3 for this purpose, taking "neutral models" implicitly into account, seems very plausible. The overall "model nature" of that definition suggests that it will nicely combine with a similar overall model formulation of "more successful." However, such a formulation will very naturally reflect the success comparison of partition theories, by suppressing virtual counterexamples. Hence, an interesting challenge seems to be to give dual formulations for both 'more successful' and 'more truthlike' of partial theories in general. But I would also not exclude the possibility that, on closer inspection, the dual formulations in ICR of both, which *prima facie* neglect the indicated partial character of theories, already take this character implicitly into account.

I would like to conclude with a straightforward example of partial predicates with a partitioning character. In SiS (Ch. 10) I show that there is a partial analogy between truth approximation and design research. In particular, in the most naïve representation, it is plausible to explicate the idea that a new prototype is an improvement of an old one in view of a set of desired properties (SiS, p. 270) in formally the same "symmetric difference" way as being "more truthlike." The set of desired or wished for properties W is here supposed to be a subset of a given set RP of "relevant properties." To be precise, RP does not contain the equally relevant "counterpart properties," that is, if RP contains P, it does not contain *not-P*. Moreover, $RP-W$ is supposed to represent the set of undesired properties. In this set-up a prototype is partitioning in the sense that 'not having a desired property' may be assumed to be equally easy (or difficult) to establish experimentally as 'having an undesired property'.

One plausible refinement results when one withdraws the assumption that a property that is not desired is automatically undesired. This results in the division of (the potentially relevant) properties into three categories; say, desired, indifferent and undesired properties, denoted by W^+, I and W^-. This may be particularly useful in view of later possibilities to negotiate again about W^+. As long as prototypes are merely described by their properties, the minimal condition of being an improvement is of course that the new prototype has all the desired properties the old one has and the old one has all undesired properties the new one has (SiS, p. 280). However, changing the topic to the market of end products, producers may each have used their own $W^+/I/W^-$ division, and may even have succeeded in realizing products 'between their W^+ and W^-'. A consumer will have his own $W^+/I/W^-$ division. For judging whether he should prefer one of the products over the other, he may well use Van Benthem's formal definition, at the end of Section 3.3, assuming that he wants to take the intentions of the producers into account. If he does not want to do so, as we may normally expect, he is well advised just to use the minimal condition mentioned above. Hence, our full-blown partial example remains rather artificial, but suggests at the same time that it will be possible to find more realistic examples.

REFERENCES

Benthem, J. van (1982). The Logical Study of Science. *Synthese* **51**, 431-472.

Benthem, J. van (1987). Verisimilitude and Conditionals. In: T. Kuipers (ed.), *What is Closer-to-the-Truth?*, pp. 103-128. Amsterdam: Rodopi.

Darden, L. (1990). Diagnosing and Fixing Faults in Theories. In: J. Shrager and P. Langley (eds.), *Computational Models of Scientific Discovery and Theory Formation*, pp. 319-353. San Mateo: Kaufmann.

Darden, L. (1991). *Theory Change in Science: Strategies from Mendelian Genetics*. Oxford: Oxford University Press.

Kuipers, T. (1982). Approaching Descriptive and Theoretical Truth. *Erkenntnis* **18** (3), 343-378.

Lakatos, I. (1976). *Proofs and Refutations: The Logic of Mathematical Discovery*. Cambridge: Cambridge University Press.

Ruttkamp, E. (2002). *A Model-Theoretic Realist Interpretation of Science*. Dordrecht: Kluwer Academic Publishers.

Thagard, P. (1988). *Computational Philosophy of Science*. Cambridge, MA: The MIT Press.

REFINED TRUTH APPROXIMATION

Thomas Mormann

GEOMETRY OF LOGIC AND TRUTH APPROXIMATION[1]

> Is it not dangerously misleading to talk as if ... truth were located somewhere in a kind of metrical or at least topological space ...? (Popper 1963, p. 232)

ABSTRACT. In this paper it is argued that the theory of truth approximation should be pursued in the framework of some kind of geometry of logic. More specifically it is shown that the theory of interval structures provides a general framework for dealing with matters of truth approximation. The qualitative and the quantitative accounts of truthlikeness turn out to be special cases of the interval account. This suggests that there is no principled gap between the qualitative and quantitative approach. Rather, there is a connected spectrum of ways of measuring truthlikeness depending on the specifics of the context in which it takes place.

1. Introduction

This paper is a remote offspring of a discussion I had with T. A. F. Kuipers on some problems of his qualitative account of truthlikeness (cf. Kuipers 1997, 1997a, Mormann 1997). At hindsight, I feel that I had not brought home the points I wanted to make, namely, (i) there is no principled difference between the qualitative and quantitative accounts of truth approximation, and (ii) Kuipers' notion of structure similarity is too weak to be really useful. This paper may be considered as a new, constructive attempt to argue for (i) and

---

[1] The first part of the title is borrowed from Miller's stimulating paper *A Geometry of Logic* (Miller 1984). Miller is one of the few authors who has explicitly noticed the importance of a "geometry of logic" for matters of truth approximation. His conception of logical geometry is informed, however, by a strictly metrical conception of geometry. Nevertheless, in the case of Boolean lattices, he comes to similar conclusions as the interval account to be presented in this paper. As the maxim of this paper shows, already Popper may be credited of having adumbrated the idea that truth can be "located somewhere in a metrical or topological space". To be honest, Popper was not ahead of his times: already in the thirties of the last century Stone had laid the foundation of what may be called a topology of logic (cf. Davey and Priestley 1990, Grosholz 1985). This topology of logic, however, seems to have been ignored by philosophers interested in matters of truthlikeness. An attempt to improve this less than optimal state of affairs is made in (Mormann 2001).

In: R. Festa, A. Aliseda and J. Peijnenburg (eds.), *Confirmation, Empirical Progress, and Truth Approximation* (*Poznań Studies in the Philosophy of the Sciences and the Humanities,* vol. 83), pp. 431-454. Amsterdam/New York, NY: Rodopi, 2005.

(ii). More precisely, I want to sketch a geometry of logic which may serve as a general framework for a theory of truth approximation. It comprises Kuipers' "qualitative" account as well as the "quantitative" account pursued, for instance, by Niiniluoto (Niiniluoto 1987). The key for this geometry of logic is provided by Kuipers' concept of structurelikeness – suitably reformulated (cf. Kuipers 2000, hereafter referred to as ICR). According to him, structurelikeness is a triadic relation $s(x, y, z)$ to be interpreted intuitively as a comparative similarity relation. That is to say, $s(a, b, c)$ is to be read as "b is at least as similar to c as a is." Kuipers takes the relation s as a primitive notion of the "refined version" of his qualitative account of truthlikeness. Although the relation of structurelikeness plays a central role in the refined acount, its characterization remains incomplete. This is explicitly acknowledged by Kuipers. He is well aware of the fact that his "three minimal conditions" (see (2.1)) are not sufficient, and he welcomes any proposal. Hence, this paper may be read as an attempt to fulfil his request. Indeed, we will formulate more than a dozen axioms for the relation of comparative similarity. This shows that this concept is a versatile and non-trivial notion.

More precisely, I want to show that the framework of interval structures[2] offers a useful conceptual setting for explicating the notion of comparative similarity for the purposes of a general theory of truth approximation. This framework may be used as a unifying scheme that comprises Kuipers' qualitative accounts as well as quantitative accounts. The conceptual unification achieved by the interval account shows that there is more between the qualitative and quantitative account than one might have thought. Depending on the specific features of the situation the interval approach may offer taylor-made comparative similarity notions.

Whether my proposal is useful or not, is to be judged, of course, by the reader. In any case, I believe that the theory of truth approximation should make greater efforts to overcome what may be called the dialectics of definitions and counter-examples: author A proposes an allegedly watertight "correct" definition of truthlikeness, B brings forward a more or less contrived counter-example invented to show that A's proposal does not keep its promise, then A responds with an improved definition etc. *ad infinitum*. Examples and counter-examples are important but the theory of truth approximation should not be restricted to them. What is needed at least as urgently, are efforts to improve the existing conceptual frameworks. Kuipers' enduring efforts not withstanding, the present level of mathematical sophistication does not suffice to deal with the tricky concept of truthlikeness, or so I want to argue. One reason is that theories of truth approximation are inextricably framed in some

---

[2] The notion of interval as it is understood here, is much more general than the standard notion of an interval in a vector space or a metrical space.

kind of "logical geometry" and up to now, we do not have a clear understanding of what this "logical geometry" or "geometry of logic" may be. All too often, we rely on geometric intuitions that are determined by Euclidean prejudices. The geometry of logic, however, does not fit the standard Euclidean metrical framework. This is evidenced by the deficits and shortcomings of the various definitions of a "distance from the truth" that have been proposed in the last decades. The main aim of this paper is not to criticise these proposals – by inventing some new counter-examples, say. Rather, I'd like to be constructive proposing a new framework in which a general geometry of logic can be developed that is useful for matters of truth approximation.

The outline of this paper is as follows: in the section 2 we introduce the basic notions of the theory of interval structures. This theory provides a general framework of the more specific topics of truth approximation treated in the following sections. The qualitative accounts of truthlikeness are shown to be special cases of the interval account. Moreover, the basic notion of convex structure is introduced which will be used in later sections. In section 3 we present a series of axioms for interval structures that lead to a characterisation of the real numbers $\mathbf{R}$ as a special interval structure. This suggests that there is no principal gap between qualitative and quantitative structures. In section 4 it is shown that distributive lattices come along with a ready-made convex interval structure. In section 5 this structure is used for defining for any Boolean lattice $(B, \leq)$ new order relations $\leq_b$ ($b \in B$) that can be used for truth approximation that may be considered to be best possible in some sense. In section 6 we close with some general remarks on the nature of a geometry of logic, its relation to Euclidean geometry, and its role in the theory of truth approximation.

The proofs of many of the results presented in this paper are too lengthy to be given here, they may be found in the indicated literature or they are elementary so that the reader can work them out for himself.

2. Interval Structures

In this section we present the basic notions of the theory of interval structures (cf. van de Vel 1993, Coppel 1999) that will be used in the following sections. The theory of interval structures has turned out to be a versatile tool in many areas of mathematics. For instance, the theory of convex structures and many areas of geometry, functional analysis and other disciplines may be studied under the heading of interval structures. Interval structures are essentially equivalent with the structures defined by the structurelikeness relations

Kuipers uses in his refined account of truthlikeness (cf. ICR)). According to him, the triadic relation $s(x, y, z)$ of structurelikeness is assumed to have the following properties:

(2.1)   *Kuipers' Minimal Properties of Structurelikeness.* Let X be a set of structures, and $s(x, y, z) \subseteq X \times X \times X$ a triadic relation defined on X. The relation s is called a relation of *structurelikeness* if it satisfies the following requirements:

(M1)	$s(x, x, x)$	(s is centered)
(M2)	$s(x, y, x) \Rightarrow x = y$	(s is centering)
(M3)	$s(x, y, z) \Rightarrow s(x, x, y)$ & $s(y, x, x)$ &	(s is conditionally
	$s(y, y, z)$ & $s(x, y, y)$ & $s(x, z, z)$ &	left/right reflexive)
	$s(y, z, z)$	

In this definition, nothing depends on the fact that X is a set of structures, hence we may consider the relation s simply as a comparative similarity relation defined on a set X, be it a set of structures or not. Thus, for any X and any triadic relation s on X satisfying (2.1) the structure (X, s) may be called a similarity structure. General similarity structures are too weak to be useful. In order to develop a contentful theory of similarity structures, we need more specific axiomatic requirements for s. A rich source for this purpose can be tapped by observing that every similarity relation gives rise to the definition of intervals in the following way:

(2.2)   *Definition.* Let (X, s) be a similarity structure, and denote the power set of X by PX. For $x, y, z \in X$ the *interval* $[x, z]$ is defined as $[x, z] := \{y; y \in X$ and $s(x, y, z)\}$. The set $[x, z]$ is called the interval with the endpoints x and z. This recipe gives us an *operator* $I: X \times X \rightarrow PX$ defined by $I(x, z) := [x, z]$. The operator I is called the interval operator defined by s. Intervals are denoted by $I(x, z)$ or simply by $[x, z]$. The structure (X, I) is called an *interval structure.*

Every similarity relation $s(x, y, z) \subseteq X \times X \times X$ defines an interval operator $I: X \times X \rightarrow PX$, and viceversa, every interval operator I defines a similarity relation by $s(x, y, z) := \{(x, y, z); y \in I(x, z)\}$. Hence, both concepts are interdefinable. From now on, we stick to the notion of interval structures (X, I). This is a matter of taste only, everything we will have to say about interval structures (X, I) could be expressed in the language of similarity structures (X, s) as well.

A useful source for finding more specific conditions for interval operators, but in no way the only one, is to study the interval operators of Euclidean, and more generally, of metrical spaces. It should be noted, however, that the follo-wing conditions on interval operators are all strictly qualitative ones, i.e. they do not depend on the real numbers $\mathbb{R}$ or on the cardinality of the underlying set X. All interval structures considered in the following are assumed to satisfy the following basic axioms:

(2.3)   *Basic Axioms for Interval Structures.* Let (X, I) be an interval structure. The operator I is assumed to have the following properties:

(I1)	$a, b \in I(a, b)$	(Extensive Law)
(I2)	$I(a, b) = I(b, a)$	(Symmetry Law)
(I3)	$a, b \in I(c, d) \Rightarrow I(a, b) \subseteq I(c, d)$	(Convexity)

As is easily seen, the standard intervals of Euclidean space E satisfy (I1)–(I3). More generally, if (X, d) is a metrical space, i.e. a set X endowed with a real-valued function $d: X \times X \to \mathbb{R}$ satisfying the familiar metrical axioms,[3] then X can be rendered an interval structure (X, I_d) satisfying (I1)–(I3) by defining

$$(2.4) \quad I_d(x, z) := \{y;\ d(x, y) + d(y, z) = d(x, z)\}$$

The interval operator I_d is called the *geodesic* interval operator of the metrical space (X, d). Although (I1)–(I3) are stronger than (M1)–(M3) they are still too weak to yield a useful theory. Further restrictions have to be imposed in order to obtain reasonable operators. A first step in this direction is done by the following definition:

(2.5)   *Definition.* Let (X, I) be an interval structure. Then (X, I) is a *geometric* interval structure if I satisfies the following three conditions:

(G1)	$I(b, b) = \{b\}$	(Idempotent Law)
(G2)	$c \in I(a, b) \Rightarrow I(a, c) \subseteq I(a, b)$	(Monotone Law)
(G3)	$c, d \in I(a,b) \Rightarrow (c \in I(a, d) \Rightarrow d \in I(c, b))$	(Inversion Law)

An intuitive interpretation of these axioms[4] is obtained by considering the case of the Euclidean plane whose standard metric d is easily seen to have a geodesic interval operator I_d which satisfies (G1)–(G3). More generally, one can prove that the geodesic interval operator of every metric space is geometric in the sense of (2.5). At this stage it may be useful to point out that there exist many more interval structures than just those derived from metrical spaces. For instance, Kuipers' "naive" account of truth approximation can be

---

[3] A function $d: X \times X \to \mathbb{R}$ is called a metric if it is satisfies the following conditions for all $x, y, z \in X$: (i) $d(x,y) \geq 0$, (ii) $d(x,y) = 0$ iff $x = y$, (iii) $d(x, y) = d(y, x)$, (iv) $d(x, y) + d(y, z) \geq d(x, z)$. For a discussion of the various kinds of metrics and related concepts, especially adapted to matters of truth approximation, see (Niiniluoto 1987, Ch. 1).

[4] Kuipers' minimal conditions and the axioms formulated in (2.3) and (2.5) are related as follows: (I1) $\Rightarrow$ (M1), (M3), (G2) = (M2).

436   *Thomas Mormann*

cast in the framework of interval structures in a natural way, even if the resulting interval structure may not be very Euclidean. According to Kuipers' model theoretic or structuralist account (cf. ICR), empirical theories x, y, z are represented as subsets of some basic set U of models or structures. The true theory is denoted by t, the comparative relation of truthlikeness is denoted by MTL(x, y, t) to be read as "y is at least as truthlike as x is." Then the "naive" account asserts:

(2.6)   *Definition.* Let $x, y \subseteq U$ be theories. Then, the theory y is at least as truthlike as x (MTL(x, y, t))) iff (1) $y - t \subseteq x - t$ and (2) $t - y \subseteq t - x$ obtain. For later use we note that (1) and (2) are equivalent with the condition $\Delta(y, t) \subseteq \Delta(x, t)$; here, $\Delta(u, v)$ is the symmetric difference of u and v, defined by $\Delta(u, v) := (u - v) \cup (v - u)$.

The deficiencies of this account of truthlikeness are well known. Sometimes, however, it seems to me that its virtues are underestimated, for instance it is easily shown that it satisfies most of the adequacy conditions Niiniluoto recently formulated (Niiniluoto 1998, see also section 5). Of course, even if (2.6) may be considered as a reasonable sufficient condition it is certainly too strong to be necessary. According to (2.6) most theories turn out to be incomparable.[5]

Nothing in (2.6) depends on the fact that t is to be the true theory. Hence, MTL may be read as a triadic comparative similarity relation between theories conceived as sets of models or structures. In section (7.2.3) of (ICR) Kuipers gives a list of formal properties of MTL(x, y, z) which shows that MTL is a similarity relation which satisfies (2.1). The following proposition shows that this characterization can be improved:

(2.7)   *Proposition.* Let PU be the power set of the basic set U of structures or models. An interval operator I_P on PU is defined by

$$I_P(x, z) := \{y; x \cap z \subseteq y \subseteq x \cup z\} \ = \ \{y; \Delta(y, z) \subseteq \Delta(x, z).$$

I_P is the interval operator defined by MTL(x, y, z), i.e., MTL(x, y, z) iff $y \in I_P(x, z)$. Moreover, I_P satisfies the conditions (I1)–(I3) and (G1)–(G3).

This proposition may be proved either directly by carrying out some tedious calculations or by observing that it follows from general results given in (van de Vel 1993, Ch. I, § 4). It may be interpreted as the assertion that Kuipers' "naive" qualitative account of truth approximation is a special case of

---

[5] This shortcoming, however, can be overcome. As is shown in (Mormann 2001) for a countable first order language the set U of structures comes along with a ready-made interval structure. This structure is even induced by a metric. Thereby the realm of compatible theories is considerably enlarged. The proof relies heavily on Stone's representation theorem.

the interval account, namely, one that is based on similarity structures of the kind (PU, I_P).

Although (PU, I_P) is an interval structure that seems to be quite well-behaved in that its interval operator satisfies (I1)–(I3) and (G1) and (G3) the resulting interval geometry is very different from the familiar Euclidean one. Later we will see that I_P and similar operators lack many properties "really" geometric interval operators such as Euclidean ones have. Here we are content to mention just one: while for a Euclidean interval $[a, c]$ every $b \in [a, c]$ cuts the interval into two subintervals $[a, b]$ and $[b, c]$ with $[a, b] \cup [b, c] = [a, c]$ and $[a, b] \cap [b, c] = \{b\}$. This does not hold for the intervals of I_P as the reader may check for himself. Intuitively this means that Euclidean intervals are "slim" while those of I_P are "fat." These and other differences render the geometry of interval spaces such as (PU, I_P) quite different from standard geometry. This shows that the domain of intended applications of the theory of interval structures is much larger than standard geometry (cf. van de Vel 1993). The embedding of the "naive" approach into the interval approach reveals that the underlying interval operator I_P has relational properties that go far beyond the minimal conditions (M1)–(M3). This suggests that (M1)–(M3) can be complemented and strengthened by more specific structural requirements not only in the case of the "naive" account but also for the "refined" version. This is indeed the case but in this paper we will not go into the details of this topic.

After having shown that the theory of interval structures comprises the qualitative account of truth approximation let us come back to a more theoretical topic, to wit, the relation of interval structures and convex structures (or convexities). Although the notion of convexity occasionally has been mentioned in the literature on truth approximation (cf. Kuipers 1997, Goldstick and O'Neill 1988), this concept has been used only in an intuitive and metaphorical way. I think, it deserves a more thorough-going treatment. Convex structures have turned out to be a very versatile tool in many realms of mathematics (cf. van de Vel 1993, Coppels 1999) and the theory of truth approximation may be well advised to take notice of this fact.

Convex structures naturally show up in the framework of the theory of interval structures. Given an interval structure (X, I) a subset $C \subseteq X$ is said to be (interval) *convex* provided $I(x, y) \subseteq C$ for all $x, y \in C$. The class of all convex subsets of X is denoted by $\mathbb{C}X$ and called the class of convex sets of (X, I). $\mathbb{C}X$ is easily seen to have the following properties:

(2.8) *Definition and Lemma.* Let (X, I) be an interval structure. Then the class $\mathbb{C}X$ of convex subsets of X has the following properties:

(1) The empty set $\varnothing$ and the universal set X are convex.

(2) $\mathbb{C}X$ is stable under arbitrary intersections, i.e. if $D \subseteq \mathbb{C}X$, then $\cap\, D$ is in $\mathbb{C}X$.

(3) $\mathbb{C}X$ is stable for nested unions, that is, if $D \subseteq \mathbb{C}X$ is totally ordered by inclusion then $\cup\, D$ is in $\mathbb{C}X$.

The first two clauses of (2.8) render $\mathbb{C}X$ a closure system (cf. van del Vel 1993, p. 4). As is well-known, a closure system gives rise to a closure operator co: $PX \to PX$ defined by

$$(2.9) \quad \mathrm{co}(A) := \cap\{C \,;\, A \subseteq C \in \mathbb{C}X\}$$

The set $\mathrm{co}(A)$ is called the convex hull of A, it is the smallest convex set containing A. If $\mathrm{co}(A) = A$, the set A is convex. A set X endowed with a closure operator co which satisfies the requirements (2.8)(1)–(3) is called a convex structure or a convexity and is denoted by (X, co).

Any interval structure (X, I) defines a convex structure (X, co), co defined as in (2.9). For all x, y we have $I(x, y) \subseteq \mathrm{co}(\{x, y\})$. These sets are equal iff I is convex, i.e., I satisfies the axiom (I3). Hence, for convex interval structures (X, I) we have a quite smooth relation between interval structures and convex structures: we could have started with a convex structure (X, co) defining an interval structure (X, I) by $I(a, b) := \mathrm{co}(\{a, b\})$, or we could have gone the other way round. Hence, one may switch back and forth between the terminologies of interval and convex structures.

(2.10) *Examples.* The standard convex structure of Euclidean space E is induced by the standard interval structure which is the geodesic interval structure of the standard metric.

Let (X, d) be a metric space. A convex structure (X, co_d) is obtained by taking the convexity operator co_d defined by the geodesic interval operator I_d.

Let $(X, \leq)$ be a partially ordered set. A subset $C \subseteq X$ is convex provided $y \in C$ whenever $x \leq y \leq z$ and $x, z \in C$.

Convex structures have immediate applications for truth approximation. Assume that theories X, Y are represented as subsets of a basic set U that is endowed with a convexity operator co.

(2.11) *Definition. Qualitative Truthlikeness via Convexity.* Let (U, co) be a convex structure. Assume that operator co is defined by an interval operator I, and let x, y, $t \subseteq U$ be theories, t being the true theory. Then y is said to be at least as truthlike as x (to be denoted by $\mathrm{CTL}(x, y, t)$) iff the following two conditions are satisfied:

(1) $x \cap t \subseteq y$

(2) $y - (x \cap t) \subseteq \mathrm{co}((x - t) \cup (t - x))$.

The characterization of truthlikeness by CTL is quite similar to that of Kuipers' refined qualitative account using MTL_r (cf. Kuipers 2000, 10.2.1). Actually, (2.11) is somewhat weaker: since the interval operator I satisfies (G1), Kuipers' requirement (Ri) implies the condition (2.11)(1), and, similarly, his (Rii) implies (2.11)(2). Hence we get $MTL_r \Rightarrow CTL$. One easily constructs examples that show that the convexity condition is more plausible than Kuipers' which essentially[6] requires that any interval $[\alpha, \beta]$ with $\alpha \in X$ and all $\beta \in t$ contains at least one $\delta \in y$. This condition may not be satisfied if, intuitively spoken, the theory y is "between" x and t but is "too small" to be met by all intervals $[\alpha, \delta]$. If this happens, we may still be inclined to consider y as being at least as truthlike as x. This intuition is not confirmed by Kuipers' original definition. It is confirmed, however, by (2.11). I do not know cases for which (2.11) yields intuitively less acceptable results than Kuipers'. This does not mean, however, that (2.11) were fully satisfying. It may be a sufficient condition but surely it is not necessary.[7] This shortcoming it shares with all approaches of truth approximation based on some notion of convexity. They have difficulties in handling cases when two theories to be compared are "lying on different sides of the truth" (cf. Niiniluoto 1991).

A convexity approach in truth approximation gains further plausibility when we consider multiple comparisons of the following kind. Let $CTL(x, y, z, t)$ denote the relation "z is at least as truthlike as x and y." Then we may use the convex structure on U to define the following sufficient condition for $CTL(x, y, z, t)$:

(2.12) *Definition. Multiple Comparison of Truthlikeness.* Let $x, y, z, t, \subseteq U$ be theories, t being the truth. Then the relation $CTL(x, y, z, t)$ obtains if

$$(x \cup y) \cap t \subseteq z \subseteq \mathrm{co}(x \cup y \cup t).$$

A simple example of this case is the following: if x, y and t are represented by points of the logical space U, then $CTL(x, y, z, t)$ obtains if z is inside the triangle xyt. This is in line with our geometric intuition since the points inside the triangle xyt are considered as at least as similar to t as the vertices x and y.

Summarizing we may conclude that qualitative accounts such as Kuipers' nicely fit into the interval approach. They may even be enhanced if one takes into account that comparative similarity structures lead to convex structures in

---

[6] I leave out the technical complication caused by Kuipers' relation of comparability.

[7] A definition of type (2.11) using only the first clause has been proposed by Goldstick and O'Neill (1988). For a discussion of the advantages and disadvantages of this type of truthlikeness based on the concept of convexity, see Niiniluoto (1991).

a natural way. As it seems, convexity is a "qualitative" concept that provides the right degree of flexibility to deal with matters of qualitative truth approximation. In the next section we show that this is only one side of the coin. We pursue the conceptual explication of interval structures and convexities in a different direction showing that also the quantitative structure of real numbers $\mathbb{R}$ can be defined in terms that ultimately rest on the notions of interval and convexity.

3. From Similarity to Numbers

In this section we sketch a quick way to obtain the "quantitative" structure of real numbers $\mathbb{R}$ from a purely "qualitative" base of an interval operator $I(x, z)$ or a comparative similarity relation $s(x, y, z)$. This shows that there is no principled gap between "the qualitative" and "the quantitative." Rather, there is a connected spectrum of structures embracing qualitative as well as quantitative accounts. Up to now, solely the extreme ends of this spectrum have been used for matters of truth approximation, to wit, Kuipers' qualitative ones on the one end, and metrical approaches on the other characterized by the usage of real valued distance functions.

The construction of the quantitative from the qualitative is mathematically non-trivial, it may be considered as one of the main achievements of 19th century mathematics. It may be traced back to the work of mathematicians such as von Staudt, Steiner, Pasch and others. It found its classical formulation in Hilbert's epoch-making work *The Foundations of Geometry* (1899) and keeps on being a living topic of contemporary mathematical research (cf. van de Vel 1993, Coppel 1999). It goes without saying that we cannot give a detailed account here. The aim of this section is much more modest. We just sketch a series of more and more specific interval structures that end up in one that encapsulates the algebraic structure of real numbers $\mathbb{R}$. The structures in this series are characterized by a series of axioms in such a way that it does not seem possible to draw a line beyond that the character of the structures would switch from qualitative to quantitative. This may be considered as a "proof" that there is no principled difference between the two realms.

In the following we heavily rely on the presentation given in Coppel (1999). For the convenience of the reader we strictly follow Coppel's system of terminology, even if we repeat some axioms that have been mentioned already earlier.

(3.1)   *Axioms for Interval Structures.* Let (X, I) be an interval structure, and consider the following conditions for the interval operator I[8]:

- (L1) $I(a, a) = \{a\}$.
- (L2)   If $b \in I(a, c)$, $c \in I(b, d)$ and $b \neq c$ then $b \in I(a, d)$.
- (L3)   If $c \notin I(a, b)$, and $b \notin I(a, c)$, then $I(a, b) \cap I(a, c) = \{a\}$.
- (L4)   If $c \in I(a, b)$ then $I(a, b) = I(a, c) \cup I(c, b)$.
- (P)   If $c_1 \in I(a, b_1)$ and $c_2 \in I(a, b_2)$, then $I(b_1, c_2) \cap I(b_2, c_1) \neq \varnothing$.
- (C)   If $c \in I(a, b_1)$ and $d \in I(c, b_2)$, then $d \in I(a, b)$ for some $b \in I(b_1, b_2)$.
- (D)   If $a, b \in X$ and $a \neq b$, then $I(a, b) \neq \{a, b\}$.
- (S)   If $a, b \in X$ and if C is *a* convex subset of $I(a, b)$ such that $a \in C$, there exists a point $c \in I(a, b)$ such that either $C = I(a, c)$ or $C = I(a,c) - \{c\}$.

Drawing some elementary diagrams the reader may familiarize himself with the intuitive meaning of these axioms. These conditions are independent of each other (cf. Coppel 1999, p. 46). Moreover, each axiom taken for itself, isolated from the rest, is not particularly strong, and allows for many strange interval structures. For instance, the trivial interval operator I defined by $I(a,b) := \{a,b\}$ satisfies all axioms except (D). The collection (3.1) may be considered as the starting point for the study of a large number of different types of interval structures and their relations. For our purposes we only mention the following four:

(3.2)   *Definition.* Let (X, I) be an interval structure.

- (1) (X, I) is called a *convex geometry* if the interval operator I satisfies (L1) and (C).
- (2)   A convex geometry (X, I) is called a *linear geometry* if the interval operator satisfies (L2), (L3), (L4) and (P).
- (3)   A linear geometry (X, I) is called *dense* if the operator I satisfies (D).
- (4)   A linear geometry is called *complete*, if the axiom (S) is satisfied.

It should be noted that these requirements are all purely qualitative in Kuipers' sense, they neither refer to the cardinality of X nor do they depend on real valued distances or something like that. All have an intuitively appealing geometric meaning as becomes evident when we check that the standard Euclidean interval structure is indeed a dense linear geometry in the sense of (3.2). Complementarily, the list (3.2) may be used to chart the difference between interval structures of Euclidean type and those of type (PU, I_P) (as defined in 2.7). The latter do not satisfy, for instance, the axioms (L2)–(L4).

---

[8] The reader should bear in mind that all these conditions for interval operators $I(x, z)$ can readily be interpreted as conditions of comparative similarity relations $s(x, y, z)$.

Let us start with introducing a concept that is essential for any sort of geometry, to wit, lines. We will show that linear geometries really deserve their name in that they have lines that have many properties "real" geometric lines are expected to have. For any interval structure (X, I), lines are defined as "extensions" of intervals in the following way:

(3.3) *Definition.* Let (X, I) be an interval structure, and a and b distinct points. The line $<a, b>$ is defined to be the set of all points c such that either $c \in I(a, b)$ or $a \in I(b, c)$ or $b \in I(a, c)$.

$$< a, b > \quad := \quad \underline{\qquad\qquad \overset{a}{} \qquad\qquad \overset{b}{} \qquad\qquad}$$

Without further assumptions the lines of interval structures may have strange properties. For instance, they may branch or overlap in unexpected ways. As the reader may check (PU, I_P) has only one big line, namely PU itself. In order to obtain lines with familiar properties that resemble the standard lines of Euclidean geometry further axioms have to be imposed on the operator I. For linear geometries we mention the following result:

(3.4) *Proposition* (Coppel 1999, III, 1, Proposition 4, p. 50). Let (X, I) be a linear geometry. Given two distinct points $x, y \in X$, the points of the line $<x, y>$ may be totally ordered so that $x \leq y$ and so that, for any points $a, b \in <x, y>$ with $a \leq b$, the interval $I(a, b)$ consists of all c such that $a \leq c \leq b$. Morerover, this total ordering is unique. We define $a \leq b$ if either of the following two conditions is satisfied:

(1) $a \in [x, b]$ and either $y \in [x, b]$ or $b \in [x, y]$.
(2) $x \in [a, y]$ and either $b \in [a, y]$ or $y \in [a, b]$.

Thus, the lines of a linear geometry satisfy at least one of the more important requirements one expects to be satisfied for honest lines.

There is more in a geometry, however, than only lines. Lines are only the first family in a series of subsets that comprise planes, hyperspaces, etc. For vector spaces these subsets are called affine subspaces. In the following we will show that linear geometries possess a class of affine subsets which share many properties with the class of affine subspaces of linear vectorspaces. In particular, affine subsets may be used to define the concept of dimension for linear geometries. As we shall see, dimensional restrictrions may be used for an elegant characterization of those interval structures which lead to quantitative structures such as the real numbers $\mathbb{R}$. Lines may be used to define a convex structure on X by a closure operator based on lines instead of intervals:

(3.5) *Definition.* Let (X, I) be a convex geometry. A subset A of X is called an *affine set* if for $x, y \in A$ always $<x, y> \subseteq A$. Analogously as for intervals, this defines a convex hull operator. It is called the affine hull operator and denoted by 'aff'. For any $A \subseteq X$ we have $\text{co}(A) \subseteq \text{aff}(A)$, i.e the affine convexity is coarser than the interval convexity.

The concept of an affine subset can be used to define the concept of dimension of convex geometries. Let $S \subseteq X$ be a subset of the convex structure (X, co) and $e \in S$. Then e is called an *extreme point* of S iff $e \notin \text{co}(S - \{e\})$. For instance, for the standard Euclidean convexity, the extreme points of a triangle ABC are its vertices A, B, and C. A set is called *independent* if all its elements are extreme points. An independent set B is called a base of C iff $\text{co}(B)=\text{co}(C)$. Now we define the concept of dimension by stipulating that the convex interval structure (X, I) has *affine dimension n* if all bases of X are finite sets $\{x_1, ..., x_{n+1}\}$ satisfying $\text{aff}(\{x_1, ..., x_n\}) = X$. As is easily checked, this definition yields the correct result for Euclidean space E: points have the dimension 0, lines the dimension 1 etc. Now we have collected all the concepts that are necessary for stating the following remarkable result:

(3.6)   *Theorem.* If (X, I) is a complete, dense linear geometry with $\dim X > 2$, then every line $<a, b>$ of X has the structure of the real numbers $\mathbb{R}$ considered as an ordered field, i.e., the line $<a, b>$ can be endowed with pointwise operations of addition $+$ and multiplication $\bullet$ which obey the familiar properties of addition and multiplication of real numbers $\mathbb{R}$.

The proof of this theorem is far beyond the scope of this paper.[9] Nevertheless, some comments on this theorem may be in order:

(i)   Theorem (3.6) shows that a purely qualitative approach suffices to obtain the quantitative structure of real numbers $\mathbb{R}$.

(ii)   The requirement (S) of completeness is a reformulation and generalization of Dedekind's completion axiom.

(iii)   The dimensional restriction is necessary. For $\dim X \leq 2$, there are so called non-Desarguesian geometries for which (3.6) does not hold.

(3.6) gives some hints that a problem that hitherto seemed to be untractable by non-metrical accounts of truth approximation may be treated by qualitative accounts as well. To wit, how to assess likeness to truth if the competing theories x and y are "on different sides" of the truth. Consider the following oversimplified case: assume that the theories x and y whose truthlikeness has to be determined are elements of a conceptual space (U, I) of theories which satisfies the assumptions of (3.6). Let us make the further assumption that x, y, t are collinear. Then the following possibilities can obtain:

---

[9] The essential content of theorem (3.6) can already be found in Hilbert (1899).

(3.7)   (1) t _________ x _________________ y

(3.7)   (2) x _________ t _________________ y

The case (3.7)(1) is unproblematic: already if (U, I) is a linear geometry due to (3.4) we have a linear order $\leq$ on the line $<t, x>$ such that $t \leq x \leq y$ or $y \leq x \leq t$, i.e x is more truthlike than y. More interesting is (3.7)(2). In this case x and y are on different sides of the truth, so to speak. By (3.6) we may bestow the line $<t, y>$ with the additive structure of the real numbers $\mathbb{R}$, t playing the role of 0. Thereby we may compare the truthlikeness of x and y although they lie on different sides of the truth by either constructing the negative x^* of x or the negative y^* of y thereby reducing the case (3.7)(2) to the situations

(3.8) t _________ x^* _________ y or y^* _________ x _________ t

which can be handled already for linear geometries. Here, the negative x^* of x is defined by $x + x^* = x^* + x = 0$. That is to say, if x and y are on different sides of the truth t we say x is more truthlike than y if the relations $y \leq x^* \leq t$ or, equivalently, $y^* \leq x \leq t$, obtain.

Admittedly this is only a very abstract sketch of how linear geometries may be used for matters of truth approxiomation but the general lesson of (3.1) should be clear: there are more possibilities for truth approximation using comparative similarity relations or interval operators than the minimal conditions (M1)–(M3), or the metric account may suggest. Truth approximation is neither obliged to be content with the meager meal of the minimal conditions nor must it order the heavy dish of some quantitative approach based on unrealistic metrical assumptions. Rather, there is a rich supply of intermediate dishes that allow to formulate specific accounts of truth approximation for different domains. Summarizing we may say that the theory of truth approximation is well advised to subscribe to a moderate pluralism: for some areas, metric methods of truth approximation may be available, for others only weak "qualitative" methods may be appropriate, for still others interval structures of an intermediate kind may turn out to work best.

4. Lattices and Interval Operators

Up to now, we have drawn our inspiration for developing a theory of interval structures mainly from the standard geometry of Euclidean spaces and similar structures. This may lead the reader to conjecture that interval structures may be helpful for describing some or even all of the standard geometric structure in a somewhat unusual way but that is all. Already the reformulation of Kuipers' naive qualitative account in the framework of the interval structures indicates that this is wrong. Interval structures such as (PU, Ip) show that the theory of interval structures comprises more than standard geometry. In the following sections we are going to explore interval structures of type (PU, Ip) in some more detail. These interval structures arise from logical structures such as lattices. These structures usually are considered as unrelated to any sort of geometry. This is wrong. As will be shown in this section, any well behaved lattice comes along with a natural geometric interval structure. As will be shown in the next section, for Boolean lattices as they arise in propositional logic, these interval structures can be used to define an "objective", i.e., language independent notion of truthlikeness. First let us recall some basic definitions of lattice theory[10]:

(4.1)   *Definition.* An ordered structure $(L, \leq)$ is a lattice iff any finite subset M of L has an infimum inf(M) and a supremum sup(M) with respect to $\leq$, i.e. a greatest lower bound and a least upper bound. The infimum and the supremum of the empty set $\varnothing$ is denoted by 1 and 0, respectively. 1 and 0 are called the top and the bottom element. As usual, if $M = \{x, y\}$, inf(M) is denoted by $x \wedge y$ and sup(M) by $x \vee y$. A lattice is distributive iff $x \vee (y \wedge z) = (x \vee y) \wedge (y \wedge z)$. A distributive lattice is Boolean iff every x has an inverse x^* which satisfies $x \vee x^* = 1$ and $x \wedge x^* = 0$.

Lattices abound in mathematics and elsewhere (cf. Davey/Priestley 1990). Probably the best known example of a lattice is the Boolean lattice $(PX, \leq)$ of the power set PX of a set X endowed with set theoretical inclusion $\subseteq$ as order relation $\leq$. A closely related class of examples is provided by the lattices of propositions where the lattice order is defined by material implication and the logical operations of conjunction and disjunction play the role of the lattice theoretical operations of inf and sup.

As is well-known the structure of a lattice has two conceptually different but mathematically equivalent descriptions. On the one hand, the lattice structure may be characterized as a geometrical structure, i.e. as an ordered set $(L, \leq)$ whose order relation $\leq$ satisfies the conditions of (4.1), on the other hand, it may be characterized as an algebraic structure $(L, \wedge, \vee)$ where the algebraic operators $\wedge: L \times L \rightarrow L$ and $\vee: L \times L \rightarrow L$ have to satisfy the laws of idempotence, commutativity, associativity etc. (cf. Davey/Priestley 1990). The relation $\leq$ and the logical operators $\wedge$ and $\vee$ are related by the basic fact:

---

[10]   For a good introduction to lattice theory and order theory the reader may consult Davey and Priestley (1990).

(4.2) $x \leq y$ iff $x \vee y = x$ or, equivalently, $x \leq y$ iff $x \wedge y = x$

Thus, a lattice may be denoted by $(L, \leq)$ or, equivalently, by $(L, \wedge, \vee)$. It is this equivalence between a "geometric" order structure $\leq$ and a "logical" algebraic structure, encapsulated in the operations $\wedge$ and $\vee$, that lies at the bottom of the geometry of logic we are after. In the following we are going to exploit it in some more detail. As a first step, we define an interval structure for lattices:

(4.3)   *Definition.* Let $(L, \leq)$ be a lattice. Then an interval operator I_L can be defined on L by

$$I_L(x, z) := \{y;\ x \wedge z \leq y \leq x \vee z\}$$

At this point it may be convenient to point out the following: although the intrinsic interval operator of a lattice shares many features with its cousins of standard Euclidean geometry, there are also important differences. For instance, in Euclidean geometry the endpoints of an interval are essentially unique, i.e. from $[a, b] = [c, d]$ one concludes that $a = c$ and $b = d$, or $a = d$ and $b = c$. This does not hold for lattice theoretical intervals, as is easily seen by constructing a counter example. Or, Euclidean intervals are decomposable in the sense that for every $b \in [a, c]$ one has $[a, b] \cup [b, c] = [a, c]$ and $[a, b] \cap [b, c] = [b]$. This does also not hold for lattice intervals either. Hence one must be careful not to overstate the similarities between the two kinds of geometry. The geometry of lattices has many unusual features, one should not expect too great a resemblance with standard geometry.

Before we go on let us come back once more to Kuipers' "naive" qualitative account. As has been shown in (2.7), it is nothing but truth approximation based on the lattice interval operator I_P defined on $(PU, \leq)$. Actually, it does not use the full conceptual strength available. This becomes evident when we observe that the definition of I_P does not use the set theoretical structure of PU at all. That is to say, the fact that the theories are conceptualized as sets is irrelevant for the truth approximation of the naive account. The only relevant fact is that $(PU, \leq)$ is a Boolean lattice, the internal structure of the elements of PU, i.e. the fact that they are sets, does not play any role. Thus, in principle, Kuipers' naive qualitative account of truth approximation has nothing to do with the set theoretical representation of theories as sets of structures. It could have been carried out for any Boolean lattice or even any distributive lattice whatsoever. This does not hold for the refined account: It works with a similarity relation $s(x, y, z)$ defined on U *and* the interval structure induced by the Boolean structure defined for *PU*. Thereby it goes beyond a pure lattice theoretical account.

The interval structure of a lattice L encapsulates the intrinsic geometry of L. Hence, it should play an important role in matters of truth approximation, or so *I* want to argue. More precisely, the intrinsic geometry may be used to replace or at least amend the often vague "intuitive" arguments in favor or against some proposed definition of truthlikeness by precise structural arguments. It helps cut down considerably the unmanageable profusion of possible metrics or distance functions thereby offering a substantial contribution for defining a good measure for truthlikeness. An analogy with the corresponding problem in Euclidean space E may make this clear. For Euclidean space there exists a profusion of metrics or distance functions. Most of them are useless. Take the worst case, the trivial metric $d_t: E \times E \to \mathbb{R}$ defined by $d_t(x, y) = 1$ iff $x \neq y$, and 0 iff $x = y$. The same holds for most metrics that can be defined on E. Hence, we should accept a metric as admissible only if it is compatible with the underlying structure of E, in our case the topological structure or the interval structure. This is a special case of a general principle applied everywhere in mathematics and other sciences, where structural considerations play a role. For matters of truth approximation this means that any distance function employed for measuring truthlikeness should pass the test that it is compatible with the logical and geometrical structure of the domain on which it is defined. This idea will be put to work in the next section applying it to the classical case of truth approximation of Boolean propositional logic.

5. Base Point Orders and Language Invariance

In the previous section we have shown that lattices $(L, \leq)$ have an interval structure derived in a natural way from the order relation $\leq$. In this section we show that for Boolean lattices the interval structure can be used to construct new order relations $\leq_b$ on L. These relations can be used to define "objective" concepts of truthlikeness. Although these "base-point-order" concepts of truth approximation are not optimal in every respect they provide a framework in which more refined concepts of truth approximation may be developed. The following definition is basic for the rest of this section (cf. van de Vel 1993, Ch. 5.1*ff*):

(5.1) *Definition.*[11] Let (X, I) be an interval structure. For $b \in X$ a binary relation $\leq_b$ on X is defined by $x \leq_b y := I(b, x) \subseteq I(b, y)$. The relation $\leq_b$ is called the basepoint quasiorder of (X, I) at b.

---

[11] After having finished this paper I came to read the contribution of Burger and Heidema (this volume). It may be interesting to note that the relation of their "T-modulated Boolean algebra BT" (2.2.4) is just the opposite order relation as my "base-point order $\leq_T$. Moreover, their paper

The relations $\leq_b$ are clearly reflexive and transitive, i.e., they are quasiorders. For the basepoint quasiorders $\leq_b$ of distributive lattices we can prove more (cf. van de Vel 1993, 5.3.4. Proposition, p. 95):

(5.2) *Proposition.* Let L be a distributive lattice. The basepoint quasiorders $\leq_b$ enjoy the following properties:

(1)   For each b the relation $\leq_b$ is a partial order on L such that $\leq_b$ and $\leq_c$ are mutually inverse orders on the interval $[b,c]$.

(2)   For $x \leq_b z$ one has $[x, z] = \{y; x \leq_b y \leq_b z\}$ and the partial order $(L, \leq_b)$ is a $\wedge$-semilattice.

(3)   If b has a complement b^*, then $(L, \leq_b)$ is a distributive lattice and the lattice operations are given by

$$x \wedge_b y := (b \vee x) \wedge (b \vee y) \wedge (x \vee y) \quad \text{and}$$
$$x \vee_b y := (b^* \vee x) \wedge (b^* \vee y) \wedge (x \vee y).$$

(4)   The bottom element 0_b and the top element 1_b of $(L, \leq_b)$ are b and b^*, respectively. If L is a Boolean lattice we get $x \wedge_b x^* = b$ and $x \vee_b x^* = b^*$. Hence, the lattices $(L, \leq_b)$ are all Boolean lattices. In particular, we have $(L, \leq_0) = (L, \leq)$.

Proposition (5.2) may be used to tackle the problem of truth approxiomation in a conceptually new way which may be characterized by the slogan "truth determines truthlikeness" in the sense that the true proposition t defines its own order relation $\leq_t$ which determines truthlikeness. Our point of departure is the following definition:

(5.3) *Definition.* Let $(L, \leq)$ be a (finite) Boolean lattice whose elements are to be interpreted as propositions. Let $x, y, t \in L$. Then one may define

     x is at least as close to t as y is iff $x \leq_t y$.

In the following I'd like to argue that this definition is the base of a natural and useful concept of truth approximation for Boolean languages in the sense that it respects the intrinsic geometric structure of L and satisfies the most important intuitive adequacy conditions. Moreover it is language independent and gives rise to a well-behaved metric. As a first step in the entreprise of showing that (5.3) is a reasonable concept, the reader may check for himself that for $t = 0$ definition (5.3) gives the standard concept of approximation for contradiction 0, to wit, logical strength.

---

contains some nice diagrams (Fig. 6 and 7) showing how in some cases these new relations look like.

Recently, Niiniluoto has formulated the following list of "nice properties" that may be taken as a list of adequacy conditions a good notion of truth approximation should satisfy (Niiniluoto 1998, p. 3):

(5.4) Adequacy Conditions for the concept of truthlikeness.

> (TR1) t is more truthlike than any other theory a.
> (TR2) If a and b are true, and if b entails a, but not vice versa, then b is more truthlike than a.
> (TR2)' If a and b are false, and if b entails a, but not vice versa, b need not be more truthlike than a.[12]
> (TR3) If a is false, then $a \vee t$ is more truthlike than a.
> (TR4) A false theory may be more truthlike than a true theory.
> (TR5) A false theory may be more truthlike than another false theory.

(5.5) *Theorem.* The truth approximation defined by the base point order $\leq_t$ (5.3) satisfies (TR1)–(TR3) and (TR5).

The proof of (5.5) is an elementary calculation using the properties of $\leq_t$ given in (5.2). Actually the basepoint orders $\leq_t$ are not new at all, since one can easily prove the following lemma:

(5.6) *Lemma.* Let L be a Boolean lattice, and t, x, y elements of L. Then the basepoint order $\leq_t$ and the symmetric difference operator Δ (see (2.6)) are related by $x \leq_t y \Leftrightarrow \Delta(x, t) \leq \Delta(y, t)$.

Lemma (5.6) should be taken as evidence that Δ provides us with the correct measure of truthlikeness, since base point order relations are not ad hoc construals but reflect the natural geometry, not only of Boolean lattices but of lattices in general. The reader may wonder, whether (5.5) does not contradict the well-known theorem that Δ cannot render a false theory more truthlike than another false theory. Indeed, *this* claim is refuted by an elementary example given below. A closer look reveals, however, that the more careful formulations of that theorem assume that the true theory t is *complete* in the sense that for every proposition c either c or c^* is true (cf. Miller 1974, p. 169).[13] In the example we are going to discuss t will be not complete. Hence (5.5) does not contradict the classical results.[14]

---

[12] (TR2)' has been added by me. A special case of it appears in (Niiniluoto 1987) as the requirement "(M5) (Falsity and logical strength)" (p. 233).

[13] See for example Niiniluoto (1987, p. 233) or (1998, p. 9).

[14] Harris' more general claim that the "comparison theory" C (in our case t) need not be complete in order that C-approximation does not satisfy (TR5) is false. Let $C^* \vDash \{c^*; c \in C\}$. Then Harris erroneously infers from $c \in C^*$ that $c^* \in C$ (Harris 1974, p. 163). This is justified only if C is

Already Popper observed that "the whole truth" may be too big a class to serve as an adequate target for truth approximation. In my opinion, the point is not that the complete truth is "too big," rather one should say that the "whole truth" T has a rather special geometric *gestalt* making its approximation particularly difficult: in the propositional approach the whole truth T is a complete theory represented by an atom of a Boolean lattice, in the model theoretic approach it corresponds to a singleton, i.e., as a set having only one element (cf. ICR, Ch. 8.5). Nevertheless, the special case of a complete theory has attracted so much attention among the "friends of verisimilitude" (Niiniluoto 1998, p. 25) perhaps due to the fact that many of them subscribe to some Tarskian theory of truth for which the whole truth may appear as a quite natural concept. Be this as it may, with respect to the broader problem of comparing theories according to their comparative similarity there is no reason to restrict one's attention to complete theories. The following elementary example shows that without the assumption that t is complete some false theories may well be more false than others:

(5.7)   *Example.* Let $(L, \leq)$ be the standard Boolean algebra generated by two independent propositions h (it is hot) and w (it is windy). Consider the following elements:

$$a = hw \text{ or } h^*w^*, \quad b = hw, \quad t = h^*$$

Take t to be the true theory. Obviously, t is not complete since (h^*w) and $(h^*w)^*$ are false. Obviously, the theories a and b are false and b is logically stronger than a, i.e., $b \leq a$. Calculating the base point order $\leq_t$ yields $a \leq_t b$. That is to say, b is less truthlike than a. Hence, logical strength of false theories does not covary with truthlikeness, and the base point definition of truthlikeness (5.3) satisfies (TR2′) and (TR5). (For a more elaborated "structuralist" account of truth approximation for non-complete true theories see (ICR, Ch. 8.2)). In order to cope with (TR4) one has to introduce a more powerful structure than just an order relation $\leq_t$. This can be done as follows (cf. Grätzer 1998, p. 225 and van de Vel 1993, p. 119):

(5.8)   *Definition and Proposition.* Let $(L, \leq)$ be a finite Boolean lattice. Define the height function $h: L \to \mathbb{R}$ by $h(x) :=$ the length of a maximal chain $0 < \ldots < x$. Then on $(L, \leq)$ a metric d is defined by

$$d(x, y) := h(x \vee y) - h(x \wedge y)^{15}$$

---

complete. Tichý implicitly presupposes that t is complete (cf. Tichý 1974, p.157). In (Miller 1974, p. 169) the requirement that t has to be complete, is explicitly stated.

[15] It can be shown that d satisfies the requirements of an admissible metric in the sense of Miller (1984, 1994). In particular, the lattice interval structure and the geodesic interval structure defined

The appeal of base point orders is further enhanced by the following pleasing result:

(5.9) *Proposition.* Let d_t be the metric on $(L, \leq_t.)$ defined by the height function. Then measuring truthlikeness by d_t satisfies the requirements (TR1)–(TR5).

The satisfaction of (TR1)–(TR3) and (TR5) is inherited from the corresponding result for the base point order $\leq_t$. An example showing that (TR4) holds is constructed along the lines of (5.7) taking, for example, the tautology 1 as a true theory far away from the truth, i.e., $d_t(1, t) \geq d_t(a, t)$. Finally, with respect to language independence, one can prove the following result:

(5.10) *Theorem.* Let $f\colon (L, \leq) \rightarrow (L', \leq')$ be a Boolean isomorphism of the finite Boolean algebras L and L'. Let $t \in L$, and denote by d_t the metric induced by the height function on $(L, \leq_t)$. Then the order $\leq_t$ and the metric d_t are objective, i.e., invariant under the translation f in the sense that

$$x \leq_t y \Leftrightarrow f(x) \leq_{f(t)} f(y) \quad \text{and} \quad d_t(x, y) = d_{f(t)}(f(x), f(y))$$

Hence one may say that with respect to the general structural requirements (TR1)–(TR5) and the requirement of language independence the base point orders $\leq_t$ and their metrics d_t score quite well. They may even be considered as best possible: if we insist that truth approximation has to be invariant under all Boolean isomorphisms it is easily seen that the base point order approach is the only one that satisfies this requirement. The reason is that every Boolean isomorphism is determined by its values on the constituents, i.e., the atoms of B. If we consider all Boolean isomorphisms as admissible every permutation of the atoms of B defines an admissible isomorphism. In other words, just any set theoretical map which maps atoms to atoms defines an admissible Boolean isomorphism. This implies that there cannot exist a non-trivial metric or some other non-trivial structure preserving map on the set of atoms that is invariant

---

by d (see (2.4)) coincide. This may be considered as evidence that d is a "natural" metric for L. Of course, for finite Boolean lattices the metrics d are essentially equivalent to the logical probabilities of the lattices $(L, \leq)$. This, however, should be considered as a mere coincidence. For this contention I offer two evidences: first, that height functions are defined for a much larger class of lattices than Boolean lattices, to wit, (semi)modular lattices (cf. Grätzer 1998). Hence, height functions should be conceived as expressing the intrinsic geometric structure of lattices which is totally unrelated to any considerations of probability. Secondly, with the help of Stone's representation one can show that countable Boolean lattices $(B, \leq)$ may be endowed with a metric adapted to the order relation $\leq$ (Mormann 2001). This entails, for instance, that the Lindenbaum algebras of countable languages (of classical logic) have "logically adequate" metrics. As it seems, these metrics do not arise from any sort of logical probability.

with respect to just every permutation. Hence, the price for total objectivity, i.e., total language invariance, is that the geometric structure on the set of atoms of B must be trivial. Hence, the base point order approach is the best possible.

Of course, if we weaken the requirement of language invariance and admit that truthlikeness need not be invariant under all isomorphisms things look different. We might hope to endow the set of atoms of B with some structure that can cope with the more fine-grained aspects of truth approximation which necessarily escape the base point approach.[16] In a radical way, this is done in (Niiniluoto 1987). Technicalities aside, he proceeds as follows: he defines a metric on the set B of atoms (constituents) which then is extended to a metric on a larger set $D(B)$.[17] Not surprisingly, this metric turns out not to be language independent. This strategy can be pursued only if the base set B has a plausible non-trivial metric. Here, Kuipers' "qualitative" approach may get along with less: instead with a metric, it may start with some "qualitative," less demanding structure on B, and use it for defining a qualitative structure on a larger set $D(B)$. Kuipers' "refined" approach may be reconstructed along these lines. A simple case in question is (2.11), i.e. truth approximation by convexity.

Evidently, Miller's Δ-approach, Kuipers' "naive approach" and measuring truthlikeness using base point orders $\leq_t$ is essentially one and the same thing. Let us call it the classical approach. Then we may state that both Niiniluoto's quantitative as well as Kuipers' refined approach may be considered as refinements of the classical approach. Hence, the classical approach should not be considered as wrongheaded from the outset but rather as a good first approximation, and the theorists of truth approximation should know to esteem the virtues of a good approximation.

6. Concluding Remarks

The general aim of this paper was to gather some evidence for the thesis that the theory of truth approximation should be pursued in the framework of some kind of geometry of logic. In many quarters of philosophy, logic and geometry

---

[16] This reciprocity between objectivity and structure corresponds to a similar reciprocity in geometry that long ago was exhibited by Klein in his *Erlanger Programm*: the larger the transformation group of a geometry the more restricted the class of invariant properties. For matters of truth approximation this means, the larger the class of admissible translations the coarser the relations of truth approximation that are translation invariant.

[17] The classical pattern of this strategy is the Hausdorff metric on the space $F(X)$ of closed subsets of a metrical space X (see Coppel 1999, p. 185*ff*).

are still used to be considered as worlds apart.[18] Hence, the first obstacle a logic of geometry has to overcome is to dissolve this deeply entrenched prejudice. I have attempted to do this by showing that, contrary to common wisdom, geometry and logic share a common ground. In its simplest form, it may be conceived of as the equivalence between an algebraic and a geometric conceptualization of lattices. This equivalence makes it that logical structures come along with ready-made geometric structures that can be used for matters of truth approximation. Admittedly, these geometric structures differ from those we are accostumed with, namely, Euclidean ones. Hence, the geometry of logic is not Euclidean geometry. This result should not come out as a big surprise. There is no reason to assume that the conceptual spaces we use for representing our theories and their relations have an Euclidean structure. On the contrary, this would appear to be an improbable coincidence.

The bits and pieces of interval geometry and the theory of convex structures presented in this paper may not yet form a coherent whole that deserves to be called a geometrical theory of truth approximation. But at least, I hope to have shown, they point at a direction where one may find such a theory.

University of the Basque Country UPV/EHU
Department of Logic and Philosophy of Science
P.O. Box 1249
20080 Donostia-San Sebastián, Spain

REFERENCES

Coppel, W.A. (1999). *Foundations of Convex Geometry*. Cambridge: Cambridge University Press.

Davey, B.A. and H. Priestley (1990). *Introduction to Lattices and Order*. Cambridge: Cambridge University Press.

Goldstick, D. and B. O'Neill (1988). Truer. *Philosophy of Science* **55**, 583–597.

Grätzer, G. (1998). *General Lattice Theory*. Second Edition. Basel: Birkhäuser.

Grosholz, E. (1985). Two Episodes in the Unification of Logic and Topology. *The British Journal for the Philosophy of Science* **36**, 147–157.

---

[18] At the latest this is obsolete since Stone proved his famous representation theorems in the 30s of the 20th century which revealed the duality between "logical" Boolean lattices and "geometric" or "topological" Boolean spaces. Since then, mathematics has discovered deep connections between logic on the one hand and geometry and topology on the other which render the traditional apartheid untenable, even for philosophers. For a first introduction into this area see Grosholz (1985).

Harris, J. H. (1974). Popper's Definitions of Verisimilitude. *The British Journal for the Philosophy of Science* **25**, 160–166.

Hilbert, D. (1971[1899]). *The Foundations of Geometry.* La Salle: Open Court.

Kuipers, T.A.F. (1997) The Dual Foundation of Qualitative Truth Approximation. *Erkenntnis* **47**, 145–179.

Kuipers, T.A.F. (1997a). Comparative versus Quantitative Truthlikeness Definitions: Reply to Thomas Mormann. *Erkenntnis* **47**, 187–192.

Kuipers, T.A.F. (2000/ICR). *From Instrumentalism to Constructive Realism, On Some Relations between Confirmation, Empirical Progress, and Truth Approximation.* Dordrecht: Kluwer.

Miller, D. (1974). Popper's Qualitative Theory of Verisimilitude. *The British Journal for the Philosophy of Science* **25**, 166–177.

Miller, D. (1984). A Geometry of Logic. In: H.J. Skala, S. Termini, and E. Trillas (eds.), *Aspects of Vagueness*, pp. 91–104. Dordrecht: Reidel.

Miller, D. (1994). *Critical Rationalism: A Restatement and Defence.* Chicago: Open Court.

Mormann, T. (1997). The Refined Qualitative Theory of Truth Approximation Does Not Deliver: Remark on Kuipers. *Erkenntnis* **47**, 181–185.

Mormann, T. (2001). Truthlikeness on Countable First Order Languages (Unpublished manuscript).

Niiniluoto, I. (1987). *Truthlikeness.* Dordrecht: Reidel.

Niiniluoto, I. (1991). Discussion: Goldstick and O'Neill on 'Truer Than'. *Philosophy of Science* **58**, 491–495.

Niiniluoto, I. (1998) Verisimilitude: The Third Period. *The British Journal for the Philosophy of Science* **49**, 1–29.

Popper, K. ([1963] 1989). *Conjectures and Refutations.* 5th revised edition. London: Routledge and Kegan Paul.

Tichý, P. (1974). On Popper's Definitions of Verisimilitude. *The British Journal for the Philosophy of Science* **25**, 155–160.

Tichý, P. (1976). Verisimilitude Redefined. *The British Journal for the Philosophy of Science* **27**, 25–43.

van de Vel, M.L.J. (1993). *Theory of Convex Structures.* Amsterdam: North Holland.

Theo A. F. Kuipers

TOWARD A GEOMETRICAL THEORY OF TRUTH APPROXIMATION

REPLY TO THOMAS MORMANN

The contribution by Thomas Mormann also escapes the risk of the phenomenon that I indicate in my reply to Burger and Heidema, viz. losing track by confronting philosophy of science with advanced techniques in logic and mathematics. Whereas Heidema and his collaborators, in much of their previous work, and Zwart (1998/2001) have illustrated the fruitfulness of looking at truth approximation from the algebraic point of view, Mormann, following David Miller, makes an impressive plea for the geometric point of view for this purpose. Quite convincingly he argues in the last two sections that, assuming that theories fit into so-called interval structures, it is possible and hence advisable to define "objective" concepts of truthlikeness, that is concepts that naturally arise from these interval structures.

More generally, Mormann shows that *prima facie* different approaches to truthlikeness have more in common than one would expect. Whereas, for example, Zwart (1998/2001) draws a sharp distinction between what he calls "content" and "likeness" theories, such as my naive and refined account, respectively, Mormann points out that there is much conceptual continuity between them. Note that this is in line with my claim that the naïve theory is an extreme or idealized special case of the refined approach (ICR, p. 165, see also Kuipers, forthcoming). However, in contrast to my claim (e.g. ICR, pp. 258-60) that Niiniluoto's quantitative approach and my qualitative one are worlds apart, Mormann also argues that the quantitative approach can be reconstrued in a qualitative way.

Below I would like to come back to our dispute about the quantitative approach and further discuss the possibility of including theoretical terms and their reference in Mormann's geometrical approach.

In: R. Festa, A. Aliseda and J. Peijnenburg (eds.), *Confirmation, Empirical Progress, and Truth Approximation* (*Poznań Studies in the Philosophy of the Sciences and the Humanities,* vol. 83), pp. 455-457. Amsterdam/New York, NY: Rodopi, 2005.

The Reach of the Qualitative and the Quantitative Approach

As stated, and assuming that I understand Mormann correctly, his basic claim with respect to the quantitative versus qualitative approach to truthlikeness is that the former can be reconstrued in a qualitative way. Without entering into his geometric argument (culminating in Theorem 3.6) I would like to endorse Mormann's general claim "that there is no principled difference between the two realms" (p. 440). At issue is the meaningfulness of "going quantitative" in specific cases. Let me quote the most relevant passage of ICR (p. 258) in this respect:

> Let us therefore turn to the fundamental problem of a quantitative approach, the use of a distance function between structures. Although Niiniluoto's proposals for definitions of quantitative truthlikeness based on distances between structures [presented in Ch. 12 of ICR] are impressive, the problem is that apart from some exceptional cases, see below, there usually is nothing like a natural real-valued distance function between the structures of scientific theories, let alone something like a quantitative comparison of theories based on such a distance-function. And even in cases where it is technically easy to define a distance function, as in some paradigm examples of the structuralist approach, e.g., between the potential models of classical particle mechanics (CPM), as far as they share the same domain and time interval, such a distance function seems never to be used by scientists. The main reason obviously is that as soon as there are two or more real-valued functions involved indicating quite different quantities, and hence expressed in quite different units of measurement, any 'overall' distance function has equally many fundamentally arbitrary aspects. For we have to add in one way or another functions with values in meters (m) for position, kilograms (kg) for mass and newtons ($kg{\cdot}m/sec^2$) for force. One should not be misled by the fact that scientists frequently compare models quantitatively, even in terms of distances. However, what they do in such cases is quantitatively comparing one (type of) function, hence one aspect of such models, but that is not at stake. If we want to compare e.g., classical particle mechanics with special or general relativity theory we have to take full potential models into account.

Hence, I would agree with Mormann when he writes in closing Section 3:

> Summarizing, we may say that the theory of truth approximation is well advised to subscribe to a moderate pluralism: for some areas, metric methods of truth approximation may be available, for others only weak 'qualitative' methods may be appropriate, for still others interval structures of an intermediate kind may turn out to work best. (p. 444)

However, despite Mormann's formally correct point, and despite his plausible plea for a moderate pluralism in these matters, I would like to stick to the above arguments for reservations with respect to the quantitative approach as an approach to be generally advised, for those arguments are not really touched by the formal point. That is, in line with Mormann's suggestion, we should go as far as is appropriate.

However, there may be one way to question my attitude. Recall that I started this reply by agreeing with Mormann's final plea for specific concepts of truthlikeness that naturally arise from the relevant interval structures. If

Niiniluoto, or somebody else, wants to plea for the quantitative approach as an approach to be generally advised and were to advance convincing arguments for claiming, for example, that physicists would profit from designing interval structures that are specific enough to define real-valued distances between models of particle mechanics, then I would be convinced. As long as this is not the case, I would suggest we try to be no more specific than appropriate in specific cases and try to make general statements about truthlikeness and truth approximation on the basis of assumptions that are as weak as possible.

Theoretical Terms and Their Reference in Mathematical Perspective

Mormann's account is on the one hand very attractive: it streamlines various *prima facie* totally different approaches and opens new ways (see Section 5) for dealing with standing problems such as how to allow false theories that "outdistance" true ones (TR4) and how to minimize language dependence as much as possible. In this respect, Mormann's geometric approach and the algebraic approaches of Burger and Heidema (this volume) and of Zwart (1998/2001) are very promising. However, as far as I know, neither type of approach yet touches the problem of reference of theoretical terms. From the philosophy of science point of view this is a very important point, e.g. see Ruttkamp's contribution to the companian volume in which she favors a basically epistemological approach to reference, rather than a semantic-cum-metaphysical one. To be sure, from my reply to her, it is not only clear that I think that my formal, but semantic-metaphysically oriented treatment of this problem in Ch. 9 of ICR is formally a step ahead, but also that it still leaves much to be desired, semantically as well as metaphysically. I consider it as a challenge to both types of mathematical approach to try to deal with this problem in a way that is formally as well as philosophically adequate.

REFERENCES

Kuipers, T. (*forthcoming*). Empirical and Conceptual Idealization and Concretization. The Case of Truth Approximation. Forthcoming in (English and Polish editions of) Liber Amicorum for Leszek Nowak.

Zwart, S. (1998/2001). *Approach to The Truth. Verisimilitude and Truthlikeness*. Dissertation Groningen. Amsterdam: ILLC Dissertation Series 1998-02. Revised version: *Refined Verisimilitude, Synthese Library*, vol. 307. Dordrecht: Kluwer Academic Publishers.

Isabella C. Burger and Johannes Heidema

FOR BETTER, FOR WORSE:
COMPARATIVE ORDERINGS ON STATES AND THEORIES

ABSTRACT. In logic, including the designer logics of artificial intelligence, and in the philosophy of science, one is often concerned with qualitative, comparative orderings on the states of a system, or on theories expressing information about the system. States may be compared with respect to normality, or some preference criterium, or similarity to some given (set of) state(s). Theories may be compared with respect to logical power, or to truthlikeness, or to how well they capture certain information. We explain a number of these relations, study their properties, and unravel some of their interrelationships.

1. Introduction

Ordering relations on semantic entities like states (worlds) and on syntactic entities like theories are ubiquitous and play crucial roles in logic, artificial intelligence, and in the philosophy of science. Often a relation on states induces a relation on theories, or conversely. Comparative orderings may concern normality, preference, deduction, abduction, verisimilitude, similarity to given information, epistemic entrenchment, scientific progress, etc. These concerns play a significant part in the broad-ranging work done by Theo Kuipers in the philosophy of science.

In this paper we discuss some of the relevant comparative orderings on states and on theories and their relationships. *Section* 2 firstly explicates a classificatory ordering on states in the form of an "epistemic state" (*Section* 2.1) which employs information to yield a linear "pile" of classes of states. In *Section* 2.2 we explore some variants of the classical inference relation and its inverse, the classical conjecture relation, on theories, as well as blends of inference and conjecture relations induced by information, represented in the form of epistemic states. *Section* 2.3 generalizes some of the results in the previous section and places them in the context of a simple treatment of knowledge and belief.

Section 3 firstly expounds comparative relations on states (*Section* 3.1) which order states according to their structural similarity to a given state (or, more generally, set of states). In *Section* 3.2 these structural orderings on states

In: R. Festa, A. Aliseda and J. Peijnenburg (eds.), *Confirmation, Empirical Progress, and Truth Approximation* (*Poznań Studies in the Philosophy of the Sciences and the Humanities*, vol. 83), pp. 459-488. Amsterdam/New York, NY: Rodopi, 2005.

(expressing some sort of compatibility with given information) are employed to induce relations between theories comparing their similarity to the given information. Some of the relations in this section can be seen as expressing (aspects of) comparative truthlikeness of theories. The difference in the contents of *Section* 2 and *Section* 3 relates to what Zwart (1998) calls "content" and "likeness" approaches.

The context for our development is the case of languages for which the structures in which we are interested may be represented by elements of $2^{\mathbb{A}}$, where 2 is the two-element Boolean algebra $\{0,1\}$ with $0 < 1$ and $\mathbb{A} = \{A_1, A_2, ...\}$ is the set of atoms of the language. Examples are propositional logics and predicate languages with a set of symbols for constants, in a situation where we are only interested in interpretations for which the function interpreting constant symbols is a bijection to the domain – a frequent situation in artificial intelligence. This means that a *possible state,* a *possible world,* or an *interpretation* may be thought of in any of four equivalent ways:

1. a truth valuation v: $\{A_1, A_2, ...\} \rightarrow \{0, 1\}$;

2. the corresponding set $\{(\pm)A_1, (\pm)A_2, ...\}$ of literals (i.e. atoms A_i without or with a negation sign) – where A_i is taken when $v(A_i) = 1$ (true) and $\neg A_i$ is taken when $v(A_i) = 0$ (false);

3. the corresponding vector $v(A_1) v(A_2)...$ of zeros and ones (written without parentheses or commas);

4. the corresponding subset $v^{-1}(1)$ of $\mathbb{A}$ consisting of those A_i's which are true under the valuation $v,$ and of which v is the characteristic function.

We shall think of a possible state as a vector of zeros and ones; and we shall denote the set $2^{\mathbb{A}}$ of all possible states by $\mathbf{W}$. Illustrative examples employ the language with two atoms $A_1 = p$ and $A_2 = q$, with its four possible states 11, 10, 01 and 00 (where the first bit indicates the value of p, the second that of q).

A *theory* is a deductively closed set of sentences; and a possible state that satisfies a theory, i.e. makes all its sentences true, is a *model* of that theory. We shall denote the set of models of a theory X by the corresponding bold letter $\mathbf{X}$, or by *Mod*(X). This means that $\mathbf{W} - \mathbf{X}$ denotes the set of *non-models* of a theory X. The term *proposition* will, with benign ambiguity, indicate any of the following: a logical equivalence class of sentences; a representative sentence from such a class; the set of models of a (class of equivalent) sentence(s); the corresponding element of the Lindenbaum-Tarski algebra of the language.

Let us from now on (for the sake of brevity) call any set of possible states, (i.e. any subset of $\mathbf{W}$, or any element of the power set $\mathcal{P}\mathbf{W}$ of $\mathbf{W}$) a *configuration.* Note that in the case of the set $\mathbb{A}$ of atoms being *finite,* every configuration is the set of models of some theory (i.e. every configuration is *effable* or, *describable*, by a theory). Moreover, every theory may then be contracted to a single proposition which entails all its elements. Thus, when $\mathbb{A}$ is finite, there is a one-to-one correspondence between the set of all theories and the set of all configurations. This means that for any theory X (which may now be thought of as a single proposition), $\mathbf{W} - \mathbf{X} = Mod(\neg X)$. On the other hand, in the case of $\mathbb{A}$ being *infinite,* there are strictly more configurations than theories (Brink and Heidema 1989).

2. Classificatory Closeness

So we have a language suitable for expressing a certain type of information about possible worlds, or possible states of some system that interests us – the latter being represented by the set $\mathbf{W}$ of all truth valuations of the atoms of the language (as described above). Information can be represented and employed rationally in many ways. We firstly explore *epistemic states* which yield sortings (or categorizations) of the states into classes which (linearly) vary in their degree of acceptability with respect to information as expressed by some theory. We then alter the classical entailment relation and its inverse by employing information represented as epistemic states. A whole spectrum of information-dependent orderings – ranging from cautious to bold – is obtained by merging these alterations in different ways. Lastly, we generalize some of these orderings by employing epistemic states differentiating between knowledge and belief.

2.1. *Epistemic States*

An *epistemic state* (of an agent, if you like) is a total preorder (i.e. a connecting, reflexive, transitive relation) $\leq_{ep}$ on the set $\mathbf{W}$. So, an epistemic state partitions the set of states into equivalence classes (where 'equivalent' means 'comparable both ways') on which a linear order is induced. The higher up a state lies in the ordering $\leq_{ep}$, the higher its degree of preference (or, normality). The simplest epistemic state would have only one partition class – that of all possible worlds. An epistemic state with two classes dichotomizes the worlds: 'has the property' versus 'does not have the property'; 'normal' versus 'abnormal'; 'good' versus 'bad'; 'accepted' versus 'rejected'. An epistemic state with more partition classes realizes finer discriminations, and hence carries more information of a certain nature. One could consider richer epistemic states, consisting not just of a single total preorder on worlds, but of a list of such orders, representing comparisons of worlds on different criteria. For the purposes of this paper we stay with a single preorder.

When we have information about the system in the form of a theory T, and we want to construct an epistemic state representing this information only, we obtain an epistemic state $\leq_{ep}$ which partitions $\mathbf{W}$ into at most two classes: $\mathbf{T}$ (the set of models of T) and $\mathbf{W} - \mathbf{T}$ (the set of non-models of T). This means that for any two states x and y, $x \leq_{ep} y$ if and only if $x \notin \mathbf{T}$ or $y \in \mathbf{T}$.

If T is tautologous (i.e. logically true), then $\mathbf{T} = \mathbf{W}$. This yields an epistemic state with just one class – consisting of all the worlds, all of them equally "good" – a totally undifferentiated "cloud of unknowing," which carries no information. On the other hand, when T is contradictory (i.e. logically false), $\mathbf{T} = \varnothing$, and we obtain again an epistemic state with just one class. This time, however, the epistemic state carries "too much information" in the sense that all states are now "equally bad."

Thus, constructing an epistemic state built on a given theory T is easy. What about the converse route – extracting information in linguistic form (i.e. a theory T) from a given epistemic state? Well, as discussed above, when having an epistemic state with one class only, it may represent two cases: $\mathbf{T} = \mathbf{W}$ or $\mathbf{T} = \varnothing$. Should we exclude these two cases and restrict ourselves to epistemic states with at least two classes with the top one representing the "good" states, there may exist another stumbling-block: the configuration consisting of the top class of states might not be describable by a theory T. To overcome these obstacles, let us assume for the rest of *Section* 2 that we have a language available which is suitable for expressing the distinguishing properties of worlds sorted by an epistemic state into (unless mentioned otherwise) at least two classes and that we are able to describe the "good class" by a theory T, i.e. the top class is the set $\mathbf{T}$ of models of a theory T. This means then that the smaller the top class is, the logically stronger T will be. In the case of the top class being $\mathbf{W}$, we may then take T as any logically true theory. An empty top class (should we allow this) will yield a contradictory theory T.

This links up nicely with the theory of semantic information of Bar-Hillel and Carnap (1953) when we consider an epistemic state with two classes – the top class represents the "good," "accepted" states, while the bottom class consists of the "bad," "excluded," or "rejected" states. This theory must be distinguished from the "information theory" of the communications engineer, i.e. Hartley, Weaver and Shannon's mathematical theory of communication, which is about the statistics of message strings – not primarily about their meaning. Simply put, Carnap and Bar-Hillel tell us that the smallest atom of information that we can have – a *content-element* of semantic information – is the information that one particular possible world is *excluded*. In the case of the set $\mathbb{A}$ of atoms of our language being finite, we may think of this bit of semantic information as syntactically represented by the negation of the state description of the excluded world. To illustrate, consider the propositional logic generated by the two atomic symbols p and q. Entailment is depicted in the Lindenbaum-Tarski algebra $\mathcal{B} = (B, \vDash, \wedge, \vee, \neg, \bot, \top)$ of this language (Figure 1), where B is the set of propositions of the language (i.e. logical equivalence classes of sentences). Note that $\top$ denotes the equivalence class of tautologies; $\bot$ denotes the equivalence class of contradictions; and $p + q = \neg(p \leftrightarrow q) = (p \wedge \neg q) \vee (\neg p \wedge q)$. If X lies below Y in the diagram and X and Y are connected by a line (either directly or transitively), then $X \vDash Y$. The four state descriptions $p \wedge q$, $p \wedge \neg q$, $\neg p \wedge q$ and $\neg p \wedge \neg q$ correspond to the four possible worlds 11, 10, 01 and 00. The four content-elements of this language are $\neg p \vee \neg q$, $\neg p \vee q$, $p \vee \neg q$ and $p \vee q$, which are the four co-atoms of the Lindenbaum-Tarski algebra of this language – i.e. the four logically weakest non-tautological propositions of the language (Figure 1). They state the exclusion of, respectively, the worlds 11, 10, 01 and 00. Every proposition is the conjunction of the content-elements entailed by it.

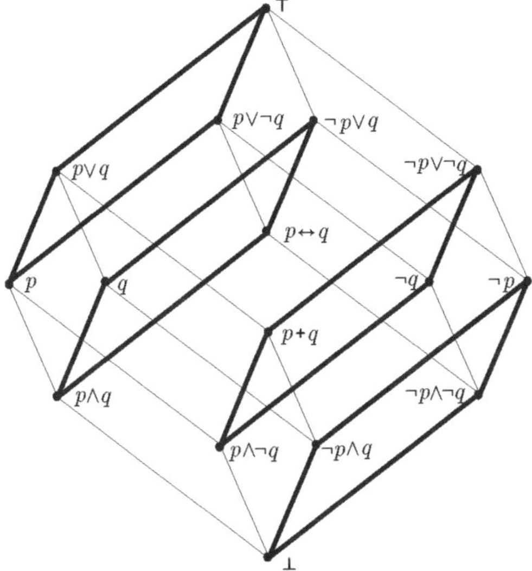

Fig. 1.The Lindenbaum-Tarski algebra $\mathcal{B}$ generated by $\{p, q\}$

The semantic information expressed by any theory logically equivalent to the proposition p with its two content-elements, $p = (p \vee q) \wedge (p \vee \neg q)$, is equivalently captured by the epistemic state of Figure 2. In general, however, an epistemic state may have more than two partition classes, and then carries more information than can be expressed by the theory describing the top class.

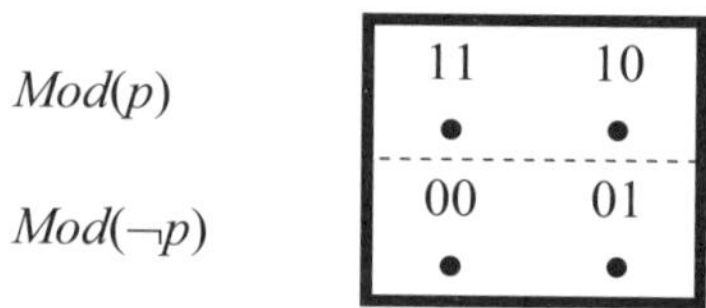

$$Mod(p)$$

$$Mod(\neg p)$$

Fig. 2. An epistemic state $\leq_{ep}$ induced by the proposition p

When the quest for more information has the aim of targeting the one actual world (state) among the possible worlds, then obtaining information can be seen as shoving worlds down into the "excluded" class and at each such step gaining one content-element of information. The quest ends with what in this context may be called complete information, when one world is left and all the others are excluded.

2.2. *Inference and Conjecture*

At the heart of logic lie the relations of *inference* (entailment, consequence) and its inverse, *conjecture* (abduction, antecedence) – whether in their classical or newer non-classical guises. Intuitively we may think of these relations on theories as dynamic, as processes: inference is a *cautious* process, since the truth of the premises ensures the truth – or at least the plausibility, i.e. the truth in all the most normally occurring of the relevant states – of the conclusion.

Conjecture (abduction), on the other hand, is a *daring* process; it moves from a given theory to a logically stronger one, which may be false (Aliseda-Llera 1997; Hendricks and Faye 1999; Hintikka 1998; Josephson and Josephson 1994; Kuipers 1999; Paul 1993). Exactly how daring depends, i.a. on the extra information available beyond the given set of propositions. Some extra information is needed to justify the conjectural move as rational.

In what follows we sketch a spectrum of comparative relations on theories – varying in degrees of cautiousness and boldness. They are obtained by merging an inferential and a conjectural relation to a single new relation, but with varying relative strengths of the two components in the blend.

2.2.1. *Classical Entailment*

The archetypal case where from theory X we cautiously infer (or deduce) theory Y is the case 'X semantically entails Y', i.e. $X \vDash Y$ or $Mod(X) \subseteq Mod(Y)$. The entailment relation $\vDash$ is cautious in the sense that it preserves truth, but may lose logical information in the sense that the consequent may be logically weaker than the antecedent. Inversely, we may also consider the classical case that from X we may daringly conjecture (or abduct) Y, i.e. $Mod(X) \supseteq Mod(Y)$, in which case truth may be lost, but logical information is preserved. We shall denote this relation (i.e. the converse of the entailment relation) which underlies classical abduction, by $\vDash^{-1}$, i.e. $X \vDash^{-1} Y : \Leftrightarrow Y \vDash X$. Entailment is depicted in the Lindenbaum-Tarski algebra $\mathcal{B} = (B, \vDash, \wedge, \vee, \neg, \bot, \top)$ of our illustrative language (Figure 1). The conjecture relation is obtained by just turning this picture upside-down. These two relations are completely independent of any (extra) data or information we may have available – they are the totally ignorant person's way of inferring and conjecturing.

2.2.2. *Entailment Relative to a Theory*

There is a simple way of taking information given by a set T of propositions into account: expand the relation $\vDash$ to $\vDash_T$:

$$X \vDash_T Y : \Leftrightarrow T \cup X \vDash Y.$$

As an illustration, we again consider Figure 1. Expanding $\vDash$ with respect to a theory T (representable by a single proposition), means collapsing the Lindenbaum-Tarski algebra modulo the congruence induced by T. Figure 3 illustrates $\vDash_T$ on the propositions of our p, q language, where T is taken as p. The annotations depict the four congruence classes modulo p. Thus, each annotation represents a set of four propositions contained in one of the four parallelograms drawn with thick lines in Figure 1. The logically strongest proposition in each class (i.e. the minimum element with respect to $\vDash$ in each class) is taken as the representative.

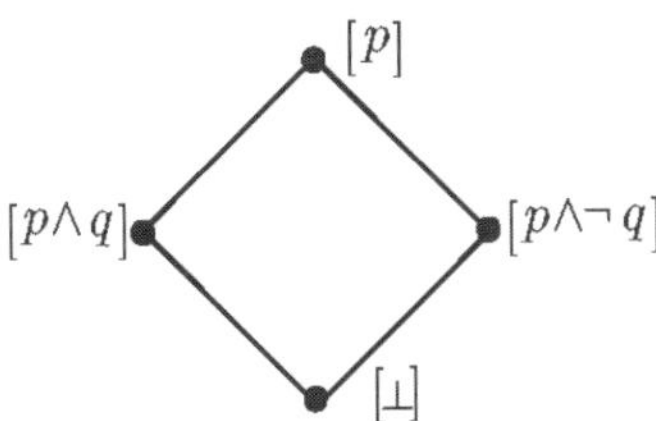

Fig. 3. The relation $\vDash_T$ with $T = p$ – annotations represent congruence classes modulo p

When T is tautologous, i.e. $T = \top$, the T-expanded relation $\vDash_T$ is just entailment; on the other hand, when $T = \bot$, all theories are equivalent under $\vDash_T$. Similarly we may expand $\vDash^{-1}$ to $\vDash_T^{-1}$, the inverse of $\vDash_T$ (i.e. $X \vDash_T^{-1} Y : \Leftrightarrow Y \vDash_T X$). This crude way of employing information T will turn out to be a very special case of our more sophisticated general ways of using information to alter $\vDash$ and $\vDash^{-1}$.

The relation $\vDash$ is *sound* in the sense that it preserves truth. This is no longer the case with its expanded version $\vDash_T$ (e.g., $\top \vDash_P p$ where p may be false even though $\top$ is obviously true). The relation $\vDash_T$ has however the property of being *T-sound* which will be explained in terms of the following definitions:

Definition 1. We call a theory X

> *T-true* iff $\mathbf{T} \cap \mathbf{X} = \mathbf{T}$;
> *T-false* iff $\mathbf{T} \cap \mathbf{X} = \varnothing$;
> *T-uncommitted* iff X is neither *T*-true nor *T*-false.

The entailment relation may be regarded as *T-sound* in the sense that no *T*-uncommitted theory is ever better than (above) a *T*-true theory, and no *T*-false theory is ever better than any *T*-true or any *T*-uncommitted theory. The relation $\vDash_T$ is however *T*-sound in a much stronger sense: in $\vDash_T$ *all* the *T*-true theories are above *all* the *T*-uncommitted theories, which are above *all* the *T*-false theories. However, all forms of soundness are lost in the conjecture relation $\vDash^{-1}$ and its *T*-expansion $\vDash_T^{-1}$ in the latter, *all* the *T*-false theories are above *all* the *T*-uncommitted theories, which are above *all* the *T*-true theories.

2.2.3. *Defeasible Inference*

In this subsection we describe how epistemic states on $\mathbf{W}$ induce plausible inference relations on theories. Defeasible inference, as in non-monotonic logics, describes the process of drawing plausible conclusions which cannot be guaranteed absolutely on the basis of the available knowledge, i.e. inference where one needs to go beyond the definite knowledge, but without making blind guesses. This may be done by employing a "default rule," justifying shrinking the set of models of X to that of X' – a theory logically stronger than X – in strengthening the process $X \vDash Y$. In other words: $X \mathrel{|\sim} Y$ (read as 'X defeasibly entails Y') if and only if $X' \vDash Y$ (where $X' \vDash X$). For the purposes of this paper, we shall require that a default rule should be expressible as an epistemic state, i.e. a total preorder $\leq_{ep}$, on the set $\mathbf{W}$ of all possible states (*Section* 2.1). Remember, the higher up a state lies within $\leq_{ep}$, the higher its degree of normality (or, preference). Within this framework of preferential model semantics (Shoham 1988), $Mod(X')$ is taken as the set of *maximal* models of X in $\leq_{ep}$ (where a state $w \in \mathbf{W}$ is "maximal" with respect to $\leq_{ep}$ in $\mathbf{X} \subseteq \mathbf{W}$ if $w \in \mathbf{X}$ and there is no $w' \in \mathbf{X}$ strictly above w in $\leq_{ep}$). Formally,

$$X \mathrel{|\sim} Y : \Leftrightarrow \mathbf{X'} \subseteq \mathbf{Y} \Leftrightarrow \max_{\leq_{ep}} Mod(X) \subseteq Mod(Y),$$

where $\max_{\leq_{ep}} Mod(X)$ denotes the set of maximal models of X with respect to $\leq_{ep}$.

Jumping from X to the logically stronger X' requires a (judicious) conjectural action, but that is where abduction ends – once X' is obtained, we proceed with classical inference. Referring the reader again to Figure 1, finding the defeasible consequences of a theory (i.e. a proposition) X in this language, means going down in the diagram to a proposition X' (according to an epistemic state $\leq_{ep}$), and then taking the filter above X' (i.e. the set consisting of X' and everything connected to X' higher up in $\models$). To illustrate, take $X = p \vee q$; and take $\leq_{ep}$ as the epistemic state induced by p, as depicted in Figure 2. The set of maximal models of X corresponds to the proposition p. Thus, Y is a defeasible consequence of $X = p \vee q$ if and only if Y is a consequence of $X' = p$; if and only if Y lies in the top thick-line parallelogram in Figure 1 (representing the filter above p).

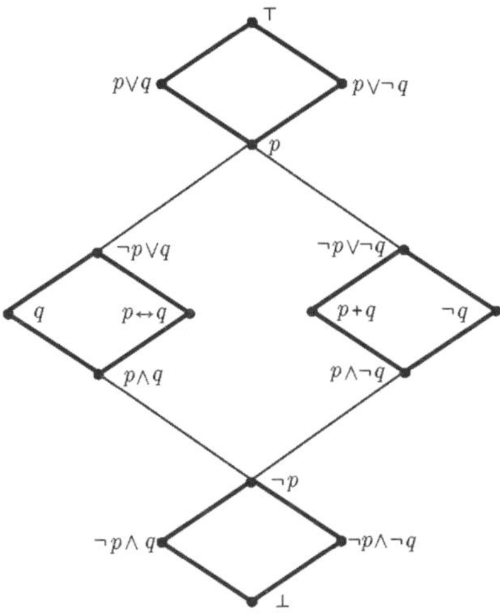

Fig. 4. B ordered by $\models_{T,\models}$ with $T = p$

When $\vert\!\sim$ employs an epistemic state which consists of only one partition class – that of all possible states, then $X = X'$ and $\vert\!\sim$ is thus $\models$. Employing an epistemic state with two classes in which the top class represents the set of models of a theory T, yields $\vert\!\sim$ as a refinement of the T-expanded relation $\models_T$. The refinement occurs only on those propositions which are inconsistent with T – they are ordered by $\models$ in $\vert\!\sim$. If we would refine $\models_T$ by "blowing up" the equivalence classes in $\models_T$ by means of $\models$, but retaining the $\models_T$-ordering between the classes, then it gives us a relation $\models_{T,\models}$, which is a refinement of any *epistemic entrenchment ordering related to T* (cf. Antoniou, 1997, p. 192, for a definition) on the propositions of the language. (In this paper we shall not discuss the one-one-correspondence between epistemic states and epistemic entrenchment orderings on propositions.) In general, the more (non-empty) partition classes there are in the epistemic state (and hence, the fewer states there are in every class), the higher the degree of boldness in the conjectural step of $|\sim$. In Figure 4 we illustrate $\models_{T,\models}$ for the p, q language where T is the proposition p. The equivalence classes in $\models_T$ (cf. Figure 3) are obtained by collapsing each of the four classes drawn with thick lines, but retaining the ordering between the classes (indicated by the thin lines). Collapsing all the classes except the bottom one gives us the relation $|\sim$ induced by the epistemic state depicted in Figure 2.

2.2.4. *Balancing Inference and Conjecture*

This and the following subsection tell how T can induce orderings on propositions (theories) which are mergers of entailment and (T-endorsed) abduction. The semantic information captured in the epistemic state given in Figure 2 may be viewed as corresponding to a theory logically equivalent to the proposition p. In general, every theory T corresponds to an epistemic state with two classes which sorts the states in $\mathbf{W}$ into the *accepted set Mod(T)* (the top class) and the *rejected set* $\mathbf{W} - Mod(T)$ (the bottom class). Suppose now that we have a (fixed) theory T, setting the norm for sorting the possible states into an accepted and a rejected class. T may represent a belief set, background information, or the (part of the) truth that is available to us now, or a hypothesis or theory that is under investigation and that we are trying to corroborate or refute by testing some of its consequences or even some conjectures that it may suggest. Against this fixed bench-mark T-sorting, we now want to compare two theories X and Y as to how well they sort the possible states into two classes. Looking at the scenario depicted in Figure 5, it seems natural to conclude that Y is "better" than X relative to the T-sorting, since the Y-sorting (represented by the thick line) lies "closer" to the T-sorting (i.e. the middle line) than the X-sorting (i.e. the thin line). (The convention in Figure 5 is that for, e.g., the X-sorting line, $Mod(X)$ lies above and $\mathbf{W} - Mod(X)$ below the line.) In formal terms this can be expressed by the following:

Definition 2. For any theories X, Y and T,

$$X \sqsubseteq_T Y : \Leftrightarrow (\mathbf{X} \cap \mathbf{T}) \subseteq (\mathbf{Y} \cap \mathbf{T}) \text{ and } [\mathbf{Y} \cap (\mathbf{W} - \mathbf{T})] \subseteq [\mathbf{X} \cap (\mathbf{W} - \mathbf{T})],$$

in which case we say that "*X sorts at best as well as Y does (relative to the T-sorting as the norm).*"

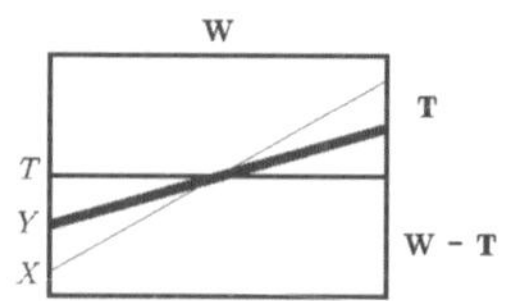

Fig. 5. *Y* sorts closer to *T* than *X*

In the special case of T being a theory in a finitely generated language (and hence, every configuration corresponds to a single proposition X), $\sqsubseteq_T$ is equivalently expressed by:

$$X \sqsubseteq_T Y : \Leftrightarrow X \wedge T \vDash Y \wedge T \text{ and } Y \wedge \neg T \vDash X \wedge \neg T$$
$$\Leftrightarrow (X \leftrightarrow T) \vDash (Y \leftrightarrow T).$$

This definition (which is studied mathematically in depth – in the context of Boolean algebras – by Burger and Heidema 2002) has previously been proposed by Miller (1978) and Kuipers (1987b). (Kuipers refers to this definition as "the naive definition" – cf. Kuipers 1992a; 1992b; 1997; 2000 (chapter 7) and 2001, chapter 9.) The proposal was launched in the context of the study of *verisimilitude*, i.e. *truthlikeness* (Brink 1989; Festa 1987; Oddie 1986; Kieseppa 1996; Kuipers 1987a; Niiniluoto 1987 and 1998; and Zwart 1998), but is unsatisfactory from that point of view, since it has some properties (leading to, e.g., Tichý's famous "child's play objection") which are contrary to our intuitions about truthlikeness. We refer the reader to Zwart (1998) for an excellent discussion on many of the properties (philosophical as well as algebraic) of this naive relation. What we suggest is that the relation need not be read as "X is at best as *close to the* (part of the) *truth* expressed by T as Y is," nor something similar in terms of *being true*. Nor has the relation primarily to do with logical strength of theories. It compares the *similarity of the X-sorting to the T-sorting* (or, the *compatibility* of X with T), expressed syntactically by $X \leftrightarrow T$ and semantically by $[\mathbf{X} \cap \mathbf{T}] \cup [(\mathbf{W} - \mathbf{X}) \cap (\mathbf{W} - \mathbf{T})]$, with the *similarity of the Y-sorting to the T-sorting* (or, the *compatibility* of Y with T), expressed correspondingly in Y. It says that the *sorting-information* in Y is equal to or better than that in X, given that in T as the norm. (In *Section* 3.1 we compare the compatibility of state x with state t to the compatibility of state y with t in a similar way.)

For a proposition T, the relation $\sqsubseteq_T$ induces a Boolean algebra $\mathcal{B}_T = (B, \sqsubseteq_T, \sqcap_T, \sqcup_T, \neg, \neg T, T)$, which has T as its top, $\neg T$ as its bottom, the same complementation $\neg$ (negation) as the Lindenbaum-Tarski algebra $\mathcal{B} = (B, \vDash, \wedge, \vee, \neg, \bot, \top)$, and meet and join operations which are related to those in $\mathcal{B}$ ($\wedge$ and $\vee$) as follows:

$$X \sqcap_T Y = (X \wedge Y) \vee [(X \vee Y) \wedge \neg T];$$
$$X \sqcup_T Y = (X \vee Y) \wedge [(X \wedge Y) \vee T].$$

The original Lindenbaum-Tarski algebra $\mathcal{B}$ (e.g., Figure 1) and the new T-*modulated* Boolean algebra $\mathcal{B}_T$ are isomorphic under the isomorphism $m_T : B \to B$, $m_T(X) := (X \leftrightarrow T)$, which is its own inverse, and hence also establishes an isomorphism in the opposite direction (Burger and Heidema 1996). The underlying Boolean group structure in $\mathcal{B}$ plays a very interesting role in this modulation processes, and may clarify many of the interesting philosophical properties of the relation $\sqsubseteq_T$, but we shall not elaborate on it in this paper. Full details appear elsewhere (Burger and Heidema 2002). To illustrate, we show $\mathcal{B}_P$ for the p, q language in Figure 6. Note that in $\mathcal{B}_P$ the four state descriptions of the four possible worlds sit ordered in a way (take a sneaking peek at Figure 13) which obviously contains the same information as the p-sorting of Figure 2. Hence one may see the ordering $\sqsubseteq_T$ on theories as a natural extension of the original epistemic ordering on the states.

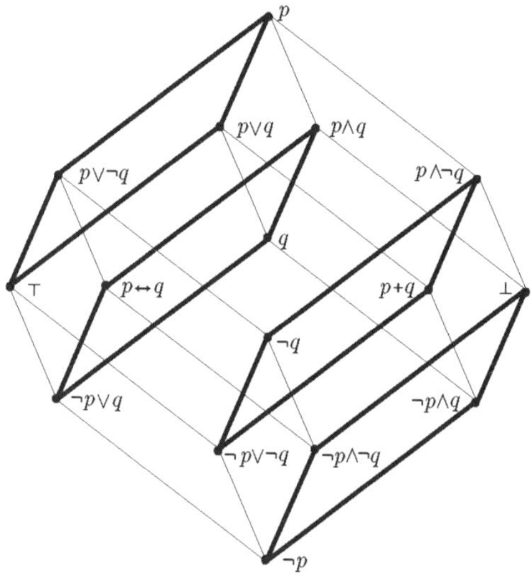

Fig. 6. $\mathcal{B}_P$ generated by $\{p, q\}$

Now for a closer look at the role played by the thick and thin lines in the diagrams of Figure 1 and Figure 6. Consider the definition of $\sqsubseteq_T$ as expressed in a finitely generated language: $X \sqsubseteq_T Y : \Leftrightarrow X \wedge T \vDash Y \wedge T$ and $Y \wedge \neg T \vDash X \wedge \neg T$. The first part, $X \wedge T \vDash Y \wedge T$, or equivalently $X \wedge T \vDash Y$, holds when $X \vDash Y$. The second part, $Y \wedge \neg T \vDash X$, holds when $Y \vDash X$. (Note that the first part is equivalent to Kuipers' (1992a; 1998) 'instantial clause'; the second part to his 'explanatory clause'.) Thus the first part expands entailment, while the second part expands the inverse relation $\vDash^{-1}$. Since the parts are equitably combined by 'and', the compatibility relation $\sqsubseteq_T$ may be seen as an entwinement of inference and conjecture: $\sqsubseteq_T = \vDash_T \cap (\vDash_{\neg T})^{-1}$. Formally this entwinement balances the two processes of cautious deduction and daring abduction, but the relation $\sqsubseteq_T$ can be viewed as dynamic – it becomes more boldly abductive the logically stronger the theory T becomes. We shall now explore this mixed nature of $\sqsubseteq_T$ in more detail.

The "Janus"-property (with respect to deduction and abduction) of $\sqsubseteq_T$ is best revealed when observing the ordering inside the congruence classes modulo T and inside the congruence classes modulo $\neg T$. To illustrate we treat the special case of $T = p$ in the p, q language, and ask the reader to follow what we say in the diagram of Figure 1 for $\vDash$ and Figure 6 for $\sqsubseteq_T$. When for a pair of propositions (X, Y), $X, Y \in B$, $Mod(X)$ and $Mod(Y)$ differ by a single state, we say that (X, Y) is a *step*. If (X, Y) is a step and $X \vDash Y$, then (X, Y) is an *inferential step*, while in the opposite case, when $Y \vDash X$, (X, Y) is a *conjectural step*. The 32 (pairs of) steps of the p, q language are represented by the 32 lines in Figure 1 and by the same 32 lines in Figure 6, although now arranged differently. Whereas in Figure 1 every step upward is inferential and every step downward is conjectural, this is no longer the case in $\mathcal{B}_T$. In Figure 6, depicting $\sqsubseteq_P$, we call a step upward *p-enhancing*, since it goes from a sorting to another one which is more similar to the p-sorting. A step downward in $\sqsubseteq_P$ is *p-diminishing*. Some p-enhancing steps are inferential, i.e. they go up in both $\vDash$ and $\sqsubseteq_P$. These are represented by the *thin* lines (circumscribing the congruence classes modulo $\neg T$) going in the same direction in $\vDash$ and $\sqsubseteq_P$. The other p-enhancing steps are conjectural, i.e. they go up in $\sqsubseteq_P$ but down in $\vDash$. They are represented by the *thick* lines (circumscribing the congruence classes modulo T), which go in opposite directions in $\vDash$ and $\sqsubseteq_P$.

So, if $X \sqsubseteq_T Y$, then there is at least one chain of steps, each either inferential or conjectural (but all T-enhancing), linking X to Y. Moreover, whenever $X \sqsubseteq_T Y$, $X \wedge Y$ and $X \vee Y$ both lie between X and Y in $\sqsubseteq_T$. Thus the relation $\sqsubseteq_T$ classifies each inferential or conjectural step that we may take from anywhere as either T-enhancing or T-diminishing. This may guide, illuminate and facilitate the logical dynamics in many ways. One example: Suppose T is any theory, belief, hypothesis, diagnosis, assumption, etc., and suppose that we want to corroborate T by testing and verifying logical consequences of $T \cup B$, where B is background information. One surmises that T-enhanced verified consequences of $T \cup B$ represent stronger corroboration of T than T-diminished verified consequences of $T \cup B$ (Burger and Heidema 2000).

In the definition of $X \sqsubseteq_T Y$ the inferential component, $X \wedge T \vDash Y \wedge T$, and the conjectural component, $Y \wedge \neg T \vDash X \wedge \neg T$, have equal weight. However, in spite of its conjectural component, the relation $\sqsubseteq_T$ stays T-sound: no T-false theory or T-uncommitted theory is ever better than any T-true theory, and no T-false theory is ever better than any T-uncommitted theory. This is due to the inferential component. (Note that the ordering of the congruence classes modulo T, relative to each other, are the same in $\vDash$ and $\sqsubseteq_T$.) The conjectural

 Isabella C. Burger and Johannes Heidema

component is responsible for ordering the elements inside the congruence classes modulo T.

The logically stronger T becomes, the bolder the relation $\sqsubseteq_T$ becomes. We explain what we mean in terms of the finitary case: If $T = \top$, then $X \sqsubseteq_T Y$ iff $X \vDash Y$. In this case, the congruence classes modulo T are singleton sets; and there is just one congruence class modulo $\neg T$, namely B, the set consisting of all propositions. $\top$ is the only T-true proposition; $\bot$ is the only T-false proposition; and every step upward (T-enhancing step) is inferential (i.e. there are only thin lines in the diagram of $\sqsubseteq_\top$). When the logical power of T is strengthened (i.e. when T moves down in the Lindenbaum-Tarski algebra B), the congruence classes modulo T become larger (and hence, those modulo $\neg T$ smaller) – yielding a larger class of T-true (and hence, also of T-false) propositions. This means that the number of T-enhancing conjectural steps

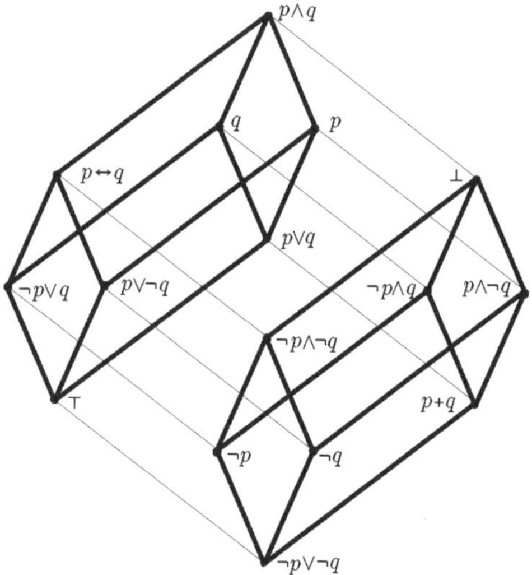

Fig. 7. $B_{p \wedge q}$ generated by $\{p, q\}$

(moving upwards by means of a thick line) increases, while the number of T-enhancing inferential steps (moving upward by means of a thin line) decreases. The very special case when we know exactly which world is the actual world, i.e. the case when T has one model only, is, in our illustrative language, represented by any one of the four propositions $p \wedge q$, $p \wedge \neg q$, $\neg p \wedge q$ or $\neg p \wedge \neg q$ lying just above $\bot$ in B (Figure 1). In Figure 7 we illustrate the Hasse diagram of B_T for $T = p \wedge q$. The congruence classes modulo $p \wedge q$ are circumscribed by the two parallelepipeds drawn with thick lines. The top one consists of all T-true propositions, the lower one of all T-false ones (there are no T-uncommitted propositions) and both these classes are daringly ordered by $\vDash^{-1}$. The thin lines connecting them are T-enhancing inferential steps. When T is contradictory, i.e. $T = \bot$, $X \sqsubseteq_T Y$ iff $X \leftrightarrow \bot \vDash Y \leftrightarrow \bot$ iff $\neg X \vDash \neg Y$ iff $Y \vDash X$ – meaning that all upward steps are conjectural and that there are only thick lines. In this case, there is only one congruence class modulo T, namely B; all the propositions are T-true as well as T-false.

In classical logic contradiction is catastrophic: once a logically false proposition can be deduced, every proposition is entailed (cf. Figure 1). Yet there are situations – belief revision, theory change, database update, truth maintenance, and paraconsistent reasoning – in which one needs to extract consistent information from inconsistent data in a way that is optimal with respect to, e.g., logical strength. In these situations, one may not want everything to be plausible from $\bot$. In $\sqsubseteq_T$, not every theory is a T-enhancement of $\bot$ (as discussed above): An increase of information leads to a decrease of the "logical power" of $\bot$ in $\sqsubseteq_T$ in the sense that the logically stronger the data T becomes, the higher $\bot$ moves up in $\sqsubseteq_T$. When, in the finitely generated case, T represents complete information (e.g., Figure 7) $\bot$ becomes innocuous: $\bot$ has only itself and T as T-enhancements. In this way we may view $\sqsubseteq_T$ as a logical system that can to some extent handle reasoning with inconsistency.

2.2.5. *Favoring Inference or Conjecture*

In $\sqsubseteq_T$ deduction and abduction are balanced. Tipping the scale to favour either the inferential or conjectural component, yields interesting variants of the relation $\sqsubseteq_T$. Instead of combining the two components by 'and', one could combine the two aspects "lexicographically" (i.e. by the "dictionary" ordering). This means that, when comparing two theories X and Y, we give one component precedence over the other one and that the less favoured one only becomes relevant when X and Y are equivalent with respect to the favoured component. Rather than formally defining the two induced lexicographically variants of $\sqsubseteq_T$, we discuss these variants by means of diagrams for $T = p$ in our illustrative language.

The diagram on the left in Figure 8 shows the *cautious* variant, $\sqsubseteq_T^c$, i.e. the relation in which preference is given to the *inferential* component. Since all propositions in the thick-line parallelograms (circumscribing the congruence classes modulo p) in Figure 1 and Figure 6 are equivalent with respect to the inferential component, $\sqsubseteq_T^c$ may be seen as the resultant of the following process: Firstly let $\vDash_T$ order the equivalence classes modulo T (modulo p, in this case – yielding the ordering depicted in Figure 3). Then, while retaining the $\vDash_T$-ordering between the classes, "blow up" each class by ordering the elements in the class by $\vDash^{-1}$. In other words, regarding T-soundness, the relation $\sqsubseteq_T^c$ behaves like the relation $\vDash_{T,\vDash}$ (as discussed in *Section* 2.2.3): all *T*-true theories are above all *T*-uncommitted theories, which are above all *T*-false theories. The difference between the two relations $\sqsubseteq_T^c$ and $\vDash_{T,\vDash}$ lies in the ordering inside the congruence classes modulo *T*. In $\sqsubseteq_T^c$ when two theories are equivalent relative to *T*, preference is given to the logically stronger one.

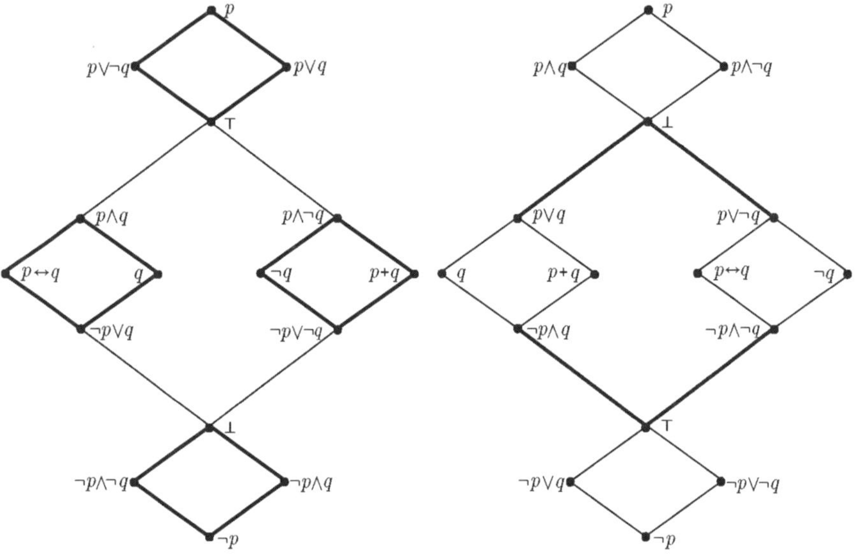

Fig. 8. *B* ordered by $\sqsubseteq_p^c$ (left) and $\sqsubseteq_p^d$ (right)

On the other hand, in the *daring* variant, $\sqsubseteq_T^d$, as depicted by the diagram on the right in Figure 8, the *conjectural* component has precedence over the inferential component. To obtain the relation $\sqsubseteq_T^d$, we firstly employ $\vDash_{\neg T}^{-1}$ to order the congruence classes modulo $\neg T$ (i.e. the classes circumscribed by thin-line parallelograms in Figure 1 and Figure 6), and then, while retaining the $\vDash_{\neg T}^{-1}$-ordering on the classes, "blow up" each class by ordering the elements in every class by $\vDash$. This yields a much bolder or daring model for the comparison of theories with respect to a theory *T* than $\sqsubseteq_T^c$ or even $\sqsubseteq_T$: some *T*-false propositions are regarded as better than some *T*-uncommitted propositions, and even better than some *T*-true propositions (compare, e.g., the positions of $\top$ and $\bot$). Thus, there is no overall *T*-soundness whatsoever in $\sqsubseteq_T^d$. The only conservatism noticeable, lies in the ordering of the theories inside the congruence classes modulo $\neg T$ (i.e. the four classes circumscribed by thin lines) in Figure 8 (diagram on the right) – which is done according to entailment, and which has as consequence that *T* is the top element and $\neg T$ is the bottom element.

2.3. *Belief and Knowledge*

The relation $\sqsubseteq_T$ is built on an epistemic state with two classes – sorting the possible states into the models and non-models of T, where T may represent, e.g., information about a system or a set of beliefs. In this section we consider an epistemic state $\leq_{ep}$ with more than two classes and study the analogue of $\sqsubseteq_T$ for this more general case. This enables us to distinguish logically weaker – and hence more secure – *knowledge* from logically stronger – and hence bolder – *beliefs*. (There is, of course, an extensive literature on knowledge and belief. For our present purposes we propose a very simple construal of these notions.) We now call the (non-empty) top equivalence class the set of *accepted* states, and the bottom class the set of *rejected* states. The union of the classes between the top and the bottom class will be the set of *undecided* states (Figure 9). Given an epistemic state with a language in which every configuration is effable, the theory R corresponding to the accepted class of states is called its *belief set*, while the theory S corresponding to the set **W-** (*rejected class*) of states is called its *knowledge set*. Note that $R \vDash S$. Looking semantically at our beliefs **R**, they should allow all the states which were found empirically ("the accepted instances" or "realized possibilities" as they are called by Kuipers 1992a). When we look syntactically at our knowledge S, we may think of it as – what Kuipers (1992a) calls – "the strongest accepted law," which means semantically that **S** excludes only those states which we must reject for empirical reasons (i.e. Kuipers' "undesired possibilities" or "counter examples"). We can now generalize the $\sqsubseteq_T$-relation:

Definition 3. For any theories X, Y, R and S with $R \vDash S$,

$$X \sqsubseteq_{[R,\,S]} Y : \Leftrightarrow (\mathbf{X} \cap \mathbf{R}) \subseteq (\mathbf{Y} \cap \mathbf{R}) \text{ and } [\mathbf{Y} \cap (\mathbf{W} - \mathbf{S})] \subseteq [\mathbf{X} \cap (\mathbf{W} - \mathbf{S})],$$

in which case we say that "X sorts at best as well as Y does (relative to the $[R, S]$-sorting as the norm)." Figure 9 illustrates this in two ways.

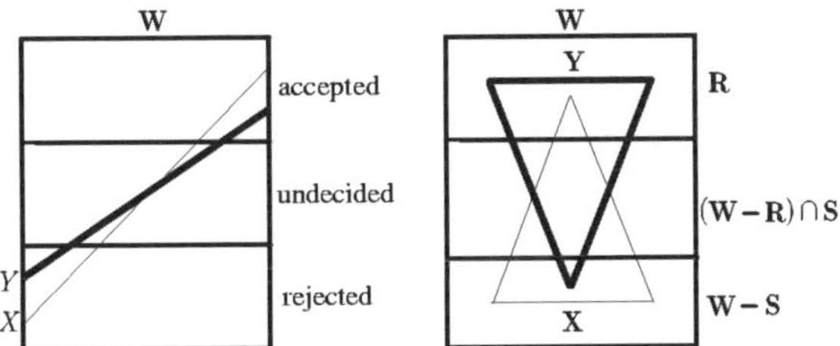

Fig. 9. General epistemic state; Y sorts better than X with respect to $[R, S]$

This definition, which does not refer explicitly to epistemic states, is also studied by Kuipers (1992a; 2000) in the context of verisimilitude. Kuipers calls it "the asymmetric naive success-dominance definition." What we propose is that the relation $\sqsubseteq_{[R, S]}$, similar to the relation $\sqsubseteq_T$, could also be seen as an ordering on theories comparing the similarity of the sortings induced by the theories to a benchmark $[R, S]$-sorting (i.e. an epistemic state differentiating between knowledge and belief as described above). Rather than dwelling further on the philosophical issues, we briefly point out a few properties of the relation $\sqsubseteq_{[R, S]}$. To simplify things, we restrict this discussion to a finitely generated language.

We generalize the relation $\sqsubseteq_T$ by replacing the role of T by a similar role, but this time played by the interval $[R, S] = \{U \in B \mid R \vDash U \text{ and } U \vDash S\}$ in B between belief and knowledge.

Propositions whose model sets differ only on the set of undecided states are equivalent in the new ordering. The equivalence classes are also intervals in B and they become the elements of the new Boolean algebra, say $\mathcal{B}_{[R, S]}$ which has $[R, S]$ as top and $[\neg S, \neg R]$ as bottom (Burger and Heidema 1997; 2000). The latter paper also explains how the new $\sqsubseteq_{[R, S]}$-relation may be employed to deliver a specific procedure for belief change in the context of the so-called AGM approach (Alchourrón, Gärdenfors and Makinson 1985).

To come again to a classification of (at least some) inferential and conjectural steps as being either enhancing or diminishing, we define a lexicographical ordering on the set of propositions B as follows: Construct the quotient Boolean algebra $\mathcal{B}_{[R, S]}$ and then pick some $T \in [R, S]$ which interests you. Now "blow up" each element of $\mathcal{B}_{[R, S]}$ by seeing it no longer as a single element, but as the class of elements of B that it is, and order them by $\sqsubseteq_T$, but retain the $[R, S]$-ordering between the classes. We call B with the resulting order $\mathcal{B}_{[R, T, S]}$ This is of course no longer a Boolean algebra. Rather than give all the mathematical details, we illustrate in Figure 10 the results for the p, q language with $[R, S] = [p \wedge q, p \vee q]$ and $T = p$. Those inferential and conjectural steps which give a comparable pair in $\mathcal{B}_{[R, T, S]}$ (diagram on the left), may now be classified as $[R, T, S]$-*enhancing* or $[R, T, S]$-*diminishing*. If you consider the belief $R = p \wedge q$ and the knowledge $S = p \vee q$ corresponding to the epistemic state depicted in Figure 10 (diagram on the right), then the Boolean algebra $\mathcal{B}_{[R, S]}$ is obtained when collapsing the four classes drawn with thick lines in the diagram of $\mathcal{B}_{[R, T, S]}$ to four single elements. The four propositions within each of the classes are ordered amongst themselves by $\sqsubseteq_p$ (as in Figure 6).

If in the total (lexicographic) picture Y lies higher than X, then this says that Y sorts better than X given a norm, which now consists of:

i) the general background of an epistemic state which accepts 11, rejects 00, and is undecided about 10 and 01; and

ii) superimposed on (but not replacing) that, the more specific state p, which decides to accept 10 (with 11) and reject 01 (with 00).

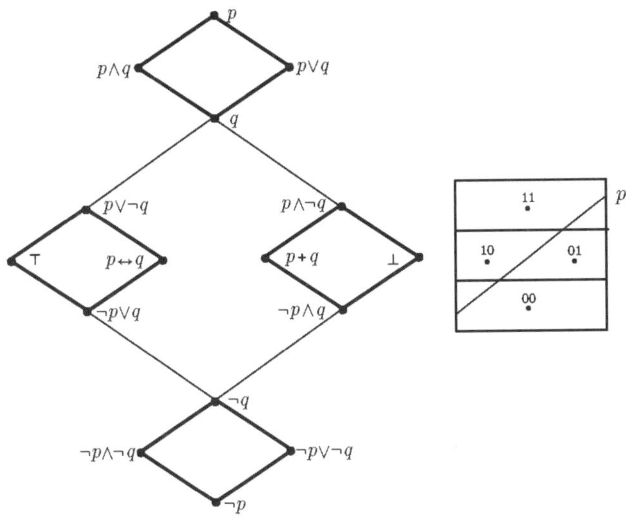

Fig. 10. $\mathcal{B}_{[R,\,T,\,S]}$ (left) and $\leq_{ep}$ (right) with $R = p \wedge q$, $T = p$, $S = p \vee q$

3. Structural Similarity

In this section we firstly explore ways in which a theory T (representing information about the system) can induce orderings on **W** which are more refined than epistemic states with two classes. These orderings, expressing some sort of structural or compositional (in terms of the atoms A_i) compatibility with the information T, are then employed to induce data-compatibility or data-similarity power orderings on theories via orderings on their model configurations.

3.1. *Similarity between States*

Sorting the possible states into an epistemic state $\leq_{ep}$ consisting of the two classes **T** and **W** – **T** is a very coarse way of differentiating between the states. Sometimes this $\leq_{ep}$ provides us with all the necessary information that we need.

But suppose we need to differentiate not only between the models of T and non-models of T, but also amongst the models of T or amongst the non-models of T. What is a rational way of doing that? A case in point is a theory T

equivalent to the proposition $p \wedge q$ in our illustrative language. According to this given information and the induced epistemic state, the state 11 is accepted and the states 10, 01 and 00 are rejected. One intuitively feels though, knowing that the actual state is 11, that, amongst the *non-models* of $p \wedge q$, the state 00, which differs from 11 on every atom and hence describes a state which is the exact opposite of the actual world, should be regarded as worse than both the states 10 and 01 which agree with 11 on one atom. On the other hand, in the case when T is equivalent to the proposition $p \vee q$, amongst the *models* of $p \vee q$, one might consider 11 as "better" than both 10 and 01, since both 10 and 01 agree with the non-model 00 on one atom. One thus might want to refine $\leq_{ep}$ to an order on **W** in which either the elements of **W** − **T** or the elements of **T** are no longer equivalent – an order which reflects that some non-model states "come closer to being a model of T" than others (or, that some model states "come closer to being a non-model of T" than others). The Boolean algebra $\mathbf{W} = \mathbf{2}^{\mathbb{A}}$ with its product order $\leq_{pr}$, induced by the ordering 0 < 1 of the Boolean algebra **2**, plays an instrumental role in our proposals of refining $\leq_{ep}$:

For every $x, y \in \mathbf{W}$,

$$x \leq_{pr} y : \Leftrightarrow x_i \leq y_i \text{ (in } \mathbf{2}\text{) for all } i.$$

Figure 11 depicts $\leq_{pr}$ on $\mathbf{W} = \{11,10,01,00\}$.

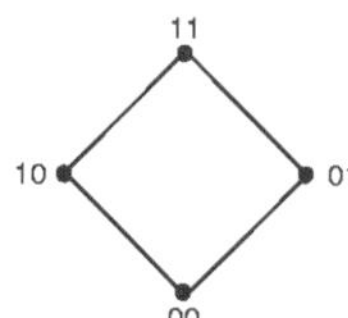

Fig. 11. The product ordering, $\leq_{pr}$, on $\mathbf{W} = \{11,10,01,00\}$

We now want to think of the elements of **W** also in another way, namely as *comparison* vectors – expressing the degree of compatibility or agreement or similarity of two states of the world:

Definition 4. Given any two elements t and x of **W**, we define $t(x) \in \mathbf{W}$ by

$$[t(x)]_i := \begin{cases} 1 & \text{if } t_i = x_i, \\ 0 & \text{if } t_i \neq x_i, \end{cases} \quad \text{for any } i.$$

The vector $t(x)$ $(= x(t))$ expresses the compatibility or structural similarity of two states x and t: the more 1's in $t(x)$, the closer state t and x resemble each other (Heidema and Labuschagne 1990). (Note that when $\mathbb{A}$ is finite, then the number of 0's in $t(x)$ is the *Hamming distance* between t and x.)

If $t = x$, then $[t(x)]_i = 1$ for every i and one may schematically think of $t(x)$ as the vector 11... – expressing *maximum compatibility* between the vectors t and x. On the other hand, $t(x) = 00...$ expresses *maximum incompatibility* – the case when $t_i \neq x_i$ for every i. Every other vector in between these extreme cases represents a partial compatibility between t and x. In other words: the higher up $t(x)$ lies within $(\mathbf{W}, \leq_{pr})$, the better the concordance of t and x; and hence, if $t(x) \leq_{pr} t(y)$, then x *is at most as compatible with t as y is*. Note that in the relation $t(x) \leq_{pr} t(y)$ single states are compared in exactly the same way as configurations are compared in *Definition* 2 (*Section* 2.2.4). In another paper (Burger and Heidema 2002) we explain the following in detail: For fixed $t \in \mathbf{W}$ we define a new ordering on $\mathbf{W} : x \sqsubseteq_t y : \Leftrightarrow t(x) \leq_{pr} t(y)$. Then $t(\cdot) : \mathbf{W} \to \mathbf{W}$, x a $t(x)$ is a Boolean isomorphism from $(\mathbf{W}, \leq_{pr})$ onto $(\mathbf{W}, \sqsubseteq_t)$. Also, for $t, x \in \mathbf{W}$, $t(x)$ is given by the derived Boolean operation $\leftrightarrow$ in $(\mathbf{W}, \leq_{pr})$: $t(x) = t \leftrightarrow x = (t \wedge x) \vee (\neg t \wedge \neg x)$. This means $(B, \sqsubseteq_T)$ is similar to $(\mathbf{W}, \sqsubseteq_t)$ in abstract Boolean terms.

The idea of *compatibility between two possible states* can now be expanded to the notion of *compatibility between a possible state and a configuration*.

Definition 5. Let $\mathbf{T} = Mod(T)$. For every $x \in \mathbf{W}$ we define

$$\mathbf{T}(x) := \{t(x) \mid t \in \mathbf{T}\}.$$

The set $\mathbf{T}(x)$ of comparison vectors represents x in so far as compatibility with the data T is concerned. Roughly speaking, the higher the set $\mathbf{T}(x)$ lies within $(\mathbf{W}, \leq_{pr})$, the higher the compatibility of the state x with the data T or the configuration $\mathbf{T}$.

In Table 1 we illustrate the set $\mathbf{T}(x)$ for each of the four possible states of our illustrative language for the two cases $T = p$ (i.e. $\mathbf{T} = \{11, 10\}$) and $T = p \vee q$ (i.e. $\mathbf{T} = \{11, 10, 01\}$).

Table 1. $\mathbf{T}(x)$ when $T = p$ and $T = p \vee q$

x:	11	10	01	00
$\mathbf{T}(x)$ when $T = p$:	{11, 10}	{10, 11}	{01, 00}	{00, 01}
$\mathbf{T}(x)$ when $T = p \vee q$:	{11, 10, 01}	{10, 11, 00}	{01, 00, 11}	{00, 01, 10}

It is easily seen that 11... $\in \mathbf{T}(x)$ if and only if $x \in \mathbf{T}$ – meaning that x is maximally compatible with one model of T. To discuss the dual situation, we define the notion 'anti-state'.

Definition 6. For every state $x \in \mathbf{W}$, the *anti-state* of x is defined as the state x^* such that $[x(x^*)]_i = 0$ for every i.

Thus x^* is the state which differs from x with respect to every coordinate. This means $00... \in \mathbf{T}(x)$ if and only if $x^* \in \mathbf{T}$ – meaning that x is maximally incompatible with one model of T. (Please note: *not maximally compatible* does not mean the same as *maximally incompatible*.) Consider now the case $T = p \vee q$, i.e. $\mathbf{T} = \{11, 10, 01\}$. The possible state 11 is then overall (of the four possibilities) the most compatible with T since $11 \in \mathbf{T}$, but $(11)^* = 00$ is not a model of T. Next in line are the states 10 and 01, which are incomparable (to each other) with respect to $T = p \vee q$: both are models of T – yielding $11 \in \mathbf{T}(10)$ and $11 \in \mathbf{T}(01)$ – but the anti-states of both are also models of T – hence $00 \in \mathbf{T}(10)$ and $00 \in \mathbf{T}(01)$. Thus both are maximally compatible with a model of T as well as maximally incompatible with a model of T. The state 00 is the least compatible with T: $11 \notin \mathbf{T}(00)$ and $00 \in \mathbf{T}(00)$.

For the case $T = \top$ there is of course always one state $t \in \mathbf{T}$ which is maximally compatible with a state x (namely $t = x$); and one state $t \in \mathbf{T}$ which is maximally incompatible with x (namely $t = x^*$).

Our aim is to refine the coarse total preorder $\leq_{ep}$ on $\mathbf{W}$ – induced by T – to an order which will express, for any two states x and y, the relation 'x is at most as compatible with T as y is'. To achieve this, we firstly need the following two notions – defined in terms of any preorder (i.e. a reflexive, transitive relation) $\leq$ on $\mathbf{W}$:

Definition 7. For any configuration $\mathbf{X} \subseteq \mathbf{W}$, we define

$$\nabla\mathbf{X} := \{w \in \mathbf{W} \mid (\exists x \in \mathbf{X})(w \leq x)\}, \text{ the } \textit{downset of } \mathbf{X};$$
$$\triangle\mathbf{X} := \{w \in \mathbf{W} \mid (\exists x \in \mathbf{X})(x \leq w)\}, \text{ the } \textit{upset of } \mathbf{X}.$$

∇ and $\triangle$ are closure operations on $\mathcal{P}\mathbf{W}$. In terms of ∇ this means:

(1) $\mathbf{X} \subseteq \nabla\mathbf{X}$
(2) $\nabla(\nabla\mathbf{X}) = \nabla\mathbf{X}$
(3) If $\mathbf{X}, \mathbf{Y} \subseteq \mathbf{W}$ and $\mathbf{X} \subseteq \mathbf{Y}$, then $\nabla\mathbf{X} \subseteq \nabla\mathbf{Y}$.
     (Similarly for $\triangle$.)

In what follows we shall discuss downsets. Most of the time, reference to upsets shall be done in parentheses. The configuration $\nabla\mathbf{X}$ ($\triangle\mathbf{X}$) consists of all elements of $\mathbf{X}$ together with everything which lies lower down (higher up) than an element of $\mathbf{X}$ within $(\mathbf{W}, \leq)$. When X is the theory describing the configuration $\mathbf{X}$, (i.e. $\mathbf{X} = Mod(X)$), and $\nabla\mathbf{X}$ ($\triangle\mathbf{X}$) is also effable, we denote by ∇X ($\triangle X$) the theory corresponding to $\nabla\mathbf{X}$ ($\triangle\mathbf{X}$), i.e. $\nabla\mathbf{X} = Mod(\nabla X)$ ($\triangle\mathbf{X} = Mod(\triangle X)$). For a more detailed discussion on "downsets" and "upsets" and their relationships to "negative" and "positive" sentences respectively, we refer the reader to Burger and Heidema (1994).

Let us go back to the idea of viewing the elements of $(\mathbf{W}, \leq_{pr})$ as comparison vectors. For a given set T of propositions, every $x \in \mathbf{W}$ is represented by the configuration $\mathbf{T}(x) \subseteq \mathbf{W}$ of comparison vectors. To every one of these sets we may then apply $\triangledown$ ($\triangle$) – defined in terms of any preordering $\leq$ on $\mathbf{W}$. Let us agree upon picking $\leq$ as the *product ordering* $\leq_{pr}$ for the rest of *Section* 3.1. We then obtain the downsets (upsets), $\triangledown\mathbf{T}(x)$ ($\triangle\mathbf{T}(x)$). In the finite case, these configurations correspond to single propositions. In agreement with our previous conventions on notation, we shall denote this proposition by $\triangledown T(x)$ ($\triangle T(x)$). Table 2 shows the propositions $\triangledown T(x)$ and $\triangle T(x)$ corresponding to respectively the downsets $\triangledown\mathbf{T}(x)$ and upsets $\triangle\mathbf{T}(x)$ of the sets $\mathbf{T}(x)$ for the two cases $T = p$ and $T = p \vee q$ listed in Table 1. These sets were obtained by means of the product ordering depicted in Figure 11.

Table 2. $\triangledown T(x)$ and $\triangle T(x)$

	x:	11	10	01	00
$T = p$	$\triangledown T(x)$:	$\top$	$\top$	$\neg p$	$\neg p$
	$\triangle T(x)$:	p	p	$\top$	$\top$
$T = p \vee q$	$\triangledown T(x)$:	$\top$	$\top$	$\top$	$\neg p \vee \neg q$
	$\triangle T(x)$:	$p \vee q$	$\top$	$\top$	$\top$

As mentioned previously, we want to refine the very coarse preorder $\leq_{ep}$ (induced by T) on the possible states of the world. With the first part of the following definition we accomplish that.

Definition 8. Let $\mathbf{T} = Mod(T)$, where T is any theory. For all $x, y \in \mathbf{W}$ we define

$$x \preceq_T^c y : \Leftrightarrow \triangledown\mathbf{T}(x) \subset \triangledown\mathbf{T}(y) \text{ or } x = y;$$
$$x \preceq_T^d y : \Leftrightarrow \triangle\mathbf{T}(y) \subset \triangle\mathbf{T}(x) \text{ or } x = y;$$

where downsets and upsets are taken with respect to the product ordering, $\leq_{pr}$, on $\mathbf{W}$. If $x \preceq_T^c y$ ($x \preceq_T^d y$, respectively), we say that x *is at most as c-compatible (d-compatible) with T as y is*. The superscripts c and d have the connotations 'cautious' and 'daring' respectively, as will become clear in *Section* 3.2, where the relations $\preceq_T^c$ and $\preceq_T^d$ on $\mathbf{W}$ induce cautious and daring relations on theories. The intuition is roughly the following: larger sets of worlds correspond to weaker theories, which are more cautious, since they are more likely to be true; while smaller sets of states correspond to logically stronger theories, which are more daring, since they are more likely to be false.

So we have two new partial orders (i.e. reflexive, transitive, antisymmetric relations) on $\mathbf{W}$. The relation $\preceq_T^c$ is a refinement of $\leq_{ep}$ in the sense that the non-models of T (i.e. the states in $\mathbf{W} - \mathbf{T}$) are no longer equivalent in $\preceq_T^c$ – in fact, no two are equivalent, some may be comparable, but some may be incomparable. However, for every $x \in \mathbf{W} - \mathbf{T}$ and every $y \in \mathbf{T}$, $x \preceq_T^c y$; and all the models of T are incomparable in $\preceq^c_T$. This means that in $\preceq^c_T$ the refinement of $\leq_{ep}$ is done inside the two partition classes of $\leq_{ep}$, but it retains the order of the two classes – every element in $\mathbf{T}$ is above every element of $\mathbf{W} - \mathbf{T}$ in $\preceq^c_T$. (If we require that the elements of $\mathbf{T}$ be equivalent with respect to $\preceq^c_T$, we can have a weaker version $\preceq^c_T{}'$ of $\preceq^c_T : x \preceq^c_T{}' y : \Leftrightarrow \nabla\mathbf{T}(x) \subseteq \nabla\mathbf{T}(y)$.) If T is either tautological or contradictory, $\preceq^c_T$ is the identity relation on $\mathbf{W}$. The ordering $\preceq^c_T$ is different from the product ordering $\leq_{pr}$ except in the very special case $\mathbf{T} = \{11...\}$ – the case when $\mathbf{T}(x) = \{x\}$ for all $x \in \mathbf{W}$ and hence $\nabla\mathbf{T}(x) \subset \nabla\mathbf{T}(y)$ if and only if $x \leq_{pr} y$. In fact if $\mathbf{T}$ is any singleton set, say $\mathbf{T} = \{t\}$, then $(\mathbf{W}, \preceq^c_T)$ is isomorphic to the Boolean algebra $(\mathbf{W}, \leq_{pr})$ under the isomorphism which maps a state x to the state $t(x)$. Thus, when $\mathbf{T} = \{t\}$, $(\mathbf{W}, \preceq^c_T)$ has t as top element and t^* as bottom element.

In the relation $\preceq^d_T$ it is not the case that every model of T is above every non-model of T. Now we have a separation between the sets $\mathbf{W} - \mathbf{T}^*$ and $\mathbf{T}^*$, where $\mathbf{T}^* = \{t^* \mid t \in \mathbf{T}\}$, i.e. all the anti-states of models of T. All the elements of $\mathbf{T}^*$ are incomparable in $\preceq^d_T$ and are below every element of $\mathbf{W} - \mathbf{T}^*$ – meaning that a state x which is maximally incompatible with an element of $\mathbf{T}$, is always placed below every state y which has some agreement with every element of $\mathbf{T}$ (i.e. a state y for which $00... \notin \mathbf{T}(y)$). Note that $\preceq^d_T$ can equivalently be defined as: $x \preceq^d_T y : \Leftrightarrow \nabla\mathbf{T}^*(y) \subset \nabla\mathbf{T}^*(x)$ or $x = y$; and we may also define a weaker version $\preceq^d_T{}'$ of $\preceq^d_T$ in which all the states in $\mathbf{T}^*$ will be equivalent: $x \preceq^d_T{}' y : \Leftrightarrow \Delta\mathbf{T}(y) \subseteq \Delta\mathbf{T}(x)$.) Just as with $\preceq^c_T$, if T is either tautological or contradictory, $\preceq^d_T$ is the identity relation on $\mathbf{W}$. If $\mathbf{T} = \{t\}$, then $\preceq^d_T = \preceq^c_T$ (and both are the product ordering $\leq_{pr}$ when $\mathbf{T} = \{11...\}$).

We can now employ the relations $\preceq^c_T$, $\preceq^c_T{}'$, $\preceq^d_T$ and $\preceq^d_T{}'$ in various ways to obtain orderings on $\mathbf{W}$ which express some degree of compatibility with the data T (Burger and Heidema 1995). One example: employing lexicographically first $\preceq^c_T{}'$ and then $\preceq^d_T$ (the latter only to order the states *inside* the two classes $\mathbf{T}$ and $\mathbf{W} - \mathbf{T}$) yields an ordering which differentiates not only between models and non-models of T, but also discriminates amongst the models as well as amongst the non-models of T. Figures 12 and 13 show $\preceq^c_T$ and $\preceq^d_T$ for, respectively, $T = p \vee q$ and $T = p$. If $T = p \wedge q$, both $\preceq^c_T$ and $\preceq^d_T$ correspond to the diagram in Figure 11.

Fig. 12. $\preceq^c_T$ (left) and $\preceq^d_T$ (right) on $\mathbf{W} = \{11, 10, 01, 00\}$ for $T = p \vee q$

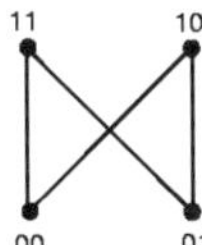

Fig. 13. $\preceq_T^c$ and $\preceq_T^d$ on **W** = {11, 10, 01, 00} for $T = p$

3.2. *Similarity between Sets of States*

One may see the orderings $\sqsubseteq_T$, $\sqsubseteq_T^c$, and $\sqsubseteq_T^d$, on theories as natural extensions to configurations of the original epistemic ordering on the states; compare, e.g., the positions of the four possible states 11, 10, 01 and 00 in the epistemic state (Figure 2) with the corresponding state descriptions $p \wedge q$, $p \wedge \neg q$, $\neg p \wedge q$ and $\neg p \wedge \neg q$ in $\sqsubseteq_p$ (Figure 6), $\sqsubseteq_p^c$ and $\sqsubseteq_p^d$ (Figure 8). In this section we employ the *structural similarity* orderings on states – expressing some sort of compatibility with a theory T – to induce orderings on theories via power orderings on configurations. Depending on the context for applications, one may employ and combine the four structural similarity relations, $\preceq_T^c$, $\preceq_T^d$, $\preceq_T^c{}'$ and $\preceq_T^d{}'$ in various ways to induce many different and interesting relations on theories. When T represents the available information, these orderings on theories express the "closeness" or "similarity" to this data T – in a mode which ranges on a spectrum from cautious to daring. Note that T can range from complete information (i.e. when **T** is a singleton set **T** = {t}) to no information (i.e. when $T = \top$); and even contradictory information $T = \bot$.

In what follows we firstly define downsets and upsets of configurations in terms of the relations $\preceq_T^c$ and $\preceq_T^d$. The ways in which these downsets and upsets are combined in defining a relation, give rise to a whole spectrum of orderings on theories (via configurations). In this paper we explore two such relations – a cautious one ($\rightrightarrows_T^c$) and a daring one ($\rightrightarrows_T^d$). 'Verisimilitude' or 'truthlikeness' describes the very special and ideal case when we have complete information, **T** = {t}. The cautious relation $\rightrightarrows_T^c$ on theories may be seen as a lexicographical version of the verisimilar power ordering of Brink and Heidema (1987, 1989), which is also equivalently described in terms of positive and negative contents of theories by Burger and Heidema (1994). (Note that Kuipers' "refined definition of truthlikeness of theories" can be understood as a variant of the verisimilar power ordering of Brink and Heidema – as pointed out by Kieseppä 1996). So, with the relations $\rightrightarrows_T^c$ and $\rightrightarrows_T^d$ we generalize verisimilitude; but also entailment and its converse.

Definition 9. For any configuration $\mathbf{X} \subseteq \mathbf{W}$, we define

$$\nabla_T^c \mathbf{X} := \{w \in \mathbf{W} \mid (\exists x \in \mathbf{X})(w \preceq_T^c x)\};$$
$$\triangle_T^c \mathbf{X} := \{w \in \mathbf{W} \mid (\exists x \in \mathbf{X})(x \preceq_T^c w)\};$$
$$\nabla_T^d \mathbf{X} := \{w \in \mathbf{W} \mid (\exists x \in \mathbf{X})(w \preceq_T^d x)\};$$
$$\triangle_T^d \mathbf{X} := \{w \in \mathbf{W} \mid (\exists x \in \mathbf{X})(x \preceq_T^d w)\}.$$

By employing downsets and upsets as defined above, we then have the following two partial orderings on theories:

Definition 10. For any theories X, Y and T,

$$X \rightrightarrows_T^c Y :\Leftrightarrow$$

    (i) $\nabla_T^c \mathbf{X} \subseteq \nabla_T^c \mathbf{Y}$ and

    (ii) (if $\nabla_T^c \mathbf{X} = \nabla_T^c \mathbf{Y}$, then $\triangle_T^c \mathbf{Y} \subseteq \triangle_T^c \mathbf{X}$) and

    (iii) (if $\nabla_T^c \mathbf{X} = \nabla_T^c \mathbf{Y}$ and $\triangle_T^c \mathbf{Y} = \triangle_T^c \mathbf{X}$, then $\mathbf{X} \subseteq \mathbf{Y}$).

$$X \rightrightarrows_T^d Y :\Leftrightarrow$$

    (i) $\triangle_T^d \mathbf{Y} \subseteq \triangle_T^d \mathbf{X}$ and

    (ii) (if $\triangle_T^d \mathbf{Y} = \triangle_T^d \mathbf{X}$, then $\nabla_T^d \mathbf{X} \subseteq \nabla_T^d \mathbf{Y}$) and

    (iii) (if $\triangle_T^d \mathbf{Y} = \triangle_T^d \mathbf{X}$ and $\nabla_T^d \mathbf{X} = \nabla_T^d \mathbf{Y}$, then $\mathbf{Y} \subseteq \mathbf{X}$).

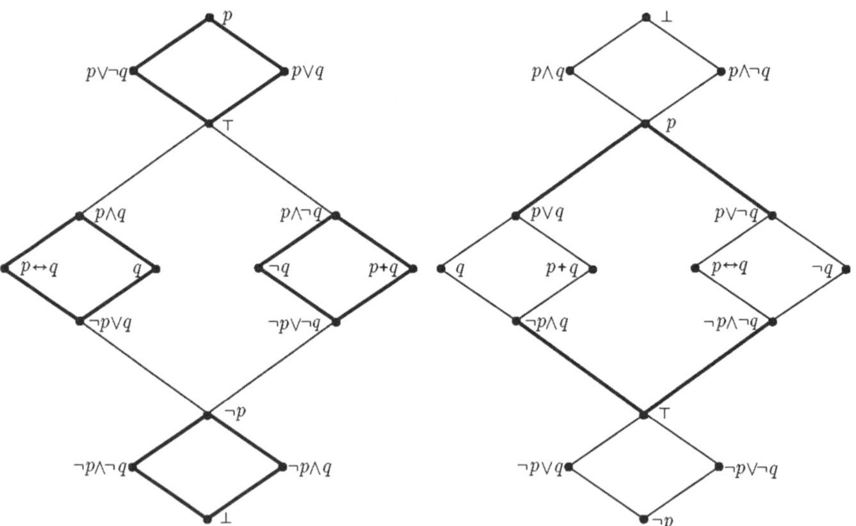

Fig. 14. B ordered by $\rightrightarrows_p^c$ (left) and $\rightrightarrows_p^d$ (right)

The relation $\rightrightarrows_T^c$ is cautious in the sense that it has the strict T-soundness property: all T-true theories are above all T-uncommitted theories, which are above all T-false theories. This is also the case in the relation $\vDash_T$, but $\rightrightarrows_T^c$ is a refinement of $\vDash_T$. The refinement occurs when step (ii) (and if necessary, step (iii)) is lexicographically applied to order the elements inside the congruence classes modulo T. Remember, when $T = \top$ or $T = \bot$, then $\preceq^c_T$ is just the identity relation on states – meaning that $\nabla^c_T \mathbf{X} = \mathbf{X}$ and hence, $X \rightrightarrows^c_T Y$ iff $X \vDash Y$. We illustrate $\rightrightarrows^c_T$ in the p, q language for $T = p$ (diagram on the left in Figure 14) and $T = p \wedge q$ (diagram on the left in Figure 15).

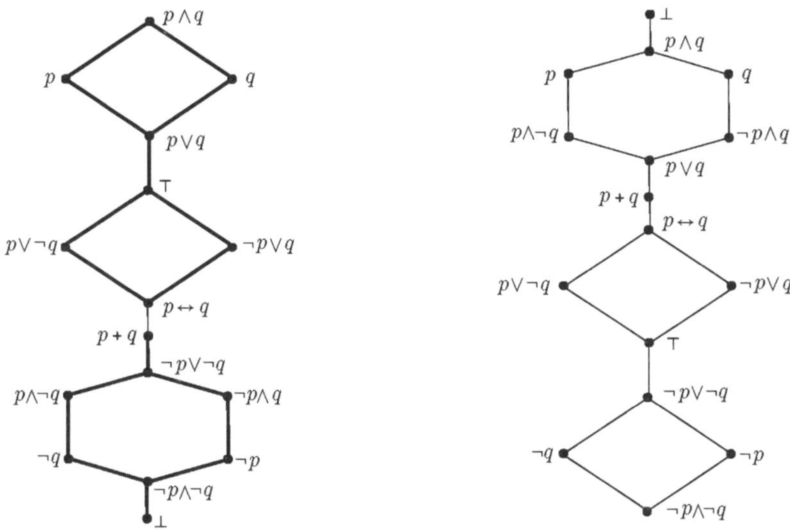

Fig. 15. *B* ordered by $\rightrightarrows^{\bar{c}}_{p \wedge q}$ (left) and $\rightrightarrows^{d}_{p \wedge q}$ (right)

The sets circumscribed by thick lines in $\rightrightarrows^c_T$ are the congruence classes modulo T. There is no need to apply step (iii) in the case $T = p$. However, when $T = p \wedge q$, the four propositions $\top$, $p \vee \neg q$, $\neg p \vee q$ and $p \leftrightarrow q$ have the same downsets ∇^c_T as well as the same upsets $\triangle^c_T$, and are therefore ordered by $\vDash$. We note that the relation $\rightrightarrows^c_T$ has none of the counterintuitive properties of the relation $\sqsubseteq_T$ (*Definition 2, Section* 2.2.4) with regard to truthlikeness as pointed out by various authors. Just one example: When we know the truth to be $p \wedge q$, viewing $\neg p \wedge \neg q$ as closer to the truth than $\neg p$ (in $\sqsubseteq_{p \wedge q}$, Figure 7) seems counterintuitive. But in $\rightrightarrows^c_{p \wedge q}$ (Figure 15, diagram on the left) these two propositions are reversed: $\neg p$ is closer to $p \wedge q$ than $\neg p \wedge \neg q$.

Instead of working lexicographically, we can define product orderings in which downsets and upsets have equal weight. In the case of complete information, the product version $\rightrightarrows^c_T{}'$ of the relation $\rightrightarrows^c_T$, $X \rightrightarrows^c_T{}' Y :\Leftrightarrow \nabla^c_T \mathbf{X} \subseteq \nabla^c_T \mathbf{Y}$ and $\triangle^c_T \mathbf{Y} \subseteq \triangle^c_T \mathbf{X}$, is equivalent to the verisimilar ordering of Brink and Heidema (1987; 1989) (and Burger and Heidema 1994).

The boldness of the relation $\rightrightarrows^d_T$ is clearly visible when looking at Figures 14 and 15 – depicting $\rightrightarrows^d_T$ for $T = p$ and $T = p \wedge q$, respectively. T-soundness is totally lost; and whenever $T = \top$ or $T = \bot$, $X \rightrightarrows^d_T Y$ iff $Y \vDash X$ iff $X \vDash^{-1} Y$. In

Figure 14 (diagram on the right) the four classes within the thin lines are the congruence classes modulo $\neg p$; in Figure 15 (diagram on the right) it is not possible to indicate the 8 congruence classes modulo $\neg p \lor \neg q$ in the same way.

For any configuration $\mathbf{X}$, the configuration of anti-states of $\mathbf{X}$ is the set $\mathbf{X}^* := \{x^* \mid x \in \mathbf{X}\}$ (cf. *Definition 6, Section* 3.1). It can be shown that for any two states x and y, $x \preceq_T^c y$ iff $y^* \preceq_T^d x^*$, and hence that $(\nabla_T^c \mathbf{X})^* = \triangle_T^d (\mathbf{X}^*)$ and $(\triangle_T^c \mathbf{X})^* = \nabla_T^d (\mathbf{X}^*)$ – so that we have the following relationship between the two orderings $\rightrightarrows_T^c$ and $\rightrightarrows_T^d$ as illustrated in Figures 14 and 15:

For all configurations $\mathbf{X}, \mathbf{Y} \subseteq \mathbf{W}$,

$$\mathbf{X} \rightrightarrows_T^c \mathbf{Y} \text{ iff } \mathbf{Y}^* \rightrightarrows_T^d \mathbf{X}^*.$$

4. Conclusion

In this paper we have given an overview of some of the many ways in which information may be employed in ordering states as well as theories. Some of these orderings, especially their philosophical properties, are also studied by others. The orderings on theories induced by epistemic states (Section 2) have, because of the classificatory nature of epistemic states, a rich underlying algebraic structure which was not discussed in depth in this paper, but which may provide hints for applications in the designer logics of Artificial Intelligence. Some of these orderings, however, have counterintuitive properties in the context of truthlikeness. On the other hand, the power orderings on configurations (and hence on theories) induced by data-similarity orderings on states as discussed in Section 3, might lack some of the convenient algebraic properties of those defined in Section 2, but they are more applicable within the context of truthlikeness. The many other interesting versions of these orderings is a topic for another paper.

Rand Afrikaans University
Department of Mathematics
P.O. Box 524
Auckland Park
2006 South Africa

University of South Africa
Department of Mathematics
South Africa

REFERENCES

Alchourrón, C., P. Gärdenfors and D. Makinson (1985). On the Logic of Theory Change: Partial Meet Contraction and Revision Functions. *Journal of Symbolic Logic* **50**, 510-530.

Aliseda-Llera, A. (1997). *Seeking Explanations: Abduction in Logic, Philosophy of Science and Artificial Intelligence* (ILLC Dissertation Series, 1997-04). Amsterdam: University of Amsterdam.

Antoniou, G. (1997). *Nonmonotonic Reasoning.* Cambridge, MA: The MIT Press.

Bar-Hillel, Y. and R. Carnap (1953). Semantic Information. *The British Journal for the Philosophy of Science* **4**, 147-157.

Brink, C. (1989). Verisimilitude: Views and Reviews. *History and Philosophy of Logic* **10**, 181-201.

Brink, C. and J. Heidema (1987). A Verisimilar Ordering of Theories Phrased in a Propositional Language. *The British Journal for the Philosophy of Science* **38**, 533-549.

Brink, C. and J. Heidema (1989). A Verisimilar Ordering of Propositional Theories: The Infinite Case. *Technical Report Series of the Automated Reasoning Project* (TR-ARP, **1/89**). Canberra: Australian National University.

Burger, I.C. and J. Heidema (1994). Comparing Theories by their Positive and Negative Contents. *The British Journal for the Philosophy of Science* **45**, 605-630.

Burger, I.C. and J. Heidema (1995). Power-Order Semantics II: Cautious and Bold Reasoning (Summary of talk given at the 10th International Congress of Logic, Methodology and Philosophy of Science, August 1995, Florence, Italy). In: E. Castagli and M. Konig (eds.), *Volume of Abstracts*, p. 214. Florence: Comune di Cesena.

Burger, I.C. and J. Heidema (1996). Boolean Algebras and Generalized Entailment I: Element-Induced Isomorphisms of Boolean Algebras (Summary of talk given at the Annual Congress of the SAMS, November 1996, University of the Western Cape, South Africa). *Notices of the South African Mathematical Society* **28**(2), 105.

Burger, I.C. and J. Heidema (1997). Boolean Algebras and Informed Entailment (Summary of talk given at the Joint Mathematical Conference of the American Mathematical Society, the South African Mathematical Society and the Southern African Mathematical Science Foundation, June 1997, University of Pretoria, South Africa). *Conference Programme*, 60.

Burger, I.C. and J. Heidema (2000). Epistemic States Guiding the Rational Dynamics of Information. In: R. Mizoguchi and J. Slaney (eds.), *PRICAI 2000: Topics in Artificial Intelligence* (Proceedings of the 6th Pacific Rim International Conference on Artificial Intelligence, Melbourne, Australia, August/September 2000 (LNAI, **1886**)), pp. 275-285. Berlin: Springer-Verlag.

Burger, I.C. and J. Heidema (2002). Merging Inference and Conjecture by Information. *Synthese* **131**(2), 223-258.

Festa, R. (1987). Theory of Similarity, Similarity of Theories, and Verisimilitude. In: Kuipers (1987a), pp. 145-176.

Heidema, J. and W.A. Labuschagne (1990). Data-Dependence of Semantic Information. *South African Computer Journal* **3**, 39-44.

Hendricks, V.F. and J. Faye (1999). Abducting Explanation. In: L. Magnani, N.J. Nersessian and P. Thagard (eds.), *Model-Based Reasoning in Scientific Discovery*, pp. 271-292. New York: Kluwer Academic/Plenum.

Hintikka, J. (1998). What is Abduction? The Fundamental Problem of Contemporary Epistemology. *Transactions of the Charles S. Peirce Society* **34**(3), 503-533.

Josephson, J.R. and S.G. Josephson (1994). *Abductive Inference.* Cambridge: Cambridge University Press.

Kieseppä, I. (1996). *Truthlikeness for Multidimensional, Quantitative Cognitive Problems.* Dordrecht: Kluwer Academic.

Kuipers, T.A.F., ed. (1987a). *What is Closer-to-the-Truth? A Parade of Approaches to Truthlikeness. Poznań Studies in the Philosophy of the Sciences and the Humanities,* vol. 10. Amsterdam: Rodopi.

Kuipers, T.A.F. (1987b). A Structuralist Approach to Truthlikeness. In: Kuipers (1987a), pp. 79-99.

Kuipers, T.A.F. (1992a). Naive and Refined Truth Approximation. *Synthese* **93**, 299-341.

Kuipers, T.A.F. (1992b). Truth Approximation by Concretization. In: J. Brzeziński and L. Nowak (eds.), *Idealization III: Approximation and Truth. Poznań Studies in the Philosophy of the Sciences and the Humanities,* vol. 25, pp. 159-179. Amsterdam: Rodopi.

Kuipers, T.A.F. (1997). The Dual Foundation of Qualitative Truth Approximation. *Erkenntnis* **47**(2), 145-179.

Kuipers, T.A.F. (1998). Pragmatic Aspects of Truth Approximation. In: P. Weingartner, G. Schurz and G. Dorn (eds.), *The Role of Pragmatics in Contemporary Philosophy* (Proceedings of the 20th International Wittgenstein Symposium, August 1997), pp. 288-300. Vienna: Hölder-Pichler-Tempsky.

Kuipers, T.A.F. (1999). Abduction Aiming at Empirical Progress or Even Truth Approximation. *Foundations of Science* **4**(3), 307-323.

Kuipers, T.A.F. (2000/ICR). *From Instrumentalism to Constructive Realism. On Some Relations between Confirmation, Empirical Progress, and Truth Approximation. Synthese Library,* vol. 287). Dordrecht: Kluwer Academic.

Kuipers, T.A.F. (2001/SiS). *Structures in Science. Heuristic Patterns Based on Cognitive Patterns.. Synthese Library,* vol. 301. Dordrecht: Kluwer Academic.

Miller, D. (1978). On Distance from the Truth as a True Distance. In: J. Hintikka, I. Niiniluoto and E. Saarinen (eds.), *Essays in Mathematical and Philosophical Logic,* pp. 415-435. Dordrecht: D. Reidel.

Niiniluoto, I. (1987). *Truthlikeness. Synthese Library,* vol. 185. Dordrecht: D. Reidel.

Niiniluoto, I. (1998). Verisimilitude: The Third Period. *The British Journal for the Philosophy of Science* **49**, 1-29.

Oddie, G. (1986). *Likeness to the Truth (University of Western Ontario Series in Philosophy of Science),* vol. 30. Dordrecht: D. Reidel.

Paul, G. (1993). Approaches to Abductive Reasoning: An Overview. *Artificial Intelligence Review* **7**, 109-152.

Shoham, Y. (1988). *Reasoning about Change: Time and Causation from the Standpoint of Artificial Intelligence.* Cambridge, MA: The MIT Press.

Zwart, S.D. (1998). *Approach to the Truth: Verisimilitude and Truthlikeness.* ILLC Dissertation Series, 1998-02. Amsterdam: University of Amsterdam.

Theo A. F. Kuipers

LOGIC IN SERVICE OF PHILOSOPHY OF SCIENCE
REPLY TO ISABELLA BURGER AND JOHANNES HEIDEMA

In order to localize the paper by Isabella Burger and Johannes Heidema, and to express my appreciation for it, I start by quoting the first part of the introduction I wrote for the proceedings of the special symposium on "Current interfaces between logic and philosophy of science," held at the 10th Logic, Methodology and Philosophy of Science (LMPS) international congress in Florence, 1995.

> Logic and philosophy of science have had many alliances since the rise of modern logic. The Wiener Kreis was even based on the idea of permanent cooperation. However, the ties have as yet not culminated in a strong interdiscipline, as for instance logic and linguistics have in logical semantics. To be sure, logicians and philosophers of science feel to belong to the same family, as is for instance clear from the very happening of the 10th LMPS international congress. But, curiously enough, besides the section General Methodology, there is not yet a section called Logical, or Formal, Methodology.
>
> One reason for the non-existence of an interdiscipline maybe Suppes' well known complaint that the logical study of science started to imitate metamathematics rather than mathematics. Another reason certainly is a general one, always threatening interdisciplinary cooperation. As Zandvoort (1995) has demonstrated, successful cooperation usually has an asymmetric means-end nature: one program plays the role of guide program, it poses and determines the leading problems, the other plays the role of supply or service program by helping to reformulate and solve these problems. However, when the potential supply discipline is well developed, as in the case of logic, the temptation is strong to create a problem area vaguely suggested by the guide discipline and to proceed by developing appropriate tools for that area. This may well be a useful strategy leading to interesting results, but not necessarily for the intended guide discipline, i.c. philosophy of science. In sum, whereas philosophy of science has the lasting problem of keeping in touch with science, the logical study of science has the problem of keeping in touch with science or at least with philosophy of science. (Kuipers 1997, p. 379)

The paper by Burger and Heidema is not only evidently a paper of a logical nature, it is also strongly guided by problems in philosophy of science or, more generally, by epistemological problems. As is clear, from the beginning to the very end, they succeed in keeping in touch with these problems. Incidentally, the same is true for the paper by Thomas Mormann, which may even be seen

In: R. Festa, A. Aliseda and J. Peijnenburg (eds.), *Confirmation, Empirical Progress, and Truth Approximation* (*Poznań Studies in the Philosophy of the Sciences and the Humanities,* vol. 83), pp. 489-492. Amsterdam/New York, NY: Rodopi, 2005.

as a "geometric" variation on the theme of Burger and Heidema's section on structural similarity. Regarding Burger and Heidema, I would like to make some remarks about "naive sorting," followed by some thoughts about the definition of a theory being false.

The Janus Character of Improving Naive Sorting

In ICR (p.152) I merely write about a "tension," whereas Burger and Heidema aptly write about the Janus character of trying to improve (naive) "sorting," where 'sorting' is their favorite term for what a theory does. If T is the target (set of models of a) theory, and X (the set of models of) a given theory, the naive challenge is to narrow down the symmetric difference between X and T. This may be done by trying to apply successfully, without knowing T (!), two operations in an arbitrary order: enlarging the set of models, with the safe aspect that no T-models are lost, but the unhappy aspect that *non-T*-models are introduced, and shrinking the resulting set, with the safe aspect of not introducing *non-T*-models, but the unhappy risk of losing T-models. Note that a similar Janus character applies in the refined, structural similarity approach. Presupposing my favorite type of structural similarity, of which it is not evident that it fits in the approach by Burger and Heidema, starting from a certain theory, a similar problem situation is generated by idealization followed by concretization, or vice versa.

I agree of course with the general claim (below *Definition* 2) that it makes a great deal of sense to generalize the terminology of truthlikeness, for example, 'Y is closer to the truth (expressed by) T than X', to the sorting terminology, in particular 'Y sorts closer to T than X'. However, I am surprised about the fact that an asymmetric (T-biased) general formulation is given, whereas the naive definition evidently is symmetric between T and X. For example, 'Y's sorting is between that of X and T' would be neutral between X and T.

Some Reflections on the Definition of a "False Theory"

It may be interesting to compare my own definition of a false theory with that of Burger and Heidema in Section 2.2.2. They follow the standard definition in model theory (cf. Przełęcki 1969, p. 19). Relative to a certain favorite or target theory T, they call an arbitrary theory X T-true iff all T-models are X-models, T-false iff no T-model is an X-model, and T-uncommitted otherwise, that is, when some T-models are X-models and some are not. Although in ICR I essentially use the same definition of being T-true, I call a theory already false when it is not T-true, that is, when some T-models are not X-models. Of course, one may ask, what's in a definition? But the point is that the standard logical terminology is not very attractive from the nomic point of view in philosophy of science. According to this view, the target set $Mod(T)$ contains all and only the nomic possibilities. This nicely fits in the definition of 'T-true', a theory X is T-true when the claim "all T-models are X-models" is true, with the consequence that T is the strongest true theory, for which reason I call it "the truth." However, in this view it is then also plausible to let being T-false correspond with the situation that this claim is false: a theory (about nomic possibilities) is false when it does not leave room for all nomic possibilities. Of course, it is possible to formalize the intended nomic interpretation within a modal language, but it is much easier to leave the intended interpretation informal. Unfortunately, Burger and Heidema do not pay attention to the nomic interpretation. If I am right, the only place where they touch upon the distinction between actual and nomic truth approximation is when, near the end of Section 2.2.4, they deal with "[t]he very special case when we know exactly which world is the actual world, i.e. the case when T has one model only." This suggests that calling a theory X T-uncommitted when it shares some models with T, without including them all, is essentially based on the "actualist" interpretation: if T contains more than one model, we do not yet know "exactly which world is the actual world." But in the nomic interpretation, T will contain more than one model and is nevertheless the strongest truth there is to know about the target. Hence, from the philosophy of science point of view, we need actual and nomic definitions: nomically true or false versus actually (or instantially) true or false.

Note that the problem of how to define what a "false theory" is, is not specific to the nomic point of view. The standard definition is more generally a problem for the structuralist approach. Przełęcki (1991) considers besides the standard "uni-referential conception of a theory's intended interpretation" also a "multi-referential" one, that is, a structuralist one. In line with structuralist practice, he carefully talks about the "empirical claim" of the theory, that is, "all intended applications are X-models," as being either true or false. In ICR I "officially" did the same in terms of nomic possibilities, by talking about the (weak) claim of the theory. However, I certainly forgot to do so all the time, and for good reason. Scientists are used to talking about true or false theories in some generic sense and about theories being true or false in particular cases. Explicating that terminology should do justice to that practice in a way that is as simple as possible. In general, although in the logical study of science logic should certainly determine the rules of the game, it should try to keep the terminology of the field intact, whenever possible. For a further elaboration of this point, see my reply to Zwart's "fourth problem," which was later written than the present reply.

REFERENCES

Przełęcki, M. (1969). *The Logic of Empirical Theories*. London: Routledge and Kegan Paul.

Przełęcki, M. (1991). Is the Notion of Truth Applicable to Scientific Theories? In: G. Schurz and G. Dorn (eds.), *Advances in Scientific Philosophy, Essays in Honour of Paul Weingartner.* (*Poznań Studies in the Philosophy of the Sciences and the Humanities*, vol. 24), pp. 283-294. Amsterdam: Rodopi.

Kuipers, T. (1997). Logic and Philosophy of Science: Current Interfaces (introduction to the proceedings of a special symposium with the same name). In: M.L. Dalla Chiara, K. Doets, D. Mundici and J. van Benthem (eds.), *Logic and Scientific Methods*, vol. 1 (10th LMPS International Congress, Florence, August, 1995), pp. 379-381. Dordrecht: Kluwer Academic Publishers.

Zandvoort, H. (1995). Concepts of Interdisciplinarity and Environmental Science. In: T.A.F. Kuipers and A.R. Mackor (eds.), *Cognitive Patterns in Science and Common Sense* (*Poznań Studies in the Philosophy of the Sciences and the Humanities*, vol. 45), pp. 45-68. Amsterdam/Atlanta: Rodopi.

REALISM AND METAPHORS

J. J. A. Mooij

METAPHOR AND METAPHYSICAL REALISM

ABSTRACT. This article discusses a number of metaphors about the nature of science, in connection with three types of metaphysical realism: minimal, moderate and essentialistic realism. From the beginning, Kuipers accepts the first and rejects the third type of realism, but it is only later on that he endorses the second type. It is argued that this makes his remarks on essentialistic realism somewhat misleading; and his moderate realism is compared with some realist positions taken by other philosophers. It is further argued that the metaphors of the mirror, the net and the map correspond to essentialistic, minimal and moderate realism, respectively. This explains why the map metaphor is by far the most suitable one in the context of Kuipers' theory of science. This does not imply, however, that it is the (only) true one and that the others are false.

0. Introductory Remarks

On the evening of April 20, 1978, some hours after the successful defense of his doctoral dissertation *Studies in Inductive Probability and Rational Expectation*, Theo Kuipers told me that he would never write a book again: he did not feel up to that ordeal, and one book in his lifetime had to be enough. I do not know how or when he changed his mind. In any case, we now have even two books from him, and impressive volumes at that. A wonderful outcome; perhaps it was a good thing after all that these books were not planned from the beginning. I am very happy to have the opportunity to comment on them and to take part in this festive enterprise!

I restrict myself to topics from *From Instrumentalism to Constructive Realism*. In the last section, Theo Kuipers discusses some well-known metaphors about the nature of (natural) science. These are, mainly, the metaphors of the mirror, the fish net, and the map.In evaluating the adequacy of these metaphors, Kuipers distinguishes between their relevance to theories and to vocabularies, in both cases referring to some important characteristics that should be captured by an adequate metaphor.

Basically, all these metaphors are about the relation between science and the natural world. Therefore, in assessing them one should keep that relationship in mind. How is it defined? This brings us to the very first sections

In: R. Festa, A. Aliseda and J. Peijnenburg (eds.), *Confirmation, Empirical Progress, and Truth Approximation* (*Poznań Studies in the Philosophy of the Sciences and the Humanities,* vol. 83), pp. 495-505. Amsterdam/New York, NY: Rodopi, 2005.

of the book, where the author introduces a number of ontological and epistemological positions, from which he makes a provisional choice.

1. Metaphysical Realisms: Essentialistic, Minimal and Moderate

Starting with the assumption that there is a natural world that is independent of human beings (answer to Question 0: ontological realism), Kuipers suggests that we should also endorse the epistemological claims to have knowledge about it (Q1: epistemological realism), to have knowledge beyond the observable (Q2: scientific realism) and to have knowledge even beyond reference claims concerning theoretical terms (Q3: theoretical realism or theory realism). In the process he rejects a number of more or less skeptical positions. However, he is adamantly opposed to a further ontological and metaphysical position, called "essentialistic realism," and defined as the view that there is "a correct or ideal conceptualization of the natural world" (Question 4: Kuipers 2000, pp. 7-8). Instead, he opts for "constructive realism". Given the presentation of the options, this must be a version of theoretical realism. And so it is; in the last chapter, where Kuipers summarizes the conclusions about constructive realism, it is characterized as an epistemological position and a version of theory realism (pp. 317, 319).

This I find somewhat misleading because it suggests that no further metaphysical commitments are called for after the positive answers to Q1, Q2 and Q3 where 'true' always means 'true in some objective sense'. (p. 4) I believe that Kuipers' anti-skeptical epistemological positions imply more than the minimal metaphysical realism of Q0: the bare assumption that the natural world exists. They also imply that the natural world has certain characteristics. If science can deliver objective knowledge about the observable and non-observable aspects of the world, then there must be, not only the world, but *a way the world is*; it cannot be completely indefinite or malleable. Perhaps there need not be one and only one way the world is, independent of any conceptualization. However, certain conceptualizations must be accurate and correct.

To be sure, the problem is partly a matter of phrasing and interpreting Question 4. Skip the words 'or ideal' and the question looks somewhat more innocent, because there could be a number of different *correct* conceptualizations, whereas there obviously can be at most one *ideal* conceptualization. As the question stands, the phrase 'correct or ideal' is ambiguous because it may mean 'correct, or even ideal' but also 'correct, that is, ideal'. The first leaves room for a certain relativity whereas the latter does not, and it is clear that Kuipers cannot accept any absolutist ontology.

The ambiguity filters from the phrasing of the question Q4 into the discussion of the answer. The weaker meaning is reflected in the statement that a positive answer to Q4 implies the view that "the natural world must have essences of a kind." The stronger interpretation is evident in the further statement that "there must be natural kinds, not only in some pragmatic sense, but in the sense of categories in which entities in the actual or the nomic world *perfectly* fit." Moreover, Kuipers explicitly accepts the uniqueness of an ideal conceptualization when he refers to the putative challenge of science to "uncover the ideal conceptualization." Very surprising is the final argument against a positive answer to Q4, viz. that "there is no reason to assume that there comes an end to the improvement of vocabularies." This is begging the question. If there are correct or ideal conceptualizations, then there may come an end to the improvement of vocabularies; just as (in Kuipers' view) there may come an end to the improvement of *theories,* viz. by finding the strongest true exemplar.

My impression is that Kuipers can only escape from something like non-minimal metaphysical realism by arguing against the maximal version. A question like "Are there correct conceptualizations of the natural world?" would deserve a positive answer, I think. Even the more specific notion of a natural kind is not suspect, given the claims to possible knowledge that have been made in the answers to the questions Q1, Q2 and Q3. Indeed, the denial of the existence of natural kinds in more than a pragmatic sense is, like the denial of the existence of a natural world (as characterized by Kuipers), "highly implausible and not taken seriously by natural scientists" (Kuipers 2000, p. 3) – or so it seems to me. Strangely enough, in introducing Q0 Kuipers did refer to the possibility that the world "brings *one or more* conceptualizations with it." (p. 3; my italics) The issue is set aside because it is irrelevant to the intended meaning of Q0, but it could easily have been taken up in connection with Q4.

As a matter of fact, in the course of the book a stronger metaphysical position than the positive answer to Q0 comes to the fore. At the beginning of Chapter 7 Kuipers in fact accepts a claim about the way the world is. He endorses the view that the truth about a certain part of reality is partly determined by "the nature of reality itself" (p. 140). Surely, the nature of reality is, according to him, only effective within the conceptual boundaries of a specific vocabulary; therefore, he rejects "extreme metaphysical realism", which is the view "that reality carries with it its ideal vocabulary". This "extreme" metaphysical realism is identical to the "essentialistic" realism as defined in the answer to Q4. But Kuipers now accepts a kind of moderate metaphysical realism of which no hint could be found in the discussion of Q4 and which he connects with the positive answers to Q2 and Q3 (p. 141). This type of realism is "moderate" in the sense that it lies between minimal realism (Q0) and extreme or essentialistic realism (Q4). It accepts not only that there is an external world, but also that external reality has a nature, that is, has certain features. And within moderate metaphysical realism, although it is not essentialistic, the acceptance of "essences of a kind" would not seem to be problematic at all.[1]

This moderate metaphysical realism is close to a number of other realist positions. To begin with, there is Niiniluoto's position as defined and defended in his recent book, *Critical Scientific Realism*. Surely, Niiniluoto is unwilling to call his "critical scientific realism" a version of metaphysical realism. But this unwillingness seems to arise mainly from his unwillingness to choose between "internal" and "metaphysical" realism as defined by Hilary Putnam; by no means does he want to endorse the metaphysical pole of this "false dichotomy".[2] Kuipers shows a similar hesitation, but in the end he accepts a kind of moderate metaphysical realism. Anyhow, apart from terminological niceties, Niiniluoto's position between bare ontological realism and strong, "Putnamian" or essentialistic realism is close to Kuipers' position.[3]

This similarity was perhaps to be expected. More interestingly, Kuipers' position seems also close to the "external" realism endorsed by John R. Searle in Chapters 7 and 8 of his book *The Construction of Social Reality*. Searle understands by external realism the view that the natural world has countless features that exist independently of our representations of them (Searle 1995, p. 153). His considered definition is as follows: "Realism is the view that there is a way that things are that is logically independent of all human representations" (p. 155). This might seem to be rather extreme indeed. But in the meantime he had denounced the common mistake "to suppose that realism

---

[1] The same metaphysical view reappears in the final chapter, for instance at the end of section 13.5 (p. 327). It is possible, of course, that moderate metaphysical realism was already implied in the positive answer to Q0. Accordingly, Kuipers' moderate metaphysical realism could be meant as a restatement of the ontological realism proposed in the answer to Q0. I feel, however, that it should have been brought into the open, preferably in connection with (something like) the question Q4, as an intermediate position. That Kuipers' answer to Q4 is not meant to imply any substantial metaphysical (realist) commitment also appears from his suggestion that instead of the name 'contructive realism' the name 'nominalistic realism' would have been adequate. (p. 334: Ch. 1, n. 4)

[2] Niiniluoto (1999, pp. viii and 211-218). Moreover, according to his own definition, a metaphysical realist "asserts the existence of some abstract entities and principles (such as Plato's ideas, Leibniz's monads, Hegel's objective spirit) which are not accessible to the scientific method" (p.7; cf. also p. 36-37, n. 22). Apparently, Niiniluoto does not want to belong to that company either.

[3] See pp. 10-13 and, in particular, Chapters 5-7. As a whole, Niiniluoto's book is a thorough critical survey of realist and anti-realist positions. On bare Ontological Realism, see p. 2 and Ch. 2. Putnam himself opted in 1976 for internal realism.

is committed to the theory that there is one best vocabulary for describing reality, that reality itself must determine how it should be described" (p. 155). This he rejects; he accepts conceptual relativity; and thus there is here no strong essentialistic but only moderate realism after all.[4]

Even the distance from the realist view of Rom Harré in his *Varieties of Realism* is perhaps not so great. Kuipers mentions Harré together with Richard Boyd as adherents of a very strong conception of natural kinds. Nonetheless, Harré disavows the principle that most theoretical statements are true or false by virtue of the way the world is.[5] He repeatedly calls his realism a "modest realism" because it is not a "truth realism" but a "referential realism." It is true that in arguing for referential realism he appeals to real essences, and he endorses Kripke's and Putnam's belief that "science is about real essences."[6] But his own definition of a real essence is a rather weak one, for he takes "real essences as the constitutions of material beings, expressed in revisable hypotheses derived from the best current theories" (Harré 1986, p. 117). Is this really objectionable from Kuipers' point of view? Harré goes on to suggest that real essences are only relevant to some sciences. The upshot is that the world *as a whole* has no ideal conceptualization, and that in so far as there is one it is defined from a specific point of view.[7]

All in all, Kuipers and Harré are allies in the fight against relativistic and skeptic positions rather than opponents with regard to the nature of reality. I even think that Kuipers' central notion of truth approximation corresponds in a way to Harré's central notion of trustworthiness; both are connected with reliability. "Trustworthiness is not just a moral quality. It is also an epistemic standard," says Harré (1986, p. 13). I feel that it is in the spirit of Kuipers' study to say: "Truth approximation is not just an epistemic notion. It is also a moral injunction." I would like to know whether this feeling is close to the truth.

Of course, besides being allies, Harré and Kuipers are opponents in several respects, in particular as to the relevance of logic and of a logical framework in the philosophy of science. Rather than defending a logical approach, Harré

---

[4] Searle's statement that if we had never existed, most of the world would have remained unaffected (Searle 1995, p. 153), is not too different from a similar claim by Kuipers in Section 13.5. With Kuipers, too, this must mean not only that there would have been a world just the same, but also that many or most of its features would have been exactly as they are now. Concerning the justification of external realism, Searle argued that it is a background condition of intelligibility.

[5] Cf. Harré (1986, esp. Ch. 5).

[6] Harré (1986, pp. 109-111); however, he adds the proviso that besides the theoretical context there is always the practical context. On Kripke and Putnam, see pp. 115-117.

[7] See also pp. 106-107 for the relativity of natural kinds; one should distinguish, though, between the natural kinds themselves and the relevant classificatory material practices.

defends the use of metaphor. "Metaphor and simile are the characteristic tropes of scientific thought," he says (1986, p. 7). Boyd has said similar things. Is it strange that adherents of a strong conception of natural kinds are so strongly in favor of metaphor? It seems that Kuipers does not pay much attention to the role of metaphors *within* science. On the other hand, he is positively interested in metaphors *about* science. Has that to do with an essentialistic view of science, with "logical essentialism," as Harré would have it?

Apparently Harré's argument in favor of simile and metaphor is not connected with his acceptance of real essences. After having divided science and its objects into three levels of research, he argues that the description of entities of the second realm typically depends on simile and that of the entities of the third realm on metaphor (1986, ch. 3). Thus the need of metaphor would arise particularly and/or typically from the extreme abstraction of the entities of the third realm. It seems that the notion of a real essence does not play any role here. But Harré subscribes to Richard Boyd's notion of "epistemic access" as the creative function of metaphor (1986, p.77), so let us see what can be found in Boyd's elaborate essay "Metaphor and Theory Change."

Again we meet with a version of realism which is less essentialistic than one might expect. For in defining "the task of arranging our language so that our linguistic categories 'cut the world at its joints'," Boyd adds that the joint metaphor is misleading because the relevant notion of "joint" may depend on context or discipline.[8] But it is true that in the remaining part of Boyd's essay one does not hear much about this dependence; and in his defense of the role of metaphor in science (not so much as an exegetic or heuristic device but as constitutive of a theory), Boyd strongly appeals to the real essences of natural kinds.

Boyd argues that theory-constitutive metaphors in presenting a terminology offer at the same time specific strategies for research. They need not be successful, of course; the "informed guess" that suggested the metaphor in the first place, does not always work out. It does work out if the characteristics suggested by the metaphor prove to be there. In that case the metaphor plays an excellent role in creating new referential terms that give us "epistemic access" to a region of objects or phenomena. In rejecting the idea that the reference of a scientific term necessarily depends on a definition of one kind or another, Boyd appeals to the causal theory of reference as proposed by Kripke and Putnam. He adds that the role of metaphor is another outstanding example

---

[8] Boyd (1979, p. 358, cf. n. 2). But see also Kuhn's comments in Kuhn (1979, pp. 417-419). Fundamentally, Boyd characterizes this task as "the task of accommodation of language to the causal structure of the world." Kuhn makes the astonishing comment: "Does it obviously make better sense to speak of accommodating language to the world than of accommodating the world to language?" (p. 418)

of a nondefinitional reference-fixing strategy. Constitutive metaphors refer to whatever has the relevant characteristics and in that way the necessary disambiguation takes place. Therefore they can only succeed if the putative objects have real essences of the appropriate kind; if not, the metaphor is doomed to float freely, as it were, without any anchoring into the real world. And so it seems that if there were no real essences at all, no metaphor could be a successful constitutive metaphor according to Boyd. As one might expect, this mode of "reference fixing" is best suited to the introduction of terms referring to natural kinds "whose real essences consist of complex relational properties, rather than features of internal constitution" (Boyd 1979, p. 358). But he also suggests that the reference fixing in the case of theoretical terms and even all general terms is similar to the reference fixing in the case of metaphor.[9]

2. The Mirror, the Net and the Map

Back to Kuipers' section on metaphors. It is clear that the metaphors he takes into consideration are not of the constitutive but of the exegetic type. They serve to explain or illuminate the nature of science and its relationship to the world as he has defined them already, not to start or to control a theory of the scientific enterprise. Let me call his metaphors "K-metaphors" and the metaphors Boyd is talking about "B-metaphors." B-metaphors occur *in* science, K-metaphors are *about* science. B-metaphors are based on metaphysical realism, K-metaphors refer to metaphysical realism. B-metaphors are right or adequate in so far as the relevant research is successful, K-metaphors are right or adequate in so far as they suggest the characteristics that have been described. Despite the differences between B- and K-metaphors there are striking similarities as well. Just as a B-metaphor is an attempt to indicate a (or perhaps: the) real structure of the world's joints, a K-metaphor is an attempt to indicate a (or perhaps: the) real structure of science. And although in the first case the relevant structure is as yet largely unknown whereas in the second case it has been elaborated already, in both cases the referential connection between metaphor and the primary subject is determined by such a structure.

---

[9] See in particular the sections "Accommodation and reference fixing," pp. 364-372, esp. 369-372, and "Epistemic access: the basis of a theory of reference," pp. 377-401. Boyd's view on the role of metaphors in science is by no means the only view on this subject, of course. Among other contributions are those of Mary Hesse, W.H. Leatherdale and Marx W. Wartofsky. For Kuhn's criticism of the notion of "epistemic access," see his (1979, pp. 411-412).

Now, in view of Kuipers' moderate metaphysical realism the "mirror" metaphor is certainly inadequate. It suggests that the world has characteristics that need only to be noticed and copied. Apparently no creative conceptual activity is called for. Thus the upshot is extreme or essentialistic realism. On the other hand, the "net" metaphor is inadequate because it suggests that the joints we find are of our own making; here conceptualization is all important. The upshot is a kind of minimal realism, something like the positive answer to Q0 only: if you cast out a net, you may catch something.

The "map" metaphor is better than both. It seems appropriate from the point of view of moderate metaphysical realism. Despite the imperfections that he indicates, Kuipers' own discussion suggests the same. As a matter of fact, its metaphysical import proves to be rather stable. This even appears from Stephen Toulmin's illuminating discussion of this metaphor in Chapter 4 of his book *The Philosophy of Science*. Toulmin's comments show that his so-called instrumentalism is only of a restricted kind because it refers mainly to his interpretation of experimental laws in Chapter 3. It does not apply to his view of theories. As appears precisely from his thorough defense of the "map" analogy and his criticism of the "net" analogy as used by Eddington, Toulmin is here a moderate realist. Surely, the "map" analogy *can* be used to defend an instrumentalist position because it can be used to suggest that theories serve only to deliver itineraries. However, as used by Toulmin it has a very different import because it is used to clarify the way theories do represent physical reality.[10] Part of this also follows from his discussion of the question whether sub-microscopic entities exist. His answer that they do exist if there is anything to show for them (1953, pp. 121-122; 1960, p. 136) amounts to scientific realism, at least; moreover, it seems in line with Kuipers' views on the dialectics of observation. And so Kuipers' moderate metaphysical realism is not only close to the so-called essentialistic realism of Harré and Boyd and the external realism of Searle, but also to the so-called instrumentalism of Stephen Toulmin. It is representative of a wide class of realist positions indeed. The "map" metaphor seems to be the appropriate, if not the optimal metaphor of this position: suggestive, to the point, not too misleading.[11]

Even the "map" metaphor is not perfect, of course. Like any metaphor it *is* misleading in certain respects. To summarize Kuipers' discussion: whereas it is strong on constructivity, selectivity and improvability, it is weak on the

---

[10] See, for instance, Toulmin (1953/67, pp. 108-110, 114-116; 1960, pp.121-122, 128-129).

[11] Like Harré, Toulmin appeals to the trustworthiness of theories (Toulmin 1953, pp. 99-100; 1960, p. 111). For him, moreover, this is something relevant to the success of the "map" metaphor. I think he is right. Once more, the "mirror" and the "net" score significantly lower, and again for opposite reasons. Briefly stated, the first suggests too much trustworthiness, the second too little.

nomic world and on dialectical progress. And so Kuipers invites his readers to propose a really satisfactory metaphor. But is there anything better, after all?

I doubt it. In this context, where the relation between science and the natural world is the central issue, a better metaphor must still be taken from the sphere of representation, even if the nomic world would now have to be properly included. Other contexts may allow for quite different adequate metaphors in which (for instance) science is characterized as an "art," or a "factory," or (like fiction in the view of Henry James), a "house with a million windows".[12] These will not do here. The same applies to the metaphor of the "lamp". In a famous book on the history of modern literary criticism, M.H. Abrams discussed the shift from the metaphor of the "mirror" to the metaphor of the "lamp" in the first decades after 1800. Whereas neo-classic criticism often presented literature as a mirror of life, romantic criticism typically praised it for the light it brought into the world; instead of truthful imitation, creative imagination and expression became the norm.[13] To a certain extent the metaphor of the lamp is a suitable metaphor with respect to science as well. In any case, it highlights the role and the impact of the scientific mind in the discovery of the secrets of the universe. But though perhaps more suitable than the metaphor of the mirror, it is not better than the metaphor of the map.It still suggests that no specific conceptual methods are called for.

In order to outdo the "map" metaphor, one has to look for representational metaphors with a (potential) reference to physically-possible worlds. Examples would be "diagram" or "blueprint," but they lack most of the rich variability that is inherent in the "map" metaphor. That there are evidently so many types of maps (cf. Toulmin 1953, pp. 102-107; 1960, pp. 114-119) is another strong feature of this metaphor, mirroring as it does the rich variability of science. Moreover, isn't something like a dialectical progress involved in the hierarchy of maps with data becoming more and more abstract? It may even be possible to stretch the applicability of the "map" metaphor to the mapping of possibilities besides actualities.

---

[12] See the preface (written in 1908) to *The Portrait of a Lady* ([1881] 1984): "The house of fiction has in short not one window, but a million – a number of possible windows not to be reckoned, rather; every one of which has been pierced, or is still pierceable, in its vast front, by the need of the individual vision and by the pressure of the individual will" (James 1984, pp. 45-46). Taken as a metaphor about science, the reasons would be different, of course, and 'a million' might become hyperbolic. For an example, see the remark by P.W. Atkins: "We have looked through the window on to the world provided by the Second Law, and have seen the naked purposelessness of nature" (1986, p. 98).

[13] This is, of course, a very rough summary of Abrams' detailed exposition; other metaphors and issues were involved in the process. One of those who played a role in the establishment of Romantic criticism was John Stuart Mill, with two essays on poetry written in 1833. See Abrams (1953, pp. 23-26 and *passim*).

504   J. J. A. Mooij

3. Epilogue

Whether the "map" metaphor is *true* and the "mirror" and "net" metaphors are *false* is, however, a very different question. For one thing, it has often been argued that metaphorical statements cannot be true anyhow because they are always false or neither true nor false; instead, the relevant distinctions would be that they can be more or less illuminating, (in)adequate, (in)appropriate, misleading and the like. There are strong reasons for such a view. Elsewhere I have indicated a number of specific procedures in human communication (the "so to speak" procedure, the procedure of metaphor deletion, and the "no truth – no sense" mechanism) which seem to corroborate such a view. However, I have also argued that this view is nonetheless not quite convincing because there are also important, maybe overriding reasons why the possibility to characterize certain metaphorical statements as true or false should be maintained (Mooij 1993). I still believe so. But whereas this point of view might well lead to the assessment of the "map" metaphor as true, it is far from evident that the "net" metaphor and even the "mirror" metaphor must then be false, for they, too, capture certain characteristics of theories and the nature of science. In certain respects science *is* (metaphorically) a net or a mirror – and even an art, a factory or a house with countless windows.[14] Therefore, such contentions may be true, although useless or even highly misleading in many contexts. In itself, this is no different from literal statements. For instance, the statement that the Thirty Years' War has come to an end is certainly true; but it may be useless, and in suggesting that this recently happened, it can be very misleading indeed. In general, the relevance of noticing any properties of persons or things is strongly dependent on the situation; many true ascriptions are (or would be) ineffective or worse.

Thus I do not feel obliged to retract my earlier view that Popper's and Hempel's statement that theories are nets is a true metaphorical statement (Mooij 1993, p. 78). At the time, Theo Kuipers suggested to me that the net metaphor is proper to vocabularies, not to theories. Apparently, he still holds that opinion, at least in the sense that the net metaphor is more adequate and less misleading with respect to vocabularies than with respect to theories.[15] I agree. But this relative inadequacy does not necessarily make the metaphor false. However, it does make it, in the context of Kuipers' discussion of three

---

[14] But not a black hole, I think, and not even a myth.

[15] Cf. Niiniluoto, in his description of the rationalist and Kantian traditions: "Concepts are like nets that we throw out to the world" (1999, p. 110). Theo Kuipers informed me that there is a printing error in Table 13.2 on p. 333. The adequacy of the net/spectacles metaphor in the category "referentially improvable" for vocabularies should be given as + instead of –, as also appears from the text.

types of metaphysical realism and his acceptance of the moderate type, an unsuitable metaphor. Apart from any further shortcomings, it suggests the wrong kind of realism.

University of Groningen
Faculty of Arts
c/o Goeman Borgesiuslaan 32
9722 RJ Groningen
The Netherlands

REFERENCES

Abrams, M.H. (1953). *The Mirror and the Lamp: Romantic Theory and the Critical Tradition.* Oxford: Oxford University Press. (Many reprints).

Ankersmit, F.R. and J.J.A. Mooij, eds. (1993). *Knowledge and Language*, vol. III: *Metaphor and Knowledge.* Dordrecht: Kluwer Academic Publishers.

Atkins, P.W. (1986). Time and Dispersal: The Second Law. In: Flood and Lockwood (1986), pp. 80-98.

Boyd, R. (1979). Metaphor and Theory Change: What is 'Metaphor' a Metaphor For? In: Ortony (1979), pp. 356-408.

Flood, R. and M. Lockwood, eds. (1986). *The Nature of Time.* Oxford: Blackwell.

Harré, R. (1986). *Varieties of Realism.* Oxford: Blackwell.

James, H. (1984). *The Portrait of a Lady.* Edited with an Introduction by G. Moore and Notes by P. Crick. London: Penguin Books.

Kuhn, T.S. (1979). Metaphor in Science. In: Ortony (1979), pp.409-419.

Kuipers, T.A.F. (2000/ICR). *From Instrumentalism to Constructive Realism.* Dordrecht: Kluwer Academic Publishers.

Mooij, J.J.A. (1993). Metaphor and Truth: A Liberal Approach. In: Ankersmit and Mooij (1993), pp. 67-80.

Niiniluoto, I. (1999). *Critical Scientific Realism.* Oxford: Oxford University Press.

Ortony, A., ed. (1979). *Metaphor and Thought.* Cambridge: Cambridge University Press.

Searle, J.R. (1995). *The Construction of Social Reality.* New York: The Free Press.

Toulmin, S. (1953). *The Philosophy of Science. An Introduction.* London: Hutchinson. Also: New York: Harper & Row, 1960.

Theo A. F. Kuipers

MODERATE REALISM AND METAPHORS

REPLY TO HANS MOOIJ

When I was preparing my dissertation in the years 1973-7, Hans Mooij and Aart Stam were the ideal supervisors. The more problematic stories about supervisors I heard, the more I understood how exceptionally good my situation as a "promovendus" (Ph.D. student) was. Reading Mooij's paper reminded me again of this fact, but now in a somewhat embarrassing way. Had I shown him my draft of ICR, he would have redrawn my attention to Toulmin's discussion of the map metaphor in his seminal book. Now my only face-saving option is to confess that my ideal world includes Hans Mooij as permanent supervisor.

Moderate Realism

Let me start, however, with what I see as Mooij's main critical point. According to him, I do not make sufficiently clear the extent to which my constructive realism goes beyond minimal metaphysical realism, without becoming essentialist realism by assuming an ideal conceptualization of the world. He writes in Section 1:

> If science can deliver objective knowledge about the observable and non-observable aspects of the world, then there must be, not only the world, but *a way the world is*; it cannot be completely indefinite or malleable. Perhaps there need not be one and only one way the world is, independent of any conceptualization. However, certain conceptualizations must be accurate and correct.

I certainly agree that I assume some in-between, moderate, metaphysical realism, but I cannot agree with his suggestion that this should be based on a notion of 'correct conceptualizations' that is weaker than the notion of (being part of) the ideal conceptualization. Certainly, if he means by a correct conceptualization merely that all its terms are referring to something in – or some aspect of – the world, then I could agree. But I am afraid Mooij means more than this minimal condition for a satisfactory vocabulary, for the

In: R. Festa, A. Aliseda and J. Peijnenburg (eds.), *Confirmation, Empirical Progress, and Truth Approximation* (*Poznań Studies in the Philosophy of the Sciences and the Humanities,* vol. 83), pp. 506-510. Amsterdam/New York, NY: Rodopi, 2005.

reference condition does not exclude terms like 'grue' and one merely referring to an "undetached rabbit part," to vary Quine's famous example. Instead of postulating that there are correct vocabularies in a stronger sense, my moderate metaphysics regarding the claim that "there must be … *a way the world is*" is very explicitly of a hybrid metaphysical-epistemological nature as, for example, expressed in the concluding passage of Section 13.5 ("The metaphysical nature of scientific research", ICR, p. 327):

> In sum, the scientist studies objective features of THE NOMIC WORLD by choosing a domain and a vocabulary. Although many choices, partly of a conceptual constructive nature, have to be made, nothing like radical social constructivism is the result, for THE NOMIC WORLD determines what we will find, given our choices. That is, although constructive realism recognizes several relativistic features, it does not at all lead to radical relativism: given previous choices, we cannot find what we want. Of course, we can frequently find what we want, but only by adapting our choices of domain and vocabularies in an appropriate way, that is, in agreement with the nature of THE NOMIC WORLD.[16]

Hence, in my view the "way the world is" is not something that can be expressed in terms of (in-)correct vocabularies but is of a dispositional nature, and hence, metaphorically, of a stimulus-response nature: imposing *this* vocabulary on *that* domain yields those true and false statements. I also think that this is very much the kind of moderate metaphysical realism that is subscribed to by Niiniluoto, Searle and, for that matter, Toulmin (1953, p. 128; 1960, p. 115). Hence, being in their company, as Mooij claims I am, is what I agree about, but I do not think that Boyd and Harré belong to the same group, simply because, as Mooij points out, they at least subscribe to some kind of correct, if not ideal, language.

I can further illustrate my view with one particular point in his argumentation. He writes: "If there are correct or ideal conceptualizations, then there may come an end to the improvement of vocabularies; just as (in Kuipers' view) there may come an end to the improvement of *theories,* viz. by finding the strongest true exemplar." (p. 497) This comparison of vocabularies and theories in terms of improvement seems inadequate to me. Given a domain and a vocabulary the improvement of theories about that domain in that vocabulary has its limits, set by the thus implied (unknown, but operative) truth. However, given only a domain, there is no boundary for correct vocabularies, at least not for referring ones, for, as already indicated, they can be crated at will. On the other hand, if correct vocabularies in a stronger sense were to exist and their improvement were to come to an end, they would be

---

[16] Incidentally, the closing phrase "in agreement with the nature of THE NOMIC WORLD" is an essentialist slip of my pen. It would have been better to write: "in agreement with – and using the provisional findings about – THE NOMIC WORLD".

combinable in some way or other and lead to the ideal vocabulary for that domain. Certainly this would be a serious option if the world were designed, but we may assume that it isn't.

Metaphors

Turning to the general topic of metaphors, I like Boyd's distinction, referred to by Mooij, between constitutive (C-) and exegetic (E-)metaphors. In principle, both can occur in science or be about science. Although I certainly have interest in both kinds of metaphors and analogies in science (SiS, p. 11) and whereas I do not remember having read or heard about stimulating C-metaphors about science, my specific interest in metaphors in the last chapter of ICR certainly concerns only E-metaphors about science, viz. exegeses of my meta-view of science as developed in ICR. As indicated above, Hans Mooij points out, in his characteristically polite way, that I was essentially re-inventing the wheel by my arguments in favor of the map metaphor relative to the mirror and the net metaphor. In particular, rereading Toulmin's excellent plea for the map metaphor and against the net metaphor convinced me that he had already said the most essential things in this respect. And indeed, as suggested by Mooij, in the relevant chapter ("Theories and Maps") Toulmin is much less the anti-realist than he is usually supposed to be, probably on the basis of the preceding chapter ("Laws of Nature"). Even the latter division can be questioned: rereading Toulmin's claims about determining the "scope" of both laws and theories, it is tempting to reinterpret his analysis in terms of what I mean by dialectically establishing a domain in combination with a vocabulary (exceeding the "domain vocabulary") and the true theory (ICR, p. 332):

> More precisely, in the short-term dynamics of 'science in the making' three things are established in dialectical interaction: the domain as a (unproblematically conceptualized) part or aspect of THE NOMIC WORLD, the (extra) vocabulary and the true theory about that domain as seen through that vocabulary.

Hence, Toulmin's claims seem very much compatible with the constructive realist attitude developed in ICR. However, unfortunately, the precise nature of this dialectical interaction has not yet been elaborated. Only some anticipations were set out regarding domain variation (ICR, p. 207). A detailed elaboration of "truth approximation by domain variation" is not merely a technical challenge, it will certainly improve our philosophical-methodological insights. For a start, see my reply to Zwart.

Of course, I was very happy with Mooij's support of the map metaphor, by references and arguments. It inspired me to combine the map metaphor with the language of possibilities, leading to the following quasi-metaphorical characterizations of four kinds of research (cf. SiS, Ch. 1).

- Theoretical research:
    mapping nomic (physical, chemical, biological,…) (im-)possibilities
- Experimental research:
    realizing and mapping (by definition, nomic) possibilities
- Technical/design research:
    mapping and realizing intended possibilities
- Historical research:
    mapping (naturally or artificially) realized possibilities

I speak of quasi-metaphorical characterizations because it is hard to conceive of the language of possibilities, let alone that of their realization, as metaphorical. The combination with mapping, however, in particular when applied to theoretical or nomological research, seems to satisfy all the criteria I suggested in ICR (pp. 330-2) for an adequate metaphor for that kind of research.[17] According to that exposition, we should distinguish sharply between the vocabularies and theories formulated with them to characterize a given domain (beyond its characterization in terms of the "domain vocabulary"). In the terms of the map metaphor, a vocabulary corresponds to a *mapping method*, determining which aspects of the domain will be mapped in what manner. A specific theory corresponds to the result of a particular attempt to map the domain, the *resulting map*. In this way the map metaphor satisfies in general three of the four joint criteria for a metaphor for vocabularies and theories: both have to be and are constructive, selective and referentially improvable. Moreover, theories as resulting maps satisfy the theory-specific criterion that theories should be substantially improvable, for many mistakes can be made in the application of a mapping method leading to a mapping product (cf. Toulmin, 1953, p. 127; 1960, p. 114).

The three joint features and the theory-specific one remain when one speaks more specifically of mapping possibilities, introducing the non-metaphorical notion of possibilities. However, a metaphor for theories should also include the nomic target of theories, viz. nomic possibilities. By speaking of "mapping nomic (im-)possibilities" I include this aspect by definition, leading to a second non-metaphorical aspect of the metaphor. Last but not least, an adequate metaphor should highlight the already indicated dialectical interaction in establishing three things: "the domain as a (unproblematically conceptualized) part or aspect of THE NOMIC WORLD, the (extra)

---

[17] I thank Hans Mooij for revealing the mistake regarding the net-metaphor in Table 13.2 (ICR, p. 333), reported in his Note 16. Moreover, I certainly agree with Mooij's closing statement that this metaphor suggests the wrong kind of metaphysics, viz. minimal realism.

vocabulary and the true theory about that domain as seen through that vocabulary." (ICR, p. 332). This dialectical feature seems also to be covered, like the nomic feature, not so much by the metaphorical mapping but by its specific target of nomic possibilities. Since a vocabulary determines the "conceptual possibilities," and the combination of a vocabulary and a given domain determines the "nomic possibilities," and hence "the true theory," there are two independent variables and a dependent one, which leaves room for a dialectical determination of an attractive triple.

In sum, instead of finding a satisfactory pure metaphor, asked for in the closing sentences of ICR, we have obtained a highly satisfactory partial metaphor. Guided by this partial metaphor, I would like to confirm that I see truth approximation not as "just an epistemic notion. It is also a moral injunction" (p. 499), not only for scientists, but also for judges and other professionals for whom the (relevantly restricted and conceptualized) truth makes all the difference. However, this certainly is not a moral injunction for all professionals – it does not hold for artists, for example.

REFERENCE

Toulmin, S. (1953). *The Philosophy of Science. An Introduction*. London: Hutchinson. Also: New
    York: Harper & Row, 1960.

Roberto Festa

ON THE RELATIONS BETWEEN (NEO-CLASSICAL) PHILOSOPHY OF SCIENCE AND LOGIC

ABSTRACT. In this paper I consider a number of metaphilosophical problems concerning the relations between logic and philosophy of science, as they appear from the neo-classical perspective on philosophy of science outlined by Theo Kuipers in ICR and SiS. More specifically, I focus on two pairs of issues: (A) the *(dis)similarities* between the goals and methods of logic and those of philosophy of science, w.r.t. (1) the role of theorems within the two disciplines; (2) the falsifiability of their theoretical claims; and (B) the *interactions* between logic and philosophy of science, w.r.t. (3) the possibility of applying logic in philosophy of science; (4) the possibility that the two disciplines are sources of challenging problems for each other.

The outcomes obtained in ICR and, more generally, the "neo-classical approach" to philosophy of science developed in ICR and SiS,[1] not only raise specific philosophical problems, related to different epistemological and methodological issues, but also a number of metaphilosophical problems about the nature of philosophy of science, and its relations with other disciplines.

A main metaphilosophical problem concerns the relations between logic and philosophy of science, as they appear from the neo-classical perspective. I focus on two pairs of issues:

(A) the *(dis)similarities* between the goals and methods of logic and those of philosophy of science, w.r.t.:
    (1) the role of theorems within the two disciplines;
    (2) the falsifiability of their theoretical claims;

(B) the *interactions* between logic and philosophy of science, w.r.t.:
    (3) the possibility of applying logic in philosophy of science;

---

[1] To be precise, the term 'neo-classical philosophy of science' is introduced in SiS, starting from the second clause of the subtitle ("An Advanced Textbook in Neo-Classical Philosophy of Science"). Although this term is not used in ICR, the definition of the "neo-classical approach" (SiS, p. ix) should allow us to use it also w.r.t to ICR. Hence, in the following, I will use 'neo-classical (approach to) philosophy of science' w.r.t. the general philosophical perspective developed both in ICR and SiS, and 'ICR-approach', 'ICR-view', 'ICR-analysis', etc. when referring to specific approaches, views and analyses developed in ICR.

In: R. Festa, A. Aliseda and J. Peijnenburg (eds.), *Confirmation, Empirical Progress, and Truth Approximation* (*Poznań Studies in the Philosophy of the Sciences and the Humanities,* vol. 83), pp. 511-521. Amsterdam/New York, NY: Rodopi, 2005.

512   *Roberto Festa*

(4) the possibility that the two disciplines are sources of challenging problems for each other.

All the issues (1)-(4) have been considered, more or less extensively, by Johan van Benthem in his programmatic paper "The logical study of science" (1982) in which a number of interesting views and desiderata related to (1)-(4) are stated.[2] I find it particularly stimulating to ask whether, and to what extant, the neo-classical approach agrees with van Benthem' views and contributes to the realization of his desiderata.

1. The Role of Theorems in Logic and in Philosophy of Science

According to van Benthem (1982, p. 433), an important "difference in 'mentality' between logicians and many formal philosophers of science" is revealed by the circumstance that "logicians want *theorems* where these philosophers often seem content with *definitions*." In this connection, he suggests the desideratum that such a difference in mentality should be reduced, so that theorems also play a significant role within philosophy of science.[3]

It seems to me that ICR *does* contribute to the realization of the above desideratum, since the formal definitions that occur in ICR are very often functional to the proof of theorems which play a crucial role in the ICR-account of the scientific enterprise. I am thinking, in particular, of two pairs of theorems – or, more precisely, of families of theorems – viz., the Success/Forward and the Projection/Upward Theorems.[4] The Success and the Forward Theorems establish appropriate links between empirical success and truth approximation; for instance, the (basic version of the) Success Theorem guarantees that "more observational truthlikeness" implies "at least as much

---

[2] See especially Sections 1 and 4; see also van Benthem (1983), especially Section 2. More precisely, issues (1) and (2) are discussed in (1982), Section 1.2, "Logicians and philosophers of science"; issue (3) in (1982), Section 4.1, "What is 'Application'?"; issue (4) in (1982), Section 1.3, "Logica Magna", and in (1983), Section 2, "Logical semantics and the general philosophy of science." For van Benthem's present views on the relations between logic and philosophy of science the reader is referred to the last pages of his contribution to the present volume.

[3] Indeed he points out that "notions should be tied up with 'regulative' theorems (already existing or created explicitly for the purpose)" (van Benthem 1982, p. 457). See also p. 434.

[4] Each family of theorems includes a *basic* version (referring to the case of purely observational theories, where all the incorrect models of a theory are treated in the same way) and more sophisticated versions, i.e., *stratified* versions (concerning proper theories, with a stratification of different, observational and theoretical, levels), and/or *refined* versions (concerning the case where the possibility that one incorrect model of a theory may be more similar, or "more structurelike," to a target model than another, is taken seriously into account). In ICR different versions of the Success and the Forward Theorems (pp. 158-161, 164, 213, 216, and 260-262) and of the Projection and Upward Theorems (pp. 213-214, 216-218, and 276) are presented.

success." Taken together, these two theorems lead to the conclusion that the evaluation methodology – governed by the rule of success RS, which prescribes the selection of the theory, if any, that has so far proven to be the most empirically successful – is functional for observational truth approximation. On the other hand, the Projection and Upward Theorems establish appropriate links between different kinds of truthlikeness, at different levels: for instance, the Projection Theorem roughly states that more theoretical truthlikeness implies more observational truthlikeness. Taken together, the Success/Forward and the Projection/Upward Theorems lead to the conclusion that the evaluation methodology based on RS is functional for theoretical truth approximation; in turn, this conclusion supports the main epistemological thesis of ICR, concerning the possibility of the transition from instrumentalism to constructive empiricism, from this to referential realism and, finally, from referential realism to constructive realism.[5]

2. The Falsifiability of Theoretical Claims in Logic and in Philosophy of Science

According to van Benthem (1982, p. 434, note 5), Popper's idea that our theoretical claims should be falsifiable can be applied to philosophy itself, including logic and philosophy of science. In this connection, he remarks that, while the most famous research programs of modern logic, such as Frege's and Hilbert's "made claims which were *falsifiable*, and indeed they were falsified – witness, e.g., Gödel Incompleteness Theorems" (p. 434), the desideratum of falsifiability has been violated by the most famous research programs in philosophy of science.

Whatever one thinks of the working style prevailing in contemporary philosophy of science, it seems pretty clear that the neo-classical approach satisfies the desideratum of falsifiability to a very broad extent. Below I argue that both the above mentioned role of theorems in ICR, and the falsifiability of

---

[5] On the methodological and epistemological implications of the Success/Forward and the Projection/Upward Theorems see ICR (especially pp. 11, 142, 161-167, 171, 186, 189-190, 208-209, 213-214, 218-219, 228-229, 233, 236, 260-262, 276, and 286).

The above mentioned theorems do not exhaust the list of theorems introduced in ICR, which also includes theorems concerning the properties of quantitative confirmation (ICR, pp. 53-55 and 63-64) and its qualitative consequences (ICR, pp. 57-58); the qualitative consequences of Popperian corroboration (ICR, pp. 69-70); the conceptual foundations of (refined and/or stratified) nomic truthlikeness (ICR, pp. 177-185, 250-253, and 272-275); the application of the concepts of truthlikeness and truth approximation to the strategy of idealization and concretization (ICR, pp. 268-270 and 285); and the principles of quantitative nomic truth approximation (ICR, pp. 309-314).

its theoretical claims, are direct and related consequences of the neo-classical view of the goals and methods of philosophy of science.

In agreement with Carnap's and Hempel's classical philosophy of science, the neo-classical approach assumes that a main target of philosophy of science is the development of a network of *explicative (research) programs*, directed at *concept explication*, i.e., to "the construction of a simple, precise, and useful concept [or *explicatum*], which is, in addition, similar to a given informal concept" (SiS, p. 8). Kuipers characterizes the strategy underlying the explication of intuitive concepts as follows:

> – From the intuitive concept to be explicated one tries to derive conditions of adequacy that the explicated concept will have to satisfy, and evident examples and counter-examples that the explicated concepts has to include or exclude." (p. 8)

Note that the activity of concept explication can be extended to the "explication of intuitive judgments, i.e., intuitions, including their justification, demystification or even undermining" (SiS, pp. 9-10): for instance, ICR provides both an explication of the intuitive concepts of empirical success and truthlikeness and of the intuition that the choice of empirically successful theories is functional for truth approximation.

The result of concept and intuition explication is given by a so-called *conceptual (meta-)theory*, viz., by a theory that "is intended to provide a perspective, a way of looking, at a certain domain without making a general empirical claim" (SiS, p. 58). However, different (not strictly empirical) kinds of claims can be associated to a conceptual theory: for instance, to the conceptual meta-theory obtained as the result of concept explication, one should attach the "(quasi-)empirical meta-claim" that the theory "roughly captures an intuitive concept or distinction" (p. 58). The conceptual theories obtained as a result of the explicative programs carried out in neo-classical philosophy of science include the theories on the notions of explanation and reduction of laws by subsumption (SiS, Ch. 3, see especially p. 92), on the qualitative and quantitative notions of confirmation (ICR, Ch. 2 and 3), and the structuralist (meta-)theory of empirical theories (SiS, Ch. 12, see especially p. 341).

The above characterization of (the conceptual theories resulting from) explicative programs clearly suggests that the notions of *success* and *progress* can be legitimately applied to explicative research. Indeed, one can evaluate the success of the conceptual theory resulting from a provisional explication in terms of evident *examples* and evident *non-examples* of the explicated concept, and of the so-called *conditions of adequacy*, to be satisfied by the explicated concept: "successes of conceptual theories … should be interpreted as satisfactorily treated cases and realized general conditions of adequacy" (SiS, p. 248). Taking into account that, as a norm, a conceptual theory has to face its own anomalies, i.e., the "problems arising from cases and aspects that are wrongly dealt with" (SiS, p. 248), a plausible definition of (conceptual) explicative progress – which can also be applied to two "falsified" provisional explications – seems to be the following:

> – Provisional explication *Y* is better than provisional explication *X*, roughly speaking, if and only if *Y* treats more evident examples and non-examples properly and/or fulfills more conditions of adequacy. (SiS, p. 264)[6]

The above characterization of the notions of explicative success and progress allows a more precise localization of the role of theorems and of the desideratum of falsifiability in the neo-classical approach to philosophy of science. Indeed, it seems to me that a main role of theorems in philosophy of science is related to two basic features of explicative programs: (1) in many cases one cannot immediately see that a given *explicatum* applies to the available evident examples, does not apply to the evident non-examples, and, finally, satisfies the accepted conditions of adequacy: one has to prove this by appropriate theorems; (2) the explication of intuitions – for instance of the intuitive judgment that more truthlikeness implies more empirical success – amounts to showing that the corresponding formal judgement is true: once again, very often this cannot be seen, but has to be proven.

As far as the desideratum of falsifiability is concerned, it seems to me that falsifiability is a proper feature of the conceptual theories developed in explicative research. Indeed, explicative research is not characterized, in general, by a fixed set of "data," viz., examples, non-examples, and conditions of adequacy: hence, it may well happen that one finds new, convincing data that can be shown – or proven – to falsify a provisionally accepted explication. Moreover, in principle, one cannot rule out that some items in the initial stock of "data" are subsequently shown – perhaps by discovering appropriate theorems – to falsify a provisionally accepted explication.[7]

3. The Application of Logic in the Philosophy of Science

Before wondering about the usefulness, or indispensability, of the *application of (modern) logic* in philosophy of science, one should make sufficiently clear what the "application of logic" in other fields amounts to. For instance,

---

[6] On progress in explicative research, see SiS, pp. 248 and 263-64.

[7] A good example of falsification of a provisional explication can be found in the development of Kuipers' research program on truthlikeness. Indeed, Kuipers' first definition of refined truthlikeness (Kuipers 1987b) was falsified by van Benthem (1987, Section 5), who proved that it did not satisfy an apparently plausible condition of adequacy. In response to this falsification, Kuipers proposed a new definition (ICR, Ch. 10.2.1 and especially note 7 to Ch. 10, p. 344).

according to a "weak sense" of the expression, specified by van Benthem (1982, p. 457), 'applying logic' just means 'using logical tools' where "these 'tools' may be anything from methods and theorems to mere notions or notations." In particular, as far as the application of logic in philosophy of science is concerned, "sometimes, the only 'tool' would even be that esoteric (though real) quality called 'logical sophistication'" (p. 457). Here it should be noticed that very often the "esoteric logical sophistication able to produce [the intended theorems] requires a very down-to-earth training in technical logic" (p. 457).

It seems to me that at least this "weak sense" of "applying logic" fits ICR rather well. In fact, in spite of the conceptually very important implications of the Success/Forward and the Projection/Upward Theorems, their proof only presupposes an elementary knowledge of logic and (naive) set theory.[8] One might even say that some "esoteric logic sophistication" is involved just in the task of obtaining interesting results by using only elementary logic.

A "strong sense" of "applying logic" in philosophy of science might refer to the usage of the advanced, up-to-date methods and theorems of modern mathematical logic. However, the results obtained in ICR suggest that the application of such methods is not indispensable when carrying out fruitful explicative programs in philosophy of science. On the other hand, the discussion of some of the issues treated in ICR and in several contributions to the *Essays* seems to indicate that more promising results can be obtained by applying some of the "new logics" developed in the last 20 years, very often under the general label of "philosophical logic," or even creating new logical tools and theories.[9]

---

[8] These remarks apply also to other theorems proven in ICR. However, I do not wish to suggest that for *any* philosophically relevant theorem one can provide an elementary proof: for instance, this is certainly not the case for some theorems proven within the so-called "theory of inductive probability" (cf. Kuipers 1978 and Festa 1993), which are also indicated in ICR (Ch. 4, pp. 85 and 80, respectively).

[9] See the contributions by Aliseda, Batens, van Benthem, Burger and Heidema, Meheus, and Schurz in this volume, and the contributions by Grobler and Wiśniewski, Kamps, and Ruttkamp in the companion volume. The "new logics" applied – and in some cases developed – by these authors include Beth's semantic tableaux method (Aliseda), ampliative adaptive logic (Batens and Meheus), dynamic-epistemic logics for information update and partial logic (van Benthem), belief revision (Burger and Heidema), relevance logic (Schurz), erotetic logic (Grobler and Wiśniewski), computational (classical) logic (Kamps), and non-monotonic logic (Ruttkamp). On the potential usefulness of some of these new logics see also the last pages of van Benthem's contribution to this volume.

4. Philosophy of Science and Logic as Sources of Problems for Each Other

It seems to me that the possibility that philosophy of science is a source of important problems for logic is related, first of all, to the general circumstance that, if logic is intended in a sufficiently wide sense, then (neo-classical) philosophy of science can be seen as a main province of the logical continent.

More precisely, suppose that logic is construed – in agreement with van Benthem (1982, p. 435) – as *logica magna*, viz. as "the study of reasoning, wherever and however it occurs." This means that "in principle, an ideal logician is interested both in that *activity*, and its *products*, both in its *normative* and its *descriptive* aspects, both in *inductive* and *deductive* argument [and looks] for stable patterns" (p. 435) underlying the activities and products of the various forms and areas of reasoning.[10] This definition implies that "ideal logicians" should look, among other things, for "stable patterns" underlying the activities and products of *scientific* reasoning. If logic is conceived in this extended sense, then the subject matter of neo-classical philosophy of science belongs to the field of logic proper – and, more precisely, of the *logic of empirical sciences*: indeed, a main target of ICR and SiS is the identification of the stable structures and patterns underlying the activities and products – or, roughly equivalently, the processes and results – of scientific reasoning.

There is also a more specific reason – related to a characteristic feature of the neo-classical approach – for the possibility that philosophy of science is a source of interesting problems for logic: I am referring to the special attention that, within the neo-classical approach, is devoted to *both* terms of the above mentioned pairs of concepts, i.e., "activities/products" and "processes/results." Indeed, Kuipers makes it explicit that the core of the neo-classical approach is "the idea that there is system in knowledge and knowledge production" (SiS, p. x) and that its "primary aim is to show in some detail the products that scientists are after, and the means by which they try to obtain them" (SiS, p. 3). Moreover, Kuipers assumes that a crucially important use-value of the cognitive structures underlying scientific activities and products is related to the circumstance that "they may play a heuristic role in actual research" (SiS, p. xii; see also ICR, p. 12).[11]

---

[10] See especially van Benthem (1982), Section 1.3 and, for a more recent and extensive treatment of this issue, van Benthem (1999).

[11] He supports this assumption – whose importance is witnessed also by the subtitle of SiS, i.e., "Heuristic Patterns Based on Cognitive Structures" – by numerous and detailed illustrations of the ways in which the heuristic patterns driving scientific research are based on specific cognitive structures: see ICR (especially pp. 10-12, 113-115, 122, 164, and 270-272) and SiS (especially pp. ix-x, xii, 32, 103, 118, 125, 231-232, 240, and 275).

Kuipers' attention to the cognitive heuristic patterns of scientific research leads him to confront sympathetically a recent research tradition – going under the name of computational philosophy of science – the basic goal of which is the design of "computer programs that deal with discovery, evaluation and revision of hypotheses" (SiS, p. 287).[12] The pursuit of this goal is legitimated by the idea – which overthrows an entrenched opinion of classical philosophy of science – that "discovery ... is accessible for methodological analysis" (p. 287) and that "the discovery process ... [is] methodologically examinable and perhaps partially programmable" (p. 289). Since this assumption also plays a crucial role in Kuipers' approach, he can point out that "there is not much competition between computational and classical philosophy of science, as soon as one is willing to reject the dogma of the impossibility of fruitful philosophical-methodological analysis of the Context of Discovery" (p. 303). As a matter of fact, the rejection of this dogma discloses the possibility of a fruitful cooperation between neo-classical and computational philosophy of science: for instance, one can explore the idea that the cognitive patterns identified by neo-classical philosophy of science "could provide the heuristics for computer programs" (p. 301) useful for the (re)production of scientific discoveries.[13] It seems clear that a fruitful exploration of this idea requires, among other things, the creation of new logical tools, such as the semantic tableaux method to instrumentalist abduction.[14] This possibility clearly shows that highly innovative logical research may be driven by questions stemming from (the neo-classical approach to) philosophy of science.

It may be interesting to notice that the "dynamic switch" of neo-classical philosophy of science – from scientific products and results to scientific activities and processes – is in agreement with a more general trend involving the whole logical continent. In this connection, van Benthem (1999, p. 25), points out that, while the traditional "focus of logical research" was and, to a large extent, still is "on static *products* of logical activities, such as statements or proofs, instead of those activities *themselves*," the situation is changing rapidly so that "in much current literature there is a 'Dynamic Turn' putting (both physical and cognitive) activities at centre stage as a primary target of research."[15]

---

[12] On computational philosophy of science, see SiS (Ch. 11).

[13] On the conceptual connections between neo-classical and computational philosophy of science, see SiS (Ch. 11.1.4).

[14] For a survey of the studies in this area see SiS (pp. 302-303). See also the contributions to the Section 2 of this volume, by Atocha Aliseda, Joke Meheus, and Diderik Batens.

[15] On the dynamic turn of the recent logical research, see van Benthem (1999), especially Sections 2.4 and 6.1.

Let me conclude with a brief remark on the possibility that logic – and, more generally, mathematics – supplies intriguing problems to philosophy of science. Indeed, this possibility is far from obvious, and one may still share with van Benthem (1983, p.302), the complaint that "philosophers of science have often shied away from applying their insights to the sacred strongholds of the purest sciences, such as mathematics or logic" although "after all, there is nothing particularly immoral about asking whether, e.g., ... Kripke's research program has entered upon a 'degenerative phase'." One immediately recognizes the Lakatosian flavor of this question; in fact Lakatos (1976) is the first splendid, but still rather isolated, example of epistemological analysis of mathematical research programs. Since Kuipers' notion of an explicative research program can be seen as a sophisticated non-empirical variant of Lakatos' concept of research program, I would not be surprised if Kuipers' notion turned out to be fruitfully applicable in the investigation of significant research programs developed in logical and mathematical disciplines.[16]

From the above remarks one can draw, at least, the general conclusion that the similarities and interactions between philosophy of science and logic depend strongly on the way in which the two disciplines are conceived. In particular, it seems to me that both the similarities and the interactions are deemed to increase if philosophy of science is practised within the neo-classical approach to philosophy of science developed by Theo Kuipers in IRC and SiS.

University of Trieste
Department of Philosophy
Androna Campo Marzio 10
34123 Trieste
Italy
e-mail: festa@units.it

REFERENCES

Benthem, J. van (1982). The Logical Study of Science. *Synthese* **51**, 431-472.

Benthem, J. van (1983). Logical Semantics as an Empirical Science. *Studia Logica* **42**, 299-313.

---

[16] Although Kuipers notices that "the concepts of 'logical consequence' and 'probability' have given rise to very successful explicative programs in the borderland between philosophy and mathematics" (SiS, p. 8), so far he did not engage in the investigation of specific logical or mathematical research programs. However, van Bendegem is certainly right in suggesting that the "quasi-empirical" account of mathematical arguments developed in his contribution to volume 2 of the *Essays* is in the spirit of Kuipers' neo-classical approach.

                    Roberto Festa

Benthem, J. van (1987). Verisimilitude and Conditionals. In: Kuipers (1987a), pp. 103-128.

Benthem, J. van (1999). Wider Still and Wider ... Resetting the Bounds of Logic. *European Review of Philosophy* **4**, 21-44.

Festa, R. (1993). *Optimum Inductive Methods. A Study in Inductive Probability Theory, Bayesian Statistics and Verisimilitude*. Dordrecht: Kluwer.

Kuipers, T.A.F. (1978). *Studies in Inductive Probability and Rational Expectation*. Dordrecht: Reidel.

Kuipers, T.A.F. Ed. (1987a). *What is Closer-to-the-truth? A Parade of Approaches to Truthlikeness*. (*Poznań Studies in the Philosophy of the Sciences and the Humanities*, vol. 10). Amsterdam: Rodopi.

Kuipers, T.A.F. (1987b). A Structuralist Approach to Truthlikeness. In: Kuipers (1987a), pp. 79-99.

Lakatos, I. (1976). *Proofs and Refutations. The Logic of Mathematical Discovery*. Cambridge: Cambridge University Press.

Theo A. F. Kuipers

UNINTENDED CONSEQUENCES AND THE CASE OF ABDUCTION

REPLY TO ROBERTO FESTA

In his essay Roberto Festa elaborates the issues of (dis)similarities and possible interactions between logic and philosophy of science put forward by Johan van Benthem, in 1982, primarily in view of ICR, but also in view of SiS. Hence, the essay provides a natural transition from Volume 1, which mainly deals with topics of ICR, to Volume 2, which mainly deals with topics of SiS. In particular, as Festa makes clear, for the comparison of logic and philosophy of science it is plausible to emphasize the dominant research perspective in both ICR and SiS, viz. explicative research. The nature of this type of research is only explicitly thematized in SiS (pp. 8, 9, 18, 58-9, 248, 263-4). Festa is certainly right in stressing that the (plausible) definition of progress in explicative research presupposes the falsifiability of provisional explications, while the specific argumentation that an evident example or non-example is treated properly or that a condition of adequacy is, or is not, fulfilled may amount to proving a more or less deep theorem, or characterizing a more or less interesting countermodel. These points illustrate the claim in ICR (p. 130-1) and SiS (p. 248) that (explicative) research in philosophy should satisfy, *mutatis mutandis*, the three general principles of testability (PT), separate evaluation (PSE) and improvement (PI), where the latter presupposes the principle of comparative evaluation (PCE). Hence, Festa's essay pleases me very much. In this reply I would like to concentrate on the role of "unintended consequences," neglected by Festa, in speaking of progress in the empirical sciences, logic, and philosophy (of science). Moreover, I shall illustrate it by the recent finding of the straightforward abductive nature of Inference to the Best Theory (IBT), as opposed to the standard conception of Inference to the Best Explanation (IBE).

In: R. Festa, A. Aliseda and J. Peijnenburg (eds.), *Confirmation, Empirical Progress, and Truth Approximation* (*Poznań Studies in the Philosophy of the Sciences and the Humanities,* vol. 83), pp. 521-525. Amsterdam/New York, NY: Rodopi, 2005.

                    Theo A. F. Kuipers

Unintended Consequences in the Empirical and Non-Empirical Sciences

Although Festa refers to several paragraphs and passages in SiS dealing with concept explication, including the definition of progress in explicative research (SiS, p. 264), surprisingly enough he neglects my remark about unintended explications:

> However, it is also considered to be very important that the proposed explication turns out to give rise to unintended explications, that is, to satisfactory explications of related concepts and intuitions. This type of success is the analogue of the extra, i.e., predictive or external, success of explanatory programs. Again the question is whether this form of success is formally defensible as a necessary condition for progress, but the fact remains that in practice this type of explicative success plays an important role. (SiS, p. 18)

In ICR, Subsection 7.5.1, I deal extensively with the last mentioned question regarding explanatory programs in view of the basic version of the structuralist theory of truthlikeness and conclude:

> In sum, ad hoc repair of a theory will seldom be a real improvement without unexpected extra success. In other words, comparative HD-evaluation of an ad hoc repair will either lead to unexpected extra successes of the new theory or extra successes of the old theory that could have been, but were not, explicitly expected before. Hence, besides some qualifications, the intuitions of Popper and Lakatos with respect to ad hoc repairs and novel facts are largely justified. [However, I]nstead of a ban on ad hoc changes, they can be allowed, provided they are subjected to comparative HD-[evaluation] with the original theory. (ICR, pp. 168-9)[1]

Hence, the first question is whether there is a similar story to tell about ad hoc improvements of explications. Unfortunately, the detailed argumentation in Subsection 7.5.1 referred to leans heavily on the unknown but fixed character of the postulated target, the truth, that is, the set of nomic possibilities determined by the domain and the vocabulary. In contrast – although less extreme than in the case of the material version of explicative research, that is, design research (SiS, pp. 282-3) – explicative research is guided by a more or less known target which may or may not be changed. Hence, the prospects for an analogous way of reasoning are not impressive.

However this may be, the intuition stated in the first quote remains: an explication of a concept generating related but unintended explications is more impressive than an explication that does not. Fortunately, close reading of this plausible specification of the intuition suggests that the above quoted comparison with ad hoc repairs of empirical theories may not be adequate. In the case of ad hoc repairs we are talking about a given theory and a revised version that solves a given problem of the former. In the case of unintended

---

[1] Note that I inserted 'However' at the beginning of the last sentence in order to stress the deviating moral. Moreover, I replaced 'HD testing' by 'HD evaluation', since using the former expression here must have been a slip of the falsificationist pen.

explications we are primarily comparing two explications of the same concept in terms of their side effects. Hence, suppose we have two explications of a given concept that are equally successful in the sense suggested by my definition of explicative progress (SiS, p. 264), quoted by Festa. Suppose, moreover, that the one generates one or more explications of, apparently, related concepts and the other does not: as far as we know, of course. In other words, the one leads to conceptual unification whereas the other does not. In several contributions and replies in both volumes there have been hints about non-empirical merits of empirical theories that may be taken into account in addition to empirical ones. Here we seem to come across a variant of this type of merits of concept explication in addition to the straightforward merits in terms of evident examples and non-examples and conditions of adequacy. However, in the empirical case it was, at least sometimes, possible to argue that non-empirical merits are functional for truth approximation in an indirect way, viz. in the case of a certain type of aesthetic criteria (Kuipers 2002). In the present case of conceptual unification such an argument seems less likely, roughly for the same reasons as we have met before: the target is more or less known and changeable. However, if we take the set of all concepts that become related by an explication of the concept we started with, we can of course apply a generalized version of the definition of explicative progress. Then we will find that the resulting set of explications of the set of concepts unified in this sense, starting from an explication of a given concept, is better than an alternative explication of this given concept that is equally successful with respect to this concept but has nothing to offer with respect to the concepts that are related to it by the first explication. However, to get a fair comparison one has to bring in the best available, presumably isolated, explications of these concepts. Even if all the resultant one-by-one comparisons show equal success, we are inclined to say that the unified explication of a set of concepts is superior to the sum total of disconnected explications of them. However, it is again difficult to see how this could be functional for truth approximation. Hence, unification remains a non-empirical merit of concept explication.

So far we have assumed that one explication of a given concept was compared with another existing explication. However, the unifying merit of explications of concepts generating unintended explications may be generalized in the sense that such explications are better than any (already existing or not yet existing) explication that is equally successful with respect to an initial concept but does not generate related explications.

Note that a similar story can be told about theoretical unification. If one theory has links with other theories whereas an (existing or hypothetical) empirically equally successful competitor remains isolated, this certainly is considered to be a merit of the former. David Atkinson argues in his contribution to the companion volume that the main merit of string theory is this kind of theoretical unification. He even suggests that the research program underlying string theory may have to be seen as exemplifying a type of program that differs from the four standard types characterized in SiS, that is, descriptive, explanatory, design and explicative programs. A research program may merely aim at unification of (domains and) theories, that is, aim at a unification that seems to have, for practical reasons, no prospects of empirical evaluation.

Inference to the Best Theory as a Case of Abduction

The general merit of conceptual unification suggested above certainly applies to the structuralist explication of the notion of truthlikeness presented in detail in ICR, relative to equally successful but isolated explications of this notion and of its relatives as generated by this explication. In ICR I summarize its merits in this respect as follows:

> … the structuralist theory of truthlikeness has generated a number of unexpected and unintended results. The most important ones are:
>
> – a justification of fundamental methodological rules, in particular the rule to choose the most successful theory (Subsection 7.3.3.),
>
> – an explanation of the success of the natural sciences in terms of truth approximation (Subsection 7.3.3.),
>
> – an explanation and justification of the non-falsificationist behavior of scientists (Section 6.2., 6.3. and 7.3.),
>
> – a corrective explication of so-called 'inference to the best explanation' (Subsection 7.5.3, [called "inference to the best theory," see also Kuipers 2004]),
>
> – an explication of Popper's bad luck, i.e., the convincing failure of his at first sight very plausible definition of truthlikeness (Section 8.1.),
>
> – an explication of the correspondence theory of truth as an intralevel intuition (Section 8.2.),
>
> – … an explication of dialectical concepts [Section 8.3] (ICR, p. 198).

Since the appearance of ICR in 2000, two other unintended explications can be added to this list, viz.

> – an explication of the intuition of many scientists and philosophers that aesthetic criteria are indirectly functional for empirical progress and even truth approximation (Kuipers 2002),
>
> – an explication of the intuition of several philosophers that 'inference to the best explanation' is a kind of abduction in the sense of Peirce (Kuipers 2004).

The last mentioned explicative success is of a special kind. As included in the first list, the structuralist explication of truthlikeness generated a corrective explication of the idea of the "inference to the best explanation" (IBE), leading to "inference to the best theory" (IBT). The last success may also be seen as a proof of the claim that the generated explication of IBE, viz. IBT, satisfies the plausible condition of adequacy that an explication of IBE should make it an abductive rule of inference in the paradigmatic sense of Peirce.

I would like to give a rough idea of the basic versions of these related explications. In ICR I argued that IBE should not be explicated as "inference to the best unfalsified theory, if any, as true," as is usually suggested, but as "inference to the best theory, if any, whether falsified or not, as the closest to the truth," i.e., as IBT. In my forthcoming paper I argue that the latter, but not the former, satisfies the crucial second premise of Peirce's scheme for abduction (Peirce 1958, p. 189):

> The surprising fact C is observed.
>
> But if A were true, C would be a matter of course.
>
> Hence, there is reason to suspect that A is true.

That is, the hypothesis of being closer to the truth[2] makes a matter of course of the, as such, surprising fact of remaining empirically more successful,[3] viz. due to the Success Theorem, see Festa's paper. In contrast, the hypothesis of being true does not do so. This may be an unintended merit of the structuralist explication of the notion of truthlikeness that can be particularly appreciated by logicians interested in abductive logic. However this may be, it at least illustrates once again that relatively simple theorems may nevertheless be very important because of their far-reaching consequences with respect to the unification of concepts and intuitions.

REFERENCES

Kuipers, T. (2002). Beauty, a Road to The Truth. *Synthese* **131** (3), 291-328.

Kuipers, T. (2004). Inference to the Best Theory, Rather Than Inference to the Best Explanation. Kinds of Induction and Abduction. In: F. Stadler (ed.), *Induction and Deduction in the Sciences*, pp. 25-51. Dordrecht: Kluwer Academic Publishers.

Peirce, C.S. (1958). *Collected Papers of Charles Sanders Peirce*, vol. 5. Cambridge, MA: Harvard University Press.

---

[2] More precisely, interpreting A as the hypothesis that the one theory is closer to the truth than the other.

[3] More precisely, interpreting C as the (surprising) comparative fact that the first theory, besides having an extra success, is empirically at least as successful as the second, and remains so after further evaluation.

BIBLIOGRAPHY OF THEO A.F. KUIPERS

Biographical Notes

Theo A.F. Kuipers (b. Horst, Limburg, NL, 1947) studied mathematics at the Technical University of Eindhoven (1964-7) and philosophy at the University of Amsterdam (1967-71). In 1978 he received his Ph.D. degree from the University of Groningen, defending a thesis on inductive logic (*Studies in Inductive Probability and Rational Expectation, Synthese Library*, vol. 123, 1978). The supervisors were J.J.A. Mooij and A.J. Stam. From 1971 to 1975 he was deputy secretary of the Faculty of Philosophy of the University of Amsterdam. In 1975 he was appointed Assistant Professor of the philosophy of science in the Faculty of Philosophy of the University of Groningen; in 1985 he became associate professor and full professor since 1988. He married Inge E. de Wilde in 1971.

A synthesis of his work on confirmation, empirical progress and truth approximation, entitled *From Instrumentalism to Constructive Realism*, appeared in 2000 (*Synthese Library*, vol. 287). A companion synthesis of his work on the structure of theories, research programs, explanation, reduction, and computational discovery and evaluation, entitled *Structures in Science*, appeared in 2001 (*Synthese Library*, vol. 301).

The works he has edited include *What is Closer-to-the-Truth? A Parade of Approaches to Truthlikeness* (*Poznań Studies in the Philosophy of the Sciences and the Humanities*, vol. 10, 1987). He also edited, with Anne Ruth Mackor, *Cognitive Patterns in Science and Common Sense. Groningen Studies in Philosophy of Science, Logic, and Epistemology* (*Poznań Studies in the Philosophy of the Sciences and the Humanities*, vol. 45, 1995).

He was one of the main supervisors of the Ph.D. theses of Henk Zandvoort (1985), Rein Vos (1988), Maarten Janssen (1990), Gerben Stavenga (1991), Roberto Festa (1992), Frank Berndsen (1995), Jeanne Peijnenburg (1996), Anne Ruth Mackor (1997), Rick Looijen (1998), Sjoerd Zwart (1998), Eite Veening (1998), Alexander van den Bosch (2001), and Esther Stiekema (2002). In one way or another, he was also involved in several other Ph.D. theses in Groningen, Amsterdam (VU and UvA), Rotterdam, Nijmegen, Utrecht, Ghent, Leuven, Lublin and Helsinki.

During the academic years 1982/3 and 1996/7 he was a fellow of the Netherlands Institute of Advanced Study (NIAS) at Wassenaar.

528

Besides working in the Faculty of Philosophy, being Dean for a number of periods, he is an active member of the Graduate School for Behavioral and Cognitive Neurosciences (BCN) of which he chaired the research committee for a number of years.

On the national level he was one of the initiators of the section of philosophy of science as well as of the Foundation for Philosophical Research (SWON) of the National Science Foundation (ZWO/NWO). During 1997-2003 he was 'the philosopher member' of the Board of the Humanities of NWO. Since 2000 he has chaired the Dutch Society for Philosophy of Science.

He is a member of the Coordination Committee of the Scientific Network on Historical and Contemporary Perspectives of Philosophy of Science in Europe of the European Science Foundation (ESF).

His research group, which is working on the program *Cognitive Structures in Knowledge and Knowledge Development*, received the highest possible scores from the international assessment committee of Dutch philosophical research in the periods 1989-93 and 1994-8.

Publications

1971

0.   *Inductieve Logica en Haar Beperkingen* (unpublished masters thesis). University of Amsterdam. 1971, 64 pp.

1972

1.   De Wetenschapsfilosofie van Karl Popper. *Amersfoortse Stemmen* **53** (4), 1972, 122-6.
2.   Inductieve Waarschijnlijkheid, de Basis van Inductieve Logica. *Algemeen Nederlands Tijdschrift voor Wijsbegeerte* **64** (4), 1972, 291-6.
3.   A Note on Confirmation. *Philosophica Gandensia* **10**, 1972, 76-7.
4.   Inductieve Logica. *Intermediair* **49**, 1972, 29-33.

1973

5.   A Generalization of Carnap's Inductive Logic. *Synthese* **25**, 1973, 334-6. Reprinted in: J. Hintikka (ed.), *Rudolf Carnap* (*Synthese Library*, vol. 73). Dordrecht: Reidel, 1977.

1976

6.   Inductive Probability and the Paradox of Ideal Evidence. *Philosophica* **17** (1), 1976, 197-205.

1977

7.   Het Verschijnsel Wetenschapsfilosofie, Bespreking van Herman Koningsveld, het Verschijnsel Wetenschap. *Kennis en Methode* **I** (3), 1977, 271-9.
8.   A Two-Dimensional Continuum of a Priori Probability Distributions on Constituents. In: M. Przełęcki, K. Szaniawski, R. Wójcicki (eds.), *Formal Methods in the Methodology of Empirical Sciences* (*Synthese Library*, vol. 103), pp. 82-92. Dordrecht: Reidel, 1977.

1978

9. On the Generalization of the Continuum of Inductive Methods to Universal Hypotheses. *Synthese* **37**, 1978, 255-84.

10. *Studies in Inductive Probability and Rational Expectation.* Ph.D. thesis University of Groningen, 1978. Also published as: *Synthese Library*, vol. 123, Dordrecht: Reidel, 1978, 145 pp.

11. Replicaties, een Reactie op een Artikel van Louis Boon. *Kennis en Methode* **II** (3), 1978, 278-9.

1979

12. Diminishing Returns from Repeated Tests. *Abstracts 6-th LMPS-Congress*, Section 6, Hannover, 1979, 118-22.

13. Boekaankondiging: G. de Brock e.a., De Natuur: Filosofische Variaties. *Algemeen Nederlands Tijdschrift Voor Wijsbegeerte* **71**.3, 1979, 200-1.

1980

14. A Survey of Inductive Systems. In: R. Jeffrey (ed.), *Studies in Inductive Logic and Probability*, pp. 183-92. Berkeley: University of California Press, 1980.

15. Nogmaals: Diminishing Returns from Repeated Tests. *Kennis en Methode* **IV** (3), 1980, 297-300.

16. a.Comment on D. Miller's "Can Science Do Without Induction?"
b.Comment on I. Niiniluoto's "Analogy, Transitivity and the Confirmation of Theories." In: L.J. Cohen, M. Hesse (eds.), *Applications of Inductive Logic*, (1978), pp. 151-2/244-5. Oxford: Clarendon Press, 1980.

1981

17. (Ed.) *Hoofdfiguren in de Hedendaagse Filosofie van de Natuurwetenschappen* (redactie, voorwoord (89) en inleiding (90-3)). *Wijsgerig Perspectief* **21** (4), (1980-) 1981. 26 pp.

1982

18. The Reduction of Phenomenological to Kinetic Thermostatics. *Philosophy of Science* **49** (1), 1982, 107-19.

19. Approaching Descriptive and Theoretical Truth. *Erkenntnis* **18** (3), 1982, 343-78.

1983

20. Methodological Rules and Truth. *Abstracts 7-th LMPS-Congress,* vol. **3** (Section 6), Salzburg, 1983, 122-5.

21. Non-Inductive Explication of Two Inductive Intuitions. *The British Journal for the Philosophy of Science* **34** (3), 1983, 209-23.

1984

22. Olson, Lindenberg en Reductie in de Sociologie. *Mens en Maatschappij* **59** (1), 1984, 45-67.

23. Two Types of Inductive Analogy by Similarity. *Erkenntnis* **21** (1), 1984, 63-87.

24. Oriëntatie: Filosofie in Polen (samenstelling, inleiding en vertaling). *Wijsgerig Perspectief* **24** (6), (1983-)1984, 216-21.

25. Empirische Mogelijkheden: Sleutelbegrip van de Wetenschapsfilosofie. *Kennis en Methode* **VIII** (3), 1984, 240-63.

26. Inductive Analogy in Carnapian Spirit. In: P.D. Asquith, Ph. Kitcher (eds.), *PSA 1984, Volume One* (Biennial Meeting Philosophy of Science Association in Chicago), pp. 157-67. East Lansing: PSA, 1984.

530

27.   Utilistic Reduction in Sociology: The Case of Collective Goods. In: W. Balzer, D.A. Pearce, H.-J. Schmidt (eds.), *Reduction in Science. Structure, Examples, Philosophical Problems* (*Synthese Library*, vol. 175, Proc. Conf. Bielefeld, 1983), pp. 239-67. Dordrecht: Reidel, 1984.

28.   What Remains of Carnap's Program Today? In: E. Agazzi, D. Costantini (eds.), *Probability, Statistics, and Inductive Logic, Epistemologia* **7**, 1984, 121-52; Proc. Int. Conf. 1981 at Luino, Italy. With discussions with D. Costantini (149-51) and W. Essler (151-2) about this paper and with E. Jaynes (71-2) and D.Costantini (166-7) about theirs.

29.   An Approximation of Carnap's Optimum Estimation Method. *Synthese* **61**, 1984, 361-2.

30.   Approaching the Truth with the Rule of Success. In: P. Weingartner, Chr. Pühringer (eds.), *Philosophy of Science – History of Science*, Selection 7th LMPS Salzburg 1983, *Philosophia Naturalis* **21** (2/4), 1984, 244-53.

1985

31.   The Paradigm of Concretization: The Law of Van der Waals. *Poznań Studies in the Philosophy of the Sciences and the Humanities*, vol. 8 (ed. J. Brzeziński), Amsterdam: Rodopi, 1985, pp. 185-99.

32.   (met Henk Zandvoort), Empirische Wetten en Theorieën. *Kennis en Methode* **9** (I), 1985. 49-63.

33.   The Logic of Intentional Explanation. In: J. Hintikka, F.Vandamme (Eds.), *The Logic of Discourse and the Logic of Scientific Discovery* (Proc. Conf. Gent, 1982), *Communication and Cognition* **18** (1/2), 1985, 177-98. Translated as: Logika wyjaśniania intencjonalnego. *Poznańskie Studia z Filozofii Nauki* **10**, 1986, 189-218.

34.   Een Beurs voor de Verdeling van Arbeidsplaatsen. *Filosofie & Praktijk* **6** (4), 1985, 205-11.

1986

35.   Some Estimates of the Optimum Inductive Method. *Erkenntnis* **24**, 1986, 37-46.

36.   The Logic of Functional Explanation in Biology. In: W. Leinfellner, F. Wuketits (eds.), *The Tasks of Contemporary Philosophy* (Proc. 10th Wittgenstein Symp. 1985), pp. 110-4. Wenen: Hölder-Pichler-Temsky, 1986.

37.   Intentioneel Verklaren van Handelingen. In: Proc. Conf. Handelingspsychologie, ISvW- Amersfoort 1985. *Handelingen*. O-nr, 1986, 12-18.

38.   Explanation by Specification. *Logique et Analyse* **29** (116), 1986, 509-21.

1987

39.   (Ed.) *What is Closer-To-The-Truth? A Parade of Approaches to Truthlikeness* (*Poznań Studies in the Philosophy of the Sciences and the Humanities*, vol. 10). Amsterdam: Rodopi, 1987, 254 pp. Introduction: 1-7.

40.   A Structuralist Approach to Truthlikeness, in 39: 79-99.

41.   Truthlikeness of Stratified Theories, in 39: 177-86.

42.   (Ed.) *Holisme en Reductionisme in de Empirische Wetenschappen, Kennis en Methode* **11** (I), 1987. 136 pp., Voorwoord: 4-5.

43.   Reductie van Wetten: een Decompositiemodel, in 42: 125-35.

44.   Fascinaties: Wetenschappelijk Plausibel en Toch Taboe. *VTI* (contactblad Ver. tot Instandhouding Int. School v. Wijsbegeerte), nr.13, juli 1987, 5-8; discussie met J. Hilgevoord in nr. 14, 1987, 6-9.

45.   A Decomposition Model for Explanation and Reduction. *Abstracts LMPS-VIII*, Moscow, 1987, vol. 4, 328-31.

46.   Truthlikeness and the Correspondence Theory of Truth. In: P. Weingartner, G. Schurz (eds.), *Logic, Philosophy of Science and Epistemology*, Proc. 11th Wittgenstein Symp. 1986, pp. 171-6. Wenen: Hölder-Pichler-Temsky, 1987.

47.   Reductie van Begrippen: Stappenschema's. *Kennis en Methode* **11** (4), 1987, 330-42.

1988

48. Voorbeelden van Cognitief Wetenschapsonderzoek. *WO-NieuwsNet* **I** (I), 1988, 13-29.
49. Structuralistische Explicatie van Dialectische Begrippen. *Congresbundel Filosofiedag Maastricht 1987*, pp. 191-7. Delft: Eburon, 1988.
50. Inductive Analogy by Similarity and Proximity. In: D.H. Helman (ed.), *Analogical Reasoning*, pp. 299-313. Dordrecht: Kluwer Academic Publishers, 1988.
51. (with Hinne Hettema), The Periodic Table – its Formalization, Status, and Relation to Atomic Theory. *Erkenntnis* **28**, 1988, 387-408.
52. Cognitive Patterns in the Empirical Sciences: Examples of Cognitive Studies of Science. *Communication and Cognition* **21** (3/4), 1988, 319-41. Translated as: Modele kognitywistyczne w naukach empirycznych: przykłady badań nad nauką, *Poznańskie Studia z Filozofii Humanistyki* **14** (1), 1994, 15-41.

1989

53. (Ed.) *Arbeid en Werkloosheid*. Redactie, inleiding, discussie thema-nummer *Wijsgerig Perspectief* **29** (4), (1988-) 1989.
54. (with Maarten Janssen), Stratification of General Equilibrium Theory: A Synthesis of Reconstructions. *Erkenntnis* **30**, 1989, 183-205.
55. *Onderzoeksprogramma's Gebaseerd op een Idee. Impressies van een Wetenschapsfilosofische Praktijk*, inaugural address University of Groningen. Assen: Van Gorcum, 1989. 32 pp.
56. How to Explain the Success of the Natural Sciences. In: P. Weingartner, G. Schurz (eds.), *Philosophy of the Natural Sciences* (Proc. 13th Int. Wittgenstein Symp. 1988), pp. 318-22. Wenen: Hölder-Pichler-Temsky, 1989.

1990

57. (Ed. with J. Brzeziński, F. Coniglione, and L. Nowak) *Idealization I: General Problems, Idealization II: Forms and Applications (Poznań Studies in the Philosophy of the Sciences and the Humanities,* vol. 16+17), Rodopi, Amsterdam-Atlanta, 1990.
58. Reduction of Laws and Concepts. In 57 I: 241-76.
59. Het Objectieve Waarheidsbegrip in Waarder. *Kennis en Methode* **XIV** (2), 1990, 198-211. (Met een reactie van Hans Radder: 212-15).
60. (met Hauke Sie), Industrieel en Academisch Onderzoek. *De Ingenieur*, nr. 6 (juni), 1990, 15-8.
61. Interdisciplinariteit en Gerontologie. In: D. Ringoir en C. Tempelman (ed.), *Gerontologie en Wetenschap*, pp. 143-9. Nijmegen: Netherlands Institute of Gerontology, 1990.
62. Het Belang van Onware Principes. *Wijsgerig Perspectief* **31** (1), 1990, 27-9.

1991

63. Economie in de Spiegel van de Natuurwetenschappen: Overeenkomsten, Plausibele Verschillen en Specifieke Rariteiten. *Kennis en Methode* **XV** (2), 1991, 182-97.
64. Realisme en Convergentie, of Hoe het Succes van de Natuurwetenschappen Verklaard Moet Worden. In: J. van Brakel en D. Raven (ed.), *Realisme en Waarheid*, pp. 61-83. Assen: Van Gorcum, 1991.
65. On the Advantages of the Possibility-Approach. In: A. Ingegno (ed.), *Da Democrito a Collingwood*, pp. 189-202. Firenze: Olschki, 1991.
66. Structuralist Explications of Dialectics. In: G. Schurz and G. Dorn (eds.), *Advances in Scientific Philosophy. Essays in honour of Paul Weingartner on the occasion of the 60-th anniversary of his birthday (Poznań Studies in the Philosophy of the Sciences and the Humanities,* vol. 24), pp.295-312. Amsterdam-Atlanta: Rodopi, 1991.
67. Dat Vind Ik Nou Mooi. In: R.Segers (ed.), *Visies op Cultuur en Literatuur. Opstellen naar aanleiding van het werk van J.J.A. Mooij*, pp. 69-75. Amsterdam: Rodopi, 1991.

532

1992

68. (Ed.) *Filosofen in Actie*. Delft: Eburon, 1992. 255 pp.
69. Methodologische Grondslagen voor Kritisch Dogmatisme. In: J.W. Nienhuys (ed.), *Het Vooroordeel van de Wetenschap*, ISvW-conferentie 23/24 februari 1991, pp. 43-51. Utrecht: Stichting SKEPSIS, 1992.
70. (with Rein Vos and Hauke Sie), Design Research Programs and the Logic of Their Development. *Erkenntnis* **37** (1), 1992, 37-63. Translated as: Projektowanie programów badawczych i logika ich rozwoju. *Projektowanie i Systemy* **15**, 1995, pp. 29-48.
71. Truth Approximation by Concretization. In: J. Brzeziński and L. Nowak (eds.), *Idealization III: Approximation and Truth (Poznań Studies in the Philosophy of the Sciences and the Humanities*, vol. 25), pp. 159-79. Amsterdam-Atlanta: Rodopi, 1992.
72. Naive and Refined Truth Approximation. *Synthese* **93**, 1992, 299-341.
73. Wetenschappelijk Onderwijs. In: *ABC van Minder Docentafhankelijk Onderwijs*, 25 jarig jubileum uitgave, pp. 133-7. Groningen: COWOG, 1992.

1993

74. On the Architecture of Computational Theory Selection. In: R. Casati & G. White (eds.), *Philosophy and the Cognitive Sciences*, pp. 271-78. Kirchberg: Austrian Ludwig Wittgenstein Society, 1993.
75. Computationele Wetenschapsfilosofie. *Algemeen Nederlands Tijdschrift voor Wijsbegeerte* **85** (4), 1993, 346-61.
76. De Pavarotti's van de Analytische Filosofie. *Filosofie Magazine* **2** (8), 1993, 36-9. Bewerking in: D. Pels en G. de Vries, *Burgers en Vreemdelingen*, t.g.v. afscheid L.W. Nauta, pp. 99-107. Amsterdam: Van Gennep, 1994. Reacties van Menno Lievers, Anthonie Meijers, Filip Buekens en Stefaan Cuypers, gevolgd door repliek TK: *Filosofie Magazine* **3** (1), 1994, 37-40.
77. Wetenschappelijk Onderwijs en Wijsbegeerte van een Wetenschapsgebied. *Universiteit en Hogeschool* **40** (1), 1993, 9-18.

1994

78. (with Andrzej Wiśniewski) An Erotetic Approach to Explanation by Specification. *Erkenntnis* **40** (3), 1994, 377-402.
79. (with Kees Cools and Bert Hamminga), Truth Approximation by Concretization in Capital Structure Theory. In: B. Hamminga and N.B. De Marchi (eds.), *Idealization VI: Idealization in Economics (Poznań Studies in the Philosophy of the Sciences and the Humanities,* vol. 38), pp. 205-28. Amsterdam-Atlanta: Rodopi, 1994.
80. Falsificationisme Versus Efficiënte Waarheidsbenadering. Of de Ironie van de List der Rede. *Algemeen Nederlands Tijdschrift voor Wijsbegeerte* **86** (4), 1994, 270-90.
81. The Refined Structure of Theories. In: M. Kuokkanen (ed.), *Idealization VII: Structuralism, Idealization, Approximation (Poznań Studies in the Philosophy of the Sciences and the Humanities*, vol. 42), pp. 3-24. Amsterdam-Atlanta: Rodopi, 1994.

1995

82. Observationele, Referentiële en Theoretische waarheidsbenadering (Reactie op Ton Derksen). *Algemeen Nederlands Tijdschrift voor Wijsbegeerte* **87** (1), 1995, 33-42.
83. Falsificationism Versus Efficient Truth Approximation. In: W. Herfel, W. Krajewski, I. Niiniluoto and R. Wojcicki (eds.), *Theories and Models in Scientific Processes (Poznań Studies in the Philosophy of the Sciences and the Humanities*, vol. 44), pp. 359-86. Amsterdam-Atlanta: Rodopi, 1995. (Extended and translated version of 80).
84. Ironie van de List der Rede. *Wijsgerig Perspectief* **35** (6), (1994-)1995, 189-90.
85. (Ed. with Anne Ruth Mackor), *Cognitive Patterns in Science and Common Sense. Groningen Studies in Philosophy of Science, Logic, and Epistemology*. With a foreword by Leszek Nowak. *Poznań Studies in the Philosophy of the Sciences and the*

533

Humanities, vol. 45. Amsterdam-Atlanta: Rodopi, 1995. With a general introduction ("Cognitive Studies of Science and Common Sense", pp. 23-34) and special introductions to the four parts.

86. Explicating the Falsificationist and the Instrumentalist Methodology by Decomposing the Hypothetico-Deductive Method. In 85: 165-86.

87. (with Hinne Hettema), Sommerfeld's *Atombau*: A Case Study in Potential Truth Approximation. In 85: 273-97.

88. Verborgen en Manifeste Psychologie in de Wetenschapsfilosofie. *Nederlands Tijdschrift voor Psychologie* **50** (6), 1995, 252.

1996

89. Truth Approximation by the Hypothetico-Deductive Method. In: W. Balzer, C.U. Moulines and J.D. Sneed (eds.), *Structuralist Theory of Science: Focal Issues, New Results*, pp.83-113. Berlin: Walter de Gruyter, 1996.

90. Wetenschappelijk en Pseudowetenschappelijk Dogmatisch Gedrag. *Wijsgerig Perspectief* **36** (4), (1995-)1996, 92-7.

91. Het Softe Paradigma. Thomas Kuhn Overleden. *Filosofie Magazine* **5** (7), 1996, 28-31.

92. Explanation by Intentional, Functional, and Causal Specification. In: A. Zeidler-Janiszewska (ed.), *Epistemology and History. Humanities as a Philosophical Problem and Jerzy Kmita's Approach to It* (*Poznań Studies in the Philosophy of the Sciences and the Humanities*, vol. 47), pp. 209-36. Amsterdam-Atlanta: Rodopi, 1996.

93. Efficient Truth Approximation by the Instrumentalist, Rather Than the Falsificationist Method. In: I. Douven and L. Horsten (eds.), *Realism in the Sciences* (*Louvain Philosophical Studies*, vol. 10, pp. 115-30. Leuven: Leuven University Press, 1996.

1997

94. Logic and Philosophy of Science: Current Interfaces. (Introduction to the proceedings of a special symposium with the same name). In: M.L. Dalla Chiara, K. Doets, D. Mundici and J. van Benthem (eds.), *Logic and Scientific Methods*, vol. 1, (*10th LMPS International Congress*, Florence, August, 1995), pp.379-81. Dordrecht: Kluwer Academic Publishers, 1997.

95. The Carnap-Hintikka Programme in Inductive Logic. In: Matti Sintonen (Ed.), *Knowledge and Inquiry: Essays on Jaakko Hintikka's Epistemology and Philosophy of Science. (Poznań Studies in the Philosophy of the Sciences and the Humanities*, vol. 51), pp. 87-99. Amsterdam-Atlanta: Rodopi, 1997. With a comment by Hintikka, pp. 317-18.

96. Boekaankondiging: A. Derksen (ed.), *The Scientific Realism of Rom Harré.* Tilburg: Tilburg University Press, 1994, *Algemeen Nederlands Tijdschrift voor Wijsbegeerte* **89** (2), 1997, 174.

97. The Dual Foundation of Qualitative Truth Approximation. *Erkenntnis* **47** (2), 1997, 145-79.

98. Comparative Versus Quantitative Truthlikeness Definitions: Reply to Thomas Mormann. *Erkenntnis* **47** (2), 1997, 187-92.

1998

99. Confirmation Theory. *The Routledge Encyclopedia of Philosophy*, vol. 2, 1998, 532-36.

100. Pragmatic Aspects of Truth Approximation. In: P. Weingartner, G. Schurz and G. Dorn (eds.), *The Role of Pragmatics in Contemporary Philosophy*, pp. 288-300. Proceedings of the 20th International Wittgenstein-Symposium, August 1997. Vienna: Hölder-Pichler-Temsky, 1998.

1999

101. Kan Schoonheid de Weg Wijzen naar de Waarheid? *Algemeen Nederlands Tijdschrift voor Wijsbegeerte* **91** (3), 1999, 174-93.

534

102.	The Logic of Progress in Nomological, Design and Explicative Research. In: J. Gerbrandy, M. Marx, M. de Rijke, and Y. Venema (eds.), *JFAK. Essays Dedicated to Johan van Benthem on the Occasion of his 50th Birthday*, CD-ROM, Amsterdam University Press, Series Vossiuspers, Amsterdam, ISBN 90 5629 104 1, 1999. (Unique) Book edition vol. 3, 1999, pp. 37-46.

103.	Zeker Lezen: Wetenschapsfilosofie. *Wijsgerig Perspectief* **39** (6), 1999, 170-1.

104.	De Integriteit van de Wetenschapper. In: E. Kimman, A. Schilder, en F. Jacobs (ed.), *Drieluijk: Godsdienst, Samenleving, Bedrijfsethiek, Liber Amicorum voor Henk van Luijk*, pp. 99-109. Amsterdam: Thela-Thesis, 1999.

105.	Abduction Aiming at Empirical Progress or Even at Truth Approximation, Leading to a Challenge for Computational Modelling. In: J. Meheus, T. Nickles (eds.), *Scientific Discovery and Creativity*, special issue of *Foundations of Science* **4** (3), 1999, 307-23.

2000

106.	*From Instrumentalism to Constructive Realism. On Some Relations Between Confirmation, Empirical Progress, and Truth Approximation* (*Synthese Library*, vol. 287). Dordrecht: Kluwer Academic Publishers, 2000.

107.	Filosofen als Luis in de Pels. Over Kritiek, Dogma's en het Moderne Turven van Publicaties en Citaties. In: J. Bremmer (ed.), *Eric Bleumink op de Huid Gezeten. Opstellen aangeboden door het College van Decanen ter gelegenheid van zijn afscheid als Voorzitter van het College van Bestuur van de Rijksuniversiteit Groningen op 24 mei 2000*, pp. 89-103. Groningen: Uitgave RUG, 2000.

108.	(with Hinne Hettema), The Formalisation of the Periodic Table. In: W. Balzer, J. Sneed, U. Moulines (eds.), *Structuralist Knowledge Representation. Paradigmatic Examples* (*Poznań Studies in the Philosophy of the Sciences and the Humanities*, vol. 75), pp. 285-305. Amsterdam-Atlanta: Rodopi, 2000. (Revised version of 51.)

2001

109.	Epistemological Positions in the Light of Truth Approximation. In: T.Y. Cao (ed.), *Philosophy of Science* (Proceedings of the 20th World Congress of Philosophy, Boston, 1998, vol. 10), pp. 79-88. Bowling Green: Philosophy Documentation Center, Bowling Green State University, 2001.

110	Naar een Alternatieve Impactanalyse. *De Academische Boekengids*, 26. Amsterdam: AUP, 2001, p. 16.

111	*Structures in Science. Heuristic Patterns Based on Cognitive Structures. An Advanced Textbook in Neo-Classical Philosophy of Science.* (*Synthese Library*, vol. 301). Dordrecht: Kluwer Academic Publishers, 2001.

112	Qualitative Confirmation by the HD-Method. *Logique et Analyse* **41** (164), 1998 (in fact 2001), 271-99.

2002

113	Beauty, a Road to The Truth. *Synthese* **131** (3), 291-328.

114	Poppers Filosofie van de Natuurwetenschappen. *Wijsgerig Perspectief* **42** (2), 2002, 17-31.

115	Quantitative Confirmation, and its Qualitative Consequences. *Logique et Analyse* **42** (167/8), 1999 (in fact 2002), 447-82.

116	Aesthetic Induction, Exposure Effects, Empirical Progress, and Truth Approximation. In: R. Bartsch e.a (ed.), *Filosofie en Empirie. Handelingen 24e NV-Filosofiedag*, 2-11-2002, pp.194-204, Amsterdam: UvA-Wijsbegeerte, 2000.

117	O dwóch rodzajach idealizacji i konkretyzacji. Przypadek aproksymacji prawdy. In: J. Brzeziński, A. Klawiter, T. Kuipers, K Łastowski, K. Paprzycka and P. Przybysz (eds.), *Odwaga Filozofowania. Leszkowi Nowakowi w darze*, pp. 117-139. Poznań: Wydawnictwo Fundacji Humaniora, 2002.

2003

2004

118   Inference to the Best Theory, Rather Than Inference to the Best Explanation. Kinds of Abduction and Induction. In: F. Stadler (ed.), *Induction and Deduction in the Sciences*, Proceedings of te ESF-workshop *Induction and Deduction in the Sciences*, Vienna, July, 2002, pp. 25-52, followed by a commentary of Adam Grobler, pp. 53-36, Dordrecht: Kluwer Academic Publishers, 2004.

119   De Logica van de G-Hypothese. Hoe Theologisch Onderzoek Wetenschappelijk Kan Zijn. In: K. Hilberdink (red.), *Van God Los? Theologie tussen Godsdienst en Wetenschap* 59-74, Amsterdam: KNAW, 2004.

2005

120   The Threefold Evaluation of Theories: A Synopsis of *From Instrumentalism to Constructive Realism* (2000) + replies to 17 contributions. In: Roberto Festa, Atocha Aliseda, and Jeanne Peijnenburg (eds.), *Confirmation, Empirical Progress, and Truth Approximation, Essays in Debate with Theo Kuipers*, Volume 1. *Poznań Studies in the Philosophy of the Sciences and the Humanities.* This volume.

121   Structures in Scientific Cognition: A Synopsis of *Structures in Science. Heuristic Patterns Based on Cognitive Structures* (2001) + replies to 17 contributions. In: Roberto Festa, Atocha Aliseda, and Jeanne Peijnenburg (Eds.), *Cognitive Structures in Scientific Inquiry, Essays in Debate with Theo Kuipers*, Volume 2. *Poznań Studies in the Philosophy of the Sciences and the Humanities.* The companion volume.

To appear

-    Inductive Aspects of Confirmation, Information, and Content. To appear in the volume of *The Library of Livings Philosophers* (Schilpp) dedicated to Jaakko Hintikka.

-    Empirical and Conceptual Idealization and Concretization. The Case of Truth Approximation. To appear in English edition Liber Amicorum for Leszek Nowak. It appeared already in the Polish edition: 117.

INDEX OF NAMES

538

Dirac, P., 351
Dorling, J., 285-8, 307
Dorn, G.J.W., 84, 488, 492, 531, 533
Douven, I., 16, 261, 274, 281, 284, 286,
	289-90, 293, 296-9, 307-13, 533
Duhem, P., 296
Dyck, M., van, 208, 217

Earman, J., 37, 84, 284, 288, 308
Eddington, A., 112, 502
Einstein, A., 112, 114, 155, 164-5, 274,
	323-5, 360
Essler, W., 530

Fagin, R., 415, 419
Faye, J., 464, 487
Feigl, H., 274
Festa, R. 14, 18-20, 38, 84, 97, 108, 216,
	259, 261, 266, 274, 307, 341, 398, 469,
	487, 511, 516, 520-3, 525, 527, 535
Fetzer, J.H., 275
Feyerabend, P., 27
Feynman, R., 358
Fine, A., 265, 274, 284
Fitelson, B., 34-5, 84, 97, 108, 122, 127
Fitzgerald, G., 284
Fizeau, A., 54, 392
Flach, P. A., 249, 252, 274, 417, 419
Flood, R., 505
Fraassen, B., van, 13, 23-4, 28, 147, 159,
	265-6, 275, 283, 285, 287-90, 293,
	295-6, 300-2, 304, 306, 308-9, 417
Franklin, R., 351
Frege, G., 513
French, P., 308

Gabbay, D., 252, 417, 419
Galavotti, M.C., 84, 108, 274
Galle, J.G., 326, 329, 331
Gärdenfors, P., 170, 189, 405, 417, 419,
	476, 487
Gellner, E., 11
Gemes, K., 94, 141, 150-1, 158-9, 161-2
Gerbrandy, M, 534
Giere, R., 78, 84, 307
Glymour, C., 144, 296-8, 300, 308
Gödel, K., 171, 513
Goldman, A.I., 367, 369-70, 372
Goldstick, D., 437, 439, 453-4
Good, I.J., 11, 93, 101, 108, 125, 127,
	308
Goodman, N., 14, 32, 129-30, 133-5,
	137-8, 148
Grätzer, G., 450-1, 453
Grobler, A., 516

Groot, A., de, 28
Grosholz, E., 431, 453
Grove, A., 36, 84
Guenthner, F., 419

Hacking, I., 13, 23-4, 293, 295-8, 300,
	303-4, 306, 308-9
Haesaert, L., 245-6
Hájek, A., 288, 308
Halpern, J., 36, 84, 419
Hamminga, B., 16, 317, 331, 336-40, 532
Hands, D.W., 336
Hanson, N.R., 256, 274
Harman, G., 259, 274
Harré, R., 295, 298, 308, 499-500, 502,
	505, 507, 533
Harris, J.H., 342-4, 352-4, 356, 385, 395,
	449, 454
Hartley, R., 462
Hartshorne, C., 275
Hausdorff, F., 452
Hegel, G., 498
Heidema, J., 15, 17-8, 398, 420, 447, 455,
	457, 459, 461, 469-71, 476, 479-80,
	482-3, 485, 487, 489-91, 516
Heilbroner, R.L., 335-6
Heller, A., 11
Helman, D.H., 531
Hempel, C.G., 13, 23, 28, 32, 42, 46, 84-
	5, 94-5, 100, 108, 120-1, 126, 173,
	175, 221, 257, 260-1, 274-5, 300, 504,
	514
Hendricks, V.F., 464, 487
Hendriks, L., 20
Herfel, W., 532
Hesse, M., 501, 529
Hettema, H., 531, 533-4
Hilbert, D., 440, 443, 454, 513
Hilgevoord, J., 530
Hilpinen, R., 249
Hintikka, J., 11, 31, 35, 37-8, 132, 148,
	162, 166, 185, 243, 246, 249, 260, 268,
	271-2, 274, 279, 417, 419, 464, 488,
	528, 530, 533, 535
Hodges, W., 59, 84, 396, 402
Hoek, W., van der, 415
Hoggar, C., 419
Holland, J., 249, 252, 274
Hooker, C.A., 307-9
Horn, A., 411
Horsten, L., 284, 308-9, 533
Howson, C., 37, 84, 286, 308
Hume, D., 308, 384
Huygens, C., 161, 392

MONOGRAPHS-IN-DEBATE

CONTENTS OF BACK ISSUES